W9-BIH-746

Trigonometry

REVISED EDITION

Arthur F. Coxford

HARCOURT BRACE JOVANOVICH, PUBLISHERS

Orlando   San Diego   Chicago   Dallas

About the Author

Arthur F. Coxford
Professor of Mathematics Education
University of Michigan
Ann Arbor, Michigan

Editorial Advisors

Brother Neal Golden, S.C.
Chairman, Department of
Mathematics and
Computer Science
Brother Martin High School
New Orleans, Louisiana

James C. Wortham
Mathematics Curriculum Specialist
Akron Public Schools
Akron, Ohio

The contributions of Brother Neal Golden, S.C. who wrote the sections on computer programming in the BASIC language in this text are gratefully acknowledged.

Picture Credits

Cover: Jerome Kresch

Page 39, The L. S. Starrett Company; 106-107, Gordon S. Smith/Photo Researchers; 159, HBJ Photo, 178, Robert A. Isaacs/Photo Researchers; 224-225, HBJ Photo; 272, Russ Kinne/Photo Researchers.

Printed in the United States of America   **ISBN 0-15-359370-9**

Contents

Chapter 1
Trigonometric
Functions

1-1 Relations and Functions

The study of <u>trigonometry</u> originated in connection with problems that involved right triangles. Recall from your previous mathematics courses that the <u>Pythagorean Theorem</u> states an important relationship for right triangles.

Pythagorean Theorem

In any right triangle, if the length of the hypotenuse is c and the lengths of the legs are a and b, then

$$a^2 + b^2 = c^2.$$

Finding the distance between two points P_1 and P_2 in the coordinate plane is the same as finding the length of $\overline{P_1P_2}$ (segment P_1P_2). As example 1 shows, $\overline{P_1P_2}$ can be viewed as the hypotenuse of a right triangle.

Example 1

Find the distance between P_1 and P_2. Write the answer in simplest form.

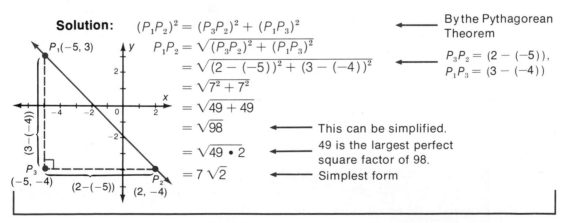

Solution:

$$(P_1P_2)^2 = (P_3P_2)^2 + (P_1P_3)^2 \qquad \longleftarrow \text{By the Pythagorean Theorem}$$

$$P_1P_2 = \sqrt{(P_3P_2)^2 + (P_1P_3)^2}$$

$$= \sqrt{(2-(-5))^2 + (3-(-4))^2} \qquad \longleftarrow \begin{array}{l} P_3P_2 = (2-(-5)), \\ P_1P_3 = (3-(-4)) \end{array}$$

$$= \sqrt{7^2 + 7^2}$$

$$= \sqrt{49 + 49}$$

$$= \sqrt{98} \qquad \longleftarrow \text{This can be simplified.}$$

$$= \sqrt{49 \cdot 2} \qquad \longleftarrow \begin{array}{l} \text{49 is the largest perfect} \\ \text{square factor of 98.} \end{array}$$

$$= 7\sqrt{2} \qquad \longleftarrow \text{Simplest form}$$

The method of Example **1** can be generalized to obtain the <u>Distance Formula</u>.

Distance Formula

The distance between two points $P_1(x_1, y_1)$ and $P_2(x_2, y_2)$ is

$$P_1P_2 = \sqrt{(x_2 - x_1)^2 + (y_2 - y_1)^2}.$$

The study of trigonometry encompasses much more than problems that deal with triangles. Trigonometry also deals extensively with the study of relations and functions.

Recall that a **relation** is a set of ordered pairs. Since the coordinates (x, y) of the points on line P_1P_2 in Example 1 are ordered pairs, line P_1P_2 is the graph of a relation. The equation $y = -x - 2$ describes this relation. You can use this equation to construct a table of ordered pairs, such as the one at the right. For this relation, no two ordered pairs have the same first element. Thus, the relation is a function.

x	−5	−3	0	2	4
y	3	1	−2	−4	−6

Definition: A **function** is a relation such that no two of its ordered pairs have the same first element.

In a relation or a function, the set of first elements is the **domain,** and the set of second elements is the **range.** In the ordered pair (x, y), x represents the elements of the domain and is often called the **independent variable.** The variable associated with the range, y, is often called the **dependent variable.** For example, in constructing the table of values above, each value for y is found by replacing x in $y = -x - 2$ with an element of the domain. This function can be described in several ways, where the domain is the set of real numbers R. Recall that "$x \in R$" means "x is an element of the set of real numbers."

1. $f = \{(x, y): y = -x - 2, x \in R\}$ ◄——— Set-builder notation
2. $f(x) = -x - 2, x \in R$ ◄——— Function notation
3. $y = -x - 2, x \in R$ ◄——— As an equation

The letter f in **1** represents the function. In **2,** $f(x)$ represents an element in the range of f. Thus, the ordered pairs of a function can be written as

$$(x, f(x))$$

as well as (x, y).

Here are some examples of functions that you have studied in your previous courses.

Name	Graph	Function	Domain/Range
Linear		$f(x) = -x + 4$	Domain: R Range: R
Quadratic		$f(x) = -3x^2 - 4x + 5$	Domain: R Range: $\{y:\ y \le 6\frac{1}{3}\}$
Exponential		$f(x) = 2^x$	Domain: R Range: $\{y:\ y > 0\}$
Logarithmic		$f(x) = \log_{10} x$	Domain: $\{x:\ x > 0\}$ Range: R

If the domain of a function is not given, it is assumed to be the set of real numbers. However, certain real numbers for which a function has no meaning have to be excluded.

Example 2
Determine the real numbers that must be excluded from the domain of each function.

Function	Exclusion
Solutions: a. $f(x) = \dfrac{9}{x+2}$	**a.** -2, since $\frac{9}{0}$ has no meaning
b. $f(x) = \dfrac{21}{x^2 - 4}$	**b.** 2 and -2, since $\frac{21}{0}$ has no meaning

Classroom Exercises

Give the domain and range of each relation. Then tell whether the relation is a function.

1. $\{(0, 1), (0, 0), (5, 3), (-2, 6)\}$
2. $\{(-1, 1), (2, 4), (-1, 7), (2, 9)\}$
3. $\{(1, 4), (2, 3), (5, 6), (8, 13)\}$
4. $\{(4, 3), (6, -3), (2, -1), (-2, -1)\}$

5.
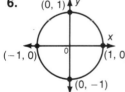

6.
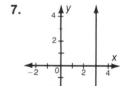

7.
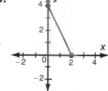

8.

Written Exercises

a

Use the Pythagorean Theorem to replace each ? with the correct length. Refer to the figure. Write answers in simplest form.

1. $a = 1$, $b = 1$, $c = \underline{?}$
2. $a = 1$, $c = 2$, $b = \underline{?}$
3. $b = 3$, $c = 5$, $a = \underline{?}$
4. $b = 2$, $a = 2$, $c = \underline{?}$
5. $b = 2$, $c = 4$, $a = \underline{?}$

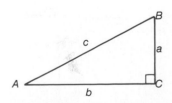

Trigonometric Functions   5

Find the distance between points P_1 and P_2. Write answers in simplest form.

6. $P_1(1, 2)$; $P_2(4, 3)$ **7.** $P_1(-4, 1)$; $P_2(-5, 4)$

8. $P_1(2, -6)$; $P_2(7, -5)$ **9.** $P_1(3, 6)$; $P_2(-1, -2)$

10. $P_1(4, -3)$; $P_2(-4, 3)$ **11.** $P_1(-3, 3)$; $P_2(0, -4)$

In Exercises 12–16 give the domain and range of each relation. Then tell whether the relation is a function.

12. $\{(1, 5), (2, 5), (-2, 3), (4, 3)\}$

13. $\{(3, 3), (2, 2), (1, 1), (-1, -1), (0, 0)\}$

14. $\{(5, -3), (3, -2), (0, 0), (3, 2), (5, 3)\}$

15. $\{(-5, 5), (-2, 2), (0, 0), (2, -2), (5, -5)\}$

16. $\{(-2, -5), (-2, 0), (-2, 5), (-2, 7)\}$

Construct a table of values for each function. Show at least five ordered pairs for each function.

17. $y = 2x - 3$ **18.** $y = x^2$ **19.** $y = x^3$

20. $y = 2^x$ **21.** $y = \dfrac{1}{x + 2}$ **22.** $y = \dfrac{x}{x - 1}$

Determine the real numbers that must be excluded from the domain of each function.

23. $f(x) = \dfrac{4}{x}$ **24.** $f(x) = \dfrac{-2}{x + 5}$ **25.** $f(x) = \dfrac{2}{|x - 3|}$

26. $f(x) = \dfrac{1}{|x| - 5}$ **27.** $f(x) = \dfrac{2x}{x^2 - 9}$ **28.** $f(x) = \dfrac{9}{25 - x^2}$

29. $f(x) = \dfrac{33}{x^2 - 2x}$ **30.** $f(x) = \dfrac{3x}{2x - 4}$ **31.** $f(x) = \dfrac{3x^2}{x^2 - 1}$

In Exercises 32–35, determine the domain and range of each function.

32. $y = 2x + 1$

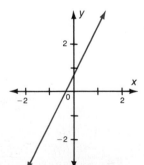

33. $y = -2$

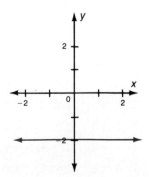

34. $y = -x^2$

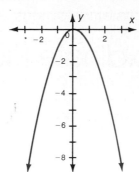

35. $y = |x|$

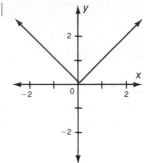

Let r be the distance from the origin to the given point.
For each point find $r, \frac{y}{r}, \frac{x}{r},$ and $\frac{y}{x}$.

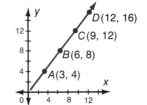

36. $A(3, 4)$

37. $B(6, 8)$

38. $C(9, 12)$

39. $D(12, 16)$

40. In the figure at the left below, triangle ABC is an isosceles right triangle with $AC = BC = s$. Use the Pythagorean Theorem to express the length AB in terms of s.

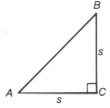

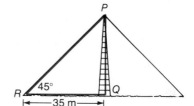

41. In the figure at the right above, a cable attached to a tower at point P is anchored in the ground 35 meters from point Q, the base of the tower. The measure of angle PRQ is 45°. Find the length of the cable and the height of point P above the ground.

42. In the figure at the left below, triangle ABC is equilateral with $AB = BC = AC = s$. From geometry you can show that triangle ADC and triangle ADB are congruent. Use this fact and the Pythagorean Theorem to find AD and DC in terms of s.

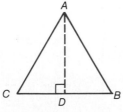

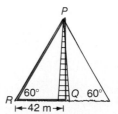

43. In the figure at the right above, a cable attached to a tower at point P is anchored in the ground 42 meters from point Q, the base of the tower. The measure of angle PRQ is 60°. Find the length of the cable and the height of point P above the ground.

1-2 Angles and the Coordinate Plane

In geometry, an angle is defined as the union of two rays with a common end-point. In trigonometry, an angle is defined in terms of a rotation. For example, angle *AOB* is formed by rotating ray *OA* about its endpoint *O* to the new position shown by ray *OB*.

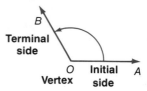

Definitions: An **angle** is formed by the rotation of a ray about its endpoint from an initial position to a terminal position.

The amount of rotation is the **measure** of the angle.

The point of rotation is the **vertex** of the angle. The initial position of the ray is the **initial side** of the angle.

The terminal position of the ray is the **terminal side** of the angle.

In geometry, the measure of an angle is always positive. In trigo-nometry, however, the measure of an angle may be positive or negative. If the rotation is in a counterclockwise direction, the measure of the angle is positive. If the rotation is in a clockwise direction, the measure of the angle is negative.

Positive Angle Measures	Negative Angle Measures
60°	−60°
420°	−420°

The measure of an angle can be given in <u>rotations</u> and in degrees.

One <u>full</u> counterclockwise rotation equals 360°. The terminal and initial sides of the angle coincide.

One-half a full counterclockwise rotation equals 180°.

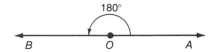

Example 1   For each rotation, find the measure of the angle in degrees. Then sketch the angle.

a. $\frac{1}{2}$ rotation, clockwise

b. $\frac{3}{4}$ rotation, counterclockwise

Solutions: a. $\frac{1}{2}(-360°) = -180°$

b. $\frac{3}{4}(360°) = 270°$

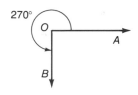

In trigonometry, it is convenient to consider angles in <u>standard position</u> in the coordinate plane.

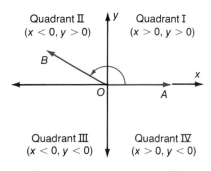

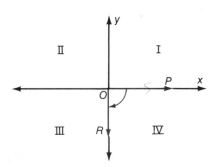

An angle is in **standard position** in the coordinate plane when its vertex is at the origin, point *O*, and its initial side coincides with the positive *x* axis. An angle in standard position is said to lie in the quadrant in which its terminal side lies. For example, ∠ *AOB*, in the figure at the left above, lies in Quadrant II. An angle whose terminal side coincides with the *x* axis or the *y* axis is a **quadrantal angle.** In the figure at the right above, ∠ *POR* is a quadrantal angle.

Angles in standard position whose terminal sides coincide are called **coterminal angles.**

Example 2 Ray *OB* is the terminal side of ∠ *AOB*, which is in standard position.

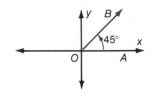

a. Find the measure of an angle coterminal with ∠ *AOB* under a counterclockwise rotation.

b. Find the measure of an angle coterminal with ∠ *AOB* under a counterclockwise rotation different from the rotation in **a.**

c. Find the measure of one angle coterminal with ∠ *AOB* under a clockwise rotation.

Solutions: **a.**

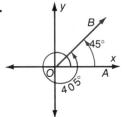

$360° + 45° = 405°$
Counterclockwise rotation

b.

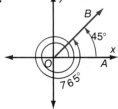

$2(360°) + 45° = 765°$
Counterclockwise rotation

c.

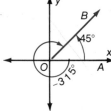

$-360° + 45° = -315°$
Clockwise rotation

Coterminal angles differ by an integral multiple of 360°.

$$405° = 1 \cdot 360° + 45°; \qquad 765° = 2 \cdot 360° + 45°; \qquad -315° = (-1) \cdot 360° + 45°$$

Thus, for any angle with measure θ (Greek letter "theta"), the measures of its coterminal angles are

$$\boldsymbol{n \cdot 360° + \theta,} \text{ where } n \text{ is an integer.}$$

Classroom Exercises

For each rotation, find the measure of the angle in degrees.

1. $\frac{1}{2}$ rotation, counterclockwise

2. $\frac{3}{4}$ rotation, clockwise

3. $\frac{2}{3}$ rotation, clockwise

4. $\frac{1}{3}$ rotation, counterclockwise

For each value of θ, find the measure of a coterminal angle under the given number of rotations.

5. $\theta = 30°$; 2 rotations, counterclockwise $= 750$
6. $\theta = -60°$; 1 rotation, clockwise
7. $\theta = -145°$; 1 rotation, counterclockwise
8. $\theta = -145°$; 2 rotations, clockwise

Written Exercises

Sketch an angle in standard position that results from the given rotation. Find the measure of the angle in degrees.

1. $\frac{1}{4}$ rotation, counterclockwise
2. $\frac{1}{2}$ rotation, counterclockwise
3. $\frac{3}{4}$ rotation, counterclockwise
4. 1 rotation, counterclockwise
5. $\frac{1}{4}$ rotation, clockwise
6. $\frac{1}{2}$ rotation, clockwise
7. $\frac{1}{6}$ rotation, clockwise
8. $\frac{1}{8}$ rotation, counterclockwise
9. $\frac{5}{12}$ rotation, counterclockwise
10. $\frac{3}{10}$ rotation, clockwise
11. $\frac{4}{5}$ rotation, counterclockwise
12. $\frac{3}{2}$ rotation, clockwise
13. $\frac{17}{12}$ rotation counterclockwise
14. $\frac{7}{6}$ rotation, clockwise
15. $\frac{2}{3}$ rotation, clockwise
16. $\frac{4}{3}$ rotation, counterclockwise

Let θ be the measure of angle AOB in standard position. Find the measure of an angle coterminal with angle AOB under the given number of rotations.

17. $\theta = 60°$; 2 clockwise
18. $\theta = -30°$; 1 clockwise
19. $\theta = 60°$; 2 counterclockwise
20. $\theta = -45°$; 1 counterclockwise
21. $\theta = -90°$; 1 counterclockwise
22. $\theta = 360°$; 1 clockwise
23. $\theta = -30°$; 3 clockwise
24. $\theta = 135°$; 2 clockwise
25. $\theta = 210°$; 2 clockwise
26. $\theta = -270°$; 1 counterclockwise

Let θ be the measure of an angle in standard position. For each value of θ, find the quadrant in which the angle lies.

27. $41°$
28. $-30°$
29. $135°$
30. $215°$
31. $-215°$
32. $420°$
33. $335°$
34. $-219°$
35. $181°$
36. $-539°$
37. $359°$
38. $-2000°$

Complete each statement in Exercises 39–42, where θ is the measure of ∠AOB in standard position. Assume that θ is in the given interval.

39. $0° < θ < 90°$; ∠AOB lies in Quadrant ? .

40. $180° < θ < 270°$; ∠AOB lies in Quadrant ? .

41. $90° < θ < 180°$; ∠AOB lies in Quadrant ? .

42. $-90° < θ < 0°$; ∠AOB lies in Quadrant ? .

Find a coterminal angle with measure θ, where $0° ≤ θ < 360°$, for each of the following.

43. 390°	**44.** 460°	**45.** 720°	**46.** 810°
47. 1000°	**48.** 1360°	**49.** −50°	**50.** −110°

51. Angle *AOB* is in standard position and $A(-3, 4)$ lies in its terminal side. Find the distance, *r*, from point *A* to the origin.

Let θ be the measure of an angle in standard position with point $P(x, y)$ in its terminal side. Let r be the distance of P from the origin. Find x and y for each value of r and θ.

52. $θ = 45°$; $r = 2$ **53.** $θ = 135°$; $r = 1$ **54.** $θ = -45°$; $r = 4$ **55.** $θ = 225°$; $r = 3$

56. For all angles in standard position with measure $θ > 0°$ that lie in Quadrant I, θ is between $n · 360°$ and $90° + n · 360°$. That is,

$$n · 360° < θ < 90° + n · 360°,$$

where *n* is a whole number. Find similar expressions for angles in Quadrants II, III, and IV.

57. Repeat Exercise 56 for $θ < 0°$.

Review Capsule for Section 1-3 _____

Simplify each radical.

Example: $\sqrt{\dfrac{25}{45}} = \dfrac{\sqrt{25}}{\sqrt{45}} = \dfrac{5}{\sqrt{9 \cdot 5}} = \dfrac{5}{3\sqrt{5}}$

The denominator, $3\sqrt{5}$, can be *rationalized*. That is,

$$\dfrac{5}{3\sqrt{5}} = \dfrac{5}{3\sqrt{5}} \cdot \dfrac{\sqrt{5}}{\sqrt{5}} = \dfrac{5\sqrt{5}}{3 \cdot 5} = \dfrac{\sqrt{5}}{3}.$$

Both $\dfrac{5}{3\sqrt{5}}$ and $\dfrac{\sqrt{5}}{3}$ are acceptable answers.

1. $\sqrt{8}$ **2.** $\sqrt{125}$ **3.** $\sqrt{52}$ **4.** $\sqrt{\dfrac{8}{25}}$ **5.** $\sqrt{\dfrac{3}{8}}$ **6.** $\dfrac{3}{2\sqrt{7}}$ **7.** $-\dfrac{2}{3\sqrt{5}}$

1-3 Sine, Cosine, and Tangent Functions

Consider the following figures where $P(x, y)$ is a point on the terminal side of an angle in standard position. When you construct a perpendicular from any point P on the terminal side of an angle to the x axis, a right triangle is formed. This triangle is called a **reference triangle.** In trigonometry, the distance r from the origin to a point P on the terminal side of an angle in standard position is the **radius vector.**

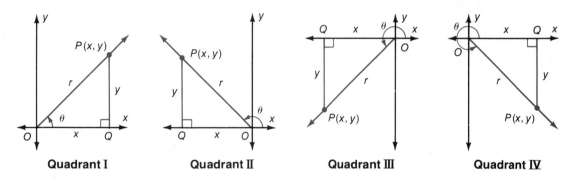

| Quadrant I | Quadrant II | Quadrant III | Quadrant IV |

The ratios $\frac{y}{r}$, $\frac{x}{r}$, and $\frac{y}{x}$ are given special names in trigonometry.

Definitions: Let $P(x, y)$ be any point (not the origin) on the terminal side of an angle in standard position with measure θ and let r be the radius vector. Then,

$$\textbf{sine } \theta \text{ (abbreviated sin } \theta) = \frac{y}{r},$$

$$\textbf{cosine } \theta \text{ (abbreviated cos } \theta) = \frac{x}{r},$$

$$\textbf{tangent } \theta \text{ (abbreviated tan } \theta) = \frac{y}{x}, x \neq 0.$$

You can find the radius vector r by applying the Pythagorean Theorem. Since $r^2 = x^2 + y^2$,

$$r = \sqrt{x^2 + y^2}. \quad \longleftarrow \text{ Distance is positive.}$$

Given the coordinates of a point P on the terminal side of an angle in standard position, you can evaluate $\sin \theta$, $\cos \theta$, and $\tan \theta$.

Trigonometric Functions **13**

Example 1 The terminal side of an angle in standard position passes through $P(4, -5)$. Sketch the reference triangle. Evaluate $\sin \theta$, $\cos \theta$, and $\tan \theta$.

Solution: Given $P(4, -5)$, $x = 4$ and $y = -5$. Thus, $r = \sqrt{4^2 + (-5)^2} = \sqrt{41}$.

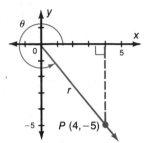

$$\sin \theta = \frac{y}{r} \qquad \cos \theta = \frac{x}{r} \qquad \tan \theta = \frac{y}{x}$$

$$\sin \theta = \frac{-5}{\sqrt{41}} \qquad \cos \theta = \frac{4}{\sqrt{41}} \qquad \tan \theta = \frac{-5}{4}$$

$$\sin \theta = -\frac{5}{\sqrt{41}} \qquad\qquad\qquad \tan \theta = -\frac{5}{4}$$

Given the value of one of the trigonometric ratios, you can find the values of the other two.

Example 2 θ is the measure of an angle in standard position that lies in Quadrant II and $\sin \theta = \frac{2}{7}$. Evaluate $\cos \theta$ and $\tan \theta$. Simplify all radicals.

Solution: Since $\sin \theta = \frac{2}{7}$, $y = 2$ and $r = 7$. Find x.

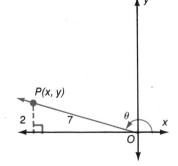

$$\sqrt{x^2 + 2^2} = 7$$
$$x^2 + 4 = 49$$
$$x^2 = 45$$
$$x = 3\sqrt{5} \text{ or } x = -3\sqrt{5}$$

Since θ is in Quadrant II, $x < 0$. Thus, $x = -3\sqrt{5}$.

Therefore, $\cos \theta = \dfrac{x}{r} = \dfrac{-3\sqrt{5}}{7}$ and $\tan \theta = \dfrac{y}{x} = \dfrac{2}{-3\sqrt{5}}$.

Thus, $\cos \theta = -\dfrac{3\sqrt{5}}{7}$ and $\tan \theta = -\dfrac{2}{3\sqrt{5}}$, or $-\dfrac{2\sqrt{5}}{15}$.

Recall that in similar triangles corresponding sides have the same ratio. Thus, as long as the measure of an angle in standard position remains the same, the ratios $\frac{y}{r}$, $\frac{x}{r}$, and $\frac{y}{x}$ will not be affected by a change in the position of point P on the terminal side of the angle.

> **Theorem 1–1** For an angle with measure θ in standard position, each of the ratios sin θ, cos θ, and tan θ is independent of the position of point P on the terminal side.

Proof: Let P_1 and P_2 be any two points on the terminal side of an angle with measure θ in standard position. Draw the reference triangle for each point. Then triangles OP_1Q_1 and OP_2Q_2 are similar and corresponding sides have the same ratio. That is,

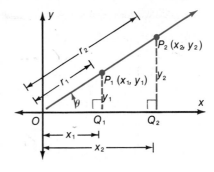

$$\frac{y_1}{r_1} = \frac{y_2}{r_2} = \sin\theta, \qquad \frac{x_1}{r_1} = \frac{x_2}{r_2} = \cos\theta,$$

and $\qquad \dfrac{y_1}{x_1} = \dfrac{y_2}{x_2} = \tan\theta.$

By Theorem 1–1, the sine, cosine, and tangent ratios depend only on the measure of the angle θ. Also, for each angle with measure θ, each ordered pair $(\theta, \sin\theta)$, $(\theta, \cos\theta)$, and $(\theta, \tan\theta)$ is <u>unique</u>. Thus, sin θ, cos θ, and tan θ are functions.

Classroom Exercises

The terminal side of an angle θ in standard position passes through the given point. Evaluate sin θ, cos θ, and tan θ.

1. $P(4, 3)$ **2.** $P(-4, 3)$ **3.** $P(-4, -3)$ **4.** $P(4, -3)$

Written Exercises

a

The terminal side of an angle with measure θ in standard position passes through the given point. Sketch the reference triangle. Evaluate sin θ, cos θ, and tan θ. Simplify all radicals.

1. $P(5, 12)$ **2.** $P(-5, 12)$ **3.** $P(-5, -12)$ **4.** $P(5, -12)$

5. $P(-6, -8)$ **6.** $P(6, -8)$ **7.** $P(6, 8)$ **8.** $P(-6, 8)$

9. $P(10, -24)$ **10.** $P(5, 10)$ **11.** $P(-6, 4)$ **12.** $P(-12, -10)$

Trigonometric Functions   **15**

θ is the measure of an angle in standard position that lies in the given quadrant. When sin θ is given, evaluate cos θ and tan θ. When cos θ is given, evaluate sin θ and tan θ. When tan θ is given, evaluate sin θ and cos θ. Simplify all radicals.

13. $\sin \theta = \frac{1}{2}$; Quadrant II

14. $\sin \theta = -\frac{1}{2}$; Quadrant III

15. $\tan \theta = -\frac{3}{5}$; Quadrant II

16. $\cos \theta = \frac{\sqrt{3}}{2}$; Quadrant IV

17. $\tan \theta = \frac{4}{5}$; Quadrant I

18. $\tan \theta = \frac{3}{5}$; Quadrant III

19. $\cos \theta = -\frac{\sqrt{2}}{2}$; Quadrant II

20. $\sin \theta = \frac{\sqrt{3}}{2}$; Quadrant II

The terminal side of an angle with measure θ in standard position passes through the given point. Sketch the reference triangle. Evaluate sin θ, cos θ, and tan θ. Simplify all radicals.

21. $P(-12, 5)$ **22.** $P(3, -4)$ **23.** $P(-2, 3)$ **24.** $P(5, 2)$

25. $P(-23, 7)$ **26.** $P(4, -5)$ **27.** $P(4, 3)$ **28.** $P(6, -3)$

θ is the measure of an angle in standard position that lies in the given quadrant. Evaluate the other two functions.

29. $\cos \theta = -\frac{3}{5}$; Quadrant III

30. $\sin \theta = \frac{3}{12}$; Quadrant I

31. $\tan \theta = -\frac{4}{3}$; Quadrant IV

32. $\sin \theta = -\frac{12}{13}$; Quadrant III

33. $\tan \theta = -1$; Quadrant II

34. $\cos \theta = -\frac{2}{9}$; Quadrant II

35. $\sin \theta = -\frac{5}{11}$; Quadrant IV

36. $\tan \theta = \frac{1}{3}$; Quadrant I

θ is the measure of an angle in standard position that lies in the given quadrant. Replace each $\underline{?}$ with $+$ or $-$ to indicate whether the values of the given functions are positive or negative.

	I	II	III	IV
37. $\sin \theta$	?	?	?	?
38. $\cos \theta$	?	?	?	?
39. $\tan \theta$	?	?	?	?

State the quadrant or quadrants in which an angle in standard position with measure θ lies under the given conditions.

40. $\cos \theta > 0$

41. $\sin \theta < 0$

42. $\tan \theta > 0$

43. $\tan \theta < 0$

44. $\cos \theta > 0$ and $\sin \theta < 0$

45. $\cos \theta < 0$ and $\sin \theta > 0$

46. $\cos \theta > 0$ and $\tan \theta > 0$

47. $\tan \theta < 0$ and $\sin \theta < 0$

48. $\cos \theta < 0$ and $\sin \theta < 0$

49. $\cos \theta < 0$ and $\tan \theta > 0$

50. $\sin \theta > 0$ and $\tan \theta > 0$

51. $\cos \theta > 0$ and $\tan \theta < 0$

1-4 Cosecant, Secant, and Cotangent Functions

There are six trigonometric functions. The reciprocals of the ratios that define the sine, cosine, and tangent functions are used to define the remaining three trigonometric functions.

Definitions: Let $P(x, y)$ be any point (not the origin) on the terminal side of an angle θ in standard position, and let r be the radius vector. Then

$$\text{\textbf{cosecant }} \theta \text{ (abbreviated csc } \theta) = \frac{r}{y}, \; y \neq 0$$

$$\text{\textbf{secant }} \theta \text{ (abbreviated sec } \theta) = \frac{r}{x}, \; x \neq 0$$

$$\text{\textbf{cotangent }} \theta \text{ (abbreviated cot } \theta) = \frac{x}{y}, \; y \neq 0.$$

Although θ is defined as the measure of an angle in standard position, it will be convenient also to let θ be the name of an angle as in the above definition. The context will make the use of θ clear.

Example 1

The terminal side of an angle in standard position passes through $P(-4, -7)$. Evaluate $\csc \theta$, $\sec \theta$, and $\cot \theta$.

Solution: Given $P(-4, -7)$, $x = -4$, and $y = -7$, find r.

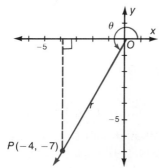

$$r = \sqrt{(-4)^2 + (-7)^2} = \sqrt{65}$$

$$\csc \theta = \frac{r}{y} \qquad\qquad \sec \theta = \frac{r}{x} \qquad\qquad \cot \theta = \frac{x}{y}$$

$$\csc \theta = \frac{\sqrt{65}}{-7} \qquad \sec \theta = \frac{\sqrt{65}}{-4} \qquad \cot \theta = \frac{-4}{-7}$$

$$\csc \theta = -\frac{\sqrt{65}}{7} \qquad \sec \theta = -\frac{\sqrt{65}}{4} \qquad \cot \theta = \frac{4}{7}$$

Given the value of one of the trigonometric functions, you can find the values of the other five.

Example 2

θ is the measure of an angle in standard position that lies in Quadrant IV and $\sec \theta = \frac{9}{7}$. Evaluate $\sin \theta$, $\cos \theta$, $\tan \theta$, $\csc \theta$, and $\cot \theta$. Simplify all radicals.

Solution: Since $\sec \theta = \frac{9}{7}$, you can let $x = 7$ and $r = 9$. Find y.

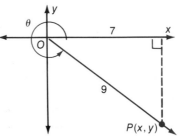

$$7^2 + y^2 = 9^2$$
$$49 + y^2 = 81$$
$$y^2 = 32$$
$$y = 4\sqrt{2} \quad \underline{\text{or}} \quad y = -4\sqrt{2}$$

Since θ is in Quadrant IV, $y < 0$. Thus, $y = -4\sqrt{2}$.

Therefore, $\sin \theta = -\dfrac{4\sqrt{2}}{9}$, $\cos \theta = \dfrac{7}{9}$, $\tan \theta = -\dfrac{4\sqrt{2}}{7}$,

$\csc \theta = -\dfrac{9}{4\sqrt{2}}$, or $-\dfrac{9\sqrt{2}}{8}$, and $\cot \theta = -\dfrac{7}{4\sqrt{2}}$, or $-\dfrac{7\sqrt{2}}{8}$.

By the definitions of the trigonometric functions,

$$\sin \theta = \frac{y}{r} \qquad \text{and} \qquad \csc \theta = \frac{r}{y}, \, y \neq 0,$$

where y is the ordinate of a point $P(x, y)$ on the terminal side of an angle θ in standard position, and r is the radius vector. Therefore, $\sin \theta$ and $\csc \theta$ are <u>reciprocals</u> of each other. Also, $\cos \theta$ and $\sec \theta$ are reciprocals of each other; and $\tan \theta$ and $\cot \theta$ are reciprocals of each other. That is,

$$\csc \theta = \frac{1}{\sin \theta}, \, \sin \theta \neq 0$$

$$\sec \theta = \frac{1}{\cos \theta}, \, \cos \theta \neq 0$$

$$\cot \theta = \frac{1}{\tan \theta}, \, \tan \theta \neq 0$$

Recall that the product of two numbers that are reciprocals of each other is 1. Thus, these <u>reciprocal</u> <u>properties</u> may also be written as:

$$\sin \theta \cdot \csc \theta = 1 \qquad \cos \theta \cdot \sec \theta = 1 \qquad \tan \theta \cdot \cot \theta = 1$$

Classroom Exercises

The terminal side of an angle θ in standard position passes through the given point. Evaluate $\csc \theta$, $\sec \theta$, and $\cot \theta$.

1. $P(8, 6)$ **2.** $P(-8, 6)$ **3.** $P(-8, -6)$ **4.** $P(8, -6)$

Written Exercises

a

The terminal side of an angle θ in standard position passes through the given point. Evaluate $\csc \theta$, $\sec \theta$, and $\cot \theta$. Simplify all radicals.

1. $P(-1, -1)$ **2.** $P(3, 3)$ **3.** $P(-2, 2)$ **4.** $P(5, -5)$

5. $P(5, 3)$ **6.** $P(-5, 3)$ **7.** $P(-5, -3)$ **8.** $P(5, -3)$

9. $P(2, -5)$ **10.** $P(1, 3)$ **11.** $P(-2, 1)$ **12.** $P(4, 2)$

θ is the measure of an angle in standard position that lies in the given quadrant. Evaluate the other five trigonometric functions of θ.

13. $\sec \theta = 4$; Quadrant IV

14. $\csc \theta = \frac{13}{12}$; Quadrant I

15. $\tan \theta = -\frac{2}{5}$; Quadrant II

16. $\cos \theta = -\frac{1}{3}$; Quadrant III

17. $\csc \theta = -\frac{3}{2}$; Quadrant III

18. $\cot \theta = -\frac{5}{12}$; Quadrant IV

19. $\sec \theta = -\frac{2\sqrt{3}}{3}$; Quadrant III

20. $\tan \theta = -1$; Quadrant II

θ is the measure of an angle in standard position that lies in the given quadrant. Replace each _?_ with $+$ or $-$ to indicate whether the values of the given functions are positive or negative.

	I	II	III	IV
21. $\csc \theta$	?	?	?	?
22. $\sec \theta$	?	?	?	?
23. $\cot \theta$	?	?	?	?

In Exercises 24–29, use the value of the given function to find the value of the reciprocal function.

24. $\sin \theta = -\frac{4}{9}$, $\csc \theta =$ _?_

25. $\cot \theta = \frac{\sqrt{6}}{5}$, $\tan \theta =$ _?_

26. $\sec \theta = -\frac{13}{5}$, $\cos \theta =$ _?_

27. $\csc \theta = 2$, $\sin \theta =$ _?_

28. $\tan \theta = -\frac{1}{5}$, $\cot \theta =$ _?_

29. $\cos \theta = -\frac{6}{7}$, $\sec \theta =$ _?_

Use the reciprocal properties to evaluate each expression.

30. $\sin 46° \cdot \csc 46° = \underline{?}$

31. $\tan 312° \cdot \cot 312° = \underline{?}$

32. $\cot 175° \cdot \tan 175° = \underline{?}$

33. $\cos 10° \cdot \sec 10° = \underline{?}$

State the quadrant or quadrants in which an angle in standard position with measure θ lies under the given conditions.

34. $\sec \theta > 0$

35. $\cot \theta < 0$

36. $\sec \theta > 0$ and $\csc \theta < 0$

37. $\sec \theta > 0$ and $\csc \theta > 0$

38. $\sec \theta < 0$ and $\cot \theta > 0$

39. $\csc \theta > 0$ and $\cot \theta < 0$

Given the values of two trigonometric functions, state the quadrant in which the angle lies. Give the values of the remaining four trigonometric functions.

40. $\sin \theta = \frac{3}{5}$; $\tan \theta = -\frac{3}{4}$

41. $\csc \theta = -\frac{13}{12}$; $\cos \theta = \frac{5}{13}$

42. $\tan \theta = -\frac{16}{63}$; $\csc \theta = -\frac{65}{16}$

43. $\sec \theta = \frac{29}{21}$; $\cos \theta = \frac{21}{29}$

44. $\sec \theta = -\frac{149}{140}$; $\csc \theta = -\frac{149}{51}$

45. $\cot \theta = -\frac{60}{11}$; $\tan \theta = -\frac{11}{60}$

Review Capsule for Section 1-5 _____

For Exercises 1–3, refer to right triangle ABC with $C = 90°$ and $A = 30°$. (HINT: Recall from geometry that the three sides of a 30°–60° right triangle are in the ratio $1 : 2 : \sqrt{3}$, where "1" corresponds to the side opposite the 30°-angle and "2" corresponds to the hypotenuse.

1. $c = 4$, $b = \underline{?}$, $a = \underline{?}$

2. $a = 1$, $c = \underline{?}$, $b = \underline{?}$

3. $b = \frac{\sqrt{3}}{2}$, $a = \underline{?}$, $c = \underline{?}$

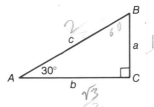

For Exercises 4–6, refer to right triangle ABC with $C = 90°$ and $A = 45°$. (HINT: Recall from geometry that the three sides of a 45°–45° right triangle are in the ratio $1 : 1 : \sqrt{2}$, where $\sqrt{2}$ corresponds to the hypotenuse.)

4. $c = \sqrt{2}$, $b = \underline{?}$, $a = \underline{?}$

5. $a = \frac{1}{\sqrt{2}}$, $b = \underline{?}$, $c = \underline{?}$

6. $b = 2$, $a = \underline{?}$, $c = \underline{?}$

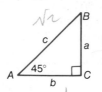

1-5 Values of the Trigonometric Functions

For an angle θ in standard position, each value of r, the radius vector, determines a unique point $P(x, y)$ on the terminal side of θ. By choosing a convenient value of r, you can use the definitions of the trigonometric functions to evaluate these functions for $\theta = 0°, 90°, 180°, 270°$, and $360°$. For these quadrantal angles, a convenient value of r is 1.

Example 1 For an angle θ in standard position, evaluate $\sin \theta$, $\cos \theta$, and $\tan \theta$ for each value of θ and r.

Solutions: a. Given $\theta = 90°$ and $r = 1$. Then $x = 0$, and $y = 1$.

Thus, $\sin 90° = \frac{1}{1} = 1$

$\cos 90° = \frac{0}{1} = 0$

$\tan 90° = \frac{1}{0}$, which is undefined.

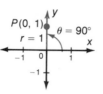

b. Given $\theta = 180°$ and $r = 1$. Then $x = -1$ and $y = 0$.

Thus, $\sin 180° = \frac{0}{1} = 0$

$\cos 180° = \frac{-1}{1} = -1$

$\tan 180° = \frac{0}{-1} = 0$

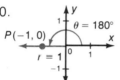

c. Given $\theta = 270°$ and $r = 1$. Then $x = 0$, and $y = -1$.

Thus, $\sin 270° = \frac{-1}{1} = -1$

$\cos 270° = \frac{0}{1} = 0$

$\tan 270° = \frac{-1}{0}$, which is undefined.

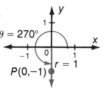

d. Given $\theta = 360°$ and $r = 1$. Then $x = 1$ and $y = 0$.

Thus, $\sin 360° = \frac{0}{1} = 0$

$\cos 360° = \frac{1}{1} = 1$

$\tan 360° = \frac{0}{1} = 0$.

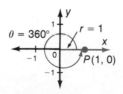

e. Since angles of $0°$ and $360°$ are coterminal, they have the same initial and terminal sides. Therefore, their trigonometric functions have the same values. Thus, $\sin 0° = 0$, $\cos 0° = 1$, and $\tan 0° = 0$.

By choosing a convenient value of the radius vector r, you can also evaluate the trigonometric functions for $\theta = 30°$, 45°, and 60°.

Recall from geometry that a 30°–60° right triangle has sides in the ratio $1 : 2 : \sqrt{3}$ and that a 45°–45° right triangle has sides in the ratio $1 : 1 : \sqrt{2}$.

Example 2

For an angle θ in standard position, evaluate $\sin\theta$, $\cos\theta$, and $\tan\theta$ for each value of θ and r.

a. $\theta = 30°$; $r = 2$ **b.** $\theta = 60°$; $r = 2$ **c.** $\theta = 45°$; $r = \sqrt{2}$

Solutions: **a.** Given $\theta = 30°$ and $r = 2$.
Then $PQ = 1$ and $OQ = \sqrt{3}$.
That is, $x = \sqrt{3}$ and $y = 1$.

Therefore, $\sin 30° = \dfrac{1}{2}$

$\cos 30° = \dfrac{\sqrt{3}}{2}$

$\tan 30° = \dfrac{1}{\sqrt{3}}$, or $\dfrac{\sqrt{3}}{3}$.

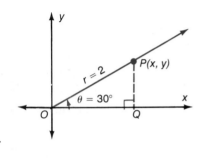

b. Given $\theta = 60°$ and $r = 2$.
Then $OQ = 1$ and $PQ = \sqrt{3}$.
That is, $x = 1$ and $y = \sqrt{3}$.

Therefore, $\sin 60° = \dfrac{\sqrt{3}}{2}$

$\cos 60° = \dfrac{1}{2}$

$\tan 60° = \sqrt{3}$.

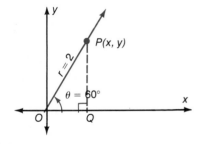

c. Given $\theta = 45°$ and $r = \sqrt{2}$.
Then $PQ = OQ = 1$.
That is, $x = 1$ and $y = 1$.

Therefore, $\sin 45° = \dfrac{1}{\sqrt{2}}$, or $\dfrac{\sqrt{2}}{2}$

$\cos 45° = \dfrac{1}{\sqrt{2}}$, or $\dfrac{\sqrt{2}}{2}$

$\tan 45° = 1$.

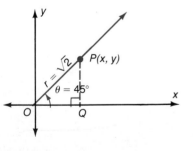

The method of Example 2 can also be used to find the function values for angles with terminal sides in Quadrants II, III, and IV.

Note that m ∠ POQ means "the measure of angle POQ."

Example 3 For each value of θ, evaluate $\sin\theta$, $\cos\theta$, and $\tan\theta$. Let $r = \sqrt{2}$.

a. $\theta = 135°$ **b.** $\theta = 225°$ **c.** $\theta = 315°$

Solutions: a. For $\theta = 135°$, m ∠ POQ = $(180° - 135°)$, or $45°$. Since $r = \sqrt{2}$ and θ is in Quadrant II, $x = -1$ and $y = 1$.

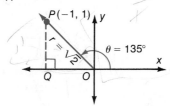

Therefore, $\sin 135° = \dfrac{1}{\sqrt{2}}$,

$\cos 135° = \dfrac{-1}{\sqrt{2}}$, or $-\dfrac{1}{\sqrt{2}}$

$\tan 135° = \dfrac{1}{-1}$, or -1.

b. For $\theta = 225°$, m ∠ POQ = $(225° - 180°)$, or $45°$. Since $r = \sqrt{2}$ and θ is in Quadrant III, $x = -1$ and $y = -1$.

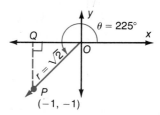

Therefore, $\sin 225° = \dfrac{-1}{\sqrt{2}}$, or $-\dfrac{1}{\sqrt{2}}$

$\cos 225° = \dfrac{-1}{\sqrt{2}}$, or $-\dfrac{1}{\sqrt{2}}$

$\tan 225° = \dfrac{-1}{-1}$, or 1.

c. For $\theta - 315°$, m ∠ POQ $-$ $(360° - 315°)$, or $45°$. Since $r = \sqrt{2}$ and θ is in Quadrant IV, $x = 1$ and $y = -1$.

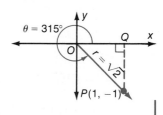

Therefore, $\sin 315° = \dfrac{-1}{\sqrt{2}}$, or $-\dfrac{1}{\sqrt{2}}$

$\cos 315° = \dfrac{1}{\sqrt{2}}$

$\tan 315° = \dfrac{-1}{1}$, or -1.

Note that for each value of θ in Example 3, m ∠ POQ $= 45°$. In each case, this angle is called the <u>reference angle</u>.

Definition: The **reference angle** of a given angle θ is the positive acute angle determined by the x axis and the terminal side of the given angle.

The figure below shows the relationship between an angle and its reference angle.

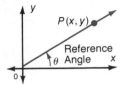
Quadrant I
Reference Angle: θ

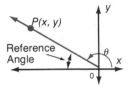
Quadrant II
Reference Angle: $180° - \theta$

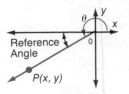
Quadrant III
Reference Angle: $\theta - 180°$

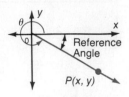
Quadrant IV
Reference Angle: $360° - \theta$

You can use the reference angle to find the values of trigonometric functions of θ where $0° < \theta < 360°$. The sign of the function value is positive or negative depending on the quadrant in which the terminal side of θ lies.

Classroom Exercises

For an angle θ in standard position, find the measure of the reference angle and evaluate sin θ, cos θ, and tan θ for each value of θ. Let $r = 2$.

1. $\theta = 120°$ **2.** $\theta = 150°$ **3.** $\theta = 135°$ **4.** $\theta = 315°$

Written Exercises

For an angle θ in standard position, evaluate sin θ, cos θ, and tan θ for each value of θ. Let $r = 2$.

1. $\theta = 210°$ **2.** $\theta = 330$ **3.** $\theta = 240°$ **4.** $\theta = 300°$

For an angle θ in standard position, evaluate sec θ, csc θ, and cot θ for each value of θ. Let $r = 1$.

5. $\theta = 0°$ **6.** $\theta = 90°$ **7.** $\theta = 180°$ **8.** $\theta = 270°$

For an angle θ in standard position, find the measure of the reference angle for each value of θ.

9. $\theta = 135°$ **10.** $\theta = 150°$ **11.** $\theta = 120°$ **12.** $\theta = 30°$

13. $\theta = 168°$ **14.** $\theta = 253°$ **15.** $\theta = 59°$ **16.** $\theta = 320°$

17. $\theta = 127°$ **18.** $\theta = 329°$ **19.** $\theta = 251°$ **20.** $\theta = 178°$

For an angle θ in standard position, find the measure of the reference angle and evaluate sec θ, csc θ, and cot θ for each value of θ and r.

21. $\theta = 30°; r = 2$　　　　**22.** $\theta = 150°; r = 2$　　　　**23.** $\theta = 210°; r = 2$

24. $\theta = 330°; r = 2$　　　　**25.** $\theta = 45°; r = \sqrt{2}$　　　　**26.** $\theta = 135°; r = \sqrt{2}$

27. $\theta = 225°; r = \sqrt{2}$　　　**28.** $\theta = 315°; r = \sqrt{2}$　　　**29.** $\theta = 60°; r = 2$

30. $\theta = 120°; r = 2$　　　　**31.** $\theta = 240°, r = 2$　　　　**32.** $\theta = 300°; r = 2$

Complete the table.

	θ	$\sin\theta$	$\cos\theta$	$\tan\theta$	$\cot\theta$	$\sec\theta$	$\csc\theta$
33.	0°						
34.	30°						
35.	45°						
36.	60°						
37.	90°						
38.	120°						
39.	135°						
40.	150°						
41.	180°						
42.	210°						
43.	225°						
44.	240°						
45.	270°						
46.	300°						
47.	315°						
48.	330°						
49.	360°						

The terminal side of an angle θ in standard position passes through the given point. Sketch the angle and evaluate sin θ, cos θ, tan θ, csc θ, sec θ, and cot θ. Then find the value of θ where 0° ≤ θ < 360°.

50. $P(2, 0)$　　　　　　**51.** $P(0, -3)$　　　　　　**52.** $P(-5, 0)$

53. $P(0, 4)$　　　　　　**54.** $P(2\sqrt{3}, 2)$　　　　　**55.** $P(-\frac{\sqrt{3}}{2}, -\frac{1}{2})$

56. $P(-3\sqrt{3}, 3)$　　　　**57.** $P(\frac{\sqrt{3}}{2}, -\frac{1}{2})$　　　　**58.** $P(3, 3\sqrt{3})$

59. $P(-\frac{1}{2}, \frac{\sqrt{3}}{2})$　　　　**60.** $P(-\frac{1}{2}, -\frac{\sqrt{3}}{2})$　　　　**61.** $P(2, -2\sqrt{3})$

62. $P(\sqrt{2}, \sqrt{2})$　　　　**63.** $P(-3, -3)$　　　　**64.** $P(2, -2)$

For an angle θ in standard position, let r be the distance from the origin to the point $P(x, y)$ on the terminal side. Find x and y for each value of θ and r.

65. $\theta = 30°$; $r = 1$

66. $\theta = 60°$; $r = 4$

67. $\theta = 90°$; $r = 5$

68. $\theta = 45°$; $r = 1$

69. $\theta = 225°$; $r = 1$

70. $\theta = 120°$; $r = 1$

71. $\theta = 0°$; $r = 7$

72. $\theta = 150°$; $r = 4$

73. $\theta = 300°$; $r = 4$

74. $\theta = 315°$, $r = 1$

75. $\theta = 210°$; $r = 1$

76. $\theta = 270°$; $r = 3$

For an angle θ in standard position where $90° < \theta \leq 360°$, find all values of θ that make the given statement true.

77. $\sin \theta = \sin 60°$

78. $\sin \theta = -\sin 30°$

79. $\cos \theta = \sin 60°$

80. $\cos \theta = -\sin 30°$

81. $\cos \theta = \sin 90°$

82. $\sin \theta = \cos 90°$

_____ Review _____

Find the distance between points P_1 and P_2. Write answers in simplest form. (Section 1–1)

1. $P_1(2, 6)$; $P_2(11, 8)$

2. $P_1(4, -2)$; $P_2(-6, 1)$

3. Tell whether the relation $\{(2, -3), (4, 2)\}$ is a function. (Section 1–1)

4. Identify the domain and range of this function. (Section 1–1)

$$\{(-2, -2), (-1, -1), (0, 0), (1, -1), (2, -2)\}$$

For each rotation, find the measure of the angle in degrees. (Section 1–2)

5. $1\frac{3}{4}$ rotations, clockwise

6. $\frac{2}{3}$ rotation, counterclockwise

In Exercises 7–12, the terminal side of an angle θ in standard position passes through the point $(-8, 6)$. Find each value. (Sections 1–3 and 1–4)

7. $\sin \theta$ **8.** $\cos \theta$ **9.** $\tan \theta$ **10.** $\cot \theta$ **11.** $\sec \theta$ **12.** $\csc \theta$

In Exercises 13–17, $\sin \theta = -\frac{5}{13}$ and θ is in Quadrant III. Find each value. (Sections 1–3 and 1–4)

13. $\cos \theta$ **14.** $\tan \theta$ **15.** $\cot \theta$ **16.** $\sec \theta$ **17.** $\csc \theta$

Find each value. (Section 1–5)

18. $\cot 90°$ **19.** $\sin 270°$ **20.** $\cos 60°$ **21.** $\tan 45°$ **22.** $\csc 330°$

1-6 Using Tables

In Section 1–5, you found exact values of the trigonometric functions of some special angles. Most values of the trigonometric functions are irrational numbers; that is, the values are infinite, nonrepeating decimals. Tables of decimal approximations for these values can be found on pages 368–372. A portion of the table is shown below. Recall that one degree equals 60 minutes ($1° = 60'$).

θ Deg.	Sin θ	Cos θ	Tan θ	
23°00′	.3907	.9205	.4245	67°00′
10′	.3934	.9194	.4279	50′
20′	.3961	.9182	.4314	40′
30′	.3987	.9171	.4348	30′
40′	.4014	.9159	.4383	20′
50′	.4041	.9147	.4417	10′
24°00′	.4067	.9135	.4452	66°00′
10′	.4094	.9124	.4487	50′
20′	.4120	.9112	.4522	40′
30′	.4147	.9100	.4557	30′
40′	.4173	.9088	.4592	20′
50′	.4200	.9075	.4628	10′
25°00′	.4226	.9063	.4663	65°00′
10′	.4253	.9051	.4699	50′
20′	.4279	.9038	.4734	40′
30′	.4305	.9026	.4770	30′
	Cos θ	Sin θ	Cot θ	θ Deg.

You can read the values of the trigonometric functions directly from the table for many values of θ. For $0° \leq \theta \leq 45°$, θ is found in the left column using the headings at the top of the table. For $45° \leq \theta \leq 90°$, θ is read in the right column using the headings at the bottom of the table.

The left column stops at 45° because every angle between 0° and 45° has a <u>complement</u> between 45° and 90°. Recall that two angles are **complements** of each other if the sum of their measures is 90°.

As you can see from the figures at the left below, placing the complement of an angle in standard position interchanges the values of x and y. Thus,

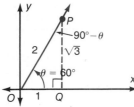

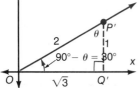

$$\left. \begin{array}{l} \sin \theta = \dfrac{y}{r} = \cos(90° - \theta) \\[1em] \cos \theta = \dfrac{x}{r} = \sin(90° - \theta) \end{array} \right\}$$ ⟵ Sine and cosine are **cofunctions.**

$$\left. \begin{array}{l} \tan \theta = \dfrac{y}{x} = \cot(90° - \theta) \\[1em] \cot \theta = \dfrac{x}{y} = \tan(90° - \theta) \end{array} \right\}$$ ⟵ Tangent and cotangent are **cofunctions.**

$$\left. \begin{array}{l} \sec \theta = \dfrac{r}{x} = \csc(90° - \theta) \\[1em] \csc \theta = \dfrac{r}{y} = \sec(90° - \theta) \end{array} \right\}$$ ⟵ Secant and cosecant are **cofunctions.**

These results suggest that any trigonometric function of θ, $0° < \theta < 90°$, is equal to the <u>cofunction</u> of its complementary angle. You use this fact when you read values from the table.

Although most values in the table are approximations, we shall use the "=" symbol to write statements involving these values because it is more convenient.

Example 1   Use the table on pages 368–372 to find each value.

a. $\sin 23°40'$ **b.** $\cos 66°20'$

Solutions: a. 1. Find $23°40'$ in the column under "θ Deg."

     2. Look directly <u>right</u> along the row until you reach the column below the heading "Sin θ."

     3. Read: $\sin 23°40' = .4014$

b. 1. Find $66°20'$ in the column over "θ Deg."

     2. Look directly <u>left</u> along the row until you reach the column above the heading "Cos θ."

     3. Read: $\cos 66°20' = .4014$

You can use the table and a reference angle to find the function values for an angle θ in standard position, where the terminal side of the angle lies in Quadrants II, III, or IV.

You can also apply the relationship regarding coterminal angles developed in Section 1–2. That is, since for every angle θ the measures of its coterminal angles are $n \cdot 360° + \theta$, where n is an integer, functions of coterminal angles will have the same values. Thus,

$$\sin (n \cdot 360° + \theta) = \sin \theta$$

$$\cos (n \cdot 360° + \theta) = \cos \theta$$

$$\tan (n \cdot 360° + \theta) = \tan \theta$$

Example 2 Use the table on pages 368–372 to find each value.

a. $\cos 600°20'$ **b.** $\sin (-40°)$ **c.** $\tan 848°$

Solutions: a. $\cos 600°20' = \cos (360° + 240°20')$

$= \cos 240°20'$ ⟵ Coterminal angle

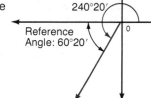

For $\theta = 240°20'$, the reference angle is $(240°20' - 180°)$, or $60°20'$.

Since $240°20'$ lies in Quadrant III,
$\cos 240°20' = -\cos 60°20'$

$= -.4950$ ⟵ From the table

b. $\sin (-40°) = \sin (-1 \cdot 360 + 320°)$

$= \sin 320°$ ⟵ Coterminal angle

For $\theta = 320°$, the reference angle is $(360° - 320°)$, or $40°$.

Since $320°$ lies in Quadrant IV,
$\sin (-40°) = -\sin 40°$

$= -.6428$ ⟵ From the table

c. $\tan 868° = \tan (2 \cdot 360° + 148°)$

$= \tan 148°$ ⟵ Coterminal angle

For $\theta = 148°$, the reference angle is $(180° - 148°)$, or $32°$.

Since $148°$ lies in Quadrant II,
$\tan 148° = -\tan 32°$

$= -.6249$

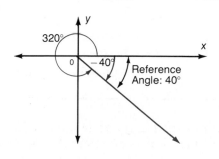

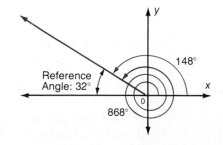

Given the value of a trigonometric function of θ, you can also use the table to find θ.

Example 3 Find θ.

a. sec $\theta = -3.994$ and
$0° < \theta < 360°$

b. tan $\theta = -.1883$ and
$270° < \theta < 360°$

Solutions: a. 1. Find 3.994 in the column above "Sec θ."

2. Look directly <u>right</u> along the row to find the column above "θ Deg."

3. Read: $\theta = 75°30'$

4. Since sec $\theta < 0$, the terminal side of θ must lie in Quadrant II or III. Thus,
$\theta = 180° - 75°30'$
$\quad = 179°60' - 75°30'$
$\quad = 104°30'$
or $\theta = 180° + 75°30'$
$\quad = 255°30'$.

b. 1. Find .1833 in the column under "Tan θ."

2. Look directly <u>left</u> along the row to find the column under "θ Deg."

3. Read: $\theta = 10°40'$

4. Since tan $\theta < 0$, and
$270° < \theta < 360°$,
$\theta = 360° - 10°40'$
$\quad = 359°60' - 10°40'$
$\quad = 349°20'$.

Classroom Exercises

Use the table on pages 368–372 to find each value.

1. cos 25°

2. sin 57°40'

3. tan 38°20'

4. cot 41°40'

5. sec 12°50'

6. csc 81°30'

Express each function in terms of the same function of the reference angle.

7. tan 518°

8. cos (−35°)

9. sin 268°

10. tan 755°

11. cot (−120°)

12. sec (−205°)

13. sin 401°50'

14. tan 123°40'

15. cos 982°

16. csc (−380°15')

17. cot (−600°40')

18. sec 220°

19. tan 640°20'

20. cos 380°50'

21. sin (−800°)

Written Exercises

Use the table on pages 368–372 to find each value.

1. sin 38°	**2.** cos 14°10′	**3.** tan 27°20′
4. csc 41°	**5.** sec 48°10′	**6.** sin 86°20′
7. sec 63°40′	**8.** cot 52°10′	**9.** cos 89°40′
10. tan 87°50′	**11.** csc 0°10′	**12.** sin 45°10′
13. tan 78°20′	**14.** csc 43°30′	**15.** cos 44°50′
16. sec 39°	**17.** cot 1°40′	**18.** sec 25°50′
19. sin 88°10′	**20.** cos 3°50′	**21.** csc 79°
22. cot 68°40′	**23.** cot 35°20′	**24.** tan 0°10′
25. sin 44°50′	**26.** cos 45°10′	**27.** csc 17°30′
28. cot 13°40′	**29.** cos 122°	**30.** tan 167°30′
31. sin 113°40′	**32.** cot 178°10′	**33.** csc 247°
34. sec 200°50′	**35.** cot 265°30′	**36.** tan 257°20′
37. sin 285°40′	**38.** csc 307°10′	**39.** cos 342°20′
40. sec 351°30′	**41.** tan 242°20′	**42.** cot 137°40′
43. csc 278°50′	**44.** sin 112°50′	**45.** sec 97°50′
46. cos 355°20′	**47.** tan 268°50′	**48.** csc 123°40′
49. cot 310°30′	**50.** sin 235°10′	**51.** sec 175°20′
52. cos (−50°)	**53.** tan 737°	**54.** sin 857°
55. cot (−138°)	**56.** csc 375°20′	**57.** sec (−38°10′)
58. cot 1293°40′	**59.** tan 708°30′	**60.** sin (−215°10′)
61. csc 1486°	**62.** cos 996°50′	**63.** sec 571°10′

Find θ.

64. sec $\theta = 1.131$ and $0° < \theta < 360°$

65. tan $\theta = -.6168$ and $270° < \theta < 360°$

66. cos $\theta = -.3961$ and $90° < \theta < 270°$

67. csc $\theta = -1.766$ and $180° < \theta < 360°$

68. sin $\theta = .9135$ and $0° < \theta < 360°$

69. cot $\theta = .7177$ and $180° < \theta < 270°$

70. tan $\theta = .2065$ and $0° < \theta < 360°$

71. cos $\theta = -.9051$ and $0° < \theta < 360°$

72. sin $\theta = -.3665$ and $0° < \theta < 360°$

1-7 Interpolation

Sometimes the measure of an angle falls between two entries in the Table. In such cases, you use a procedure called **linear interpolation.**

Example 1 Find sin 38°14′ to four decimal places.

Solution: In the Table, 38°14′ lies between 38°10′ and 38°20′.

$$
\begin{array}{cc}
\theta & \sin\theta \\
\end{array}
$$

$$
10'\left[\ 4'\left[\begin{array}{c}38°10' \\ 38°14'\end{array}\right.\right.\quad \left.\begin{array}{c}.6180 \\ \underline{\ ?\ }\end{array}\right]d\ \Bigg]\ .0022
$$

$$
38°20' \qquad\qquad .6202
$$

Since 38°14′ is $\frac{4}{10}$ of the distance between 38°10′ and 38°20′, sin 38°14′ is about $\frac{4}{10}$ of the distance between .6180 and .6202. That is,

$$\frac{4}{10} = \frac{d}{.0022}$$

$$d = \frac{.0088}{10}, \text{ or } .00088.$$

Thus, sin 38°14′ = sin 38°10′ + d

$$= .6180 + .00088$$

$$= .61888, \text{ or } .6189.$$

A similar procedure is used to find cos 138°07′.

Example 2 Find cos 138°07′ to four decimal places.

Solution: 1. An angle of 138°07′ has its terminal side in Quadrant II, Therefore, the reference angle for 138°07′ is 180° − 138°07′ or 41°53′ and cos 138°07′ = −cos 41°53′.

2. Use the table to find $\cos 41°53'$. $41°53'$ lies between $41°50'$ and $42°00'$.

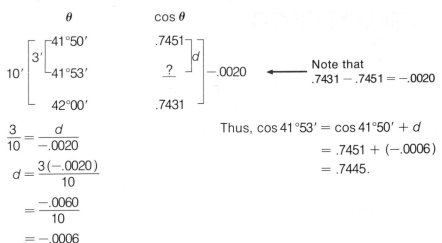

	θ		$\cos\theta$	

$$\begin{array}{ccc} & \theta & \cos\theta \\ 10' \left[3' \left[\begin{array}{c} 41°50' \\ 41°53' \end{array} \right. \right. & & \left. \left. \begin{array}{c} .7451 \\ ? \end{array} \right] d \right] -.0020 \\ & 42°00' & .7431 \end{array}$$

Note that
$.7431 - .7451 = -.0020$

$$\frac{3}{10} = \frac{d}{-.0020}$$

$$d = \frac{3(-.0020)}{10}$$

$$= \frac{-.0060}{10}$$

$$= -.0006$$

Thus, $\cos 41°53' = \cos 41°50' + d$
$= .7451 + (-.0006)$
$= .7445.$

3. Since $\cos 41°53' = .7445$, $\cos 138°07' = -.7445$.

Given a value of one of the trigonometric functions of θ, you can use the table to find θ.

Example 3 Find θ to the nearest minute if $\tan\theta = 1.1485$ and $0° < \theta < 360°$.

Solution: 1. Look in the column headed $\tan\theta$. 1.1485 falls between 1.1436 and 1.1504.

$$\begin{array}{ccc} & \theta & \tan\theta \\ 10' \left[d \left[\begin{array}{c} 48°50' \\ \theta \end{array} \right. \right. & & \left. \left. \begin{array}{c} 1.1436 \\ 1.1485 \end{array} \right] .0049 \right] .0068 \\ & 49°00' & 1.1504 \end{array}$$

$$\frac{d}{10} = \frac{.0049}{.0068}$$

$$d = 7.2, \text{ or } 7'$$

Thus, $\theta = 48°50' + d$
$= 48°50' + 7'$
$= 48°57'.$

2. Since $\tan\theta > 0$, and $0° < \theta < 360°$, the terminal side of θ lies in Quadrant I or III. Therefore, $\theta = 48°57'$ or $\theta = (180° + 48°57') = 228°57'$.

Classroom Exercises

Express each function in terms of the same function of the reference angle.

1. sin 165°48'   **2.** cos 325°15'   **3.** tan 310°23'   **4.** csc 213°12'

Written Exercises

Use the table on pages 368–372 and interpolation to find each value.

1. cos 64°45'   **2.** sin 40°36'   **3.** tan 22°12'   **4.** csc 64°23'

5. sec 42°18'   **6.** cot 75°29'   **7.** cos 51°36'   **8.** tan 34°23'

9. sin 17°20'   **10.** cos 87°30'   **11.** csc 11°33'   **12.** sec 43°17'

13. tan 125°37'   **14.** csc 217°12'   **15.** cot 275°46'   **16.** cos 145°48'

17. csc 200°53'   **18.** tan 283°22'   **19.** sin 116°32'   **20.** cot 225°28'

21. cot 162°36'   **22.** sec 149°57'   **23.** sec 99°58'   **24.** sin 357°24'

25. sin 156°45'   **26.** cos 267°54'   **27.** tan 304°14'   **28.** csc 123°14'

Find θ to the nearest minute.

29. $\sin \theta = -.7482$ and $0° < \theta < 360°$ **30.** $\cos \theta = .0766$ and $0° < \theta < 360°$

31. $\tan \theta = -.8932$ and $90° < \theta < 270°$ **32.** $\cot \theta = 1.8178$ and $0° < \theta < 270°$

33. $\csc \theta = -5.582$ and $90° < \theta < 270°$ **34.** $\sec \theta = -1.187$ and $180° < \theta < 360°$

35. $\sec \theta = 1.238$ and $0 < \theta < 180°$ **36.** $\sin \theta = -.2715$ and $180° < \theta < 270°$

37. $\cos \theta = -.7324$ and $180° < \theta < 270°$ **38.** $\tan \theta = 4.349$ and $90° < \theta < 360°$

Review Capsule for Section 1-8

Find the complement of each angle.

Example: 28°40'   **Solution:** First write 90° as: 89°60'
Then subtract.   28°40'
The complement is: **61°20'**

1. 24°   **2.** 32°20'   **3.** 68°30'   **4.** 44°50'

1-8 Solving Right Triangles

Right triangles have properties that are directly related to the trigonometric functions. In the figure at the right, θ is an angle in standard position where $0° < \theta < 90°$ and triangle OQP is a right triangle. The values of the trigonometric functions of θ can be expressed in terms of the lengths of the sides of triangle OQP where

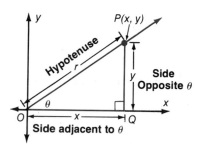

$x =$ length of side adjacent to θ
$y =$ length of side opposite θ
$r = \sqrt{x^2 + y^2} =$ length of the hypotenuse

$\sin \theta = \dfrac{\text{length of side opposite } \theta}{\text{length of hypotenuse}}$ $\qquad \csc \theta = \dfrac{\text{length of hypotenuse}}{\text{length of side opposite } \theta}$

$\cos \theta = \dfrac{\text{length of side adjacent to } \theta}{\text{length of hypotenuse}}$ $\qquad \sec \theta = \dfrac{\text{length of hypotenuse}}{\text{length of side adjacent to } \theta}$

$\tan \theta = \dfrac{\text{length of side opposite } \theta}{\text{length of side adjacent to } \theta}$ $\qquad \cot \theta = \dfrac{\text{length of side adjacent to } \theta}{\text{length of side opposite } \theta}$

To **solve a triangle** means to find the measures of the unknown sides and angles of the triangle.

Example 1

In the triangle at the right, $b = 5$, $c = 13$, and $C = 90°$. Solve the triangle. Find angles to the nearest minute.

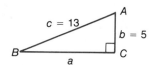

Solution: Since two sides of right triangle ABC are given, you can use the Pythagorean Theorem to find a.

$a = \sqrt{(13)^2 - (5)^2} = 12$ ⟵ Now use $\sin B$ to find B.

$\sin B = \frac{5}{13} = .3846$ ⟵ Use the table on page 370 and interpolation to find B.

$B = 22°37'$

Since angles A and B are complementary, $A = 67°23'$.
Thus, $a = 12$, $B = 22°37'$, and $A = 67°23'$.

In Example 2, the measures of a side and one acute angle are given.

Example 2   Sove this triangle. Find each side to the nearest
whole number.

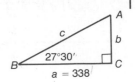

Solution: Given: $a = 338$, $B = 27°30'$, $C = 90°$.
Since angles A and B are complementary, $A = 62°30'$.
To find b, you can use $\tan B = \dfrac{b}{a}$.

Thus, $\tan 27°30' = \dfrac{b}{338}$, or $338(\tan 27°30') = b$.

Therefore, $b = 338(.5206)$ ⟵ Use the table
or a calculator.

$= 176$.

To find c, you can use $\sec B = \dfrac{c}{a}$. (You can also use $\cos B$.)

Thus, $\sec 27°30' = \dfrac{c}{338}$, or $338(\sec 27°30') = c$.

Therefore, $c = 338(1.127)$ ⟵ Use the table
or a calculator.

$= 381$.

Thus, $A = 62°30'$, $b = 176$, $c = 381$.

Classroom Exercises

Refer to the figure at the right. Replace each ? with
a ratio in terms of a, b, and c.

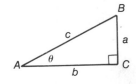

1. $\sin \theta = \underline{?}$ **2.** $\csc \theta = \underline{?}$

3. $\cos \theta = \underline{?}$ **4.** $\sec \theta = \underline{?}$

Refer to the figure at the right.

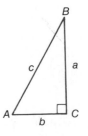

5. If you know the measure of $\angle B$ and the value of b, which
trigonometric function would you use to find c?

6. If you know the measure of $\angle A$ and the value of b, which
trigonometric function would you use to find c?

7. If you know the measure of $\angle B$ and the value of a, which
trigonometric function would you use to find b?

Written Exercises

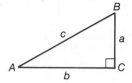

In Exercises 1–21, refer to the figure at the right to solve each right triangle. Find angles to the nearest minute and sides to the nearest whole number.

1. $c = 30$, $b = 24$
2. $a = 6$, $b = 8$
3. $a = 14$, $b = 14$
4. $c = 10$, $a = 8$
5. $c = 2$, $b = 1$
6. $a = 28$, $b = 45$
7. $c = 75$, $b = 45$
8. $c = 97$, $a = 72$
9. $c = 169$, $a = 119$
10. $b = 4.1$, $c = 7.3$
11. $c = 23.2$, $a = 15.3$
12. $b = 635$, $a = 446$
13. $A = 45°$, $b = 17$
14. $B = 30°$, $a = 12$
15. $A = 60°$, $a = 1.5$
16. $A = 61°$, $b = 18$
17. $B = 62°30'$, $c = 30$
18. $A = 35°40'$, $c = 20$
19. $B = 72°35'$, $a = 3420$
20. $A = 43°42'$, $a = 16.42$
21. $B = 62°53'$, $c = 74.37$

Review

Use the table on pages 368–372 to find each value. (Section 1–6)

1. $\sin 16°40'$
2. $\cos 217°30'$
3. $\tan 286°10'$

Use the table on pages 368–372 to find θ, where $0° < \theta < 90°$. (Section 1–6)

4. $\sin \theta = .3907$
5. $\cos \theta = .3665$
6. $\tan \theta = .4314$

Use the table on pages 368–372 for Exercises 7–8. (Section 1–7)

7. Find $\cos 38°28'$ to four decimal places.
8. If $\sin \theta = .3982$, where $0° < \theta < 90°$, find θ to the nearest minute.

Use the table on pages 368–372 for Exercises 9–10. (Section 1–8)

9. In the figure at the left below, $a = 24$, $c = 26$, and $C = 90°$. Solve the triangle. Find angles to the nearest minute.

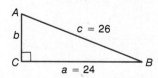

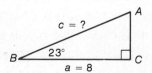

10. In the figure at the right above, $a = 8$ and $B = 23°$. Find c to the nearest tenth.

Significant Digits

To better understand about the accuracy of measurements, it is important to understand the concept of significant digits.

> **RULE:** Given a measurement:
> 1. Each nonzero digit is significant.
> 2. Any zero used to place the decimal point is not significant.
> 3. All other zeros are significant.

The position of the decimal point has nothing to do with determining the number of significant digits in a given measure. For example, 250, .25 and .025 each have two significant digits, because each has the same degree of accuracy, which is determined by the relative error.

A measure of 250 units means that the "true" measure is between 245 and 255. Thus, 250 has an **absolute error** of 5. The ratio of the absolute error to the measurement is called the **relative error.** Thus,

$$\frac{5}{250} = .02 \longleftarrow \text{ Relative error}$$

Further, a measure of .25 is between .245 and .255. Consequently, .25 has an absolute error of .005. Thus,

$$\frac{.005}{.25} = .02 \longleftarrow \text{ Relative error}$$

Lastly, a measure of .025 is between .0245 and .0255. Consequently, .025 has an absolute error of .0005. Thus,

$$\frac{.0005}{.025} = .02 \longleftarrow \text{ Relative error}$$

A similar discussion will illustrate why .000025 has two significant digits. Thus, the four zeros in this number are not significant.

Ordinarily, the zeros in a large number like 25,000,000 are used to place the decimal point. Thus, the zeros are not significant. However, there are times when such zeros are significant. In such cases, scientific notation may be used to show which zeros, if any, are significant.

$$25,000,000 = 2.5 \times 10^7 \longleftarrow \textbf{2 significant digits}$$
$$25,000,000 = 2.50 \times 10^7 \longleftarrow \textbf{3 significant digits}$$

Since the zeros in 25.0 and .250 are not used to place the decimal point, they are significant. Thus, these measures each have three significant digits.

Much of the data in applied problems in trigonometry is obtained through measurement, and measurements are approximations. Thus, the following rules are applied when multiplying, dividing, adding, or subtracting with approximations.

RULE: Perform the multiplication or division as if the numbers were exact. Then round the answer to the smallest number of significant digits which occurs in any of the numbers used.

RULE: When adding or subtracting approximations, perform the addition or subtraction as if the numbers were exact. Then round the answer to the same number of decimal places as the approximation with the least number of decimal places.

Throughout this text, attention has been paid to the concepts discussed on these pages by indicating in Examples and Exercises what degree of accuracy is required.

EXERCISES

1. Show that .000025 and 25,000 have the same relative error.

   Round each of the following to the given number of significant digits.

2. 61,458; three significant digits
3. .00215388; five significant digits
4. 6.312162; four significant digits
5. 125.51; two significant digits

   Evaluate c sin 41°55′ for each value of c.

6. $c = 3$
7. $c = 3.2$
8. $c = 3.18$

   Find each quotient. Assume that all measures are approximations.

9. $75.625 \div .284$
10. $750.0 \div 50.0$
11. $56 \div 2$

12. The sides of a quadrilateral are 118.13 feet, 147 feet, 205.3 feet, and 106.03 feet. Find the perimeter.

13. The sides of a quadrilateral are 10.2 centimeters, 5.76 centimeters, 3.42 centimeters, and 12.3 centimeters. Find the perimeter.

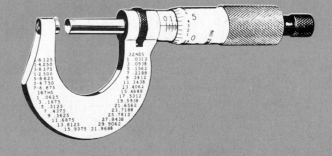

Trigonometric Functions   **39**

Trigonometric Functions

You can use a scientific calculator instead of tables to find the value of trigonometric functions. Example 1 shows how to use the calculator to change minutes (60 minutes = 1 degree) to decimal form.

Example 1   Change 37°15' to decimal form.

Solution: [3] [7] [+] [1] [5] [÷] [6] [0] [=] $\boxed{37.25}$

Example 2 shows how to use a scientific calculator to find the value of trigonometric functions. On calculators that do not have keys for the secant, cosecant, and cotangent functions, you use the corresponding reciprocal function and the reciprocal key [1/x].

Example 2   **a.** $\sin 34° = \underline{?}$ **b.** $\cot 9°40' = \underline{?}$

Solutions: a. [3] [4] [sin]

$\boxed{0.5591929}$

b. [9] [+] [4] [0] [÷] [6] [0] [=] [tan] [1/x] $\boxed{5.8708042}$

Given the value of a trigonometric function, you can use the [INV] key to find the corresponding angle measure. On calculators that do not have keys for the secant, cosecant, and cotangent functions, you use the [1/x] key, the [INV] key, and the corresponding reciprocal functions.

Example 3   **a.** $\cos \theta = .4226;\ \theta = \underline{?}$ **b.** $\csc \theta = 3.179;\ \theta = \underline{?}$

Solutions: a. [·] [4] [2] [2] [6] [INV] [cos] $\boxed{65.001154}$

b. [3] [·] [1] [7] [9] [1/x] [INV] [sin] $\boxed{18.334515}$

To change .334515 to minutes, multiply by 60 and round to the nearest ten minutes.

[·] [3] [3] [4] [5] [1] [5] [×] [6] [0] [=] $\boxed{20.0709}$

Thus, $\theta = 18°20'$.

1-9 Applications: Angle of Elevation/Depression

The trigonometric functions are often used in calculating distances that are difficult to measure directly. In many of these instances, an angle is determined by a horizontal line and a <u>line of sight</u>.

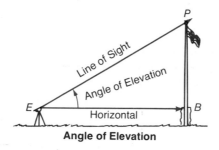

Angle of Elevation

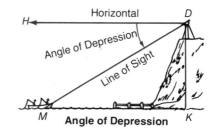

Angle of Depression

In the figure at the right above, lines *HD* and *MK* are parallel. Therefore, the measure of angle *HDM*, the **angle of depression,** equals the measure of angle *KMD*.

It is important to remember that the height of the person doing the sighting is a factor in determining the sides of the triangle used.

Example 1

A person two meters tall stands five meters from a building. The angle of elevation from where the person stands to the top of the building is 75°. Find the height of the building. Give your answer to the nearest meter.

Solution: Height of the building: $BC + 2$

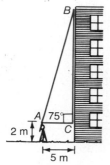

$$\tan 75° = \frac{BC}{5}$$

$$BC = 5 \tan 75°$$

$$= 5(3.7321) \longleftarrow \text{From the table or calculator.}$$

$$= 18.6605, \text{ or about 19 meters}$$

Thus, the height of the building is about $(19 + 2)$, or 21 meters.

Example 2

Find, to the nearest meter, the **line-of-sight distance,** AB, from the person to the top of the building referred to in Example 1.

Solution: $\sec 75° = \dfrac{AB}{5}$

$AB = 5 \sec 75°$ ⟵ You can also use $\cos 75° = \dfrac{5}{AB}$.

$ = 5(3.864)$ ⟵ From the table or calculator

$ = 19.32$, or about 19 meters

Written Exercises

In Exercises 1–21, find lengths to the nearest unit and angle measures to the nearest ten minutes.

1. In the figure below, the angle of depression from the top of the lighthouse to the top of the buoy is 16°. Find the distance, DC, from the cliff to the buoy.

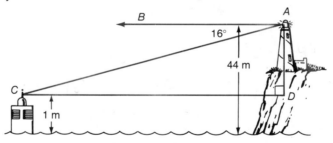

2. For the buoy and lighthouse of Exercise 1 find the line-of-sight distance, AC, from the top of the lighthouse to the buoy.

The angle of elevation from a radar antenna to an airplane is 5°50'. The antenna is 10 meters above the ground. The altitude of the plane is 112.2 m.

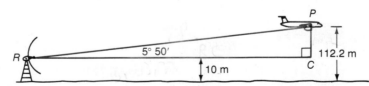

3. Find the line of sight distance, RP, from the antenna to the plane.

4. Find the distance, RC, from the antenna to C directly below the plane.

5. The angle of depression of the closest point on the ground that is visible over the nose of an airplane is called the **cockpit cutoff angle.** For a certain plane flying level at an altitude of 620 meters, the cockpit cutoff angle is 13°. Find the line-of-sight distance from the pilot to the closest visible point on the ground. Refer to the figure below.

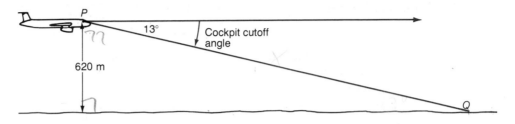

6. The angle of depression from the top of a cliff 800 meters high to the base of a log cabin is 37°20′. How far is the cabin from the foot of the cliff?

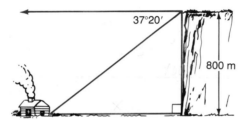

7. From the deck of a boat, the angle of elevation of the top of an offshore oil rig is found to be 31°30′. The top of the oil rig is 127 meters above the level of the platform on which it stands. Assume that the head of the person doing the sighting is level with the base of the oil rig. What is the distance between the base of the oil rig and the boat? See the figure at the left below.

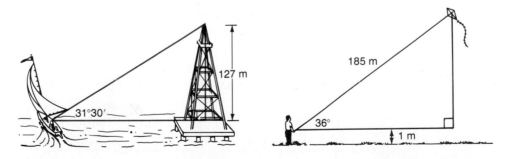

8. A kite string is 185 meters long and makes an angle of 36° with the horizontal as shown in the figure at the right above. What is the altitude of the kite? (Assume that the string is straight and that it is held one meter above the ground.)

9. A salvage ship using SONAR finds the angle of depression of wreckage on the ocean floor to be 13°10'. The charts show that in this region the ocean floor is 35 meters below the surface. How far must a diver lowered from the salvage ship travel along the ocean floor to reach the wreckage?

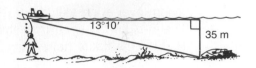

10. From the fire tower in Flatlands National Park, a forest ranger sighted a fire. To measure the angle of depression of the fire, the ranger used an instrument that was known to be 32 meters above the ground. The angle of depression from the tower to the fire was 2°10'. What was the distance between the fire and the base of the tower?

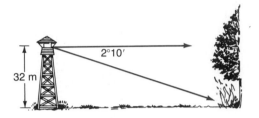

11. From the end of the shadow cast by a vertical object such as the flagpole in the figure at the right, the angle of elevation of the top of the object is the same as the angle of elevation of the sun. Find the angle of elevation of the sun when a flagpole 7.6 meters tall casts a shadow 18.2 meters long.

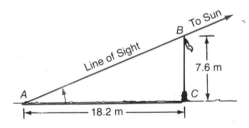

12. The Eiffel Tower in Paris casts a shadow 150 meters long when the angle of elevation of the sun is 63°26'. Find the height of the Tower. See the figure at the left below.

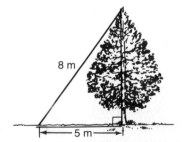

13. An 8-meter pole is leaning against a tree. The foot of the pole is 5 meters from the foot of the tree. What angle does the pole make with the tree? See the figure at the right above.

A ladder 10.4 meters long makes an angle of 68° with the ground as it leans against a building. Use this information for Exercises 14–15.

14. How far up the building does the ladder reach? (See the figure at the left below.)

15. How far is the foot of the ladder from the foot of the building? (See the figure at the left below.)

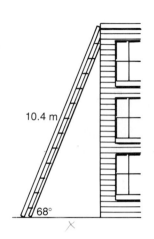

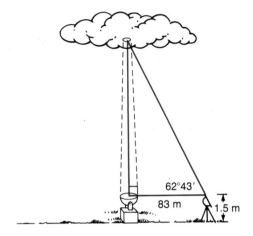

16. The altitude of the base of a cloud formation is called the ceiling. To find the ceiling one night, the famous meteorologist Dr. Gail Storm directed a spotlight vertically to the clouds. (See the figure at the right above.) Using an instrument placed 83 meters from the spotlight and 1.5 meters above the ground, Dr. Storm found the angle of elevation of the light's image on the clouds to be 62°43′. What ceiling did Dr. Storm find?

17. A ladder 12.2 meters long can be so placed that it will reach a window 10.1 meters above the ground on one side of the street. If a person tips it back without moving its foot, it will reach a window 6.4 meters above the ground on the other side. Find the width of the street.

18. The Hirsch Building and the County Hospital are 38 meters apart. From a window in the Hirsch Building, the angle of elevation of the top of the hospital is 73°. From the same window the angle of depression of the ground at the base of the hospital is 64°. Find the height of the hospital.

19. Two boats are observed from a tower 75 meters above a lake. The angles of depression are 12°30′. and 7°10′. How far apart are the boats? Refer to the figure at the right.

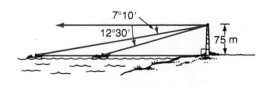

20. From a point at eye level with the base of the pedestal of the Statue of Liberty, an observer found the angle of elevation of the top of the torch to be 3°37'. From the same point, the angle of elevation of the foot of the statue was 1°51'. The distance between the top of the torch and the foot of the statue is 46 meters, and the height of the pedestal alone is 49 meters. Find the distance between the observer and the base of the pedestal. Give your answer to the nearest 100 meters.

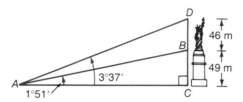

21. A certain tree grows vertically on a hill which makes an angle of 8°15' with the horizontal. When the angle of elevation of the sun is 27°20', the end of the tree's shadow is 76 meters directly downhill from the base of the tree. Find the height of the tree. Refer to the figure below.

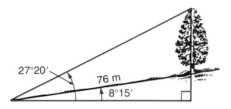

22. A tree 12.2 meters tall grows on the east bank of the Quagmire River at a point where the river is 183 meters wide. The angle of elevation of the top of the tree from the west bank is 3°20'. Find whether the west bank is higher or lower than the east bank and by how much.

23. A tree grows vertically on a hill which is inclined at an angle of 10°05'. The angle of elevation of the top of the tree is measured from downhill points A and B where A, B, and the base of the tree lie on a line and the distance between A and B is 31 meters. The angle of elevation is 30°25' at A and 20°55' at B. Find the height of the tree. Refer to the figure below.

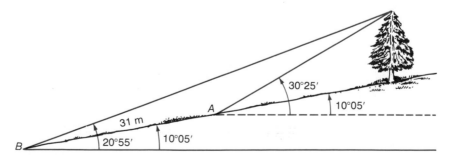

1-10 Applications: Surveying

Surveyors often need to obtain a distance that is difficult to measure directly. In that case, the distance is measured indirectly, applying trigonometry to distances and angles that <u>can</u> be measured directly. Since small errors in angle measures may result in large errors in the calculated distance, surveyors use a very sensitive angle measuring instrument called a **transit.**

Example

A bridge is to be constructed across a lake and thus the distance between points A and B must be determined. A surveyor has made the measurements shown in the figure. Find the distance between points A and B to the nearest meter.

Solution: $\tan 42°10' = \dfrac{AB}{532}$

Therefore, $AB = 532 \cdot \tan 42°10'$

$= 532\,(.9057)$

$= 482$ m.

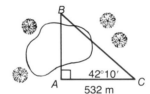

Written Exercises

In Exercises 1–10, unless stated otherwise, compute distances to the nearest meter and angles to the nearest ten minutes.

1. In order to estimate the width of a straight river, a surveyor determined the measurements shown in the figure below. Find the width of the river.

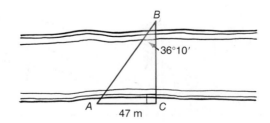

2. To find the distance *CA* across the river, as shown at the left below, a length *CB* of 50 meters was measured on one bank. The measure of ∠*B* was found to be 34°10′. Angle *C* is a right angle. How long is $\overline{CA}$?

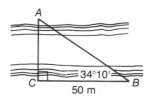

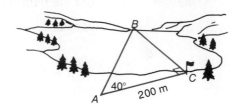

3. To find the distance *BC* between a flagpole on the shore of Silver Lake and a point on the opposite shore, a surveyor determined the measurements shown in the figure at the right above. Find the distance *BC*.

4. Points *P* and *Q* are on the north and south rims, respectively, of the Grand Canyon, with *Q* directly south of *P*. Point *R* is located 780 meters west of *P*. The measure of ∠*PRQ* is 84°40′. (See the figure at the right.) Find *PQ*, the width of the canyon.

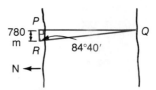

5. A surveyor wishes to determine the width of a north-south highway. While standing on the eastern edge of the highway, he notices a tree at point *B*, directly opposite him on the western edge. He walks 65 meters from his original position, *C*, directly south to a new position, *A*, and finds that the measure of angle *BAC* is 28°40′. Find the width of the highway.

6. A surveyor standing in a gully, finds that the angle of elevation of the top of one side of the gully is 15°. Her eyes are 1.6 meters above the ground and she is standing 4.2 meters from the base of this side. How deep is the gully?

7. The **angle of inclination** of a highway or railroad is the angle formed by the roadbed and the horizontal. At the steepest place, the angle of inclination of the railroad that runs to the summit of Pikes Peak is 27°. How many meters would you rise vertically in traveling 400 meters along this track?

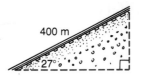

8. In traveling along a highway or railroad built on a slope, your position changes vertically as well as horizontally. The ratio of vertical distance to horizontal distance is called the **grade.** To find the grade of a proposed highway, a surveyor places a leveling instrument on the slope *AB*, and the line *CD* is sighted to two upright rods. If *AC* = 3.9 meters, *BD* = 1 meter, and *AB* = 49.21 meters, find the grade and the angle of inclination.

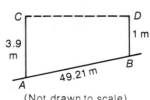

(Not drawn to scale)

9. To find the height of a tree standing at point *C* across a river from point *A*, a base line, *AB*, 80 meters long is established on one side of the river. The measure of ∠*CBA* was found to be 74°10′. The angle of elevation of the top of the tree from *A* measures 10°20′. Find the height of the tree.

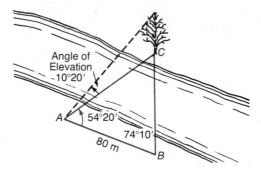

10. An exit ramp with a .0496 grade is to be constructed from Interstate 201 to an overpass 10.7 meters above the horizontal level of the road. A surveyor must locate the position on the road at which the ramp will start. Find the distance measured along the horizontal from the position of the overpass to the position at which the ramp will start. Write your answer to the nearest tenth of a meter.

1-11 Applications: Navigation

In navigation, the **course** of a ship or plane is the angle measured clockwise from north to the line of travel. Thus, in the figure below, the course of the ship is 37°20′.

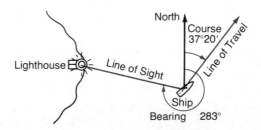

The **bearing** of a line of sight is the angle measured clockwise from north to the line of sight. Thus, the bearing of the lighthouse in the figure above is 283°.

Example 1 From a ship traveling on a course of 27° the navigator sights a lighthouse. The line of sight forms a right angle with the ship's line of travel. Find the bearing of the lighthouse.

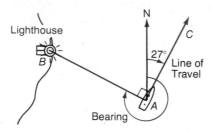

Solution: Bearing of lighthouse = 360° − m ∠ BAN. Find m ∠ BAN.

m ∠ BAN = 90° − 27° ⟵ Since ∠BAC is a right angle and ∠NAC is the ship's course

= 63°

Thus, the bearing of the lighthouse = 360° − 63°, or 297°.

Example 2 The ship of Example 1 makes a second sighting after traveling 2.2 kilometers from its first position. The course remains 27°. The new bearing of the lighthouse is 247°20′. Find, to the nearest tenth of a kilometer, the distance of the ship from the lighthouse when the first sighting was made.

Solution: First find m ∠ ACB in triangle ACB.

m ∠ ACB = Bearing − (course + 180°)

= 247°20′ − (27° + 180°)

= 40°20′

Next, use the tangent ratio to find the desired distance, AB.

$$\tan 40°20′ = \frac{AB}{2.2}$$

$$AB = 2.2 \cdot \tan 40°20′$$

$$= 2.2(.8491)$$

$$= 1.87, \text{ or about } 1.9 \text{ km}$$

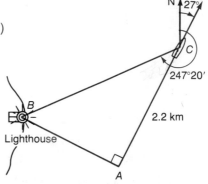

Written Exercises

In Exercises 1–4, a ship is sailing along the east coast of Maine. The ship's navigator sights a lighthouse on land. The line of sight forms a right angle with the ship's line of travel. Find the bearing of the lighthouse for the course given.

1. 0°   **2.** 54°   **3.** 207°   **4.** 300°

In Exercises 5–15, compute distances to the nearest tenth of a kilometer and angles to the nearest ten minutes.

5. The steamship *Linda Lou* is 23.3 kilometers due south of Pequanic Lighthouse. From Pequanic Lighthouse, the bearing of Wontauk Lighthouse is 90°. The distance between the lighthouses is 26.6 kilometers. Find the bearing of Wontauk Lighthouse from the *Linda Lou*.

Refer to the figure at the right below for Exercises 6–7.

6. The bearing of a lighthouse sighted by the navigator of a ship was found to be 103°20′. After the ship traveled 3.3 kilometers on a course of 13°20′, the navigator found the bearing of the lighthouse to be 147°50′. Find the distance between the ship and the lighthouse at the time of the first sighting.

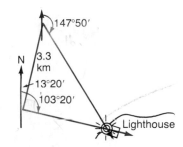

7. In Exercise 6, find the distance between the ship and the lighthouse at the time of the second sighting.

8. An airplane is 191.6 kilometers due east of radar station *A*. A second radar station is 226.6 kilometers due north of *A*. Find the bearing of the second radar station from the airplane.

9. In Exercise 8, find the distance of the second radar station from the airplane.

10. The *King Fisher* sailed 45 kilometers on course 31° and then 30 kilometers on course 121°. The ship will return to its starting point by the shortest route. Find the total distance traveled.

11. In Exercise 10, find the course of the ship when it returns to the starting point by the shortest route.

12. The bearing of a lighthouse from a ship 9.2 kilometers away is 32°30′. How far must the ship sail on course 0° for the bearing of the lighthouse to be 122°30′? Refer to the figure at the right.

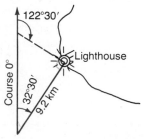

13. At 6 A.M., a ship is sailing due south at a constant speed. The navigator sights a lighthouse on a bearing of 270° at a distance of 24.2 kilometers. At 6:30 A.M., the bearing of the lighthouse is 285°. Find, to the nearest kilometer per hour, the rate at which the ship is sailing.

14. The ship of Exercise 13 continues sailing south at a constant rate. Find the bearing of the lighthouse at 9:00 A.M.

15. Two ships leave the same harbor at the same time. One sails at a constant rate of 40 kilometers per hour on course 42°15′, and the other sails at a constant rate of 51 kilometers per hour on course 132°15′. How far apart are the ships three hours after leaving the harbor?

1-12 Applications: Construction

The figure at the left below shows the names of some of the structural members in the frame of a roof.

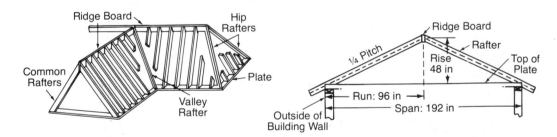

The **span** of a roof is the distance between the outside walls. For a symmetrical roof, the **run** is one half the span. The **rise** of a roof is the distance between the center of the ridge board and the level of the top of the plate. Architects and carpenters refer to the ratio

$$\frac{\text{Rise}}{\text{Span}}$$

as the **pitch** of the roof. In the figure above, a roof is shown with a pitch of $\frac{1}{4}$.

Example

Find, to the nearest inch, the rafter length for a symmetrical roof with a $\frac{1}{3}$ pitch and a span of 24 feet.

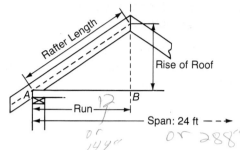

or 144" or 288"

Solution: First, find the rise and run.

$$\text{Run} = \frac{1}{2} \times \text{Span}$$
$$= \frac{1}{2} \times 288 = 144 \text{ inches} \longleftarrow 24 \text{ feet} = 288 \text{ inches}$$

Since Pitch $= \dfrac{\text{Rise}}{\text{Span}}$,   *= $\frac{1}{3} = \frac{rise}{288"}$ $\frac{3rise = 288}{3} = \frac{288}{3}$*

$$\text{Rise} = \text{Pitch} \times \text{Span}$$
$$= \frac{1}{3} \times 288 = 96 \text{ inches}$$ *rise 96"*

Then rafter length $= \sqrt{144^2 + 96^2}$
$$= \sqrt{29{,}952}, \text{ or about 173 inches}$$

Written Exercises

In Exercises 1–4, find the pitch of a symmetrical roof for the given dimensions.

1. Rise: 83 in; Span: 249 in

2. Rise: $4\frac{1}{2}$ ft; Span: 6 ft   *$p = \frac{rise}{span}$*

3. Rise: 200 cm; Run: 400 cm

4. Rise: 1.8 m; Run: 1.2 m

In Exercises 5–8, find the rise of each roof. Assume it is symmetrical.

5. Span: 488 cm; Pitch: $\frac{1}{4}$

6. Span: 2.2 m; Pitch: $\frac{1}{2}$ *$p = \frac{rise}{span}$*

7. Run: 111 in; Pitch: $\frac{1}{3}$

8. Run: 3 yd; Pitch: $\frac{2}{3}$

In Exercises 9–14, compute lengths to the nearest unit and angles to the nearest ten minutes.

9. Find the pitch of the symmetrical roof shown in the figure at the left below.

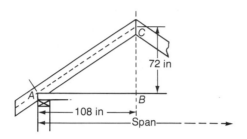

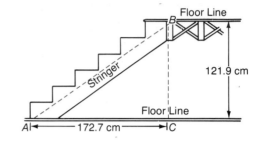

10. A piece of lumber cut to support the treads of a stair as shown in the figure at the right above is called a **stringer.** Find the length, *AB*, of the piece of lumber required for the stringer.

11. A truss for a bridge is to be constructed with measurements as shown in the figure at the right. Find the length, *CD*, of the vertical tie rod.

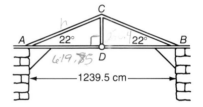

12. A steel bridge has a truss with measurements as shown in the figure at the left below. Find the measure of angle *FAD*.

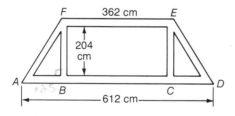

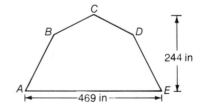

13. In the gambrel roof shown in the figure at the right above, the upper rafters have a pitch of $\frac{1}{4}$ and the lower rafters have a pitch of 1. Find the lengths *AB* and *CD*.

14. Two rafters are sometimes joined by a **collar beam** for extra strength. Find the length, *CD*, of the collar beam in the figure at the right if the roof has a pitch of $\frac{1}{3}$.

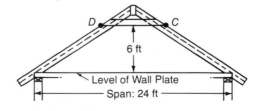

54   Chapter 1

BASIC: INTRODUCTION

The following table shows how certain algebraic expressions are written in the computer language called BASIC. In algebra, you can write 2W instead of $2 \times W$. However, in BASIC, the multiplication symbol, *, must <u>always</u> be used.

Algebraic Expression	BASIC Expression
$2l + 2w$	2*L + 2*W
$\dfrac{2y}{z-1}$	2*Y/(Z - 1) ← The slash, /, means <u>divide</u>.
$360 - 2(A + B)$	36Ø - 2*(A + B)
$(a+7)(2b)^2$	(A + 7)*(2*B)↑2 ← The ↑2 means to <u>square</u>.

The slash in the numeral Ø distinguishes it from capital letter O.

Problem: Given the lengths of the two legs of a right triangle, find the length of the hypotenuse.

The following program shows the steps for solving the above problem.

```
1Ø   READ A,B                        ← A and B represent the
2Ø   LET C = SQR(A↑2 + B↑2)            lengths of the legs.
3Ø   PRINT "HYPOTENUSE  =";C        ← SQR(A↑2 + B↑2) means √a² + b².
4Ø   GO TO 1Ø
5Ø   DATA 5, 12, 8, 9, 6.3, 4.2    ← A = 5, 8, and 6.3;
6Ø   END                             B = 12, 9, and 4.2
```

Statement 2Ø: SQR(A↑2 + B↑2) means $\sqrt{a^2 + b^2}$.

Analysis: Statement 1Ø: This tells the computer to accept values for the variables A and B from statement 5Ø, the DATA statement. It will first accept 5 for A and 12 for B. These values are stored in **memory** locations.

Statement 2Ø: The stored values of A and B are used to calculate a value for C, which is then stored in a memory location. Since A = 5 and B = 12, statement 2Ø will compute and store $\sqrt{13}$ for C.

Statement 3Ø: The PRINT statement is used to output HYPOTENUSE = 13.

Statement 4Ø: This tells the machine to "loop" back to statement 1Ø. This will cause the next values for A and B to be accepted from the DATA in statement 5Ø and stored in memory. Then using A = 8 and B = 9, statement 2Ø computes and stores a new value for C. Next, statement 3Ø outputs HYPOTENUSE = 12.Ø416.

Statement 4Ø again tells the computer to loop back to statement 1Ø. Statement 2Ø calculates C using A = 6.3 and B = 4.2, and statement 3Ø prints HYPOTENUSE = 7.57166.

Statement 4Ø again instructs the computer to loop back to statement 1Ø to read the next values for A and B. However, all of the values in statement 5Ø have been used. Therefore, the computer will now output the message OUT OF DATA AT LINE 1Ø.

Statement 6Ø: This tells the computer that all steps have been completed.

This is what the <u>printout</u> or <u>output</u> looks like.

Output: HYPOTENUSE = 13
HYPOTENUSE = 12.Ø416
HYPOTENUSE = 7.57166

OUT OF DATA AT LINE 1Ø ⟵ The computer prints this statement when the complete program is run.

Written Exercises

a

Write each algebraic expression as a BASIC expression.

1. $(x - 3)^2$

2. $\dfrac{3r^2}{v}$

3. $\dfrac{q + r}{4}$

4. $3x^2 + 5x - 9$

5. $2(l + w)$

6. $\dfrac{ab}{c + 7}$

7. $-7 + \dfrac{b}{c - d}$

8. $\sqrt{c^2 - a^2}$

9. $\sqrt{\dfrac{x^2 - a}{5z}}$

Write a program for each of the following exercises.

10. Given the degree measure of an angle, find the measure of its complement.

11. Given the degree measure of an angle, find the measure of its supplement.

12. Given the length of the radius of a circle, find the circumference.

13. Given the length of the radius of a circle, find the area.

b

14. Given the lengths of the legs of a right triangle, find the perimeter.

15. Given the lengths of the legs of a right triangle, find the area.

16. Given the lengths of one leg and the hypotenuse of a right triangle, find the area.

Chapter Objectives and Review

Objective: To know the meanings of the mathematical terms in this chapter.

1. Be sure that you know the meanings of these mathematical terms.

angle (p. 8)
    initial and terminal sides (p. 8)
    measure (p. 8)
    standard position (p. 9)
    vertex (p. 8)
angle of depression (p. 41)
angle of elevation (p. 41)
cofunction (p. 28)
complement of an angle (p. 27)
cosecant of an angle (p. 17)
cosine of an angle (p. 13)
cotangent of an angle (p. 17)
coterminal angle (p. 10)
dependent variable (p. 3)
Distance Formula (p. 3)

domain and range (p. 3)
function (p. 3)
independent variable (p. 3)
linear interpolation (p. 32)
Pythagorean Theorem (p. 2)
quadrantal angle (p. 9)
radius vector (p. 13)
reference angle (p. 23)
reference triangle (p. 13)
relation (p. 3)
secant of an angle (p. 17)
sine of an angle (p. 13)
solving a triangle (p. 35)
tangent of an angle (p. 13)
trigonometry (p. 2)

Objective: To use the distance formula to find the distance between two points. (Section 1–1)

Find the distance between points P_1 and P_2. Write answers in simplest form.

2. $P_1(1, 3)$; $P_2(4, 7)$

3. $P_1(-3, -1)$; $P_2(3, 2)$

Objective: To tell whether a relation is a function. (Section 1–1)

4. Tell whether the relation $\{(0, 1), (1, 4), (2, 4), (3, 0)\}$ is a function.

Objective: To identify the domain and range of a function. (Section 1–1)

Determine the domain and range of each function in Exercises 5–6.

5. $y = 1$

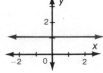

6. $y = x^2 + 1$

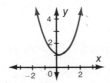

7. Determine the real numbers that must be excluded from the domain of $f(x) = \dfrac{x-3}{x^2 - 1}$.

Objectives: To find the measure of an angle, given the number of rotations and the direction of rotation. (Section 1–2)

For each rotation, find the measure of the angle in degrees.

8. $1\frac{1}{2}$ rotations, clockwise

9. $\frac{3}{8}$ rotation, counterclockwise

Objective: To find the six trigonometric ratios for any angle, given the coordinates of a point on the terminal side of the angle. (Sections 1–3 and 1–4)

The terminal side of angle θ in standard position passes through the point $(-3, 4)$. Find each value.

10. $\sin \theta$ **11.** $\cos \theta$ **12.** $\tan \theta$ **13.** $\cot \theta$ **14.** $\sec \theta$ **15.** $\csc \theta$

Objective: Given the value of one of the six trigonometric functions of an angle with measure θ and the quadrant for θ, to find the value of the other five trigonometric functions. (Sections 1–3 and 1–4)

If $\cos \theta = \frac{12}{13}$ and θ is in Quadrant I, find each value.

16. $\sin \theta$ **17.** $\tan \theta$ **18.** $\cot \theta$ **19.** $\sec \theta$ **20.** $\csc \theta$

Objective: To evaluate the trigonometric functions of the quadrantal angles and other special angles. (Section 1–5)

Find each value.

21. $\sin 90°$ **22.** $\cos 90°$ **23.** $\tan 30°$

24. $\sin 45°$ **25.** $\cos 135°$ **26.** $\cot 210°$

Objective: To use the Table of Values of the Trigonometric Functions. (Section 1–6)

Use the table on pages 368–372 to find each value.

27. $\sin 25°20'$ **28.** $\cos 228°50'$ **29.** $\tan 115°10'$

Use the table on pages 368–372 to find θ, where $0° < \theta < 90°$.

30. $\sin \theta = .2419$ **31.** $\cos \theta = .5640$ **32.** $\tan \theta = 1.8418$

Objective: To use linear interpolation with the Table of Values of the Trigonometric Functions. (Section 1–7)

33. Find $\sin 42°17'$ to four decimal places. Use the Table of Values of the Trigonometric Functions on pages 368–372.

34. If $\cos \theta = .8534$, where $0° < \theta < 90°$, find θ to the nearest minute. Use the table on pages 368–372.

Objective: To solve right triangles. (Section 1–8)

Refer to the figure at the right for Exercises 35–36. Use the table on pages 368–372 or a calculator.

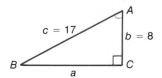

35. Find *a*.

36. Find m ∠ *A* to the nearest minute.

37. In the figure at the right, $b=7$ and m ∠ $B=41°$. Find *c* to the nearest tenth. Use the table on pages 368–372 or a calculator.

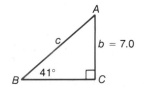

Objective: To apply right-triangle trigonometry in solving practical problems in surveying, navigation, and construction. (Sections 1–9, 1–10, 1–11, and 1–12)

38. A kite string is 90 meters long and makes a 48° angle with the horizontal. (See the figure at the right.) Find the altitude of the kite to the nearest meter. Use the table on pages 368–372 or a calculator.

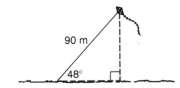

39. A surveyor has made the measurements shown in the figure. Find the distance, *CB*, across the pond to the nearest meter.

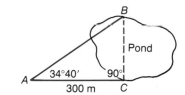

40. A pipe is to be set so as to join two horizontal pipes whose difference in level is 58 cm. The angle of rise is 21°. Find the length of the pipe to the nearest centimeter.

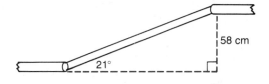

41. From a ship traveling on a course of 33°, the navigator sights a lighthouse. The line of sight to the lighthouse forms a right angle with the ship's line of travel. Find the bearing of the lighthouse.

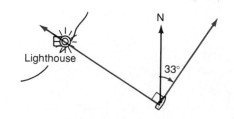

Chapter Test

Classify each statement in Exercises 1–5 as <u>True</u> or <u>False</u>.

1. An angle of 720° in standard position is a quadrantal angle.
2. An angle of 123° in standard position is coterminal with an angle of −123°.
3. The relation $\{(0, 1), (1, 2), (2, 3)\}$ is a function.
4. For an angle of 137°, the reference angle is 53°.
5. The value of $\tan 180°$ is undefined.

6. Determine the real numbers that must be excluded from the domain of the function
$$f(x) = \frac{x - 1}{25 - x^2}.$$

7. Find the degree of the angle that results from $1\frac{5}{6}$ clockwise rotations.
8. Find the distance between $(1, 7)$ and $(4, 3)$.

The terminal side of an angle θ in standard position passes through the point $(-12, 5)$. Find each value.

9. $\sin \theta$ 10. $\csc \theta$ 11. $\tan \theta$ 12. $\sec \theta$

If $\sin \theta = -\frac{8}{17}$ and θ is in Quadrant III, find each value.

13. $\cos \theta$ 14. $\cot \theta$ 15. $\sec \theta$ 16. $\csc \theta$

17. Find $\cos 228°17'$ to four decimal places.
18. If $\sin \theta = .8534$ where $0° < \theta < 90°$, find θ to the nearest minute.
19. In triangle ABC, C is a right angle, $b = 15$ and $c = 17$. Solve the triangle. Find angle measures to the nearest ten minutes.
20. A park is to be constructed in the shape of a right triangle. A surveyor has found the measures of the sides to be 72 m, 65 m, and 97 m. Find the measures of the remaining angles to the nearest ten minutes.

Chapter 2
Graphs of
Trigonometric Functions

2-1 Radian Measure

It is often convenient in mathematics to use a unit of angular measure called a <u>radian</u>. In the figure at the right, if the length of arc AB intercepted by $\angle AOB$ is equal to radius r, then measure of central angle AOB is **one radian.**

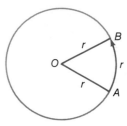

Example 1 In circle O, $r = 3$ cm, $\angle POQ$ intercepts $\overset{\frown}{QP}$, and the length of $\overset{\frown}{QP} = 7$ cm. Find to the nearest tenth, the radian measure of $\angle POQ$.

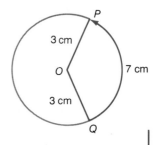

Solution: Since the radius of the circle is 3 cm, the measure of $\angle POQ$ will be one radian when the length of $\overset{\frown}{QP}$ is 3 cm. Thus, the radian measure of $\angle POQ$ is $\frac{7}{3}$, or about 2.3 radians.

Now consider circle O with radius r and ray OA as shown. When $\overset{\rightarrow}{OA}$ has made one complete counterclockwise rotation about O, the arc intercepted by the ray is the circle. Since $C = 2\pi r$, the number of r's in the circumference is 2π.

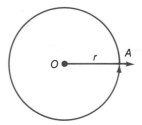

Thus, there are 2π radians in the circle. That is,

$$2\pi \text{ radians} = 360°, \text{ or} \qquad \textbf{1}$$

$$\pi \text{ radians} = 180°. \qquad \textbf{2}$$

You can use equation **2** to find the number of radians in one degree or the number of degrees in one radian.

$$1° = \frac{\pi}{180}, \text{ or} \qquad \textbf{3}$$

$1°$ is about .0174533 radians.

Similarly, **1 radian** $= \dfrac{180°}{\pi}$, or

1 radian is about 57.2958 degrees.

You can use these relationships to change degree measure to radian measure and to change radian measure to degree measure.

As with degree measure, the radian measure of an angle formed by a clockwise rotation is negative. Similarly, the radian measure of an angle formed by more than one complete rotation is greater than 2π or less than -2π, depending on the direction of the rotation.

Example 2 Change each degree measure to radian measure in terms of π.

a. 30°

b. −45°

Solutions: a. $1° = \dfrac{\pi}{180}$ radians

b. $1° = \dfrac{\pi}{180}$ radians

$30° = 30\left(\dfrac{\pi}{180}\right) = \dfrac{\pi}{6}$ radians

$-45° = -45\left(\dfrac{\pi}{180}\right) = -\dfrac{\pi}{4}$ radians

Example 3 Change each radian measure to degree measure.

a. $\dfrac{\pi}{30}$

b. $-\dfrac{4\pi}{3}$

Solutions: a. π radians $= 180°$

b. π radians $= 180°$

$\dfrac{\pi}{30} = \dfrac{180°}{30} = 6°$

$-\dfrac{4\pi}{3} = -\dfrac{4}{3}(180°) = -240°$

Classroom Exercises

Complete the table.

1.

Degrees	30°	?	60°	?	?	135°	150°	−30°	?
Radians	?	$\dfrac{\pi}{4}$	?	$\dfrac{\pi}{2}$	$\dfrac{2\pi}{3}$	?	?	?	$-\dfrac{\pi}{3}$

Written Exercises

In circle O with radius of length r, central angle POQ intercepts $\overset{\frown}{QP}$. Find, to the nearest tenth, the radian measure of $\angle POQ$ for each given value of r and length of $\overset{\frown}{QP}$.

1. $r = 8$ cm, length of $\overset{\frown}{QP} = 12$ cm

2. $r = 1$ m, length of $\overset{\frown}{QP} = 2$ m

3. $r = 10$ m, length of $\overset{\frown}{QP} = 5$ m

4. $r = 10$ cm, length of $\overset{\frown}{QP} = 40$ cm

5. $r = 20$ cm, length of $\overset{\frown}{QP} = 8$ cm

6. $r = 4$ m, length of $\overset{\frown}{QP} = 20$ m

Change each degree measure to radian measure in terms of π.

7. 360°

8. 270°

9. −90°

10. 30°

11. −60°

12. 225°

13. 120°

14. −210°

15. 240°

16. −32°

17. 27°30′

18. 6°

19. 13°45′

20. 420°

21. 0°30′

22. −390°

Change each radian measure to degree measure.

23. $\frac{\pi}{3}$

24. $\frac{\pi}{2}$

25. $-\frac{3\pi}{2}$

26. $-\pi$

27. $\frac{\pi}{6}$

28. $\frac{5\pi}{6}$

29. $\frac{2\pi}{3}$

30. $\frac{5\pi}{3}$

31. $\frac{7\pi}{6}$

32. $-\frac{7\pi}{4}$

33. 4π

34. -7π

35. $\frac{5\pi}{2}$

36. $-\frac{9\pi}{2}$

37. $\frac{\pi}{4}$

38. $\frac{3\pi}{4}$

39. $-\frac{5\pi}{3}$

40. $\frac{\pi}{12}$

41. $\frac{13\pi}{2}$

42. $-\frac{4\pi}{3}$

Sketch an angle in standard position that results from the given rotation. Find the radian measure of the angle.

43. $\frac{1}{4}$ rotation, counterclockwise

44. $\frac{1}{2}$ rotation, counterclockwise

45. $\frac{3}{4}$ rotation, counterclockwise

46. 1 rotation, counterclockwise

47. $\frac{1}{4}$ rotation, clockwise

48. $\frac{1}{2}$ rotation, clockwise

49. $\frac{1}{6}$ rotation, clockwise

50. $\frac{1}{8}$ rotation, counterclockwise

51. $\frac{5}{12}$ rotation, counterclockwise

52. $\frac{3}{10}$ rotation, clockwise

53. $\frac{4}{6}$ rotation, counterclockwise

54. $\frac{3}{2}$ rotation, clockwise

55. $\frac{17}{12}$ rotation, counterclockwise

56. $\frac{7}{6}$ rotation, clockwise

57. $\frac{2}{3}$ rotation, clockwise

58. $\frac{4}{3}$ rotation, clockwise

59. What part of a complete rotation does the minute hand of a clock make in 6 minutes?

60. Through how many degrees does the minute hand rotate in 6 minutes?

61. Through how many radians does the minute hand rotate in 6 minutes?

62. Find the length of the arc traversed in 20 minutes by the end of a 10-cm minute hand on a clock.

63. Find the radian measure of the angle formed by the hands of a clock at 4 o'clock.

Let x be the radian measure of an angle in standard position where $0 < x \le 2\pi$. The terminal side of the angle contains P. Find x.

64. $P(3, 0)$ **65.** $P(0, -5)$ **66.** $P(0, 2)$ **67.** $P(-4, 0)$

Let x be the radian measure of an angle AOB in standard position. Find the radian measure of an angle coterminal with angle AOB under the given number of rotations.

68. $x = \frac{\pi}{3}$; 2 clockwise **69.** $x = -\frac{\pi}{6}$; 1 clockwise

70. $x = \frac{\pi}{3}$; 2 counterclockwise **71.** $x = -\frac{\pi}{4}$; 1 counterclockwise

72. $x = -\frac{\pi}{6}$; 1 counterclockwise **73.** $x = 2\pi$; 1 clockwise

74. $x = -\frac{\pi}{6}$; 3 clockwise **75.** $x = \frac{3\pi}{4}$; 2 clockwise

76. $x = \frac{7\pi}{6}$; 2 clockwise **77.** $x = -\frac{3\pi}{2}$; 1 counterclockwise

78. For any angle with degree measure θ, the measures of its coterminal angles are $n \cdot 360° + \theta$ where n is an integer. Write a similar expression for the radian measures of angles coterminal with an angle whose radian measure is x.

Angles in standard position with radian measure x where $0 < x < \frac{\pi}{2}$ lie in Quadrant I. Complete each statement.

79. Angles in standard position with radian measure x where $\pi < x < \frac{3\pi}{2}$ lie in Quadrant _?_.

80. Angles in standard position with radian measure x where $\frac{\pi}{2} < x < \pi$ lie in Quadrant _?_.

81. Angles in standard position with radian measure x where $\frac{3\pi}{2} < x < 2\pi$ lie in Quadrant _?_.

82. For all angles in standard position with radian measure $x > 0$ which lie in Quadrant I, x is between $n \cdot 2\pi$ and $\frac{\pi}{2} + n \cdot 2\pi$, where n is a whole number or $n \cdot 2\pi < x < \frac{\pi}{2} + n \cdot 2\pi$, where n is a whole number. Find similar expressions for angles in Quadrants II, III, and IV.

83. Repeat Exercise 82 for $x < 0$.

2-2 Radian Measure and Tables

In the table on pages 368–372, the column headed "θ Rad." gives decimal approximations for the radian measure of angles whose degree measure is listed in the "θ Deg." column. Here is a portion of the table.

θ Deg.	θ Rad.	Sin θ	Cos θ	Tan θ	Cot θ	Sec θ	Csc θ		
27° 00′	.4712	.4540	.8910	.5095	1.9626	1.122	2.203	1.0996	63° 00′
10′	.4741	.4566	.8897	.5132	1.9486	1.124	2.190	1.0966	50′
20′	.4771	.4592	.8884	.5169	1.9347	1.126	2.178	1.0937	40′
30′	.4800	.4617	.8870	.5206	1.9210	1.127	2.166	1.0908	30′
40′	.4829	.4643	.8857	.5243	1.9074	1.129	2.154	1.0879	20′
50′	.4858	.4669	.8843	.5280	1.8940	1.131	2.142	1.0850	10′
35° 00′	.6109	.5736	.8192	.7002	1.4281	1.221	1.743	.9599	55° 00′
10′	.6138	.5760	.8175	.7046	1.4193	1.223	1.736	.9570	50′
20′	.6167	.5783	.8158	.7089	1.4106	1.226	1.729	.9541	40′
30′	.6196	.5807	.8141	.7133	1.4019	1.228	1.722	.9512	30′
40′	.6225	.5831	.8124	.7177	1.3934	1.231	1.715	.9483	20′
50′	.6254	.5854	.8107	.7221	1.3848	1.233	1.708	.9454	10′
36° 00′	.6283	.5878	.8090	.7265	1.3764	1.236	1.701	.9425	54° 00′
		Cos θ	Sin θ	Cot θ	Tan θ	Csc θ	Sec θ	θ Rad.	θ Deg.

Example 1

Find a four-place decimal approximation for the radian measure of the angle with the given degree measure.

a. 27°40′

b. 54°20′

Solutions: a. 1. Find 27°40′ in the column under "θ Deg."

2. Look directly <u>right</u> along the row to the column below the heading "θ Rad."

3. Read:
27°40′ = .4829 radians

b. 1. Find 54°20′ in the column over "θ Deg."

2. Look directly <u>left</u> along the row to the column above the heading "θ Rad."

3. Read:
54°20′ = .9483 radians

You can also use the table to find the degree measure of an angle whose radian measure is given as a decimal. When no unit of measure is indicated, always assume that the measure of the angle is in radians.

Example 2 Find the degree measure of the angle with the given radian measure.

a. .9570

b. .4741

Solutions: a. 1. Find .9570 in the column over "θ Rad."

2. Look directly <u>right</u> along the row to the column over the heading "θ Deg."

3. Read:
.9570 radians $= 54°50'$

b. 1. Find .4741 in the column under "θ Rad."

2. Look directly <u>left</u> along the row to the column under the heading "θ Deg."

3. Read:
.4741 radians $= 27°10'$

Radian measure provides an important link between angular measure and the length of an arc of a circle. This link can be determined by noting that, in a circle, the measure of the central angle in radians is directly proportional to the arc that it intercepts. Thus, in the figure at the right,

$$\frac{s}{r} = \frac{\theta}{1} \qquad (0 \le \theta < 2\pi)$$

or $\qquad s = r\theta \qquad (0 \le \theta < 2\pi).$

In this formula, θ always represents radian measure.

Example 3 In the circle at the right, central angle QOP intercepts $\overset{\frown}{QP}$. Find the length of $\overset{\frown}{QP}$ in terms of π for each given radian measure of $\angle QOP$ and value of r.

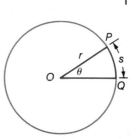

a. $\theta = \frac{\pi}{4}, r = 2$

b. $\theta = \frac{\pi}{6}, r = 1$

Solutions: a. $s = r\theta$

$\qquad = 2 \cdot \frac{\pi}{4}$

$\qquad = \frac{\pi}{2}$

b. $s = r\theta$

$\qquad = 1 \cdot \frac{\pi}{6}$

$\qquad = \frac{\pi}{6} \longleftarrow \theta$ also equals $\frac{\pi}{6}$.

Part **b** of Example 3 illustrates that if the given circle is a <u>unit circle</u> ($r = 1$), then $s = \theta$. Thus, if the measure of the central angle of a unit circle is θ, then the length of the intercepted arc is also θ.

Classroom Exercises

Use the table on pages 368–372 to find a four-place decimal approximation for the radian measure of the angle with the given degree measure.

1. 12°10′   **2.** 54°30′   **3.** 6°   **4.** 3°40′

Use the table on pages 368–372 to find the degree measure of the angle with the given radian measure.

5. .6516   **6.** 1.0530   **7.** 1.5010   **8.** .1687

9. In circle O with radius 6 cm, a central angle with radian measure 3 intercepts an arc with measure s. Find s.

Written Exercises

a

Use the table on pages 368–372 to find a four-place decimal approximation for the radian measure of the angle with the given degree measure.

1. 11°40′   **2.** 69°10′   **3.** 18°   **4.** 45°

5. 81°30′   **6.** 3°50′   **7.** 9°   **8.** 27°30′

Use the table on pages 368–372 to find the degree measure of the angle with the given radian measure.

9. .1367   **10.** 1.5213   **11.** .2153   **12.** .4480

13. 1.2479   **14.** 1.3526   **15.** .5760   **16.** .7912

In circle O with radius of length r, central angle POQ intercepts $\overset{\frown}{QP}$. Find the length of $\overset{\frown}{QP}$ in terms of π for each given radian measure of $\angle POQ$ and value of r.

17. $\theta = \frac{\pi}{3}$, $r = 20$ **18.** $\theta = \frac{\pi}{4}$, $r = 50$

19. $\theta = \frac{3\pi}{5}$, $r = 15$ **20.** $\theta = \frac{4\pi}{3}$, $r = 100$

21. $\theta = \frac{8\pi}{5}$, $r = 11$ **22.** $\theta = \frac{9\pi}{8}$, $r = 12$

23. $\theta = \frac{12\pi}{13}$, $r = 26$ **24.** $\theta = 2\pi$, $r = 7$

b

In circle O with radius of length r, central angle POQ intercepts $\overset{\frown}{QP}$. Find the length of $\overset{\frown}{QP}$ in terms of π for each given degree measure of ∠POQ and value of r. (HINT: First find the radian measure of ∠POQ.)

25. $\theta = 130°$, $r = 10$ 26. $\theta = 175°$, $r = 20$ 27. $\theta = 70°$, $r = 36$

28. $\theta = 90°$, $r = 50$ 29. $\theta = 10°$, $r = 18$ 30. $\theta = 180°$, $r = 17$

31. $\theta = 280°$, $r = 56$ 32. $\theta = 320°$, $r = 24$ 33. $\theta = 300°$, $r = 13$

34. Find, in terms of π, the radius of a circle in which a central angle of 20° intercepts an arc 9 cm long.

C

35. Find, to the nearest millimeter, the length of a driving belt running around two wheels of radii 20 cm and 10 cm respectively. The distance between their centers is 40 cm.

36. Repeat Exercise 35 for the case where the belt crosses between the circles.

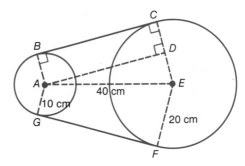

 ————————————— *Radian Measure* ——————

The three Examples below show how to use a scientific calculator having a radian mode key to obtain the trigonometric functions of radian measures directly. To place the calculator in the radian mode, you press the "Rad" key, or, on some calculators, a special switch.

Example 1   Find tan 1.3526.

Solution:
4.5100646 _{RAD}

Example 2   Find $\cos\dfrac{\pi}{3}$.

Solution:
0.5 _{RAD}

Example 3   Find sec .1367.   ◄——— Use the reciprocal key [1/x] and the cosine function.

Solution:
1.0094168 _{RAD}

2-3 Applications: Angular Velocity

Rotary motion is the motion of a body turning about an axis. The path followed by any point on an object in rotary motion is a circle with center at the axis.

As the propeller at the right revolves about the axis at O, the angular displacement of point P is the angle θ through which ray OP moves. Since the value of θ is the same for every point on the propeller, θ is also called the angular displacement of the propeller.

Example 1

In the figure above, the propeller makes $\frac{1}{2}$ revolution about the axis. Find, in radians, the angular displacement θ of point P.

Solution: Since ray OP moves through 2π radians for each revolution,

$$\theta = \frac{1}{2} \cdot 2\pi = \pi.$$

The **angular velocity** ω (Greek letter omega) of an object in rotary motion is its angular displacement per unit of time. That is, if t represents the time required for a ray to move through an angle θ, then

$$\omega = \frac{\theta}{t}.$$

1

Angular velocity may be measured in revolutions per minute (rpm), radians per minute (rad/min), or radians per second (rad/sec).

Example 2

During the interval $t = 2$ seconds, the angular displacement of a point P of an object in rotary motion is $\theta = 1.64$ radians. Find the angular velocity ω in radians per second.

Solution: $\omega = \dfrac{\theta}{t} = \dfrac{1.64 \text{ radians}}{2 \text{ seconds}} = .82$ rad/sec

Example 3

A record player turntable rotates at $33\frac{1}{3}$ revolutions per minute (rpm). Find ω, the angular velocity of the turntable, in radians per second.

X top by $\frac{2\pi}{60\ sec}$

Solution: Since 1 revolution $= 2\pi$ radians,

$$\frac{33\frac{1}{3}\text{ revolutions}}{1\text{ minute}} = \frac{2\pi \cdot 33\frac{1}{3}\text{ radians}}{1\text{ minute}}, \text{ or } 66\frac{2}{3}\pi \text{ rad/min.}$$

Since 1 minute $= 60$ seconds,

$$\frac{66\frac{2}{3}\pi\text{ radians}}{1\text{ minute}} = \frac{66\frac{2}{3}\pi\text{ radians}}{60\text{ seconds}}.$$

Thus, $\omega = 1\frac{1}{9}\pi$ rad/sec.

Since Equation **1** can be rewritten as

$$\theta = \omega \cdot t, \qquad\qquad 2$$

you can use angular velocity to find angular displacement during any interval of time. Also, since the path followed by a point P on an object in rotary motion is a circle, you can use the angular displacement to find the distance that point P travels in any interval of time.

Example 4

In the figure at the right below, points P_1 and P_2 are on a record player turntable in rotary motion at $1\frac{1}{9}\pi$ rad/sec. Find each of the following.

a. The angular displacement of P_1 and P_2 during $\frac{1}{10}$ of a second.

b. The distances s_1 and s_2 traveled during $\frac{1}{10}$ second by points P_1 and P_2 respectively

Solutions: a. $\theta = \omega \cdot t$

$\qquad = 1\frac{1}{9}\pi \dfrac{\text{rad}}{\text{sec}} \cdot \dfrac{1}{10} \text{ sec}$

$\qquad = \frac{1}{9}\pi$ radians

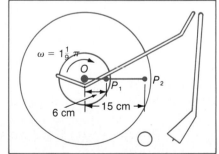

b. The path traveled by each point is a circle. Thus,

$$\begin{array}{ll} s_1 = \theta \cdot r_1 & s_2 = \theta \cdot r_2 \\ \quad = \frac{1}{9}\pi \cdot 6 & \quad = \frac{1}{9}\pi \cdot 15 \\ \quad = \frac{2}{3}\pi \text{ cm} & \quad = 1\frac{2}{3}\pi \text{ cm} \end{array}$$

Classroom Exercises

The hour hand of a clock is moved clockwise from the 12 o'clock position to the given position. Find the angular displacement θ in radians of the hour hand.

1. 4 o'clock   **2.** 10 o'clock   **3.** 7 o'clock   **4.** 1 o'clock

5. 6 o'clock   **6.** 9 o'clock   **7.** 11 o'clock   **8.** 8 o'clock

Written Exercises

In Exercises 1–4, point P is on the circumference of a rotating turntable. Find, in radians, the angular displacement of P for the given number of revolutions.

1. $\frac{1}{3}$ **2.** $\frac{3}{4}$ **3.** $\frac{2}{5}$ **4.** $\frac{7}{8}$

5–8. Find, in radians per second, the angular velocity of point P for the angular displacements of Exercises 1–4. The time interval is 3 seconds.

9. The crankshaft of an automobile engine is rotating at 3600 rpm. Find its angular velocity ω in rad/sec.

10. In the figure at the right, point P is on the circumference of a phonograph record which is on a turntable rotating at $33\frac{1}{3}$ rpm. Find the angular displacement in radians of point P during .5 seconds.

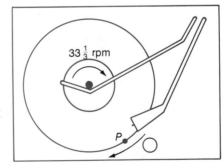

$33\frac{1}{3}$ rpm

11. Find the distance in centimeters to the nearest tenth traveled by point P in Exercise 10 during .5 seconds if the diameter of the record is 30.48 centimeters.

12. A wheel turns at 600 rad/sec. Find its angular velocity ω in rpm.

In Exercises 13–15 find the angular velocity ω in rad/min.

13. The second hand of a watch

14. The minute hand of a watch

15. The hour hand of a watch

It can be shown that the velocity of a point moving along a circular path is given by

$$v = \omega r,$$

where ω is the angular velocity of the point in rad/sec and r is the radius of the circle along which the point moves. Use this formula and the formula for angular velocity to solve Exercises 16–31.

16. The two-bladed aircraft propeller in the figure at the left below is 3.1 meters long. It rotates at 1500 rpm. Find its angular velocity ω in rad/sec, and the velocity in m/sec (meters per second) of the tip of a blade as it moves along a circular path.

17. In the figure at the right above, a seat is attached to the rim of a ferris wheel at a point P that is 7.6 meters from the axle. The ferris wheel makes one complete revolution every 20 seconds. Find, in m/sec, the velocity of point P along its circular path.

The figure at the right shows the platform of a merry-go-round. The merry-go-round rotates at 10 rpm. Find each of the following.

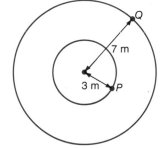

18. The angular velocity in rad/sec of point P

19. The velocity in m/sec of P along its circular path

20. The angular velocity in rad/sec of point Q

21. The velocity, in m/sec of point Q along its circular path

In the figure at the right, the small pulley is connected to the large pulley by a belt. The crankshaft is turning at 3600 rpm. In Exercises 22–25, find each of the following.

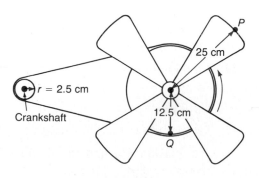

22. The angular velocity in rad/sec of the large pulley

23. The velocity in cm/sec (centimeters per second) of point P

24. The velocity in cm/sec of point Q on the belt

25. The angular velocity in rad/sec of the small pulley

26. The bolt shown in the figure at the left below advances .25 centimeters for each full revolution. The bolt is to be tightened by an air-powered impact wrench. The wrench turns at 1600 rpm and the length of the bolt is 6 centimeters. Find the time required to tighten the bolt with the wrench.

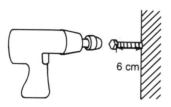

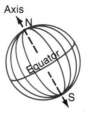

Axis

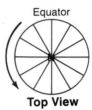

Equator

Top View

27. The earth's axis is the line determined by the north and south poles. (See the figures at the right above.) Assume that the earth is a sphere with radius 6336 kilometers that rotates about its axis once in 24 hours. Find, in kilometers per hour, the velocity of a point on the equator as it moves along its circular path.

The small gear in the figure below is turned by an electric motor at 1800 rpm. In Exercises 28–31, find each of the following.

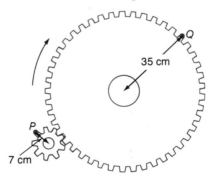

28. The angular velocity in rad/sec of the small gear

29. The velocity in cm/sec of point P along the circumference of the small gear

C **30.** The angular velocity in rad/sec of the large gear

31. The velocity in cm/sec of point Q along the circumference of the large gear

32. If t is the time required for a point to move a distance s along a circular path, then the point's velocity along the path is given by

$$v = \frac{s}{t}.$$

Use this formula and the formula for arc length (page 67) to derive the formula $v = \omega r$.

Review Capsule for Section 2-4_____

Write SAS *(for "side-angle-side"),* ASA, SSS, *or* HA *(hypotenuse-angle) to indicate why the two triangles are congruent.*

1.
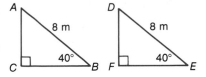

2.

3.
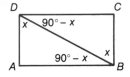

2-4 Properties of sin θ and cos θ

The properties that are developed in this section will be useful in drawing the graphs of the sine and cosine functions in Section 2–5.

Consider angles in standard position with measures θ and $-\theta$, $0 \le \theta \le \frac{\pi}{2}$, where $P(x, y)$ and $P'(x', y')$ are points on their respective terminal sides. Each point is at a distance r from the origin. The two triangles formed by segment PP' are congruent by side-angle-side ($OP = OP'$, $\theta = |-\theta|$, $OQ = OQ$). Thus, $x' = x$, and $y' = -y$. Then, by definition of the sine and cosine functions,

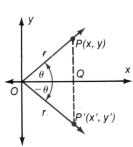

$$\textbf{sin}\,(-\theta) = \frac{y'}{r} = -\frac{y}{r} = -\textbf{sin}\,\theta \qquad \textbf{1}$$

and
$$\textbf{cos}\,(-\theta) = \frac{x'}{r} = \frac{x}{r} = \textbf{cos}\,\theta. \qquad \textbf{2}$$

Formulas **1** and **2** are true for all θ, even though the angle here is in Quadrant I. (This will be proved in Chapter 3.)

Any function having the property that for each x in the domain, $-x$ is also in the domain and $f(-x) = -f(x)$, is called an **odd function.** Thus, since sin $(-\theta) = -\sin \theta$, the <u>sine function is an odd function</u>.

Also, any function having the property that for each x in the domain, $-x$ is in the domain and $f(-x) = f(x)$, is called an **even function.** Thus, since cos $(-\theta) = \cos \theta$ the <u>cosine function is an even function.</u>

Example 1
Given $\sin\theta$, write $\sin(-\theta)$ and given $\cos\theta$, write $\cos(-\theta)$.

a. $\sin 45° = \dfrac{\sqrt{2}}{2}$ **b.** $\cos 60° = \dfrac{1}{2}$

Solutions: a. $\sin(-45°) = -\sin 45°$ **b.** $\cos(-60°) = \cos 60°$

$$= -\dfrac{\sqrt{2}}{2}$$ $$= \dfrac{1}{2}$$

Consider angles in standard position with measures θ and $180° + \theta$, where $P(x, y)$ and $P'(x', y')$ are points on their respective terminal sides. Each point is at a distance r from the origin. (See Figures 1 and 2.)

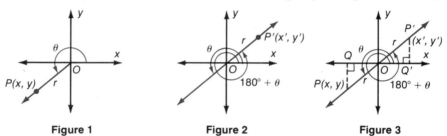

Figure 1   **Figure 2**   **Figure 3**

The two reference triangles formed by dropping perpendiculars from P and P' to the x axis are congruent by hypotenuse-angle because $OP = OP'$ and $m\angle POQ = m\angle P'OQ'$. (See Figure 3.) Thus, $x' = -x$ and $y' = -y$. Then, by definition of the sine and cosine functions,

$$\sin(180° + \theta) = \frac{y'}{r} = -\frac{y}{r} = -\sin\theta \qquad\qquad 3$$

and
$$\cos(180° + \theta) = \frac{x'}{r} = -\frac{x}{r} = -\cos\theta. \qquad\qquad 4$$

It can also be shown that

$$\sin(180° - \theta) = \sin\theta \qquad\qquad 5$$

and
$$\cos(180° - \theta) = -\cos\theta. \qquad\qquad 6$$

You are asked to derive these formulas in the Exercises.

The formulas for $(180° + \theta)$ and $(180° - \theta)$ can be used to derive the following formulas.

$\sin(360° - \theta) = -\sin\theta$ **7** $\cos(360° - \theta) = \cos\theta$ **8**

$\sin(360° + \theta) = \sin\theta$ **9** $\cos(360° + \theta) = \cos\theta$ **10**

Example 2 illustrates the derivation of formula **7**.

Example 2

Show that $\sin(360° - \theta) = -\sin\theta$.

Solution: $\sin(360° - \theta) = \sin[180° + (180° - \theta)]$ ⟵ $360° = 180° + 180°$

Let $\alpha = (180° - \theta)$. Then

$$\sin(360° - \theta) = \sin(180° + \alpha)$$
$$= -\sin\alpha \quad\quad\quad\quad\quad ⟵ \text{Formula 3}$$
$$= -\sin(180° - \theta) \quad\quad ⟵ \alpha = (180° - \theta)$$
$$= -\sin\theta \quad\quad\quad\quad\quad ⟵ \text{Formula 5}$$

The derivations of formulas **8, 9,** and **10** are asked for in the Exercises.

Formulas **3** through **10** can also be stated in terms of radian measure as shown below.

$\sin(\pi + \theta) = -\sin\theta$	**3'**	$\cos(\pi + \theta) = -\cos\theta$	**4'**
$\sin(\pi - \theta) = \sin\theta$	**5'**	$\cos(\pi - \theta) = -\cos\theta$	**6'**
$\sin(2\pi - \theta) = -\sin\theta$	**7'**	$\cos(2\pi - \theta) = \cos\theta$	**8'**
$\sin(2\pi + \theta) = \sin\theta$	**9'**	$\cos(2\pi + \theta) = \cos\theta$	**10'**

The formulas developed in this section are known as **reduction formulas,** because they enable you to express $\sin x$ and $\cos x$ where x is greater than 90° in terms of an angle θ between 0° and 90°. Proofs for Formulas 1–10 for <u>any</u> angle θ will be developed in Chapter 3.

These formulas can be used to develop tables for graphing the sine and cosine functions.

Example 3

Given that $\sin\frac{\pi}{4} = .71$, write $\sin\theta$ for each given value of θ.

Solution:

θ	$\sin\theta$	
$-\frac{\pi}{4}$	$-.71$	⟵ $\sin(-\theta) = -\sin\theta$
$\frac{3\pi}{4}$	$.71$	⟵ $\sin(\pi - \theta) = \sin\theta$
$\frac{5\pi}{4}$	$-.71$	⟵ $\sin(\pi + \theta) = -\sin\theta$
$\frac{7\pi}{4}$	$-.71$	⟵ $\sin(2\pi - \theta) = -\sin\theta$

Classroom Exercises

Given: sin 60° = .8660 and cos 60° = .5000. Complete each table.

	θ	$\sin \theta$
1.	$-60°$	?
2.	$120°$	?
3.	$240°$	?

	θ	$\cos \theta$
4.	$-60°$	?
5.	$120°$	?
6.	$240°$	?

Written Exercises

a

Given $\sin \theta$, write $\sin(-\theta)$ and given $\cos \theta$ write $\cos(-\theta)$.

1. $\sin 30° = \frac{1}{2}$

2. $\cos 45° = \frac{\sqrt{2}}{2}$

3. $\cos 30° = \frac{\sqrt{3}}{2}$

4. $\sin 60° = \frac{\sqrt{3}}{2}$

5. $\sin 52° = .7880$

6. $\cos 23° = .9205$

For Exercises 7–14, $\sin \theta = .6000$ and $\cos \theta = .8000$. Find the value of each expression.

7. $\cos(-\theta)$

8. $\sin(-\theta)$

9. $\cos(180° + \theta)$

10. $\sin(180° + \theta)$

11. $\sin(\pi - \theta)$

12. $\cos(\pi - \theta)$

13. $\sin(2\pi - \theta)$

14. $\cos(360° - \theta)$

Given: $\sin 25° = .4226$. Find the value of each expression.

15. $\sin(-25°)$

16. $\sin 155°$

17. $\sin 205°$

18. $\sin 335°$

Given: $\cos 25° = .9063$. Find the value of each expression.

19. $\cos(-25°)$

20. $\cos 155°$

21. $\cos 205°$

22. $\cos 335°$

In Exercises 23–46, use the given statement to complete the table.

Given: $\sin \frac{\pi}{8} = .3827$.

Given: $\cos \frac{\pi}{5} = .8090$.

Given: $\sin \frac{\pi}{10} = .3090$.

	θ	$\sin \theta$
23.	$-\frac{\pi}{8}$	?
24.	$\frac{7\pi}{8}$	?
25.	$\frac{9\pi}{8}$	?
26.	$\frac{15\pi}{8}$	?

	θ	$\cos \theta$
27.	$-\frac{\pi}{5}$	?
28.	$\frac{4\pi}{5}$	?
29.	$\frac{6\pi}{5}$	?
30.	$\frac{9\pi}{5}$	?

	θ	$\sin \theta$
31.	$-\frac{\pi}{10}$	?
32.	$\frac{9\pi}{10}$	?
33.	$\frac{11\pi}{10}$	?
34.	$\frac{19\pi}{10}$	?

Given: $\cos \frac{\pi}{10} = .9511$.   Given: $\sin \frac{\pi}{12} = .2588$.   Given: $\cos \frac{\pi}{12} = .9659$.

	θ	$\cos \theta$		θ	$\sin \theta$		θ	$\cos \theta$
35.	$-\frac{\pi}{10}$	?	**39.**	$-\frac{\pi}{12}$	?	**43.**	$-\frac{\pi}{12}$	?
36.	$\frac{9\pi}{10}$	?	**40.**	$\frac{11\pi}{12}$	?	**44.**	$\frac{11\pi}{12}$	?
37.	$\frac{11\pi}{10}$	?	**41.**	$\frac{13\pi}{12}$	?	**45.**	$\frac{13\pi}{12}$	?
38.	$\frac{19\pi}{10}$	?	**42.**	$\frac{23\pi}{12}$	?	**46.**	$\frac{23\pi}{12}$	?

47. Show that $\cos (360° - \theta) = \cos \theta$. (See Example 2.)

48. Show that $\sin (360° + \theta) = \sin \theta$. (See Example 2.)

49. Show that $\cos (360° + \theta) = \cos \theta$. (See Example 2.)

Derive the following reduction formulas. Follow the procedure used in this section to develop formulas 3 and 4.

50. $\sin (180° - \theta) = \sin \theta$ **51.** $\cos (180° - \theta) = -\cos \theta$

Given: $\cos \theta = \frac{5}{13}$. Find the value of each expression.

52. $\sin (-\theta)$ **53.** $\sin (180° + \theta)$ **54.** $\cos (\pi - \theta)$ **55.** $\sin (2\pi - \theta)$

Tell whether each function is <u>odd</u>, <u>even</u>, or <u>neither</u>.

56. $y = 2x$ **57.** $y = x^2$ **58.** $y = -1$ **59.** $y = x + 1$

60. $y = -x$ **61.** $y = x^3$ **62.** $y = x^2 + 1$ **63.** $y = x^4$

64. There is just one function, defined for all real numbers, that is both odd <u>and</u> even. Find it.

65. Give the equation of the line that is an axis of symmetry for all even functions.

66. Classify the tangent function as <u>odd</u>, <u>even</u>, or <u>neither</u>.

67. A portion of the graph of an <u>even</u> function is sketched at the left below. Sketch the function for $-2\pi \le x \le 2\pi$.

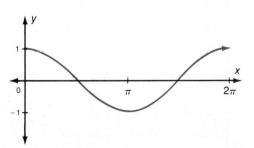   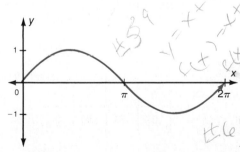

68. A portion of the graph of an <u>odd</u> function is sketched at the right above. Sketch the function for $-2\pi \le x \le 2\pi$.

2-5 Graphs of sin x and cos x

To graph the function $f(\theta) = \sin \theta$ in the coordinate plane, you begin by making a table of values for the ordered pairs $(\theta, \sin \theta)$. However, in order to graph on the customary xy axes, x will be used to represent θ and y will be used to represent $f(\theta)$, or $\sin \theta$. The values of x are given in radian measure.

To graph $y = \sin x$, use the table on pages 368-372 to make a table of values for $0 \le x \le \frac{\pi}{2}$. Then plot these points and join them with a smooth curve.

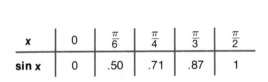

x	0	$\frac{\pi}{6}$	$\frac{\pi}{4}$	$\frac{\pi}{3}$	$\frac{\pi}{2}$
sin x	0	.50	.71	.87	1

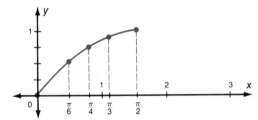

The properties developed in Section 2–4 can be used to extend both the table and the graph of $y = \sin x$.

Example 1

Use the table of values above and the graph of $y = \sin x$ for $0 \le x \le \frac{\pi}{2}$ to extend the graph of the function for $\frac{\pi}{2} < x \le 2\pi$.

Solution: 1. Extend the table and the graph for $\frac{\pi}{2} < x \le \pi$. Use the table and the property $\sin (\pi - x) = \sin x$.

x	$\frac{2\pi}{3}$	$\frac{3\pi}{4}$	$\frac{5\pi}{6}$	π
sin x	.87	.71	.50	0

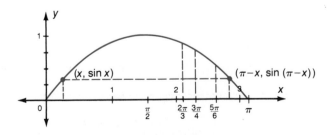

The graph is now completed for Quadrants I and II.

2. Extend the table and the graph for $\pi < x \le \frac{3\pi}{2}$. Use the table of values for $0 < x \le \frac{\pi}{2}$ and the property $\sin(\pi + x) = -\sin x$.

x	$\frac{7\pi}{6}$	$\frac{5\pi}{4}$	$\frac{4\pi}{3}$	$\frac{3\pi}{2}$
sin x	−.50	−.71	−.87	−1

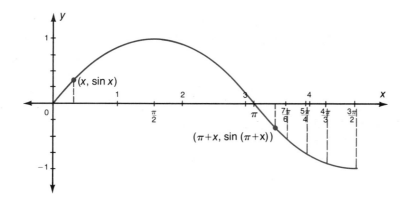

3. Extend the table and the graph for $\frac{3\pi}{2} < x \le 2\pi$. Use the table of values for $0 \le x \le \frac{\pi}{2}$ and the property $\sin(2\pi - x) = -\sin x$.

x	$\frac{5\pi}{3}$	$\frac{7\pi}{4}$	$\frac{11\pi}{6}$	2π
sin x	−.87	−.71	−.50	0

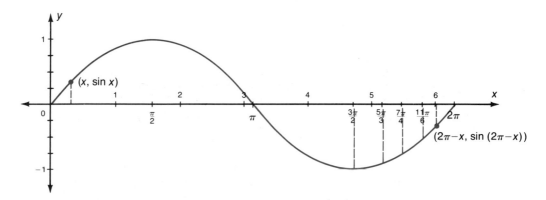

This is the graph of $y = \sin x$ for the interval $0 \le x \le 2\pi$.

The graph of Example 1 illustrates clearly the fact that the function $y = \sin x$ is positive in Quadrants I and II and that it is negative in Quadrants III and IV.

You can use Example 1 and the property sin (−x) = −sin x to draw the graph of y = sin x for −2π ≤ x ≤ 2π.

Example 2 Graph y = sin x for −2π ≤ x ≤ 2π.

Solution: Use Example 1 to sketch the graph of y = sin x for 0 ≤ x ≤ 2π. Then, since sin (−x) = −sin x, you can plot several points for (−x, −sin x).

−x	$-\frac{\pi}{4}$	$-\frac{\pi}{2}$	$-\frac{3\pi}{4}$	−π	$-\frac{5\pi}{4}$	$-\frac{3\pi}{2}$	$-\frac{7\pi}{4}$	−2π
−sin x	−.71	−1	−.71	0	.71	1	.71	0

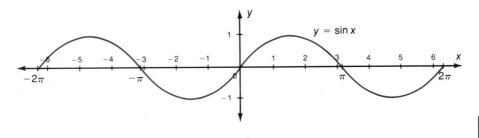

You can use a similar procedure to graph the function y = cos x.

Example 3 Graph y = cos x for −2π ≤ x ≤ 2π.

Solution: Follow the procedures of Examples 1 and 2.

1. For 0 ≤ x ≤ $\frac{\pi}{2}$, use the table on pages 368–372.

x	0	$\frac{\pi}{6}$	$\frac{\pi}{4}$	$\frac{\pi}{3}$	$\frac{\pi}{2}$
cos x	1	.87	.71	.50	0

2. For $\frac{\pi}{2}$ < x ≤ π, use the table of Step **1** and the property cos (π − x) = −cos x.

x	$\frac{2\pi}{3}$	$\frac{3\pi}{4}$	$\frac{5\pi}{6}$	π
cos x	−.50	−.71	−.87	−1

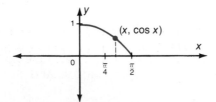

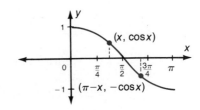

3. For $\pi < x \leq \frac{3\pi}{2}$, use the table of Step **1** and the property $\cos(\pi + x) = -\cos x$.

x	$\frac{7\pi}{6}$	$\frac{5\pi}{4}$	$\frac{4\pi}{3}$	$\frac{3\pi}{2}$
cos x	−.87	−.71	−.50	0

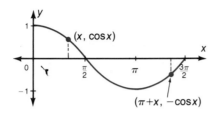

4. For $\frac{3\pi}{2} < x \leq 2\pi$, use the table of Step **1** and the property $\cos(2\pi - x) = \cos x$.

x	$\frac{5\pi}{3}$	$\frac{7\pi}{4}$	$\frac{11\pi}{6}$	2π
cos x	.50	.71	.87	1

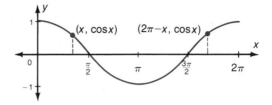

5. Use the property $\cos(-x) = \cos x$ to extend the graph from $x = 0$ to $x = -2\pi$.

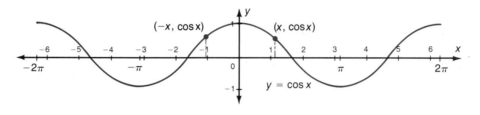

The graphs of $\sin x$ and $\cos x$ show that these functions repeat their values in the same order after every 2π units and that this is the smallest interval for which these values repeat. Thus, $\sin x$ and $\cos x$ are called periodic functions with period 2π.

Definition: If there is a smallest positive number p such that $f(p + x) = f(x)$ for every x in the domain of f, then p is the period of the function f and f is a **periodic function.**

Note also that the domain of $y = \sin x$ and $y = \cos x$ is the set of real numbers. The range of each function is the set of real numbers such that $-1 \leq y \leq 1$.

The graphs in Examples 1, 2, and 3 were drawn over restricted intervals of the domain. When the domain is not restricted, arrowheads are used to show that the graph of a function extends indefinitely. The pattern of the graph repeats over the period p.

You can use the graphs in Examples 1, 2, and 3 to sketch the graphs of functions such as $y = \sin x + D$ and $y = \cos x + D$, for $D \in R$.

Example 4 Graph $y = \sin x + 3$ for $0 \le x \le 2\pi$.

Solution: Since $y = \sin x + 3$, each point on the graph is three units above the corresponding point on the graph of $y = \sin x$.

1. Sketch the graph of $y = \sin x$ for $0 \le x \le 2\pi$.
2. Sketch a similar curve three units above the graph of $y = \sin x$.

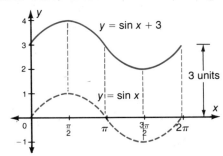

In Example 4, the graph of $y = \sin x + D$ is sketched for $D > 0$. For $D < 0$, the graph of $y = \sin x + D$ and $y = \cos x + D$ are D units below the graphs of $y = \sin x$ and $y = \cos x$, respectively. It can also be seen from Example 4 that $y = \sin x + D$, $D \in R$, and $y = \sin x$ have the same period, 2π. Similarly, the period of $y = \cos x + D$, $D \in R$, is 2π.

Classroom Exercises

Match each function with its graph.

1. $y = \cos x$ 2. $y = \sin x$ 3. $y = \cos x + 1$ 4. $y = \sin x - \frac{1}{2}$

a.
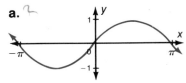

b.
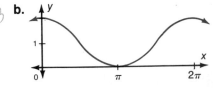

c.
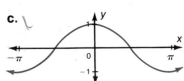

d.
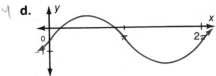

Written Exercises

a

Make a table of values. Then graph each function on the given interval.

1. $y = \sin x$, $-4\pi \le x \le -2\pi$

2. $y = \cos x$, $-4\pi \le x \le 2\pi$

Graph each function on the given interval.

3. $y = \sin x$, $2\pi \le x \le 4\pi$

4. $y = \cos x$, $\frac{\pi}{2} \le x \le \frac{5\pi}{2}$

5. $y = \cos x$, $-\frac{3\pi}{2} \le x \le \frac{3\pi}{2}$

6. $y = \sin x$, $-\pi \le x \le 3\pi$

7. $y = \cos x$, $-\frac{3\pi}{4} \le x \le \frac{3\pi}{2}$

8. $y = \sin x$, $-\frac{3\pi}{4} \le x \le 3\pi$

Graph each function for $-2\pi \le x \le 2\pi$.

9. $y = \cos x + 2$ up 2

10. $y = \sin x + \frac{1}{2}$

11. $y = \cos x - 2$ down 2

12. $y = \cos x - \frac{1}{2}$

State the period of each function graphed below.

13.

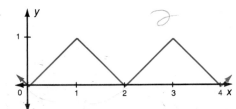

14.

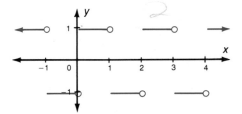

15.

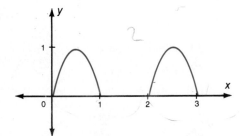

16.

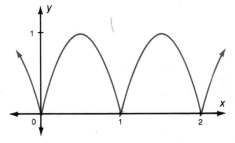

17.

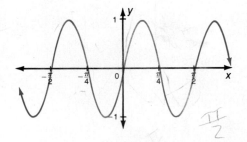

18.

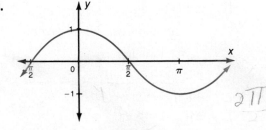

Graphs of Trigonometric Functions   **85**

The graph of $y = \sin x - \frac{1}{2}$ for $-2\pi \le x \le 2\pi$ is shown in the figure at the right below. Use this graph to find the value of y in $y = \sin x - \frac{1}{2}$ for each given value of x.

19. $x = -\frac{3\pi}{2}$

20. $x = \frac{\pi}{2}$

21. $x = 0$

22. $x = \frac{3\pi}{2}$

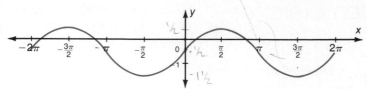

In the figure below, the functions $y = \sin x$ and $y = \cos x$ for $0 \le x \le 2\pi$ are graphed on the same set of axes. Refer to this figure to give a reasonable estimate for the value of x in the interval $0 < x < 2\pi$ that satisfies each of the following statements. Some statements may be satisfied by no values of x.

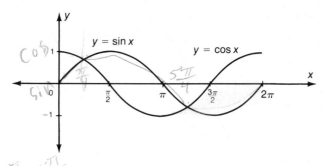

23. $\cos x = \sin x$

24. $\cos x = -\sin x$

25. $\sin x < \cos x$

26. $\sin x > \cos x$

27. $\sin x + \cos x = 1$

28. $\sin x + \cos x = 2$

In Exercise 29–38, graph each function for $0 \le x \le 2\pi$.

29. $y = 4 - \sin x$

30. $y = 3 - \cos x$

31. $y = |\sin x|$

32. $y = -|\cos x|$

33. $y = 2 + |\sin x|$

34. $y = 2 - |\sin x|$

35. $y = x + \sin x$

36. $y = x + |\sin x|$

37. $y = x - \cos x$

38. $y = \cos x - 2x$

For each function in Exercises 39–41 find a value for D such that the range of the function will be as given.

Function	Range
39. $y = \sin x + D$	$-4 < y < -2$
40. $y = \cos x + D$	$10 < y < 12$
41. $y = -2 \sin x + D$	$-1 < y < 3$

Review

In circle O with radius r, central angle POQ intercepts $\overset{\frown}{QP}$. Find, to the nearest tenth, the radian measure of $\angle POQ$ for each given value of r and length of $\overset{\frown}{QP}$. (Section 2–1)

1. $r = 12$ m, length of $\overset{\frown}{QP} = 20$ m

2. $r = 4$ cm, length of $\overset{\frown}{QP} = 7$ cm

Change each degree measure to radian measure in terms of π. (Section 2–1)

3. $-45°$ **4.** $60°$ **5.** $180°$ **6.** $-135°$

Change each radian measure to degree measure. (Section 2–1)

7. $\dfrac{\pi}{3}$ **8.** $-\dfrac{\pi}{12}$ **9.** $\dfrac{3\pi}{4}$ **10.** $\dfrac{\pi}{6}$

Use the tables on pages 368–372 to find a four-place decimal approximation for the radian measure of the angle with the given degree measure. (Section 2–2)

11. $34°20'$ **12.** $15°50'$ **13.** $78°30'$ **14.** $60°10'$

Use the tables on pages 368–372 to find the degree measure of the angle with the given radian measure. (Section 2–2)

15. .7243 **16.** .5149 **17.** 1.2217 **18.** 1.5504

19. In the figure at the right, point P is on the circumference of a wheel that is rotating 600 rpm. Find the angular velocity of point P in radians per second. (Section 2–3)

20. The radius of the wheel in Exercise 19 is 2.5 m. Find, to the nearest tenth of a meter, the distance traveled by point P in 1.4 seconds. (Section 2–3)

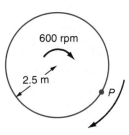

21. Given: $\sin \theta = .6000$ Find $\sin (180° - \theta)$. (Section 2–4)

22. Given: $\cos 30° = .8660$ Complete the table below. (Section 2–4)

θ	$-30°$	$150°$	$210°$	$330°$
$\cos \theta$	?	?	?	?

23. Graph $y = \cos x + 1$ in the interval $0 \le x \le 2\pi$. (Section 2–5)

2-6 Amplitude

You have seen that the range of each of the functions $y = \sin x$ and $y = \cos x$ is the set of real numbers y such that $-1 \le y \le 1$. That is,

$$-1 \le \sin x \le 1$$

and

$$-1 \le \cos x \le 1.$$

Thus, each function has a minimum value of -1 and a maximum value of 1.

When M and m are the maximum and minimum values, respectively, of a periodic function, then the **amplitude** of the function is

$$\tfrac{1}{2}(M - m).$$

Example 1

Find the amplitude of each function.

a. $y = \sin x$

b. $y = \sin x + 3$

Solutions: a. For $y = \sin x$, $M = 1$ and $m = -1$. ⟵ From Example 1, pages 80–81

$$
\begin{aligned}
\text{Amplitude} &= \tfrac{1}{2}(M - m) \\
&= \tfrac{1}{2}[1 - (-1)] \\
&= \tfrac{1}{2}(2) \\
&= 1
\end{aligned}
$$

b. For $y = \sin x + 3$, $M = 4$ and $m = 2$. ⟵ From Example 4, page 84

$$
\begin{aligned}
\text{Amplitude} &= \tfrac{1}{2}(M - m) \\
&= \tfrac{1}{2}(4 - 2). \\
&= \tfrac{1}{2}(2) \\
&= 1
\end{aligned}
$$

The amplitude may be any positive number. Example 2 will help you see the relationship between the amplitude and the value of A in the function $y = A \sin x$.

Example 2

Graph on the same set of axes for $0 \le x \le 2\pi$.

a. $y = \sin x$ **b.** $y = 3 \sin x$ **c.** $y = -\frac{1}{3} \sin x$

Solution: 1. Use the table on pages 368–372 to make a table for $0 \le x \le \frac{\pi}{2}$.

2. Use reduction formulas to extend both the table and the graph for $\frac{\pi}{2} \le x \le 2\pi$.

x	$\sin x$	$3 \sin x$	$-\frac{1}{3} \sin x$
0	0	0	0
$\frac{\pi}{6}$	$.50$	1.50	$-.17$
$\frac{\pi}{3}$	$.87$	2.61	$-.29$
$\frac{\pi}{2}$	1	3	$-.33$
$\frac{2\pi}{3}$	$.87$	2.61	$-.29$
$\frac{5\pi}{6}$	$.50$	1.50	$-.17$
π	0	0	0
$\frac{7\pi}{6}$	$-.50$	-1.50	$.17$
$\frac{4\pi}{3}$	$-.87$	-2.61	$.29$
$\frac{3\pi}{2}$	-1	-3	$.33$
$\frac{5\pi}{3}$	$-.87$	-2.61	$.29$
$\frac{11\pi}{6}$	$-.50$	-1.50	$.17$
2π	0	0	0

You can use the graphs in Example 2 to make the following comparisons.

Function	Value of A in $y = A \sin x$	Amplitude
$y = \sin x$	1	$\frac{1}{2}[1 - (-1)] = 1$
$y = 3 \sin x$	3	$\frac{1}{2}[3 - (-3)] = 3$
$y = -\frac{1}{3} \sin x$	$-\frac{1}{3}$	$\frac{1}{2}[\frac{1}{3} - (-\frac{1}{3})] = \frac{1}{3}$

Thus, for each function $y = A \sin x$, the amplitude is $|A|$. Similarly, for each function $y = A \cos x$, the amplitude is $|A|$.

By knowing the period, general shape, and amplitude of the functions defined by $y = A \sin x + D$ and $y = A \cos x + D$, you can sketch them readily.

Graphs of Trigonometric Functions **89**

Example 3

Graph the following functions on the same set of axes for $0 \le x \le 2\pi$.

a. $y = 2 \sin x$ **b.** $y = -2 \sin x + 1$

Solution: For both functions, the period is 2π and the amplitude, $|A|$, is 2. First sketch the graph of $y = 2 \sin x$ and $y = -2 \sin x$. Then, by shifting the graph of $y = -2 \sin x$ upward by 1 unit, you obtain the graph of $y = -2 \sin x + 1$.

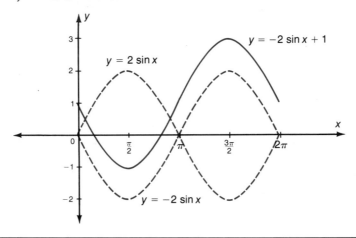

In Example 3, note that the graph of $y = -2 \sin x$ is the "mirror image" or <u>reflection</u> of the graph of $y = 2 \sin x$ with respect to the x axis. In general, for $A < 0$ the graph of $y = A \sin x$ is the reflection of the graph of $y = |A| \sin x$ with respect to the x axis. Similarly, the graph of $y = A \cos x$ for $A < 0$ is the reflection of the graph of $y = |A| \cos x$ with respect to the x axis.

Classroom Exercises

Replace each __?__, with the correct value.

1.

x	$\sin x$	$-\sin x$
0	0	0
$\dfrac{\pi}{2}$	1	-1
π	$?$	$?$
$\dfrac{3\pi}{2}$	$?$	$?$

2.

x	$\sin x$	$\frac{1}{2} \sin x + 4$
0	0	4
$\dfrac{\pi}{2}$	1	$?$
π	$?$	$?$
$\dfrac{3\pi}{2}$	-1	$?$

3.

x	cos x	−cos x
0	?	?
$\frac{\pi}{3}$	?	?
$\frac{\pi}{2}$	?	?
π	?	?

4.

x	cos x	$\frac{1}{3}$ cos x − 2
0	?	?
$\frac{\pi}{3}$	?	?
$\frac{\pi}{2}$	?	?
π	?	?

Find the amplitude of each function.

5.

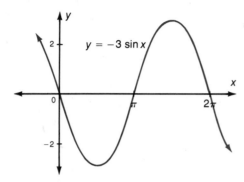

$y = -3 \sin x$

6.

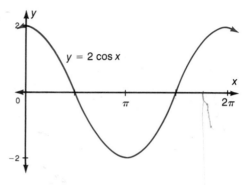

$y = 2 \cos x$

7.

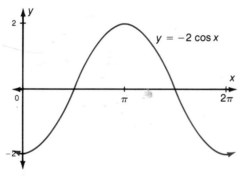

$y = -2 \cos x$

8.

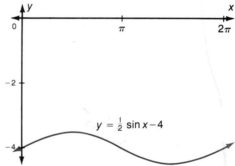

$y = \frac{1}{2} \sin x - 4$

Written Exercises

a

In Exercises 1–8 graph each function for $0 \le x \le 2\pi$. Find the amplitude of each function.

1-7 odd

9-12

13,15

17-20

1. $y = 5 \sin x$

2. $y = -5 \sin x$

3. $y = -2 \cos x$

4. $y = \frac{1}{2} \cos x$

5. $y = \sin x + 2$

6. $y = \sin x - 2$

7. $y = -1 \cos x - 1$

8. $y = -\frac{1}{4} \cos x - \frac{1}{4}$

Graphs of Trigonometric Functions   **91**

In Exercises 9–12, find the amplitude of each function. Then match each function with its graph.

9. $y = -\frac{1}{2} \sin x - 2$

10. $y = 2 \sin x + \frac{1}{2}$

11. $y = -2 \cos x + \frac{1}{2}$

12. $y = -\frac{1}{2} \cos x + 2$

a.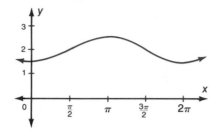

b.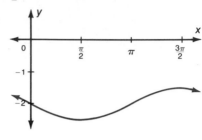

c.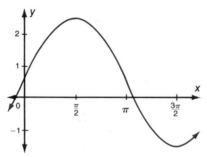

d.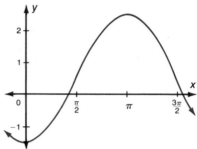

In Exercises 13–16, graph $y = \sin x$ and the given function on the same axes for $0 \le x \le 2\pi$.

13. $y = 2 \sin x - 3$

14. $y = -3 \sin x + 2$

15. $y = \frac{1}{2} \sin x - \frac{1}{2}$

16. $y = -\frac{1}{2} \sin x + \frac{1}{2}$

In Exercises 17–20, write an equation of a cosine function with the given characteristics. (NOTE: There may be more than one answer.)

17. Amplitude: 3, Period: 2π, Contains $(0, 3)$.

18. Amplitude: 2, Period: 2π, Contains $(\frac{\pi}{2}, 0)$.

19. Period: 2π, Maximum: 10, Minimum: -10, Contains $(\frac{3\pi}{2}, 0)$.

20. Period: 2π, Maximum: 6, Minimum: -2, Contains $(0, 6)$.

In Exercises 21–24, function f has equation $y = 2 \sin x + 3 \cos x$. The domain is $-2\pi \le x \le 2\pi$. First graph $y = 2 \sin x$ and $y = 3 \cos x$ on the same set of axes. Use the same domain as for f.

21. Graph f by adding the y coordinates of the first two curves.

22. Find the minimum and maximum values of f.

23. Find the amplitude of f.

24. Find the period of f.

2-7 Period

The functions $y = \sin x$ and $y = \cos x$ have the same period, 2π. However, functions of the form $y = \sin Bx$ and $y = \cos Bx$ for $B \neq 0$ and $|B| \neq 1$ will <u>not</u> have a period of 2π.

Example 1

For $0 \leq x \leq 4\pi$, graph $y = \sin x$, $y = \sin 2x$, and $y = \sin \frac{1}{2}x$ on the same set of axes. Find the period of each function.

Solution: 1. Use the table on pages 368–372 to make a table of values and to sketch the graph of each function for $0 \leq x \leq \frac{\pi}{2}$.

2. Use the appropriate reduction formulas to extend the table and the graphs for $\frac{\pi}{2} < x \leq 2\pi$.

3. Use the reduction formula $\sin(2\pi + x) = \sin x$ (see page 77) to extend the table and the graphs for $2\pi < x \leq 4\pi$.

x	$2x$	$\sin 2x$	$\frac{1}{2}x$	$\sin\frac{1}{2}x$
0	0	0	0	0
$\frac{\pi}{4}$	$\frac{\pi}{2}$	1	$\frac{\pi}{8}$	$.38$
$\frac{\pi}{2}$	π	0	$\frac{\pi}{4}$	$.71$
$\frac{3\pi}{4}$	$\frac{3\pi}{2}$	-1	$\frac{3\pi}{8}$	$.92$
π	2π	0	$\frac{\pi}{2}$	1
$\frac{5\pi}{4}$	$\frac{5\pi}{2}$	1	$\frac{5\pi}{8}$	$.92$
$\frac{3\pi}{2}$	3π	0	$\frac{3\pi}{4}$	$.71$
$\frac{7\pi}{4}$	$\frac{7\pi}{2}$	-1	$\frac{7\pi}{8}$	$.38$
2π	4π	0	π	0
$\frac{9\pi}{4}$	$\frac{9\pi}{2}$	1	$\frac{9\pi}{8}$	$-.38$
$\frac{5\pi}{2}$	5π	0	$\frac{5\pi}{4}$	$-.71$
$\frac{11\pi}{4}$	$\frac{11\pi}{2}$	-1	$\frac{11\pi}{8}$	$-.92$
3π	6π	0	$\frac{3\pi}{2}$	-1
$\frac{13\pi}{4}$	$\frac{13\pi}{2}$	1	$\frac{13\pi}{8}$	$-.92$
$\frac{7\pi}{2}$	7π	0	$\frac{7\pi}{4}$	$-.71$
$\frac{15\pi}{4}$	$\frac{15\pi}{2}$	-1	$\frac{15\pi}{8}$	$-.38$
4π	8π	0	2π	0

The graphs are shown at the top of page 94.

The graphs of $y = \sin 2x$ and $y = \sin \frac{1}{2}x$ are shown below.

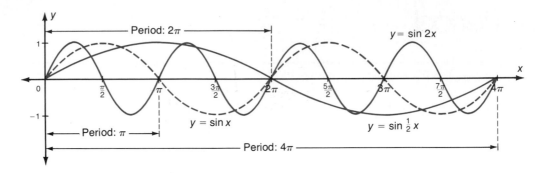

Function	Period
$y = \sin x$	2π
$y = \sin 2x$	π
$y = \sin \frac{1}{2}x$	4π

In Example 1, the amplitudes of the three functions are the same. However, the period of $y = \sin 2x$ is one half the period of $y = \sin x$ and the period of $y = \sin \frac{1}{2}x$ is twice the period of $y = \sin x$. This suggests the following rule.

For the function defined by $y = \sin Bx$ where $B \neq 0$, the period is given by

$$\frac{2\pi}{|B|}.$$

A similar rule is true for $y = \cos Bx$. This is illustrated in Example 2.

Example 2   Use the same set of axes to graph each function for $0 \leq x \leq 2\pi$. Find the period and amplitude of $y = 3 \cos 3x$.

a. $y = \cos x$ **b.** $y = 3 \cos 3x$

Solution: **1.** Sketch the graph of $y = \cos x$ for $0 \leq x \leq 2\pi$.

2. Make a table of values for $y = 3 \cos 3x$ where $0 \leq x \leq \frac{\pi}{2}$. Use the appropriate reduction formulas to extend the table for $\frac{\pi}{2} < x \leq 2\pi$.

3. Graph $y = 3 \cos 3x$ on the same set of axes as $y = \cos x$.

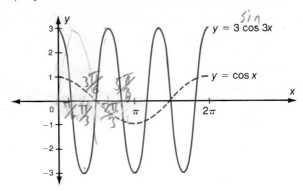

The period of $y = 3 \cos 3x$ is $\frac{2\pi}{3}$.

The amplitude of $y = 3 \cos 3x$ is $\frac{1}{2}[3 - (-3)]$, or 3.

In Example 2, the period of $y = 3 \cos 3x$ is one-third the period of $y = \cos x$. If $y = 3 \cos 3x$ is written in the more general form as $y = A \cos Bx$, the period is $\frac{2\pi}{|B|}$ and the amplitude is $|A|$. These results can be summarized as follows.

Functions of the form $y = A \sin Bx$ and $y = A \cos Bx$ where A and B are real numbers, have amplitude **|A|** and period $\dfrac{2\pi}{|B|}$.

Since $\sin (-\theta) = -\sin \theta$, the graphs of $y = A \sin Bx$ and $y = A \sin |B|x$ where $B < 0$ are mirror images of each other. Since $\cos (-\theta) = \cos \theta$, the graphs of $y = A \cos Bx$ and $y = A \cos |B|x$ are identical.

Classroom Exercises

Replace each ? with the correct value.

	x	$3x$	$\cos 3x$
1.	0	0	?
2.	$\frac{\pi}{4}$	$\frac{3\pi}{4}$	?
3.	$\frac{\pi}{2}$	$\frac{3\pi}{2}$	?
4.	$\frac{2\pi}{3}$	2π	?

	x	$\frac{1}{3}x$	$\sin\frac{1}{3}x$
5.	0	0	?
6.	2π	$\frac{2\pi}{3}$	?
7.	4π	$\frac{4\pi}{3}$	?
8.	6π	2π	?

Graphs of Trigonometric Functions **95**

Find the amplitude and period of each function.

9. $y = \sin 3x$

10. $y = -\cos \frac{1}{3}x$

11. $y = -4 \cos 4x$

Written Exercises

In Exercises 1–6, graph $y = \sin x$ and the given function on the same set of axes for $0 \le x \le 4\pi$. Find the period of each function.

1. $y = \sin 3x$

2. $y = \sin \frac{2}{3}x$

3. $y = \sin \frac{1}{3}x$

4. $y = \sin \frac{1}{6}x$

5. $y = \sin \frac{3}{2}x$

6. $y = \sin 4x$

In Exercises 7–12, graph $y = \cos x$ and the given function on the same set of axes for $0 \le x \le 4\pi$. Find the period of each function.

7. $y = \cos 2x$

8. $y = \cos \frac{1}{2}x$

9. $y = \cos \frac{1}{3}x$

10. $y = \cos 4x$

11. $y = \cos \frac{2}{3}x$

12. $y = \cos \frac{1}{6}x$

In Exercises 13–18, find the period and amplitude of each function. Then match each function with its graph.

Function	Period	Amplitude	Graph
13. $y = 2 \sin x$	?	?	?
14. $y = -2 \sin x$	?	?	?
15. $y = \sin 2x$	?	?	?
16. $y = \frac{1}{2} \sin 2x$	?	?	?
17. $y = \cos 2x$	?	?	?
18. $y = -2 \cos 2x$	?	?	?

a.

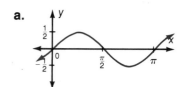

b.

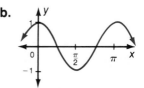

c.

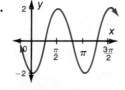

d.

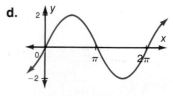

e.

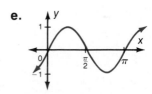

f.

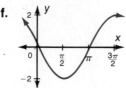

96   Chapter 2

In Exercises 19–24, find the amplitude and period of each function.

19. $y = 2 \cos \frac{1}{3} x$

20. $y = -\frac{1}{2} \cos \frac{1}{3} x$

21. $y = \frac{1}{18} \cos \frac{1}{18} x$

22. $y = \frac{7}{4} \sin \frac{3}{2} x$

23. $y = \frac{4}{5} \cos 2x$

24. $y = -5 \sin \frac{2}{3} x$

In Exercises 25–28, find a function of the form $y = A \cos Bx$ with the given properties.

25. Amplitude: 3; Period: π

26. Amplitude: $\frac{1}{2}$; Period: 3π

27. Amplitude: 12; Period: $\frac{\pi}{3}$

28. Amplitude: $\frac{2}{3}$; Period: $\frac{\pi}{2}$

In Exercises 29–33, find a function of the form $y = A \sin Bx$ with the given properties.

29. Amplitude: 4; Minimum: $(\frac{\pi}{4}, -4)$

30. Amplitude: $\frac{2}{3}$; Maximum: $(\frac{3\pi}{4}, \frac{2}{3})$

31. Amplitude: 7; Contains: $(\frac{2\pi}{3}, 0)$

32. Maximum: $(\pi, 4)$; Period: 4π

33. Maximum: $(\frac{15\pi}{4}, 2)$; Minimum: $(\frac{5\pi}{4}, -2)$

The reciprocal of the period of a periodic function is the **frequency** of the function. Thus, the frequency of $y = \sin x$ is $\frac{1}{2\pi}$. The frequency represents the number of cycles completed by a function over an interval one unit long. In Exercises 34–39, find the frequency.

34. $y = \sin 2x$

35. $y = \sin \frac{1}{2} x$

36. $y = \cos 120x$

37. $y = -4 \cos 80x$

38. $y = \sin 40x$

39. $y = -20 \cos (-20x)$

Review Capsule for Section 2-8 _____

Use the facts that $ax - b = a(x - \frac{b}{a})$ and that $d = -(-d)$ to write each expression in the form $A(x - B)$.

Example 1: $4x - 3\pi = 4(x - \frac{3\pi}{4})$ ⟵ $A = 4;\ B = \frac{3\pi}{4}$

Example 2: $6x + \frac{\pi}{3} = 6(x + \frac{\pi}{18})$

$= 6[x - (-\frac{\pi}{18})]$ ⟵ $A = 6;\ B = -\frac{\pi}{18}$

1. $2x + 2$

2. $4x - 4$

3. $2x - 2$

4. $4x - 2$

5. $3x - 2$

6. $3x + 2$

7. $2x - \pi$

8. $5x + \frac{\pi}{2}$

2-8 Phase Shift

In Section 2–5, you saw how changes in D affect the graphs of $y = \sin x + D$ and $y = \cos x + D$.

Function	Relation to the graphs of $y = \sin x$
$y = \sin x + 2$	Each point on the graph of $y = \sin x + 2$ is two units above the corresponding point on the graph of $y = \sin x$.
$y = \sin x - 3$	Each point on the graph of $y = \sin x - 3$ is three units below the corresponding point on the graph of $y = \sin x$.

Example 1 compares the position of the graphs of $y = \cos x$ and functions of the form $y = \cos(x - C)$.

Example 1

Graph $y = \cos x$ and $y = \cos(x - \frac{\pi}{4})$ for $0 < x \leq \frac{5\pi}{2}$ on the same coordinate plane. Compare the positions of the two graphs.

Solution: 1. Sketch the graph of $y = \cos x$ for $0 < x < \frac{5\pi}{2}$.

2. Make a table of values for $y = \cos(x - \frac{\pi}{4})$ where $0 < x \leq \frac{3\pi}{4}$. Use the appropriate reduction formulas to extend the table for $\frac{3\pi}{4} < x \leq \frac{5\pi}{2}$.

x	$x - \frac{\pi}{4}$	$\cos(x - \frac{\pi}{4})$
$\frac{\pi}{4}$	0	1
$\frac{\pi}{2}$	$\frac{\pi}{4}$	0.71
$\frac{3\pi}{4}$	$\frac{\pi}{2}$	0
π	$\frac{3\pi}{4}$	-0.71
$\frac{5\pi}{4}$	π	-1
$\frac{3\pi}{2}$	$\frac{5\pi}{4}$	$-.71$
$\frac{7\pi}{4}$	$\frac{3\pi}{2}$	0
2π	$\frac{7\pi}{4}$	$.71$
$\frac{9\pi}{4}$	2π	1
$\frac{5\pi}{2}$	$\frac{9\pi}{4}$	$.71$

3. Graph $y = \cos\left(x - \frac{\pi}{4}\right)$ on the same set of axes as $y = \cos x$.

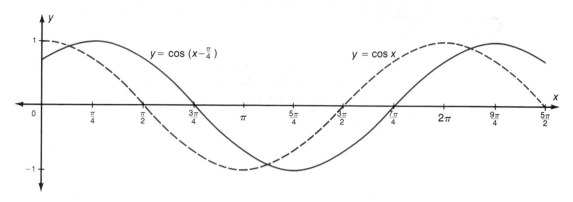

The graph of $y = \cos\left(x - \frac{\pi}{4}\right)$ has the same size, shape, amplitude, and period as the graph of $y = \cos x$, but it is shifted to the right by $\frac{\pi}{4}$ units.

In Example 1, the phase shift of $y = \cos\left(x - \frac{\pi}{4}\right)$ is $\frac{\pi}{4}$ units to the right. In general, the graph of any function defined by $y = \cos(x - C)$ is congruent to the graph of $y = \cos x$. The graph of $y = \cos(x - C)$ is C units to the right of the graph of $y = \cos x$ if $C > 0$ or $|C|$ units to the left if $C < 0$. A similar statement is true for the graphs of $y = \sin(x - C)$ and $y = \sin x$.

Example 2 Graph $y = 3 \sin\left(x + \frac{\pi}{2}\right)$ over one period. Find the amplitude, phase shift, and period.

Solution: 1. Rewrite $y = 3 \sin\left(x + \frac{\pi}{2}\right)$ in the form $y = A \sin(x - C)$. $y = 3 \sin\left(x + \frac{\pi}{2}\right)$ is the same as $y = 3 \sin\left[x - \left(-\frac{\pi}{2}\right)\right]$.

2. Use this form to find the amplitude, period, and phase shift.

$$\text{Amplitude: } 3 \quad \longleftarrow \quad |A| = 3$$

$$\text{Phase Shift: } \frac{\pi}{2} \text{ units to the left} \quad \longleftarrow \quad |C| = \frac{\pi}{2} \text{ and } C < 0$$

$$\text{Period: } 2\pi \quad \longleftarrow \quad \frac{2\pi}{|B|}; \; |B| = 1$$

3. Sketch the graph of $y = 3 \sin x$. Then sketch a curve congruent to $y = 3 \sin x$ but $\frac{\pi}{2}$ units to its left. Both graphs are shown at the top of the next page.

Graphs of Trigonometric Functions **99**

Note that the graphs of $y = 3 \sin x$ and $y = 3 \sin(x + \frac{\pi}{2})$ are identical but that the graph of $y = 3 \sin(x + \frac{\pi}{2})$ is $\frac{\pi}{2}$ units to the <u>left</u> of the graph of $y = 3 \sin x$.

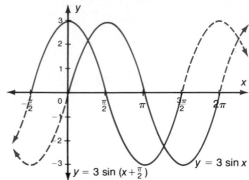

Now compare the graph of a function of the form $y = A \cos B(x - C)$ with the graph of a function of the form $y = A \cos Bx$.

Example 3 Graph $y = 2 \cos 3x$ and $y = 2 \cos 3(x - \frac{\pi}{4})$ for $0 \le x \le 2\pi$ on the same coordinate plane. Find the amplitude, period and phase shift of $y = 2 \cos 3(x - \frac{\pi}{4})$.

Solution: 1. Sketch the graph of $y = 2 \cos 3x$ for $0 \le x \le 2\pi$.

2. Make a table of values for $y = 2 \cos 3(x - \frac{\pi}{4})$.

3. Sketch the graph of $y = 2 \cos 3(x - \frac{\pi}{4})$ on the same set of axes.

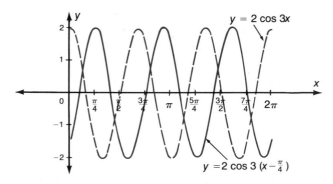

4. Use the graph of $y = 2 \cos 3(x - \frac{\pi}{4})$ to find each of the following.

Amplitude: $\frac{1}{2}[2 - (-2)] = 2$ Period: $\dfrac{2\pi}{3}$

Phase Shift: $\dfrac{\pi}{4}$ units to the right

Examples 1–3 suggest the following.

The **phase shift** for functions of the form $y = A \cos B(x - C)$ and $y = A \sin B(x - C)$ is C units to the right if $C > 0$ or $|C|$ units to the left if $C < 0$.

To sketch the graph of a function such as $y = \sin(2x + \frac{\pi}{3})$, first write the function in the form $y = \sin 2[x - (-\frac{\pi}{6})]$. Then use the amplitude, period, and phase shift of the function to sketch its graph.

Example 4 Graph $y = \sin(2x + \frac{\pi}{3})$ for $0 \le x \le 2\pi$.

Solution: **1.** Write $y = \sin(2x + \frac{\pi}{3})$ in the form $y = \sin 2[x - (-\frac{\pi}{6})]$.

2. Use this form of the equation to find the amplitude, period, and phase shift.

Function	Amplitude	Period	Phase Shift
$y = \sin 2[x - (-\frac{\pi}{6})]$	$|A| = 1$	$\dfrac{2\pi}{|B|} = \pi$	$\dfrac{\pi}{6}$ units to the left

3. Now sketch the graph of $y = \sin 2x$. The graph of $y = \sin(2x + \frac{\pi}{3})$ will be $\frac{\pi}{6}$ units to the <u>left</u> of this graph.

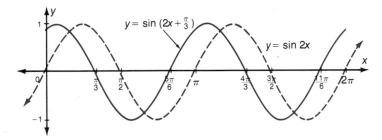

Classroom Exercises

Find the phase shift of each function.

1. $y = \cos 2(x + \frac{\pi}{2})$

2. $y = \sin(x - \pi)$

3. $y = 3 \sin(\frac{1}{2}x + \pi)$

4. $y = \cos(3x - \frac{\pi}{2})$

Written Exercises

In Exercises 1–8, graph each function for $0 \le x \le 2\pi$. Find the amplitude, period, and phase shift of each function.

1. $y = 2 \sin \left(x + \frac{\pi}{4}\right)$

2. $y = -2 \sin \left(x - \frac{\pi}{2}\right)$

3. $y = \frac{1}{2} \cos \left(x + \frac{\pi}{4}\right)$

4. $y = -\cos \left(x - \frac{\pi}{3}\right)$

5. $y = \sin 3 \left(x - \frac{\pi}{4}\right)$

6. $y = \cos 3 \left(x + \frac{\pi}{4}\right)$

7. $y = -2 \cos (2x - \pi)$

8. $y = -\frac{1}{2} \sin (3x + \pi)$

In Exercises 9–16, find the amplitude, period, and phase shift of each of the given functions.

9. $y = \frac{1}{2} \sin (3x - 2\pi)$

10. $y = -\frac{1}{2} \cos \left(\frac{1}{2}x + 4\pi\right)$

11. $y = -3 \sin \left(x - \frac{3\pi}{2}\right)$

12. $y = -3 \sin \left(4x - \frac{3\pi}{2}\right)$

13. $y = -\frac{3}{4} \cos (-3x - \pi)$

14. $y = \frac{2}{3} \sin \left(-\frac{1}{2}x + \pi\right)$

15. $y = 4 \sin (-2x - 2\pi)$

16. $y = -7 \cos (-5x + \pi)$

In Exercises 17–20, find the amplitude, period, and phase shift of each of the given functions. Then match each function with one of the four graphs below.

a.

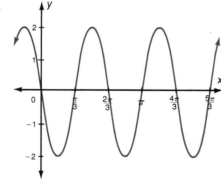

b.

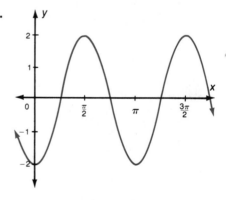

c.

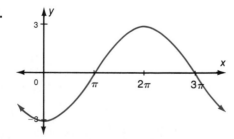

d.

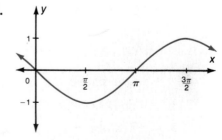

Function	Amplitude	Period	Phase Shift	Graph
17. $y = -2 \sin (2x + \frac{\pi}{2})$	?	?	?	?
18. $y = \cos (x + \frac{\pi}{2})$	?	?	?	?
19. $y = 2 \cos (3x + \frac{\pi}{2})$	?	?	?	?
20. $y = 3 \sin (\frac{1}{2}x - \frac{\pi}{2})$	?	?	?	?

b

In Exercises 21–23, find a function of the form $y = A \sin B(x - C)$ with the given properties.

21. Amplitude: 5;   Period: 2π;   Phase Shift: $\frac{\pi}{3}$ units to the right

22. Amplitude: 3;   Period: $\frac{2\pi}{3}$;   Phase Shift: $\frac{\pi}{3}$ units to the right

23. Amplitude: $\frac{2}{3}$;   Period: $\frac{\pi}{4}$;   Phase Shift: $\frac{\pi}{8}$ units to the left

In Exercises 24–26, find a function of the form $y = A \cos B(x - C)$ with the given properties.

24. Amplitude: $\frac{1}{5}$;   Period: $\frac{\pi}{5}$;   Phase Shift: 2π units to the right

25. Amplitude: $\frac{7}{3}$;   Period: $\frac{5\pi}{6}$;   Phase Shift: π units to the left

26. Amplitude: 1;   Period: $\frac{3\pi}{4}$;   Phase Shift: $\frac{3\pi}{4}$ units to the right

In Exercises 27–30, graph each function for $-2\pi \le x \le 2\pi$.

27. $y = 2 \sin (2x + \frac{\pi}{3}) - 3$ c hart

28. $y = 3 \sin (\frac{1}{2}x - \frac{\pi}{2}) + 2$

c **29.** $y = \sin x + \cos 2x$

30. $y = 2 \sin \frac{1}{2}x - \cos x$

—— Review ————————————————————

Find the amplitude of each function. Then graph the function for $0 \le x \le 2\pi$. (Section 2–6)

1. $y = -\sin x$

2. $y = 2 \cos x$

3. $y = -4 \sin x - 2$

4. $y = \frac{1}{2} \cos x + 1$

Find the period of each function. Then graph the function for $0 \le x \le 2\pi$. (Section 2–7)

5. $y = \cos 3x$

6. $y = \sin \frac{1}{3}x$

7. $y = \frac{1}{4} \sin 4x$

8. $y = -2 \cos x$

Find the amplitude, period, and phase shift of each function. Then graph each function for $0 \le x \le 2\pi$. (Section 2–8)

9. $y = 2 \cos (x - \frac{\pi}{4})$

10. $y = -\cos (x + \frac{\pi}{6})$

2-9 Addition of Ordinates

In Section 2–5 you found that the graph of $y = \sin x + 3$ could be sketched by first sketching the graph of $y = \sin x$ and then raising each point three units. You can use a similar technique to sketch the graph of a function such as $y = \sin x + \frac{1}{3} \sin 3x$.

Example 1

Graph $y = \sin x + \frac{1}{3} \sin 3x$ for $0 \le x \le 4\pi$.

Solution: 1. Sketch the graphs of $y_1 = \sin x$ and $y_2 = \frac{1}{3} \sin 3x$ on the same set of axes.

2. For each value of x, add y_1 and y_2 to find the value of y.

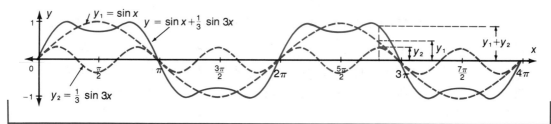

The procedure of Example 1 is called graphing by **addition of ordinates.**

Example 2

Graph $y = 2 \cos x - \cos 2x$ for $0 \le x \le 2\pi$.

Solution: 1. Sketch the graphs of $y_1 = 2 \cos x$ and $y_2 = -\cos 2x$ on the same set of axes.

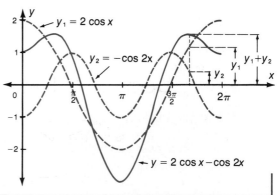

2. For each value of x, add the values of y_1 and y_2 to find y.

Classroom Exercises

Given: $y_1 = 2 \cos x$, $y_2 = \cos 2x$. Complete the table.

	x	cos x	y_1	2x	y_2	$y_1 + y_2$
1.	0	1	?2	0	?1	3
2.	$\frac{\pi}{6}$	?	?	?	?	?
3.	$\frac{\pi}{4}$	?	?	?	?	?
4.	$\frac{\pi}{3}$	?	?	?	?	?

Written Exercises

a

In Exercises 1–16, graph each function. Show one period, where $x \geq 0$.

1. $y = 2 \cos x + \cos 2x$ $0 \rightarrow 2\pi$ 1per
2. $y = 2 \sin x + \sin 2x$

3. $y = \sin x + \frac{1}{2} \cos 2x$ $0 \rightarrow 2\pi$ 1per
4. $y = \cos x + \frac{1}{2} \cos 2x$

5. $y = \sin 4x + \sin 2x$ $0 \rightarrow \pi$ 1per
6. $y = \sin 4x + \sin x$

7. $y = \sin 2x - 2 \sin x$
8. $y = \sin 2x - 2 \cos x$

9. $y - 2 \cos x - 2 \sin x$ $0 \rightarrow 2\pi$ 1per
10. $y = \frac{1}{2} \cos x - 2 \sin 2x$

11. $y = \cos 2x - \sin 3x$
12. $y = \sin 2x + \cos 3x$

b

13. $y = \sin x + \sin 2x + \sin 3x$
14. $y = 2 \sin x + \frac{1}{2} \sin 2x + \sin 4x$

15. $y = \sin (x - \frac{\pi}{3}) + \cos (x - \frac{\pi}{4})$
16. $y = \sin (x + \frac{\pi}{6}) + \cos (x + \frac{\pi}{3})$

c

17. Graph $y = \frac{1}{x} \sin x (x \neq 0)$ for $-2\pi \leq x \leq 2\pi$.

Review Capsule for Section 2-10

In Exercises 1–6, use the table to find the value of the given expression. (For further review, see pages 17–18 and 22–24.)

1. $\tan (-\frac{\pi}{6})$ **2.** $\tan (\pi - \frac{\pi}{3})$ **3.** $\cot \frac{2\pi}{3}$

4. $\cot \frac{\pi}{3}$ **5.** $\sec \frac{\pi}{3}$ **6.** $\csc \frac{\pi}{6}$

x	0	$\frac{\pi}{6}$	$\frac{\pi}{3}$
sin x	0	.50	.87
cos x	1	.87	.50
tan x	0	.58	1.7

Tidal Motion

The data collected by **oceanographers** measuring tide levels at a particular shore area over specific intervals of time can be used to define a **tide function.** On a graph of such a function, the horizontal axis represents the elapsed time t. The vertical axis represents the height of the tide above an arbitrary level. This level is assigned a value of zero.

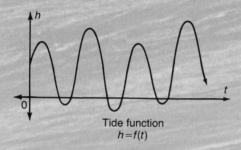

Tide function
$h = f(t)$

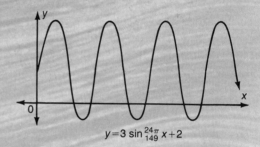

$y = 3 \sin \frac{24\pi}{149} x + 2$

Compare the graph of the tide function $h = f(t)$ at the left above with the graph of $y = 3 \sin \dfrac{24\pi}{149} x + 2$ shown at its right. Since the graphs are very similar, the periodic function $y = 3 \sin \dfrac{24\pi}{149} x + 2$ provides a good mathematical model for $h = f(t)$. That is, by replacing x with t and y with h,

$$h = 3 \sin \frac{24\pi}{149} t + 2$$

is a rule for finding the approximate height of the tide at this location as a function of time.

Tides in different parts of the world are represented by different models. Many of the models are equations of the form

$$f(t) = A \sin Bt + d$$

as in the example just given. Sometimes, however, the mathematical model is more complicated.

For example, for a shore at a spot on the west coast of North America, the graph of the tide function is shown below.

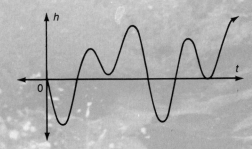

For this graph, the tide level can be interpreted as the combination of two different tidal motions that interact to produce the effect shown. The graph can be approximated by the periodic function

$$h = A_1 \sin B_1(t - C_1) + A_2 \sin B_2(t - C_2)$$

which is similar to functions found in Section 2–9 of this text.

EXERCISES

1. For the tide function $h = 3 \sin \dfrac{24\pi}{149} t + 2$, find the height of the tide for each value of t. Round your answer to the nearest whole number.

t	0	6	12	18	24
h	?	?	?	?	?

2. Find the amplitude and period of the function in Exercise 1.

2-10 Graphs of tan x, cot x, sec x, csc x

The sine and cosine functions are the basic trigonometric functions. The remaining trigonometric functions can be expressed in terms of these two. For example, by definition,

$$\sin \theta = \frac{y}{r} \quad \text{and} \quad \cos \theta = \frac{x}{r}.$$

Thus,

$$\frac{\sin \theta}{\cos \theta} = \frac{\frac{y}{r}}{\frac{x}{r}} = \frac{y}{x} = \tan \theta.$$

Therefore,

$$\tan \theta = \frac{\sin \theta}{\cos \theta}, \quad \cos \theta \neq 0.$$

The equation above shows that $\tan \theta$ is not defined for $\cos \theta = 0$, that is, for $\theta = \pm\frac{\pi}{2}, \pm\frac{3\pi}{2}, \cdots, \frac{(2n+1)\pi}{2}, n \in \mathscr{I}$.

Near these values of θ, $f(\theta) = \tan \theta$ increases or decreases without bound. Thus, amplitude is not defined for $f(\theta) = \tan \theta$. Recall that in order to graph on the customary xy axes, x is used to represent θ and y is used to represent $f(\theta)$.

Example 1 Graph $y = \tan x$ for $-\pi \leq x \leq 2\pi$.

Solution: 1. Make a table of values for $-\pi \leq x \leq 2\pi$.

2. Plot these points and join them with a smooth curve.

x	tan x
$-\pi$	0
$-\frac{3\pi}{4}$	1
$-\frac{\pi}{2}$	Undefined
$-\frac{\pi}{4}$	-1
0	0
$\frac{\pi}{4}$	1
$\frac{\pi}{2}$	Undefined

x	tan x
$\frac{3\pi}{4}$	-1
π	0
$\frac{5\pi}{4}$	1
$\frac{3\pi}{2}$	Undefined
$\frac{7\pi}{4}$	-1
2π	0

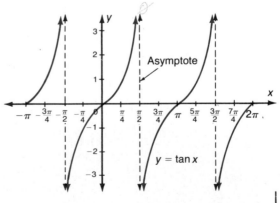

The graph in Example 1 shows that the range of $y = \tan x$ is the set of real numbers. The period of $y = \tan x$ is π, because the graph repeats at intervals of π units. That is, $\tan(\pi + \theta) = \tan\theta$. The dashed lines in the graphs are asymptotes. An **asymptote** is a line that a curve nears but does not intersect.

Unlike the sine and cosine functions, the tangent function has "breaks" in its graph. The sine and cosine are **continuous** functions. The function $y = \tan x$ is **discontinuous** at

$$x = \pm\frac{\pi}{2}, \pm\frac{3\pi}{2}, \cdots, \frac{(2n+1)\pi}{2}, n \in \mathscr{I}.$$

Also, since

$$\tan(-x) = \frac{\sin(-x)}{\cos(-x)} = \frac{-\sin x}{\cos x} = -\tan x,$$

the tangent function is an odd function.

The cotangent, secant, and cosecant functions are reciprocals of the tangent, cosine, and sine functions respectively.

$$\cot x = \frac{1}{\tan x}, \tan x \neq 0$$

$$\sec x = \frac{1}{\cos x}, \cos x \neq 0$$

$$\csc x = \frac{1}{\sin x}, \sin x \neq 0$$

Each function has the same period as its reciprocal. Also, each of these functions is discontinuous at values of x for which its reciprocal is zero.

Function	Values for which the function is not defined	Asymptotes
$y = \cot x$	$x = n\pi, n \in \mathscr{I}$	$x = n\pi, n \in \mathscr{I}$
$y = \sec x$	$x = \left(\frac{2n+1}{2}\right)\pi, n \in \mathscr{I}$	$x = \left(\frac{2n+1}{2}\right)\pi, n \in \mathscr{I}$
$y = \csc x$	$x = n\pi, n \in \mathscr{I}$	$x = n\pi, n \in \mathscr{I}$

Near those values of x for which each function is not defined, the values of the functions are unbounded. Hence, amplitude is not defined for these functions.

The graph of the cotangent function can be sketched by first sketching the graph of $y = \tan x$ and then estimating the reciprocals of $\tan x$. You can use a similar procedure to sketch graphs of the $y = \sec x$ and $y = \csc x$.

Example 2 Graph each function.

a. $y = \cot x$ **b.** $y = \sec x$ **c.** $y = \csc x$

Solutions: a.

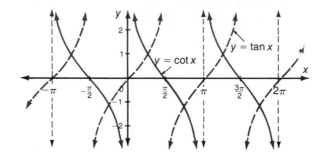

b.

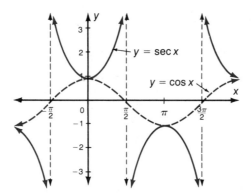

c.

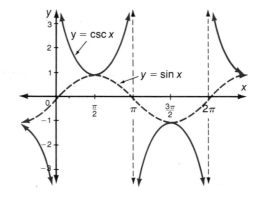

The period and phase shift of functions of the form $y = \tan B(x - C)$, $y = \cot B(x - C)$, $y = \sec B(x - C)$, and $y = \csc B(x - C)$ are found in a manner similar to that shown in the previous section.

Function	Period	Phase Shift
$y = \tan B(x - C)$	$\dfrac{\pi}{\lvert B \rvert}$	For each function, the phase shift is $\lvert C \rvert$ units to the right if $C > 0$ or $\lvert C \rvert$ units to the left if $C < 0$.
$y = \cot B(x - C)$	$\dfrac{\pi}{\lvert B \rvert}$	
$y = \sec B(x - C)$	$\dfrac{2\pi}{\lvert B \rvert}$	
$y = \csc B(x - C)$	$\dfrac{2\pi}{\lvert B \rvert}$	

Example 3 Find the period and phase shift for each function.

a. $y = \tan \left(2x - \frac{\pi}{4}\right)$ **b.** $y = \sec \left(\frac{1}{2}x + 1\right)$

Solutions: First write $\left(2x - \frac{\pi}{4}\right)$ and $\left(\frac{1}{2}x + 1\right)$ in the form $B(x - C)$

Function	Period	Phase Shift
a. $y = \tan 2\left(x - \frac{\pi}{8}\right)$	$\dfrac{\pi}{\lvert B \rvert} = \dfrac{\pi}{2}$	$C = \frac{\pi}{8}$; $\frac{\pi}{8}$ units to the <u>right</u>
b. $y = \sec \frac{1}{2}[x - (-2)]$	$\dfrac{2\pi}{\lvert B \rvert} = \dfrac{2\pi}{\frac{1}{2}}$, or 4π	$C = -2$; 2 units to the <u>left</u>

Summary of Important Properties

Function	Domain	Range	Period	Continuity
$y = \sin x$	All real numbers x	$\{y: -1 \le y \le 1, y \in R\}$	2π	Continuous
$y = \cos x$	All real numbers x	$\{y: -1 \le y \le 1, y \in R\}$	2π	Continuous
$y = \tan x$	All real numbers x, except $x = \left(\dfrac{2n+1}{2}\right)\pi, n \in \mathscr{I}$	All real numbers y	π	Discontinuous at $x = \left(\dfrac{2n+1}{2}\right)\pi, n \in \mathscr{I}$
$y = \cot x$	All real numbers x, except $x = n\pi, n \in \mathscr{I}$	All real numbers y	π	Discontinuous at $x = n\pi, n \in \mathscr{I}$
$y = \sec x$	All real numbers x, except $x = \left(\dfrac{2n+1}{2}\right)\pi, n \in \mathscr{I}$	All real numbers y, except $-1 < y < 1$	2π	Discontinuous at $x = \left(\dfrac{2n+1}{2}\right)\pi, n \in \mathscr{I}$
$y = \csc x$	All real numbers x, except $x = n\pi, n \in \mathscr{I}$	All real numbers y, except $-1 < y < 1$	2π	Discontinuous at $x = n\pi, n \in \mathscr{I}$

Classroom Exercises

Find the period and phase shift of each function. Then match each function with its graph.

Function	Period	Phase Shift	Graph
1. $y = \tan(3x - 2\pi)$	?	?	?
2. $y = \sec(x - 2\pi)$	?	?	?
3. $y = \csc(2x + 3\pi)$	?	?	?
4. $y = \cot(2x - \pi)$	?	?	?

a.
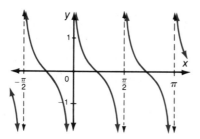

b.
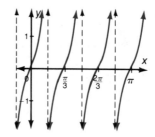

c.
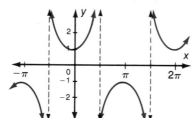

d.
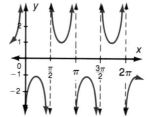

Written Exercises

In Exercises 1–10, find the period and phase shift of each function.

1. $y = \sec(x - \frac{\pi}{2})$
2. $y = \cot(2x + \pi)$
3. $y = \tan(x + \frac{\pi}{2})$
4. $y = \cot(3x + \pi)$
5. $y = \tan(3x - \pi)$
6. $y = \csc(x + \pi)$
7. $y = \sec(2x - \frac{\pi}{4})$
8. $y = \tan(\frac{1}{2}x - \frac{\pi}{2})$
9. $y = \sec(3x - 3\pi)$
10. $y = \csc(\frac{2}{5}x + \pi)$

In Exercises 11–14, find the values of x for which each function is undefined. Consider only the interval $-2\pi \le x \le 2\pi$.

11. tan x   **12.** cot x   **13.** sec x   **14.** csc x

In Exercises 15–16 graph the given pair of functions on the same set of axes for $-2\pi \le x \le 2\pi$.

15. $y = \sin (x - \frac{\pi}{4}); \ y = \csc (x - \frac{\pi}{4})$

16. $y = \cos (2x - \pi); \ y = \sec (2x - \pi)$

In Exercises 17–24, find the period and phase shift of each function. Then graph each function.

17. $y = \tan 2x$ **18.** $y = \cot \frac{1}{2}x$

19. $y = 3 \sec 3x$ **20.** $y = \tan (x + \frac{\pi}{2})$

21. $y = 2 \cot (x - \frac{\pi}{2})$ **22.** $y = 2 \sec (x + \frac{\pi}{4})$

23. $y = 2 \csc (3x + \pi) + 1$ **24.** $y = \tan (2x - \frac{\pi}{2}) - 2$

25. Show that $\tan (\pi + \theta) = \tan \theta$. (HINT: Use the fact that $\tan \theta = \dfrac{\sin \theta}{\cos \theta}$ and the reduction formulas for $\sin (\pi + \theta)$ and $\cos (\pi + \theta)$.)

26. Prove: $|\tan x| \ge |\sin x|$, for all x for which tan x is defined.

27. Prove: $|\cot x| \ge |\cos x|$, for all x for which cot x is defined.

28. Prove: $|\sec x| \ge |\tan x|$, for all x for which sec x and tan x are defined.

29. Prove: $|\cot x| \le |\csc x|$, for all x for which cot x and csc x are defined.

BASIC: TRIGONOMETRIC GRAPHS

The TAB and INT functions of BASIC can be used in writing programs that will instruct the computer to "draw" the graphs of trigonometric functions. "Built-in" functions such as COS and SIN are also used.

Problem: *Given y = sin x, draw the graph for the interval 0 ≤ x ≤ 2π.*

The following program shows the steps for solving the above problem.

```
10   PRINT TAB(37);"X  AXIS"
20   PRINT "Y  AXIS:-2";TAB(24);"-1";TAB(40);"0";TAB(56);
     "1";TAB(71);"2"
30   LET X = 0
40   LET D = 4/63
50   LET Y = SIN(X)
60   LET P = INT((Y + 2)/D) + 9
70   IF Y = 0 THEN 190
80   IF Y > 0 THEN 140
90   IF X <> INT(X) THEN 120    ◄────── "< >" means ≠.
100  PRINT TAB (P);"*";TAB(39);X
110  GO TO 200
120  PRINT TAB(P);"*";TAB(40);"!"
130  GO TO 200                       180  GO TO 200
140  IF X <> INT(X) THEN 170          190  PRINT TAB(P);"*"
150  PRINT TAB(39);X;TAB(P);"*"       200  LET X = X + .25
160  GO TO 200                        210  IF X <= 6.28 THEN 50
170  PRINT TAB(40);"!";TAB(P);"*"  220  END
```

Analysis: Statements 10 and 20: The TAB function can be used to specify print positions. Line 20 prints the scale for the *y*-axis. The 63 print positions from the ninth to the 71st correspond to the four units from −2 to 2 on the *y*-axis.

Statements 40 and 60: The value stored for D by statement 40 corresponds to the scaling of the *y*-axis. This value is used in statement 60 to compute the print position P that corresponds to sin *x*. The INT function is used to compute the greatest integer that is less than or equal to a given number. It is used in statement 60 since P must be a whole number.

Statements 70 and 80: These involve decisions. If Y = 0, then the * must be printed on the *x*-axis (line 190). If Y > 0, then either line 150 or line 170 prints the * to the right of the *x*-axis. If Y < 0, then line 90 is reached. If X <> INT(X), then X is not an integer and ! will be printed by line 120 to show the *x*-axis. (Line 170 does the same.) If X is an integer, then to show the scaling of the *x*-axis, line 100 prints X instead of !.

Statement 2ØØ: .25 is added to the current value of X and the result is stored as the new value of X. Thus each line of printout corresponds to .25 units along the *x*-axis.

Statement 21Ø: If X is still smaller than 6.28 (2π), return to line 5Ø and compute the next Y value.

Output: Turn your book so that the top of the page is at your left.

NOTE: The full range (−2 to 2) of the *y*-axis scale is too wide to be shown. Thus, the *y*-values 2 and −2 are omitted from the diagram.

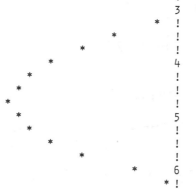

Written Exercises

In Exercises 1–10, replace line 50 in the program on page 114 to obtain graphs of the given trigonometric functions.

1. 5Ø LET Y = COS(X)
2. 5Ø LET Y = SIN(2*X)
3. 5Ø LET Y = SIN(.5*X)
4. 5Ø LET Y = 2*COS(X)
5. 5Ø LET Y = .5*COS(X)
6. 5Ø LET Y = SIN(X + 3.14)
7. 5Ø LET Y = COS(2*X - 3.14/2)
8. 5Ø LET Y = SIN(-2*X)
9. 5Ø LET Y = SIN(X) + 1
10. 5Ø LET Y = COS(X) - 1

Chapter Objectives and Review

Objective: To know the meanings of the mathematical terms in this chapter.

1. Be sure that you know the meanings of these mathematical terms.

addition of ordinates (p. 104)
amplitude (p. 88)
angular velocity (p. 70)
asymptote (p. 109)
continuous function (p. 109)
discontinuous function (p. 109)
even function (p. 75)

odd function (p. 75)
period (p. 94)
periodic function (p. 83)
phase shift (p. 101)
radian measure (p. 62)
reduction formulas (p. 77)
rotary motion (p. 70)

Objective: To find the radian measure of a central angle of a circle when the radius and length of the intercepted arc are known. (Section 2–1)

In circle O with radius r, central angle POQ intercepts $\overgroup{QP}$. Find, to the nearest tenth, the radian measure of $\angle POQ$ for each given value of r and length of $\overgroup{QP}$.

2. $r = 8$ cm, length of $\overgroup{QP} = 14$ cm

3. $r = 6$ m, length of $\overgroup{QP} = 10$ m

Objective: To change degree measure to radian measure and vice versa. (Section 2–1)

Change each degree measure to radian measure in terms of π.

4. 45°

5. −30°

6. 90°

7. 210°

Change each radian measure to degree measure.

8. $\dfrac{\pi}{18}$

9. $-\dfrac{\pi}{10}$

10. $\dfrac{3\pi}{2}$

11. $-\dfrac{5\pi}{6}$

Objective: To use the table on pages 368–372 to change degree measure to radian measure and vice versa. (Section 2–2)

Find a four-place decimal approximation for the radian measure of the angle with the given degree measure.

12. 28°30′

13. 67°40′

14. 83°10′

15. 12°50′

Find the degree measure of the angle with the given radian measure.

16. .6138

17. 1.0937

18. .2269

19. 1.3177

Objective: To solve problems involving rotary motion. (Section 2–3)

Use the figure at the right below for Exercises 20–21. In the figure, point P is on the circumference of a circular saw blade which rotates at 1800 rpm.

20. Find the angular velocity of the blade in rad/sec.

21. The diameter of the blade is 17.8 cm. Find, to the nearest tenth of a centimeter, the distance traveled by point P in .5 seconds.

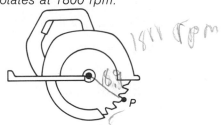

Objective: To use the reduction formulas to evaluate expressions involving the sine and cosine. (Section 2–4)

For Exercises 22–25, $\sin \theta = .8000$ and $\cos \theta = .6000$. Find the value of each expression.

22. $\cos(-\theta)$ **23.** $\cos(360° - \theta)$ **24.** $\sin(180° + \theta)$ **25.** $\sin(180° - \theta)$

Given: $\sin 30° = .5000$
Complete the table.

	θ	$\sin \theta$
26.	$-30°$	?
27.	$150°$	?
28.	$210°$	?
29.	$330°$	?

Given: $\cos \frac{\pi}{6} = .8660$
Complete the table.

	θ	$\cos \theta$
30.	$-\frac{\pi}{6}$	?
31.	$\frac{5\pi}{6}$	?
32.	$\frac{7\pi}{6}$	?
33.	$\frac{11\pi}{6}$	?

Objective: To graph functions of the form $y = \sin x + D$ and $y = \cos x + D$ on a given interval. (Section 2–5)

Graph each function on the given interval.

34. $y = \sin x, \; -\pi \le x \le 2\pi$

35. $y = \cos x, \; -\frac{\pi}{2} \le x \le \frac{5\pi}{2}$

36. $y = \cos x - 1, \; -2\pi \le x \le 2\pi$

37. $y = \sin x + 2, \; -2\pi \le x \le 2\pi$

Objective: To find the amplitude of a function of the form $y = A \sin x + D$ or $y = A \cos x + D$ and to graph the function. (Section 2–6)

Find the amplitude of each function. Then graph each function for $0 \le x \le 2\pi$.

38. $y = -2 \sin x$

39. $y = \frac{1}{2} \cos x$

40. $y = -5 \sin x + 2$

41. $y = -\frac{1}{2} \cos x + 1$

Graphs of Trigonometric Functions **117**

Objective: To find the period of a function of the form $y = A \sin Bx$ or $y = A \cos Bx$ and to graph the function. (Section 2-7)

Find the period of each function. Then graph each function for $0 \le x \le 2\pi$.

42. $y = \cos 4x$

43. $y = \sin \frac{1}{4}x$

44. $y = \frac{1}{3} \sin 3x$

45. $y = 2 \cos \frac{1}{2}x$

Objective: To find the amplitude, period, and phase shift of a function of the form $y = A \sin B(x - C)$ or $y = A \cos B(x - C)$ and to graph the function. (Section 2-8)

Find the amplitude, period, and phase shift of each function. Then graph each function for $0 \le x \le 2\pi$.

46. $y = -2 \cos \left(x - \frac{\pi}{2}\right)$

47. $y = -\sin \left(x - \frac{\pi}{3}\right)$

48. $y = \sin (2x + \pi)$

49. $y = -\frac{1}{2} \cos (4x + 8\pi)$

Objective: To use the method of addition of ordinates to graph certain functions. (Section 2-9)

Graph each function. Show one period, where $x \ge 0$.

50. $y = 2 \cos x + \sin 2x$

51. $y = \sin 4x - 2 \sin x$

Objective: To graph $\tan x$, $\cot x$, $\sec x$, and $\csc x$. (Section 2-10)

Graph each function for $-2\pi \le x \le 2\pi$.

52. $y = \tan x$

53. $y = \cot x$

54. $y = \sec x$

55. $y = \csc x$

Chapter Test

1. In circle O with radius $r = 4$ cm, central angle POQ intercepts $\overset{\frown}{QP}$ where $\overset{\frown}{QP} = 12$ cm. Find the radian measure of $\angle POQ$.

2. Change $-210°$ to radian measure.

3. Change $\frac{4\pi}{3}$ radians to degree measure.

4. Find a four-place approximation for the radian measure of an angle with degree measure $56°20'$.

5. Find the degree measure of an angle with radian measure .5760.

6. The shaft of an electric motor is rotating at 3600 rpm. Find its angular velocity in rad/sec.

7. Point P is on the circumference of a phonograph record which is on a turntable rotating at 45 rpm. The diameter of the record is 16 centimeters. Find, to the nearest tenth of a centimeter, the distance traveled by point P during .5 seconds.

8. Given: $\sin 25° = .4226$. Find the value of $\sin 205°$.

9. Given that $\sin \theta = .6000$ and $\cos \theta = .8000$, find the value of $\cos (180° + \theta)$.

10. Graph $y = \cos x$ on the interval $-2\pi \le x \le 2\pi$.

11. Graph $y = \sin x - 1$ on the interval $-2\pi \le x \le 2\pi$.

12. Find the amplitude of $y = -\frac{1}{2} \sin x + 2$.

13. Graph $y = -\frac{1}{2} \sin x + 2$ for $0 \le x \le 2\pi$.

14. Find the period of $y = 4 \cos \frac{1}{4} x$.

15. Graph $y = 4 \cos \frac{1}{4} x$ for $0 \le x \le 2\pi$.

16. Find the amplitude, period, and phase shift of $y = -\frac{1}{4} \cos (4x + 8\pi)$.

17. Graph $y = -\frac{1}{4} \cos (4x + 8\pi)$ for $0 \le x \le 2\pi$.

18. Graph $y = \tan x$ for $-2\pi \le x \le 2\pi$.

19. Graph $y = \csc x$ for $-2\pi \le x \le 2\pi$.

20. Graph $y = \sin x + \cos x$. Show at least one cycle.

learn this...

Do this First mAKE cards

Graphs of Trigonometric Functions   **119**

Cumulative Review: Chapters 1 and 2

$d = \sqrt{(x_2 - x_1)^2}$

Write the letter of the response that best answers each question.

1. Find the distance between $P(1, -3)$ and $Q(-2, 5)$. (Section 1–1)

   a. $\sqrt{5}$ **b.** 5   **c.** 73   **d.** $\sqrt{73}$

2. The domain of a function is $\{-2, -1, 0, 1, 2\}$. The function is described by the equation $y = 3x - 2$. Find the range of the function. (Section 1–1)

   a. $\{-2, 1, 4\}$ **b.** $\{-4, -2, -1, 1, 4\}$

   c. $\{-8, -5, -2, 1, 4\}$ **d.** $\{-8, -5, -2, 5, 8\}$

3. Find, in degrees, the measure of a $\frac{3}{4}$ counterclockwise rotation. (Section 1–2)

   a. $135°$ **b.** $270°$ **c.** $-135°$ **d.** $-270°$

4. Point $P(x, y)$ is on the terminal side of an angle with measure θ in standard position. P is located r units from the origin. Tell which statement is false. (Sections 1–3 and 1–4)

   a. $\sin \theta = \dfrac{y}{r}$ **b.** $\cos \theta = \dfrac{x}{r}$ **c.** $\cot \theta = \dfrac{x}{y}$ **d.** $\sec \theta = \dfrac{r}{y}$

5. Find the value of $\tan 30°$. (Section 1–5)

   a. $\dfrac{\sqrt{3}}{3}$ **b.** $\sqrt{3}$ **c.** $\dfrac{\sqrt{3}}{2}$ **d.** 2

6. Find the value of $\cos 225°$. (Section 1–5)

   a. 2   **b.** $-\dfrac{\sqrt{2}}{2}$ **c.** $\dfrac{\sqrt{2}}{2}$ **d.** $\dfrac{1}{2}$

7. Tell which angle is coterminal with $172°$. (Section 1–6)

   a. $18°$ **b.** $-172°$ **c.** $432°$ **d.** $-188°$

8. Given that $\sin 13° = .2250$, find $\sin 193°$. (Section 1–6)

   a. $.9744$ **b.** $-.2250$ **c.** $.2250$ **d.** $-.9744$

9. Given that $\sin 13° = .2250$, find $\sin 77°$. (Section 1–6)

   a. $.9744$ **b.** $-.2250$ **c.** $.2250$ **d.** $-.9744$

10. Find $\sin 29°42'$ to four decimal places. (Section 1–7)

   a. $.4894$ **b.** $.4955$ **c.** $.4890$ **d.** $.4960$

11. Find $\cos 54°21'$ to four decimal places. (Section 1–7)

   a. $.8126$ **b.** $.5809$ **c.** $.5829$ **d.** $.5828$

12. Refer to the figure at the right. Then find the value of x to the nearest meter. (Section 1–8)

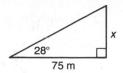

a. 39 m
b. 39.5 m
c. 40 m
d. 40.5 m

13. The angle of depression from the top of a cliff 600 meters high to the foundation of a house is 33°40′. To the nearest meter, how far is the house from the foot of the cliff? (Section 1–9)

a. 901
b. 400
c. 333
d. 499

14. To find the distance from point C on the shore to point A on Maple Island, a surveyor made the measurements shown in the figure at the right. Find the distance AC to the nearest tenth of a meter. (Section 1–10)

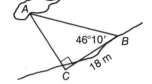

a. 17.9
b. 17.3
c. 18.1
d. 18.7

15. An airplane is 200 kilometers due east of radar station A. Radar station B is 318 kilometers due north of A. To the nearest ten minutes, find the bearing of radar station B from the airplane. (Section 1–11)

a. 147°50′
b. 57°50′
c. 327°50′
d. 237°50′

16. A house has a symmetrical roof that has a span of 4.0 meters and a pitch of $\frac{7}{20}$. Find the length of a rafter to the nearest tenth of a meter. (Section 1–12)

a. 2.4
b. 8.1
c. 4.2
d. $\sqrt{18}$

17. Tell which number you multiply by in order to change 213° to radian measure. (Section 2–1)

a. 2π
b. π
c. $\frac{180}{\pi}$
d. $\frac{\pi}{180}$

18. Find a four-place decimal approximation for the radian measure of an angle which measures 72°40′. (Section 2–2)

a. .3200
b. .3025
c. 1.2683
d. 1.2508

19. Find the degree measure of an angle which measures .8058 radians. (Section 2–2)

a. 46°10′
b. 43°50′
c. 47°10′
d. 44°50′

20. The turntable of a record player rotates at 45 rotations per minute. Find the angular velocity of the turntable. (Section 2–3)

a. 90π rad/min
b. $\frac{45\pi}{2}$ rad/min
c. 45π rad/min
d. $\frac{45}{\pi}$ rad/min

21. Tell which equality is false. (Section 2–4)

a. $\sin(-\theta) = \sin(180° + \theta)$
b. $\cos(2\pi - \theta) = \cos(-\theta)$
c. $\cos(180° + \theta) = \cos(360° - \theta)$
d. $\sin(-\theta) = \sin(360° - \theta)$

22. Tell which function has a graph that contains the points $P(\frac{\pi}{4}, \frac{\sqrt{2}}{2})$ and $Q(\frac{\pi}{2}, 1)$. (Section 2–5)

   a. $y = \cos x$ **b.** $y = \cos x + 1$

   c. $y = \sin x$ **d.** $y = \sin x + 1$

23. Find the amplitude of $y = -3 \sin(2x + \pi)$. (Sections 2–6 and 2–8)

   a. -3 **b.** 3 **c.** π **d.** $\frac{\pi}{3}$

24. Find the period of $y = 2 \cos(3x - \pi)$. (Sections 2–7 and 2–8)

   a. 2 **b.** $\frac{2\pi}{3}$ **c.** $\frac{\pi}{3}$ **d.** $\frac{\pi}{3}$

25. Find the phase shift of $y = 2 \sin(\frac{1}{2}x + \pi)$. (Section 2–8)

   a. 2π units to the left   **b.** 2π units to the right

   c. π units to the left   **d.** π units to the right

26. Find the function that has neither of the points $P(0, 3)$ and $Q(\frac{\pi}{2}, -1)$ on its graph. (Section 2–9)

   a. $y = 2 \cos x + \cos 2x$ **b.** $y = \sin x + 2 \cos 2x$

   c. $y = 2 \sin x + \sin 2x$ **d.** $y = \sin 2x + 3 \cos x$

27. Find the function that is undefined at $\frac{\pi}{2}$ and that has period 2π. (Section 2–10)

   a. $y = \tan x$ **b.** $y = \cot x$ **c.** $y = \sec x$ **d.** $y = \csc x$

Chapter 3
Trigonometric Identities

3-1 Fundamental Identities

In trigonometry as in algebra, conditional equations are true only for particular values of the variable. Thus, for $0° < \theta < 360°$,

$$\sin \theta = \tfrac{1}{2} \quad \text{only if} \quad \theta = 30° \text{ or } \theta = 150°.$$

Trigonometric identities are equations which are true for all values of the variable for which the trigonometric functions in the identity have meaning. For example, you can prove the trigonometric identity

$$\cot \theta = \frac{\cos \theta}{\sin \theta}, \text{ where } \sin \theta \neq 0,$$

by applying the definitions of the trigonometric functions.

Example 1

Prove: $\cot \theta = \dfrac{\cos \theta}{\sin \theta}$, $\sin \theta \neq 0$

Proof:

$\sin \theta = \dfrac{y}{r}$ ⟵ Definition

$\therefore r \sin \theta = y$ ⟵ The symbol $\therefore$ means "therefore."

$\cos \theta = \dfrac{x}{r}$ ⟵ Definition

$\therefore r \cos \theta = x$

$\cot \theta = \dfrac{x}{y}$ ⟵ Definition

Then $\cot \theta = \dfrac{r \cos \theta}{r \sin \theta}$

and $\cot \theta = \dfrac{\cos \theta}{\sin \theta}$.

As shown in Example 1, if $P(x, y)$ is a point on the terminal side of an angle θ in standard position, then

$$x = r \cos \theta \quad \text{and} \quad y = r \sin \theta$$

where r is the distance from the origin to point P.

You can use a similar procedure to prove the other Ratio Identity and the Reciprocal Identities. The proofs of these identities are asked for in the Exercises.

Reciprocal Identities

1. $\csc \theta = \dfrac{1}{\sin \theta}$, $\sin \theta \neq 0$

2. $\sec \theta = \dfrac{1}{\cos \theta}$, $\cos \theta \neq 0$

3. $\cot \theta = \dfrac{1}{\tan \theta}$, $\tan \theta \neq 0$

Ratio Identities

4. $\tan \theta = \dfrac{\sin \theta}{\cos \theta}$, $\cos \theta \neq 0$

5. $\cot \theta = \dfrac{\cos \theta}{\sin \theta}$, $\sin \theta \neq 0$

The three Pythagorean Identities are based on the Pythagorean Theorem. (NOTE: The symbol "$\sin^2 \theta$" means $(\sin \theta)^2$, $\cos^2 \theta$ means $(\cos \theta)^2$, and so on.)

Pythagorean Identities

6. $\sin^2 \theta + \cos^2 \theta = 1$

7. $1 + \cot^2 \theta = \csc^2 \theta$

8. $1 + \tan^2 \theta = \sec^2 \theta$

You can prove each of these identities by applying the definitions of the trigonometric functions and the Pythagorean Theorem.

Example 2 Prove Identity **6**: $\sin^2 \theta + \cos^2 \theta = 1$

Proof: By definition, $\sin \theta = \dfrac{y}{r}$ and $\cos \theta = \dfrac{x}{r}$.

$\therefore \sin^2 \theta + \cos^2 \theta = \left(\dfrac{y}{r}\right)^2 + \left(\dfrac{x}{r}\right)^2$

$\therefore \sin^2 \theta + \cos^2 \theta = \dfrac{x^2 + y^2}{r^2}$ ⟵ By the Pythagorean Theorem, $x^2 + y^2 = r^2$.

$= \dfrac{r^2}{r^2}$

$\therefore \sin^2 \theta + \cos^2 \theta = 1$

The proofs of the other two Pythagorean Identities are asked for in the Exercises.

Identities **1–8** are generally referred to as the **Fundamental Identities.** You can use the Fundamental Identities to write trigonometric expressions in terms of specific trigonometric functions.

Example 3

Write $\sec \theta - \tan \theta \sin \theta$ in terms of $\cos \theta$.

Solution: $\sec \theta - \tan \theta \sin \theta = \dfrac{1}{\cos \theta} - \dfrac{\sin \theta}{\cos \theta} \cdot \sin \theta$ ⟵ Identities 2 and 4.

$$= \dfrac{1}{\cos \theta} - \dfrac{\sin^2 \theta}{\cos \theta}$$

$$= \dfrac{1 - \sin^2 \theta}{\cos \theta}$$ ⟵ Use Identity 6.

$$= \dfrac{\cos^2 \theta}{\cos \theta}$$

$$= \cos \theta$$

Classroom Exercises

Write $\tan^2 \theta - 2 \sec \theta \sin \theta$ in the manner described.

1. In terms of $\sin \theta$ and $\cos \theta$

2. In terms of $\tan \theta$

3. In terms of $\cot \theta$

Written Exercises

In Exercises 1–2, complete each proof shown.

1. Prove Identity **1:** $\csc \theta = \dfrac{1}{\sin \theta}$, $\sin \theta \neq 0$

Proof: $\sin \theta = \dfrac{y}{r}$ (Definition)

$\therefore r \sin \theta = y$

$\csc \theta = \underline{\ ?\ }$ (Definition)

$\therefore \csc \theta = \dfrac{r}{r \cdot \sin \theta}$

$\csc \theta = \underline{\ ?\ }$

2. Prove Identity **2:** $\sec \theta = \dfrac{1}{\cos \theta}$ (Definition)

    Proof: $\cos \theta = \dfrac{?}{}$

        $\therefore \underline{\;?\;} = x$

        $\sec \theta = \dfrac{r}{x}$ (Definition)

        $\therefore \sec \theta = \dfrac{r}{?}$

        $\sec \theta = \dfrac{1}{\cos \theta}$

3. Prove Identity **3:** $\cot \theta = \dfrac{1}{\tan \theta}$, $\tan \theta \neq 0$

4. Prove Identity **4:** $\tan \theta = \dfrac{\sin \theta}{\cos \theta}$, $\cos \theta \neq 0$

5. Prove Identity **7:** $1 + \cot^2 \theta = \csc^2 \theta$

6. Prove Identity **8:** $1 + \tan^2 \theta = \sec^2 \theta$

Write in terms of cos θ.

7. $(1 - \cot^2 \theta)(\cot^2 \theta + 1)$ **8.** $(1 - \sin^2 \theta)(1 + \sec^2 \theta)$

9. $2 \sin^2 \theta - 1$ **10.** $\dfrac{\tan \theta}{\sin \theta}$

Write in terms of sin θ.

11. $\cot \theta \cdot \cos \theta$ **12.** $\cot^2 \theta$ **13.** $\tan^2 \theta \cdot \cos^2 \theta + \csc \theta$ **14.** $\dfrac{1}{\sec^2 \theta}$

Write in terms of tan θ.

15. $\dfrac{\sec \theta \cdot \sin \theta}{\tan \theta + \cot \theta}$ **16.** $\dfrac{1 - \cos^2 \theta}{\cos^2 \theta}$ **17.** $\dfrac{\sin \theta}{\cos \theta} + \dfrac{\cos \theta}{\sin \theta}$ **18.** $\csc^2 \theta - 1$

Write in terms of cos θ.

19. $\sin \theta$ **20.** $\csc \theta$ **21.** $\tan \theta$ **22.** $\cot \theta$

23. Write $\dfrac{\sin^2 \theta + \cos^2 \theta}{\sin \theta + \cos \theta}$ in terms of sec θ and csc θ.

24. Write $\dfrac{\cot \theta \tan \theta - \cos^2 \theta}{\cos \theta \tan \theta}$ in terms of sin θ.

25. Show that $\sin \theta + \cos \theta = 1$ is not an identity.

Write the remaining five trigonometric functions in terms of the given function.

26. $\sin \theta$ **27.** $\tan \theta$ **28.** $\cot \theta$ **29.** $\sec \theta$

Trigonometric Identities **127**

Review Capsule for Section 3-2 _____

Add or subtract as indicated. Then simplify.

1. $\dfrac{a}{b} - 1$

2. $\dfrac{\tan\theta}{\sec\theta} - 1$

3. $\dfrac{2b}{x^2} - x$

4. $\dfrac{2\cos\theta}{\sin^2\theta} - \sin\theta$

5. $\dfrac{\sin^2\theta}{\cos\theta} + \cos\theta$

6. $\dfrac{a}{b} + \dfrac{2}{d}$

7. $\dfrac{\tan\theta}{\sec\theta} + \dfrac{2}{\cot\theta}$

8. $\dfrac{5}{x^2-4} + \dfrac{3}{x+2}$

9. $\dfrac{5}{\csc^2\theta - 4} + \dfrac{3}{\csc\theta + 2}$

3-2 Proving Identities

You can use the Fundamental Identities to prove that other trigonometric equations are also identities. Examples 1 and 2 exhibit a basic strategy for proving identities.

Strategy 1: *Use known identities to replace the expressions on one side of the equation with equivalent expressions. Continue this procedure until the expressions on both sides of the equation are the same.*

Example 1

Prove: $\sin^2\theta = (1 - \sin^2\theta)\cdot \tan^2\theta$

Proof:

$$\begin{array}{c|c}
\sin^2\theta & (1 - \sin^2\theta)\cdot\tan^2\theta \\[4pt]
 & \cos^2\theta\cdot\tan^2\theta \\[4pt]
 & \cos^2\theta\cdot\dfrac{\sin^2\theta}{\cos^2\theta}
\end{array}$$

$\longleftarrow$ $1 - \sin^2\theta = \cos^2\theta$

$\longleftarrow$ $\tan\theta = \dfrac{\sin\theta}{\cos\theta}$

$$\sin^2\theta = \sin^2\theta$$

(handwritten annotations:) $\sin^2\theta + \cos\theta = 1$; $-\sin^2\theta \to \sin^2\theta$; $\cos^2 = 1-\sin^2\theta$

Since the last equation is obviously an identity,
$\sin^2\theta = (1 - \sin^2\theta)\cdot \tan^2\theta$ is an identity.

In Example 1, note that $1 - \sin^2\theta = \cos^2\theta$ is simply another way of writing the identity $\sin^2\theta + \cos^2\theta = 1$.

Example 2

Prove: $2 \tan^2 \theta = \dfrac{\sin \theta}{\csc \theta - 1} + \dfrac{\sin \theta}{\csc \theta + 1}$

Proof:

$2 \tan^2 \theta$ $\Bigg|$ $\dfrac{\sin \theta}{\csc \theta - 1} + \dfrac{\sin \theta}{\csc \theta + 1}$ ⟵ $\csc \theta = \dfrac{1}{\sin \theta}$

$\dfrac{\sin \theta}{\dfrac{1}{\sin \theta} - 1} + \dfrac{\sin \theta}{\dfrac{1}{\sin \theta} + 1}$ ⟵ Simplify the denominator.

$\dfrac{\sin \theta}{\dfrac{1 - \sin \theta}{\sin \theta}} + \dfrac{\sin \theta}{\dfrac{1 + \sin \theta}{\sin \theta}}$ ⟵ $\dfrac{a}{\frac{b}{c}} = \dfrac{a}{1} \cdot \dfrac{c}{b}$

$\dfrac{\sin^2 \theta}{1 - \sin \theta} + \dfrac{\sin^2 \theta}{1 + \sin \theta}$

$\dfrac{\sin^2 \theta (1 + \sin \theta)}{(1 - \sin \theta)(1 + \sin \theta)} + \dfrac{\sin^2 \theta (1 - \sin \theta)}{(1 + \sin \theta)(1 - \sin \theta)}$

$\dfrac{\sin^2 \theta + \sin^3 \theta + \sin^2 \theta - \sin^3 \theta}{(1 - \sin^2 \theta)}$ ⟵ $1 - \sin^2 \theta = \cos^2 \theta$

$\dfrac{2 \sin^2 \theta}{\cos^2 \theta}$

$2 \tan^2 \theta = 2 \tan^2 \theta$

In general, change trigonometric functions to sines and cosines, as was done in Examples 1 and 2.

Strategy 2: *When Strategy 1 does not work, use known identities to replace the expressions on both sides of the equation. Continue this procedure until the expressions on both sides of the equation are the same.*

Example 3

Prove: $\tan \theta \cdot \sin \theta = \sec \theta - \cos \theta$

Proof:

$\tan \theta \cdot \sin \theta$ $\Big|$ $\sec \theta - \cos \theta$

Identity **4** ⟶ $\dfrac{\sin \theta}{\cos \theta} \cdot \sin \theta$ $\Big|$ $\dfrac{1}{\cos \theta} - \cos \theta$ ⟵ Identity **2**

$\dfrac{\sin^2 \theta}{\cos \theta}$ $\Big|$ $\dfrac{1 - \cos^2 \theta}{\cos \theta}$

$\dfrac{\sin^2 \theta}{\cos \theta} = \dfrac{\sin^2 \theta}{\cos \theta}$ ⟵ Identity **6**

The vertical rule is used in this strategy to emphasize that each side of the equation is done independently. When you use Strategy 2, it is important not to operate "across" the equality symbol. If you were to do this, you would really be assuming that the given identity is true, and this is the very thing you are to prove.

Classroom Exercises

Prove each identity.

1. $1 - \cos^2 \theta = \dfrac{1}{\csc^2 \theta}$

2. $\dfrac{1}{1 - \sin^2 \theta} = \sec^2 \theta$

3. $1 + \tan^2 \theta = \dfrac{1}{1 - \sin^2 \theta}$

4. $\dfrac{1}{\cot^2 \theta + 1} = 1 - \cos^2 \theta$

Written Exercises

a

Prove each identity.

1. $\tan \theta = \sin \theta \cdot \sec \theta$

2. $\cot \theta = \cos \theta \cdot \csc \theta$

3. $\tan^2 \theta = \dfrac{1 - \cos^2 \theta}{\cos^2 \theta}$

4. $\sec^2 \theta = \dfrac{\sin^2 \theta + \cos^2 \theta}{\cos^2 \theta}$

5. $\tan^2 \theta = \sec^2 \theta - 1$

6. $\cot^2 \theta = \csc^2 \theta - 1$

7. $\dfrac{\cos^2 \theta}{\sin \theta} + \sin \theta = \csc \theta$

8. $\dfrac{\tan \theta}{1 - \cos^2 \theta} = \sec \theta \cdot \csc \theta$

9. $\csc \theta = \dfrac{\cot \theta}{\cos \theta}$

10. $\dfrac{1}{\sec^2 \theta} + \dfrac{1}{\csc^2 \theta} = 1$

11. $\csc^2 \theta \tan^2 \theta - 1 = \tan^2 \theta$

12. $\dfrac{\sec \theta}{\cos \theta} - \dfrac{\tan \theta}{\cot \theta} = 1$

13. $\csc^4 \theta - \cot^4 \theta = \csc^2 \theta + \cot^2 \theta$

14. $\sec^4 \theta - \tan^4 \theta = \tan^2 \theta + \sec^2 \theta$

15. $(1 - \tan \theta)^2 = \sec^2 \theta - 2 \tan \theta$

16. $(1 - \sin^2 \theta)(1 + \tan^2 \theta) = 1$

17. $\dfrac{\cot \theta}{\cos \theta} + \dfrac{\sec \theta}{\cot \theta} = \sec^2 \theta \cdot \csc \theta$

18. $2 \sin^2 \theta - 1 = 1 - 2 \cos^2 \theta$

b

19. $\sec \theta - \tan \theta \cdot \sin \theta = \cos \theta$

20. $(1 - \cos^2 \theta)(\cot^2 \theta + 1) = 1$

21. $\dfrac{1 - \sin^2 \theta}{1 + \tan^2 \theta} = \cos^4 \theta$

22. $\dfrac{\sin \theta}{\csc \theta} + \dfrac{\cos \theta}{\sec \theta} = 1$

23. $\dfrac{\sin \theta + \tan \theta}{1 + \sec \theta} = \sin \theta$

24. $\dfrac{1 - \tan^2 \theta}{1 + \tan^2 \theta} = 2 \cos^2 \theta - 1$

25. $\sec \theta - \tan \theta \sin \theta = \dfrac{1}{\sec \theta}$

26. $\dfrac{1 + \cos \theta}{\sin \theta} = \csc \theta + \cot \theta$

27. $\dfrac{\sec \theta \cdot \sin \theta}{\tan \theta + \cot \theta} = \sin^2 \theta$ 2 - 50

28. $\csc \theta + \cot \theta = \dfrac{1 + \cos \theta}{\sin \theta}$

29. $\cos^2 \theta - \sin^2 \theta = 1 - 2 \sin^2 \theta$ even

30. $\cos^2 \theta - \sin^2 \theta = 2 \cos^2 \theta - 1$

31. $\dfrac{\sec^2 \theta}{\sec^2 \theta - 1} = \csc^2 \theta$

32. $\tan^2 \theta \sin^2 \theta = \tan^2 \theta - \sin^2 \theta$

33. $(\sin \theta + \cos \theta)^2 + (\sin \theta - \cos \theta)^2 = 2$

34. $(\sin \theta + \cos \theta)(\tan \theta + \cot \theta) = \sec \theta + \csc \theta$

35. $\dfrac{\tan \theta - 1}{\tan \theta + 1} = \dfrac{1 - \cot \theta}{1 + \cot \theta}$

36. $\dfrac{1 - \tan^2 \theta}{1 + \tan^2 \theta} = 1 - 2 \sin^2 \theta$

37. $\dfrac{\cos \theta + 1}{\sin^3 \theta} = \dfrac{\csc \theta}{1 - \cos \theta}$

38. $\dfrac{\sin \theta}{1 - \cos \theta} = \csc \theta + \cot \theta$

39. $\dfrac{\tan \theta}{\sec \theta} + \dfrac{\cot \theta}{\csc \theta} = \sin \theta + \cos \theta$

40. $\dfrac{\sin \theta + \tan \theta}{1 + \sec \theta} = \sin \theta$

41. $\dfrac{2 \tan \theta}{1 + \tan^2 \theta} = 2 \sin \theta \cos \theta$

42. $\dfrac{1 + \cot \theta}{\csc \theta} = \dfrac{1 + \tan \theta}{\sec \theta}$

43. $\dfrac{1 - \cos^6 \theta}{\sin^2 \theta} = 1 + \cos^2 \theta + \cos^4 \theta$

44. $\dfrac{3 \cos^2 \theta}{\sec \theta} + \dfrac{\tan^2 \theta \cos \theta}{1 + \tan^2 \theta} = \cos \theta$

45. $\dfrac{\sin^3 \theta + \cos^3 \theta}{\sin \theta + \cos \theta} = 1 - \sin \theta \cos \theta$

46. $\cos^4 \theta - \sin^4 \theta = 2 \cos^2 \theta - 1$

47. $\dfrac{1 - 2 \sin \theta - 3 \sin^2 \theta}{\cos^2 \theta} = \dfrac{1 - 3 \sin \theta}{1 - \sin \theta}$

48. $(\tan x + \cot x)^2 = \sec^4 x \cot^2 x$

49. $(\sec x + \tan x) = \dfrac{\cos x}{1 - \sin x}$

50. $\cos^4 x - \sin^4 x - 1 = -2 \sin^2 x$

Review Capsule for Section 3-3 _____

Simplify.

Example 1: $\sqrt{27} = \underline{?}$
$$\sqrt{27} = \sqrt{9} \cdot \sqrt{3}$$
$$= 3\sqrt{3}$$

Example 2: $6\sqrt{5} - 2\sqrt{20} = \underline{?}$
$$6\sqrt{5} - 2\sqrt{20} = 6\sqrt{5} - 2 \cdot \sqrt{4} \cdot \sqrt{5}$$
$$= 6\sqrt{5} - 2 \cdot 2 \cdot \sqrt{5}$$
$$= 6\sqrt{5} - 4\sqrt{5}$$
$$= 2\sqrt{5}$$

1. $\sqrt{12}$

2. $\sqrt{50}$

3. $3\sqrt{48}$

4. $\sqrt{18} + 4\sqrt{2}$

5. $\sqrt{72} - 5\sqrt{2}$

6. $\sqrt{3} \cdot \sqrt{2}$

7. $3\sqrt{6} \cdot \sqrt{3}$

8. $2\sqrt{3} \cdot 3\sqrt{12}$

3-3 Sum and Difference Identities for Cosine

It can be readily shown that the cosine of a difference does not equal the difference of cosines. For example,

$$\cos(60° - 30°) \neq \cos 60° - \cos 30°.$$

Proof: $\cos(60° - 30°) = \cos 30° = \dfrac{\sqrt{3}}{2}$ $\qquad \cos 60° - \cos 30° = \dfrac{1}{2} - \dfrac{\sqrt{3}}{2}$

Since $\dfrac{\sqrt{3}}{2} \neq \dfrac{1}{2} - \dfrac{\sqrt{3}}{2}$, $\cos(60° - 30°) \neq \cos 60° - \cos 30°$.

The correct relationship is the <u>difference identity for cosine</u>.

Difference Identity for Cosine

9. $\cos(\alpha - \beta) = \cos\alpha \cos\beta + \sin\alpha \sin\beta$

Proof: Recall from Section 3–1 that if $P(x, y)$ is on the terminal side of an angle θ in standard position, then

$$x = r\cos\theta \qquad \text{and} \qquad y = r\sin\theta,$$

where r is the distance from the origin to point P.

In Figure 1, α (alpha) and β (beta) are angles in standard position. Points Q and P are on the terminal sides of α and β, respectively, such that $OQ = OP = 1$. Thus, Q and P are points on the circle with center at O and radius $r = 1$ (called the <u>unit circle</u>). Therefore, Q and P have coordinates as shown in Figure 1. The difference $\alpha - \beta$ is $\angle POQ$.

In Figure 2, $\alpha - \beta$ is placed in standard position and R is a point on the terminal side such that $OR = 1$. Therefore R is on the unit circle and has coordinates as shown in Figure 2.

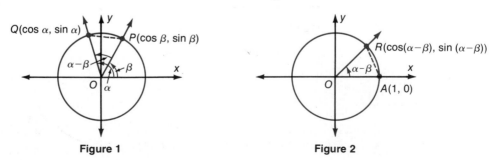

Figure 1   **Figure 2**

Use the distance formula to find PQ and AR.

$$(PQ)^2 = (\cos \alpha - \cos \beta)^2 + (\sin \alpha - \sin \beta)^2$$
$$(AR)^2 = [\cos (\alpha - \beta) - 1]^2 + [\sin (\alpha - \beta) - 0]^2$$

Angles POQ and AOR are congruent central angles of the unit circle. Thus, their chords, $\overline{PQ}$ and $\overline{AR}$, are equal in length. Since $PQ = AR$, $(PQ)^2 = (AR)^2$.

$$(\cos \alpha - \cos \beta)^2 + (\sin \alpha - \sin \beta)^2 =$$
$$[\cos (\alpha - \beta) - 1]^2 + [\sin (\alpha - \beta) - 0]^2$$

Expand the left side.

$$(\cos^2 \alpha - 2 \cos \alpha \cos \beta + \cos^2 \beta) + (\sin^2 \alpha - 2 \sin \alpha \sin \beta + \sin^2 \beta) =$$
$$(\cos^2 \alpha + \sin^2 \alpha) + (\cos^2 \beta + \sin^2 \beta) - 2(\cos \alpha \cos \beta + \sin \alpha \sin \beta) =$$
$$1 + 1 - 2(\cos \alpha \cos \beta + \sin \alpha \sin \beta) = 2 - 2(\cos \alpha \cos \beta + \sin \alpha \sin \beta)$$

Now expand the right side.

$$\cos^2 (\alpha - \beta) - 2 \cos (\alpha - \beta) + 1 + \sin^2 (\alpha - \beta) =$$
$$[\cos^2 (\alpha - \beta) + \sin^2 (\alpha - \beta)] + 1 - 2 \cos (\alpha - \beta) =$$
$$1 + 1 - 2 \cos (\alpha - \beta) = 2 - 2 \cos (\alpha - \beta)$$

Thus,

$$2 - 2(\cos \alpha \cos \beta + \sin \alpha \sin \beta) = 2 - 2 \cos (\alpha - \beta)$$
or $\qquad\qquad 2 \cos (\alpha - \beta) = 2(\cos \alpha \cos \beta + \sin \alpha \sin \beta)$
$\therefore \cos (\alpha - \beta) = \cos \alpha \cos \beta + \sin \alpha \sin \beta.$

This proof is valid for any α or β. You can use the <u>Difference Identity for Cosine</u> to prove that the following two reduction formulas are true for all values of α.

$$\cos (-\alpha) = \cos \alpha \qquad\qquad \sin (-\alpha) = -\sin \alpha$$

The proofs are asked for in the Exercises.

Since $\cos (\alpha + \beta)$ can be written as $\cos [\alpha - (-\beta)]$, the <u>Sum Identity for Cosine</u> follows directly from the Difference Identity for Cosine.

Sum Identity for Cosine

10. $\cos (\alpha + \beta) = \cos \alpha \cos \beta - \sin \alpha \sin \beta$

Proof: $\cos [\alpha - (-\beta)] = \cos \alpha \cos (-\beta) + \sin \alpha \sin (-\beta)$
$$= \cos \alpha \cos \beta + \sin \alpha(-\sin \beta) \longleftarrow \begin{array}{l} \cos (-\beta) = \cos \beta; \\ \sin (-\beta) = -\sin \beta \end{array}$$
$\therefore \cos (\alpha + \beta) = \cos \alpha \cos \beta - \sin \alpha \sin \beta$

The sum and difference identities can be used to evaluate the cosine of certain angles without using tables.

Example 1 Evaluate without tables. Write radicals in simplest form.

a. $\cos 15°$ **b.** $\cos 105°$

Solutions: a. $\cos 15° = \cos (60° - 45°)$

$= \cos 60° \cos 45° + \sin 60° \sin 45°$

$= \left(\frac{1}{2}\right)\left(\frac{\sqrt{2}}{2}\right) + \left(\frac{\sqrt{3}}{2}\right)\left(\frac{\sqrt{2}}{2}\right)$

$= \frac{\sqrt{2}}{4} + \frac{\sqrt{6}}{4} = \frac{1}{4}(\sqrt{2} + \sqrt{6})$

b. $\cos 105° = \cos (60° + 45°)$

$= \cos 60° \cos 45° - \sin 60° \sin 45°$

$= \left(\frac{1}{2}\right)\left(\frac{\sqrt{2}}{2}\right) - \left(\frac{\sqrt{3}}{2}\right)\left(\frac{\sqrt{2}}{2}\right)$

$= \frac{\sqrt{2}}{4} - \frac{\sqrt{6}}{4} = \frac{1}{4}(\sqrt{2} - \sqrt{6})$

Example 2 Without tables, evaluate $\cos (\alpha - \beta)$ if $\sin \alpha = \frac{4}{5}$, $0 < \alpha < \frac{\pi}{2}$, and $\cos \beta = -\frac{3}{5}$, $\frac{\pi}{2} < \beta < \pi$.

Solution: First find $\cos \alpha$ and $\sin \beta$

$\sin^2 \alpha + \cos^2 \alpha = 1$

$\left(\frac{4}{5}\right)^2 + \cos^2 \alpha = 1$

$\cos^2 \alpha = 1 - \left(\frac{4}{5}\right)^2$

$1 - = \frac{9}{25}$

$\therefore \cos \alpha = \frac{3}{5}$ ⟵ Since $0 < \alpha < \frac{\pi}{2}$.

$\sin^2 \beta + \cos^2 \beta = 1$

$\sin^2 \beta + \left(-\frac{3}{5}\right)^2 = 1$

$\sin^2 \beta = 1 - \left(-\frac{3}{5}\right)^2$

$1 = \frac{16}{25}$

$\sin \beta = \frac{4}{5}$ ⟵ Since $\frac{\pi}{2} < \beta < \pi$.

Then $\cos (\alpha - \beta) = \cos \alpha \cos \beta + \sin \alpha \sin \beta$ ⟵ Difference Identity

$= \left(\frac{3}{5}\right)\left(-\frac{3}{5}\right) + \left(\frac{4}{5}\right)\left(\frac{4}{5}\right) = -\frac{9}{25} + \frac{16}{25}$, or $\frac{7}{25}$.

You can use the sum and difference identities to prove some of the reduction formulas for cosine.

Example 3 Prove: $\cos (\pi - \theta) = -\cos \theta$

> **Proof:** $\cos (\pi - \theta) = \cos \pi \cos \theta + \sin \pi \sin \theta$
> $$= (-1) \cos \theta + (0) (\sin \theta)$$
> $$= -\cos \theta$$

980

Classroom Exercises

Evaluate without tables. Use the sum and difference identities for cosine. Write radicals in simplest form.

1. $\cos \left(\frac{\pi}{2} - \frac{\pi}{3} \right)$ **2.** $\cos 135°$ **3.** $\cos \frac{5\pi}{12}$ **4.** $\cos 195°$

Written Exercises

a

Evaluate without tables. Write radicals in simplest form.

1. $\cos (45° + 30°)$ **2.** $\cos \left(\frac{20\pi}{12} - \frac{3\pi}{12} \right)$

3. $\cos \frac{11\pi}{12}$ **4.** $\cos 255°$

$-1-8$

5. $\cos 345°$ **6.** $\cos \frac{19\pi}{12}$

7. $\cos \frac{23\pi}{12}$ $9-32$ **8.** $\cos 285°$

Without tables, evaluate $\cos (\alpha + \beta)$ and $\cos (\alpha - \beta)$.

9. $\sin \alpha = \frac{3}{5}$, $\cos \beta = \frac{5}{13}$; $0 < \alpha < \frac{\pi}{2}, 0 < \beta < \frac{\pi}{2}$

10. $\tan \alpha = \frac{4}{3}$, $\cos \beta = \frac{12}{13}$; $0 < \alpha < \frac{\pi}{2}, 0 < \beta < \frac{\pi}{2}$

11. $\tan \alpha = \frac{3}{4}$, $\cos \beta = \frac{5}{13}$; $0 < \alpha < \frac{\pi}{2}, 0 < \beta < \frac{\pi}{2}$

12. $\cos \alpha = -\frac{3}{5}$, $\tan \beta = -\frac{5}{12}$; neither α nor β in Quadrant II

13. $\sin \alpha = -\frac{7}{25}$, $\cot \beta = -\frac{8}{15}$; neither α nor β in Quadrant IV

Trigonometric Identities   **135**

Prove the following cofunction identities.

14. $\cos(90° - \theta) = \sin \theta$

15. $\cos \theta = \sin(90° - \theta)$ [HINT: Let $\alpha = 90° - \theta$ and $\beta = 90°$.]

Prove the following formulas.

16. $\cos(180° - \theta) = -\cos \theta$

17. $\cos(180° + \theta) = -\cos \theta$

18. $\cos(270° + \theta) = \sin \theta$

19. $\cos\left(\frac{3\pi}{2} - \theta\right) = -\sin \theta$

20. $\cos(360° + \theta) = \cos \theta$

21. $\cos(2\pi - \theta) = \cos \theta$

Express each of the following in terms of functions of θ.

22. $\cos(\theta + 30°)$

23. $\cos\left(\frac{\pi}{4} + \theta\right)$

24. $\cos(45° - \theta)$

25. $\cos(60° - \theta)$

26. $\cos\left(\theta - \frac{\pi}{3}\right)$

27. $\cos\left(\theta - \frac{\pi}{6}\right)$

28. Prove $\sin\left(\frac{\pi}{4} + \theta\right) = \cos\left(\frac{\pi}{4} - \theta\right)$

Prove the following identities.

29. $\cos(\alpha + \beta) - \cos(\alpha - \beta) = -2\sin\alpha\sin\beta$

30. $\cos(\alpha - \beta)\cos(\alpha + \beta) = \cos^2\alpha - \sin^2\beta$

31. Use the Difference Identity for Cosine to prove that $\cos(-\alpha) = \cos\alpha$.

32. Use the Difference Identity for Cosine to prove that $\sin(-\alpha) = -\sin\alpha$. (HINT: In the Difference Identity for Cosine, replace β with 90°; in the identity for Exercise 14, replace θ with $-\alpha$.)

Review Capsule for Section 3-4 _____

Find θ, where $0° < \theta < 90°$.

Example: $\cos 58° = \sin \theta$

For $0° < \theta < 90°$, any trigonometric function of θ is equal to the cofunction of its complementary angle (see page 28). Thus,

$$\cos 58° = \sin(90° - 58°) = \sin 32°.$$

Therefore, $\sin 32° = \sin \theta$.

For $0° < \theta < 90°$, the equation $\sin \theta = \sin 32°$ has one solution, **$\theta = 32°$.**

1. $\sin 40° = \cos(90° - \theta)$

2. $\cot 70° = \tan(90° - \theta)$

3. $\sin 63° = \cos \theta$

4. $\sec 12° = \csc \theta$

5. $\tan 51°20' = \cot \theta$

6. $\csc 17°40' = \sec \theta$

3-4 Sum and Difference Identities for Sine

The Difference Identity for Cosine and the fact that sine and cosine are cofunctions are the key ideas in proving the <u>Sum Identity for Sine</u>.

Sum Identity for Sine

11. $\sin(\alpha + \beta) = \sin\alpha \cos\beta + \cos\alpha \sin\beta$

Proof: $\sin\theta = \cos(90° - \theta)$ ⟵ See page 28 and Exercise 14 on page 137.

Replace θ with $\alpha + \beta$.

$$\sin(\alpha + \beta) = \cos[90° - (\alpha + \beta)]$$
$$= \cos[(90° - \alpha) - \beta]$$
$$= \cos(90° - \alpha)\cos\beta + \sin(90° - \alpha)\sin\beta$$

Since $\cos(90° - \alpha) = \sin\alpha$ and $\sin(90° - \alpha) = \cos\alpha$,

$$\sin(\alpha + \beta) = \sin\alpha\cos\beta + \cos\alpha\sin\beta.$$

Since $\sin(\alpha - \beta)$ can be written as $\sin[\alpha + (-\beta)]$, the <u>Difference Identity for Sine</u> can be readily proved.

Difference Identity for Sine

12. $\sin(\alpha - \beta) = \sin\alpha\cos\beta - \cos\alpha\sin\beta$

Proof: $\sin(\alpha - \beta) = \sin[\alpha + (-\beta)]$
$$= \sin\alpha\cos(-\beta) + \cos\alpha\sin(-\beta)$$

Since $\cos(-\beta) = \cos\beta$ and $\sin(-\beta) = -\sin\beta$,

$$\sin(\alpha - \beta) = \sin\alpha\cos\beta - \cos\alpha\sin\beta.$$

The sum and difference identities can be used to evaluate the sine of certain angles without using tables. First, express the given angle as the sum or difference of angles such as 30°, 45°, 60°, and so on.

Example 1

Evaluate. Write radicals in simplest form.

a. $\sin 105°$ **b.** $\sin 15°$

Solutions: a. $\sin 105° = \sin (45° + 60°)$

$$= \sin 45° \cos 60° + \cos 45° \sin 60°$$

$$= \left(\frac{\sqrt{2}}{2}\right)\left(\frac{1}{2}\right) + \left(\frac{\sqrt{2}}{2}\right)\left(\frac{\sqrt{3}}{2}\right)$$

$$= \frac{\sqrt{2}}{4} + \frac{\sqrt{6}}{4} = \frac{1}{4}(\sqrt{2} + \sqrt{6})$$

b. $\sin 15° = \sin (45° - 30°)$

$$= \sin 45° \cos 30° - \cos 45° \sin 30°$$

$$= \left(\frac{\sqrt{2}}{2}\right)\left(\frac{\sqrt{3}}{2}\right) - \left(\frac{\sqrt{2}}{2}\right)\left(\frac{1}{2}\right)$$

$$= \frac{\sqrt{6}}{4} - \frac{\sqrt{2}}{4} = \frac{1}{4}(\sqrt{6} - \sqrt{2})$$

Example 2

Evaluate $\sin (\alpha + \beta)$ if $\sin \alpha = \frac{3}{5}$, $\frac{\pi}{2} < \alpha < \pi$, and $\sin \beta = -\frac{5}{13}$, $\pi < \beta < \frac{3\pi}{2}$.

Solution: First find $\cos \alpha$ and $\cos \beta$.

$$\sin^2 \alpha + \cos^2 \alpha = 1 \qquad\qquad \sin^2 \beta + \cos^2 \beta = 1$$

$$\left(\frac{3}{5}\right)^2 + \cos^2 \alpha = 1 \qquad\qquad \left(-\frac{5}{13}\right)^2 + \cos^2 \beta = 1$$

$$\cos^2 \alpha = 1 - \left(\frac{3}{5}\right)^2 \qquad\qquad \cos^2 \beta = 1 - \left(-\frac{5}{13}\right)^2$$

$$= \frac{16}{25} \qquad\qquad\qquad\qquad = \frac{144}{169}$$

$$\therefore \cos \alpha = -\frac{4}{5} \quad \xleftarrow{\text{Since}} \quad \frac{\pi}{2} < \alpha < \pi. \qquad \cos \beta = -\frac{12}{13} \quad \xleftarrow{\text{Since}} \quad \pi < \beta < \frac{3\pi}{2}$$

But $\sin (\alpha + \beta) = \sin \alpha \cos \beta + \cos \alpha \sin \beta$ $\quad\longleftarrow$ Sum Identity

$$= \left(\frac{3}{5}\right)\left(-\frac{12}{13}\right) + \left(-\frac{4}{5}\right)\left(-\frac{5}{13}\right)$$

$$= -\frac{36}{65} + \frac{20}{65} = -\frac{16}{65}$$

You can use the sum and difference identities to prove some of the reduction formulas for sine.

Example 3 Prove: $\sin(\pi - \theta) = \sin\theta$.

Proof: $\sin(\pi - \theta) = \sin\pi\cos\theta - \cos\pi\sin\theta$
$$= (0 \cdot \cos\theta) - (-1)(\sin\theta) = \sin\theta$$

Classroom Exercises

Write each angle measure as the sum or difference of two measures for which sine and cosine are known.

1. $\frac{\pi}{6}$ **2.** $135°$ **3.** $\frac{5\pi}{12}$ **4.** $195°$

Evaluate. Write radicals in simplest form.

5. $\sin\frac{\pi}{6}$ **6.** $\sin 135°$ **7.** $\sin\frac{5\pi}{12}$ **8.** $\sin 195°$

Written Exercises

a

Evaluate without tables. Write radicals in simplest form.

1. $\sin(135° + 30°)$ **2.** $\sin(315° - 60°)$

3. $\sin 345°$ **4.** $\sin 285°$

5. $\sin\frac{13\pi}{12}$ **6.** $\sin\frac{\pi}{12}$

7. $\sin\frac{7\pi}{4}$ **8.** $\sin\frac{7\pi}{12}$

Without tables, evaluate $\sin(\alpha + \beta)$ and $\sin(\alpha - \beta)$.

9. $\sin\alpha = \frac{3}{5}$, $\cos\beta = \frac{5}{13}$; $0 < \alpha < 90°$, $0 < \beta < 90°$

10. $\tan\alpha = \frac{4}{3}$, $\sec\beta = \frac{13}{12}$; $0 < \alpha < 90°$, $0 < \beta < 90°$

11. $\sin\alpha = \frac{4}{5}$, $\cos\beta = \frac{12}{13}$; $\frac{\pi}{2} < \alpha < \pi$, $\frac{3\pi}{2} < \beta < 2\pi$

12. $\sin\alpha = -\frac{3}{5}$, $\cos\beta = -\frac{5}{13}$; α and β in Quadrant III.

13. $\sin\alpha = -\frac{7}{25}$, $\tan\beta = -\frac{15}{8}$; neither α nor β in Quadrant IV.

14. $\tan\alpha = \frac{4}{3}$, $\cos\beta = -\frac{12}{13}$; neither α nor β in Quadrant III.

Prove the following formulas.

15. $\sin\left(\frac{\pi}{2}+\theta\right)=\cos\theta$ **16.** $\sin\left(\theta-90°\right)=-\cos\theta$

17. $\sin\left(\pi-\theta\right)=\sin\theta$ **18.** $\sin\left(180°+\theta\right)=-\sin\theta$

19. $\sin\left(270°-\theta\right)=-\cos\theta$ **20.** $\sin\left(\frac{3\pi}{2}+\theta\right)=-\cos\theta$

21. $\sin\left(2\pi-\theta\right)=-\sin\theta$ **22.** $\sin\left(2\pi+\theta\right)=\sin\theta$

Express each of the following in terms of a function of θ.

23. $\sin\left(\theta+45°\right)$ **24.** $\sin\left(30°+\theta\right)$

25. $\sin\left(\theta-\frac{\pi}{3}\right)$ **26.** $\sin\left(\theta+\frac{\pi}{3}\right)$

27. $\sin\left(\theta-\frac{\pi}{6}\right)$ **28.** $\sin\left(\theta-\frac{\pi}{4}\right)$

Prove each identity.

29. $2\sin\left(\frac{\pi}{4}+\theta\right)\sin\left(\theta-\frac{\pi}{4}\right)=\sin^2\theta-\cos^2\theta$

30. $\sin\left(\alpha+\beta\right)+\sin\left(\alpha-\beta\right)=2\sin\alpha\cos\beta$

31. $\sin\left(\alpha+\beta\right)\cdot\sin\left(\alpha-\beta\right)=\sin^2\alpha-\sin^2\beta$

Review

Use the definitions of the trigonometric functions to prove each identity. (Section 3–1)

1. $\cot\theta=\dfrac{\cos\theta}{\sin\theta}$ **2.** $\sec^2\theta-\tan^2\theta=1$

Prove each identity. (Section 3–2)

3. $\dfrac{\cos\theta-\sin\theta}{\cos\theta}=1-\tan\theta$ **4.** $\tan\theta+\cot\theta=\sec\theta\csc\theta$

5. $\dfrac{\cot\theta+1}{\cot\theta}=1+\tan\theta$ **6.** $\tan\theta\left(\tan\theta+\cot\theta\right)=\sec^2\theta$

7. Evaluate $\cos165°$ without using tables. (Section 3–3)

8. Prove: $\cos\left(3\pi-\theta\right)=-\cos\theta$ (Section 3–3)

9. If $\sin\alpha=-\frac{5}{13}$ and $\cos\beta=\frac{3}{5}$, evaluate $\cos\left(\alpha+\beta\right)$. Assume that neither α nor β is in Quadrant IV. (Section 3–3)

10. Evaluate $\sin255°$ without using tables. (Section 3–4)

11. Prove: $\sin\left(\pi+\theta\right)=-\sin\theta$ (Section 3–4)

12. If $\sin\alpha=\frac{5}{13}$ and $\cos\beta=\frac{3}{5}$, evaluate $\sin\left(\alpha-\beta\right)$. Assume that $\frac{\pi}{2}<\alpha<\pi$, $\frac{3\pi}{2}<\beta<2\pi$. (Section 3–4)

3-5 Sum and Difference Identities for Tangent

The sum identities for sine and cosine can be used to derive a Sum Identity for Tangent.

Sum Identity for Tangent

For all angle measures α and β for which the functions have meaning,

13. $\tan (\alpha + \beta) = \dfrac{\tan \alpha + \tan \beta}{1 - \tan \alpha \tan \beta}.$

Proof: Since $\tan \theta = \dfrac{\sin \theta}{\cos \theta}$,

$$\tan (\alpha + \beta) = \frac{\sin (\alpha + \beta)}{\cos (\alpha + \beta)}. \quad \longleftarrow \quad \text{Replace } \theta \text{ with } \alpha + \beta.$$

Next, apply the sum identities for sine and cosine.

$$\frac{\sin (\alpha + \beta)}{\cos (\alpha + \beta)} = \frac{\sin \alpha \cos \beta + \cos \alpha \sin \beta}{\cos \alpha \cos \beta - \sin \alpha \sin \beta}$$

$$= \frac{\dfrac{\sin \alpha \cos \beta}{\cos \alpha \cos \beta} + \dfrac{\cos \alpha \sin \beta}{\cos \alpha \cos \beta}}{\dfrac{\cos \alpha \cos \beta}{\cos \alpha \cos \beta} - \dfrac{\sin \alpha \sin \beta}{\cos \alpha \cos \beta}}$$

$$= \frac{\tan \alpha + \tan \beta}{1 - \tan \alpha \tan \beta}$$

Since $\tan [\alpha + (-\beta)]$ can be written as $\tan (\alpha - \beta)$, the Difference Identity for Tangent can be readily proved.

Difference Identity for Tangent

For all angle measures α and β for which the functions have meaning,

14. $\tan (\alpha - \beta) = \dfrac{\tan \alpha - \tan \beta}{1 + \tan \alpha \tan \beta}.$

Proof: $\tan(\alpha - \beta) = \tan[(\alpha + (-\beta)]$

$$= \frac{\tan\alpha + \tan(-\beta)}{1 - \tan\alpha\tan(-\beta)} \quad \longleftarrow \text{Sum Identity for Tangent}$$

$$= \frac{\tan\alpha - \tan\beta}{1 + \tan\alpha\tan\beta} \quad \longleftarrow \tan(-\beta) = -\tan\beta$$

As is the case with the sum identities for sine and cosine, the sum and difference identities for tangent can be used to evaluate the tangent of certain angles without using tables.

Example 1 Evaluate $\tan 105°$. Write the answer in radical form.

Solution: $\tan 105° = \tan(60° + 45°)$

$$= \frac{\tan 60° + \tan 45°}{1 - \tan 60° \tan 45°}$$

$$= \frac{\sqrt{3} + 1}{1 - \sqrt{3} \cdot 1}, \quad \text{or} \quad \frac{1 + \sqrt{3}}{1 - \sqrt{3}}$$

The answer to Example 1 may be simplified by rationalizing the denominator. To do this, multiply the fraction by the number 1 expressed in a form that will make the denominator a perfect square.

$$\frac{1 + \sqrt{3}}{1 - \sqrt{3}} \cdot \frac{1 + \sqrt{3}}{1 + \sqrt{3}} = \frac{(1 + \sqrt{3})^2}{1^2 - (\sqrt{3})^2}$$

$$= \frac{1 + 2\sqrt{3} + 3}{1 - 3}$$

$$= -2 - \sqrt{3}$$

Example 2 Evaluate $\tan 15°$. Write the radical in simplest form.

Solution: $\tan 15° = \tan(60° - 45°)$

$$= \frac{\tan 60° - \tan 45°}{1 + \tan 60° \tan 45°}$$

$$= \frac{\sqrt{3} - 1}{1 + \sqrt{3} \cdot 1}$$

$$= \frac{\sqrt{3} - 1}{\sqrt{3} + 1} \quad \longleftarrow \text{Rationalize the denominator.}$$

$$\tan 15° = \frac{\sqrt{3}-1}{\sqrt{3}+1} \cdot \frac{\sqrt{3}-1}{\sqrt{3}-1} \qquad \longleftarrow \qquad \frac{\sqrt{3}-1}{\sqrt{3}-1} = 1$$

$$= \frac{(\sqrt{3}-1)^2}{(\sqrt{3})^2 - 1^2}$$

$$= \frac{3 - 2\sqrt{3} + 1}{3 - 1}$$

$$= \frac{4 - 2\sqrt{3}}{2}$$

$$= \frac{\cancel{2}(2 - \sqrt{3})}{\cancel{2}}$$

$$= 2 - \sqrt{3} \qquad \longleftarrow \quad \text{Simplest form}$$

The sum and difference identities can be used to prove some of the reduction formulas for tangent.

Example 3   Prove: $\tan (2\pi + \theta) = \tan \theta$

Proof: $\tan (2\pi + \theta) = \dfrac{\tan 2\pi + \tan \theta}{1 - \tan 2\pi \tan \theta}$

$$= \frac{0 + \tan \theta}{1 - 0 \cdot \tan \theta}$$

$$= \frac{\tan \theta}{1},$$

$$= \tan \theta$$

Classroom Exercises _____

Evaluate without using tables. Use the Sum and Difference Identities for tangent. Write radicals in simplest form.

1. $\tan \left(-\frac{\pi}{12}\right)$ **2.** $\tan 135°$ **3.** $\tan \frac{5\pi}{12}$ **4.** $\tan 195°$

Written Exercises

1-4, 10, 12, 13, 18
16, 19, 20
1-2

a

Evaluate without using tables. Write radicals in simplest form.

1. $\tan \left(\frac{8\pi}{12} + \frac{3\pi}{12}\right)$ **2.** $\tan (300° - 45°)$ **3.** $\tan 75°$ **4.** $\tan \frac{17\pi}{12}$

5. $\tan 345°$ **6.** $\tan \frac{19\pi}{12}$ **7.** $\tan \frac{23\pi}{12}$ **8.** $\tan 285°$

Complete each reduction formula. Use the sum and difference identities for tangent.

9. $\tan (\pi - \theta) = \underline{?}$ **10.** $\tan (180° + \theta) = \underline{?}$

11. $\tan (-\theta) = \underline{?}$ **12.** $\tan (2\pi - \theta) = \underline{?}$

b

Without using tables, evaluate $\tan (\alpha + \beta)$ and $\tan (\alpha - \beta)$.

13. $\sin \alpha = \frac{3}{5}$, $\cos \beta = \frac{5}{13}$; $0 < \alpha < \frac{\pi}{2}, 0 < \beta < \frac{\pi}{2}$

14. $\tan \alpha = \frac{4}{3}$, $\cos \beta = \frac{12}{13}$; $0 < \alpha < \frac{\pi}{2}, 0 < \beta < \frac{\pi}{2}$

15. $\tan \alpha = \frac{3}{4}$, $\cos \beta = \frac{5}{13}$; $0 < \alpha < \frac{\pi}{2}, 0 < \beta < \frac{\pi}{2}$

16. $\cos \alpha = -\frac{3}{5}$, $\tan \beta = -\frac{5}{12}$; neither α nor β in Quadrant II

17. $\sin \alpha = -\frac{7}{25}$, $\cot \beta = -\frac{8}{15}$; neither α nor β in Quadrant IV

Express each of the following in terms of functions of θ.

18. $\tan (\theta + 30°)$ **19.** $\tan \left(\frac{\pi}{4} + \theta\right)$

20. $\tan (45° - \theta)$ **21.** $\tan (60° - \theta)$

22. $\tan \left(\theta - \frac{\pi}{3}\right)$ **23.** $\tan \left(\theta - \frac{\pi}{6}\right)$

Prove each identity in Exercises 24–25.

24. $\tan (\pi + \theta) = \tan \theta$ **25.** $\tan (3\pi - \theta) = -\tan \theta$

c

26. Derive a formula for $\cot (\alpha + \beta)$ in terms of $\cot \alpha$ and $\cot \beta$.

27. Derive a formula for $\cot (\alpha - \beta)$ in terms of $\cot \alpha$ and $\cot \beta$.

Review Capsule for Section 3-6

Complete each sum or difference formula. (For further review, see pages 133, 134, 138, and 142.)

1. $\sin (\alpha + \beta) = \underline{?}$ **2.** $\sin (\alpha - \beta) = \underline{?}$

3. $\cos (\alpha + \beta) = \underline{?}$ **4.** $\cos (\alpha - \beta) = \underline{?}$

5. $\tan (\alpha + \beta) = \underline{?}$ **6.** $\tan (\alpha - \beta) = \underline{?}$

3-6 Double-Angle Identities

By replacing β with α in the sum identities for $\sin (\alpha + \beta)$, $\cos (\alpha + \beta)$ and $\tan (\alpha + \beta)$, you can derive the Double-Angle Identities.

$$\sin (\alpha + \beta) = \sin (\alpha + \alpha) \quad \longleftarrow \quad \text{Let } \beta = \alpha.$$

$$\sin (\alpha + \alpha) = \sin \alpha \cos \alpha + \cos \alpha \sin \alpha$$

$$\sin 2\alpha = 2 \sin \alpha \cos \alpha$$

$$\cos (\alpha + \beta) = \cos (\alpha + \alpha) \quad \longleftarrow \quad \text{Let } \beta = \alpha.$$

$$\cos (\alpha + \alpha) = \cos \alpha \cos \alpha - \sin \alpha \sin \alpha$$

$$\cos 2\alpha = \cos^2 \alpha - \sin^2 \alpha$$

$$\tan (\alpha + \beta) = \tan (\alpha + \alpha) \quad \longleftarrow \quad \text{Let } \beta = \alpha.$$

$$\tan (\alpha + \alpha) = \frac{\tan \alpha + \tan \alpha}{1 - \tan \alpha \tan \alpha}$$

$$\tan 2\alpha = \frac{2 \tan \alpha}{1 - \tan^2 \alpha}, \text{ where } \tan \alpha \neq \pm 1$$

The following two alternate forms of the identity for $\cos 2\alpha$ can be derived by using the identity $\cos^2 \alpha + \sin^2 \alpha = 1$.

$$\cos 2\alpha = \cos^2 \alpha - \sin^2 \alpha$$

$$= \cos^2 \alpha - (1 - \cos^2 \alpha)$$

$$= 2 \cos^2 \alpha - 1 \quad \text{(first form)}$$

$$\cos 2\alpha = \cos^2 \alpha - \sin^2 \alpha$$

$$= (1 - \sin^2 \alpha) - \sin^2 \alpha$$

$$= 1 - 2 \sin^2 \alpha \quad \text{(second form)}$$

Double-Angle Identities

15. $\sin 2\alpha = 2 \sin \alpha \cos \alpha$

16. $\cos 2\alpha = \cos^2 \alpha - \sin^2 \alpha$

  $\cos 2\alpha = 2 \cos^2 \alpha - 1$

  $\cos 2\alpha = 1 - 2 \sin^2 \alpha$

17. $\tan 2\alpha = \dfrac{2 \tan \alpha}{1 - \tan^2 \alpha}$, where $\tan \alpha \neq \pm 1$

Example 1 If $\cos \alpha = \frac{12}{13}$, where α is in Quadrant IV, evaluate each of the following.

a. $\sin 2\alpha$ **b.** $\cos 2\alpha$ **c.** $\tan 2\alpha$

Solutions: Since $\sin 2\alpha = 2 \sin \alpha \cos \alpha$, find $\sin \alpha$.

In Quadrant IV, $\sin \alpha < 0$. Thus,

$\sin \alpha = -\sqrt{1 - \cos^2 \alpha}$ $\longleftarrow$ Identity **6**

$ = -\sqrt{1 - (\frac{12}{13})^2}$

$ = -\frac{5}{13}.$

a. $\sin 2\alpha = 2 \sin \alpha \cos \alpha$

$\sin 2\alpha = 2\left(-\frac{5}{13}\right)\left(\frac{12}{13}\right) = -\frac{120}{169}$

b. $\cos 2\alpha = 2 \cos^2 \alpha - 1$

$ = 2\left(\frac{12}{13}\right)^2 - 1 = \frac{119}{169}$

c. $\tan 2\alpha = \dfrac{\sin 2\alpha}{\cos 2\alpha}$ $\longleftarrow$ Since $\sin 2\alpha$ and $\cos 2\alpha$ are known, use Identity **4**.

$ = \dfrac{-\frac{120}{169}}{\frac{119}{169}} = -\frac{120}{119}$

The Double-Angle Identities can also be used to prove other identities.

Example 2 Prove: $\tan \alpha = \dfrac{1 - \cos 2\alpha}{\sin 2\alpha}$

Proof: Simplify the right side.

$\dfrac{1 - \cos 2\alpha}{\sin 2\alpha} = \dfrac{1 - (1 - 2 \sin^2 \alpha)}{2 \sin \alpha \cos \alpha}$ $\longleftarrow$ Identity **16**
$\phantom{\dfrac{1 - \cos 2\alpha}{\sin 2\alpha}}$ $\longleftarrow$ Identity **15**

$\phantom{\dfrac{1 - \cos 2\alpha}{\sin 2\alpha}} = \dfrac{2 \sin^2 \alpha}{2 \sin \alpha \cos \alpha}$ $\longleftarrow$ $\dfrac{2 \sin \alpha \sin \alpha}{2 \sin \alpha \cos \alpha}$

$\phantom{\dfrac{1 - \cos 2\alpha}{\sin 2\alpha}} = \dfrac{\sin \alpha}{\cos \alpha}$

$\phantom{\dfrac{1 - \cos 2\alpha}{\sin 2\alpha}} = \tan \alpha$

$\sin 2\alpha = 2\sin\alpha\cos\alpha$

$\tan 2\alpha = \dfrac{2\tan\alpha}{1-\tan^2\alpha}$

$\cos 2\alpha = 2\cos^2\alpha - 1$

$\cos 2\alpha = \cos^2\alpha - \sin^2\alpha$

$\cos^2\alpha = 1 - 2\sin^2\alpha$

Classroom Exercises

If $\sin\alpha = -\frac{3}{5}$, and α is in quadrant III, evaluate each of the following.

1. $\sin 2\alpha$ **2.** $\cos 2\alpha$ **3.** $\tan 2\alpha$

Written Exercises

If $\sin\alpha = \frac{4}{5}$ and α is in Quadrant I, evaluate each of the following.

1. $\sin 2\alpha$ **2.** $\cos 2\alpha$ **3.** $\tan 2\alpha$

If $\sin\alpha = \frac{2}{\sqrt{5}}$; $0 < \alpha < \frac{\pi}{2}$, evaluate each of the following.

4. $\cos 2\alpha$ **5.** $\sin 2\alpha$ **6.** $\tan 2\alpha$

If $\tan\alpha = -\frac{5}{12}$ and α is in Quadrant II, evaluate each of the following.

7. $\sin 2\alpha$ **8.** $\cos 2\alpha$ **9.** $\tan 2\alpha$

For the given values of $\sin\alpha$ and $\cos\alpha$, evaluate $\sin 2\alpha$, $\cos 2\alpha$, and $\tan 2\alpha$.

10. $\sin 21° = .3584$, $\cos 21° = .9336$

11. $\sin 1.23 = .9425$, $\cos 1.23 = .3342$

12. $\sin 4.5 = -.9775$, $\cos 4.5 = -.2108$

Prove each identity.

13. $\tan\alpha = \dfrac{\sin 2\alpha}{1 + \cos 2\alpha}$ **14.** $\cos 2\alpha = \dfrac{1 - \tan^2\alpha}{1 + \tan^2\alpha}$

15. $\sin 2\alpha = \dfrac{2\tan\alpha}{1 + \tan^2\alpha}$ **16.** $\tan\alpha = \csc 2\alpha - \cot 2\alpha$

17. $4\csc^2 2\alpha = \sec^2\alpha(1 + \cot^2\alpha)$ **18.** $\sec 2\alpha = \dfrac{1}{1 - 2\sin^2\alpha}$

19. $\dfrac{1 + \cos 2\alpha}{\sin 2\alpha} = \cot\alpha$ **20.** $\csc 2\alpha + \cot 2\alpha = \cot\alpha$

21. $\dfrac{\cot^2\alpha - 1}{\csc^2\alpha} = \cos 2\alpha$ **22.** $\dfrac{\sin 2\alpha}{\sin\alpha} - \dfrac{\cos 2\alpha}{\cos\alpha} = \sec\alpha$

23. $(\sin\alpha - \cos\alpha)^2 = \sec^2\alpha - \tan^2\alpha - \sin 2\alpha$

24. $\csc 2\alpha = \dfrac{\sec \alpha \csc \alpha}{2}$

25. $\sec 2\alpha = \dfrac{\sec^2 \alpha}{2 - \sec^2 \alpha}$

26. $\sec 2\alpha = \dfrac{\csc^2 \alpha}{\csc^2 \alpha - 2}$

27. $\sec 2\alpha = \dfrac{\sec^2 \alpha \csc^2 \alpha}{\csc^2 \alpha - \sec^2 \alpha}$

28. $\cot 2\alpha = \dfrac{1 + \cos 4\alpha}{\sin 4\alpha}$

29. $\sin 3\alpha = 3 \sin \alpha - 4 \sin^3 \alpha$

30. $\cos 3\alpha = 4 \cos^3 \alpha - 3 \cos \alpha$

31. $\tan 3\alpha = \dfrac{3 \tan \alpha - \tan^3 \alpha}{1 - 3 \tan^2 \alpha}$

32. Derive an identity for $\cot 2\alpha$ in terms of $\cot \alpha$.

33. A vertical tree casts a shadow of 10 m at one time and 3.2 m at a later time when the angle of the line of sight to the sun and the horizontal is twice as large. Find the height of the tree.

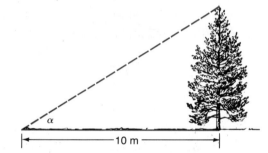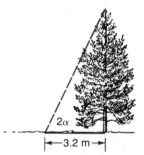

Prove each identity.

34. $\cos 4\alpha = 8 \cos^4 \alpha - 8 \cos^2 \alpha + 1$

35. $\sin 4\alpha = 4 \sin \alpha \cdot \cos \alpha (2 \cos^2 \alpha - 1)$

Review Capsule for Section 3-7 _____

Change each degree measure to radian measure in terms of π. (For further review, see pages 62–65.)

1. 60°   **2.** 360°   **3.** −135°   **4.** −810°

Change each radian measure to degree measure. (For further review, see pages 62–65.)

5. $\dfrac{3\pi}{2}$ **6.** $-\dfrac{11\pi}{6}$ **7.** $\dfrac{7\pi}{2}$ **8.** $-\dfrac{15\pi}{4}$

3-7 Half-Angle Identities

The Double-Angle Identity, $\cos 2\theta = \cos^2 \theta - \sin^2 \theta$, can be used to derive other identities.

$$\cos 2\theta = 1 - 2\sin^2 \theta \qquad\qquad \cos 2\theta = 2\cos^2 \theta - 1$$

$$2\sin^2 \theta = 1 - \cos 2\theta \qquad\qquad 2\cos^2 \theta = 1 + \cos 2\theta$$

$$\sin^2 \theta = \frac{1 - \cos 2\theta}{2} \qquad\qquad \cos^2 \theta = \frac{1 + \cos 2\theta}{2}$$

$$\sin \theta = \pm\sqrt{\frac{1 - \cos 2\theta}{2}} \quad (1) \qquad \cos \theta = \pm\sqrt{\frac{1 + \cos 2\theta}{2}} \quad (2)$$

Since $\tan \theta = \dfrac{\sin \theta}{\cos \theta}$, $\quad \tan \theta = \dfrac{\pm\sqrt{\frac{1 - \cos 2\theta}{2}}}{\pm\sqrt{\frac{1 + \cos 2\theta}{2}}}$

Thus, $\qquad\qquad\qquad \tan \theta = \pm\sqrt{\dfrac{1 - \cos 2\theta}{1 + \cos 2\theta}} \quad (3)$

If θ is replaced with $\frac{\alpha}{2}$ in equations (1), (2), and (3), the <u>Half-Angle Identities</u> result.

Half-Angle Identities

18. $\quad \sin \dfrac{\alpha}{2} = \pm\sqrt{\dfrac{1 - \cos \alpha}{2}}$

19. $\quad \cos \dfrac{\alpha}{2} = \pm\sqrt{\dfrac{1 + \cos \alpha}{2}}$

20. $\quad \tan \dfrac{\alpha}{2} = \pm\sqrt{\dfrac{1 - \cos \alpha}{1 + \cos \alpha}}$

NOTE: The sign of each Half-Angle Identity is determined by the quadrant in which $\frac{\alpha}{2}$ terminates.

Alternate forms of Identity **20** can be derived by:

1. Rationalizing the denominator:

$$\tan \frac{\alpha}{2} = \frac{\sin \alpha}{1 + \cos \alpha}, \text{ where } \cos \alpha \neq -1$$

2. Rationalizing the numerator:

$$\tan \frac{\alpha}{2} = \frac{1 - \cos \alpha}{\sin \alpha}, \text{ where } \sin \alpha \neq 0$$

You are asked to derive these two forms in the Exercises and to explain why the ± sign is not needed. The Half-Angle Identities can be used to evaluate the sine, cosine, and tangent of certain angles without using tables.

Example 1 Evaluate. Write radicals in simplest form.

a. $\sin \frac{\pi}{12}$

b. $\cos 255°$

Solutions: a. Since $\frac{\alpha}{2} = \frac{\pi}{12} = 15°$, the terminal side of $\frac{\alpha}{2}$ lies in Quadrant I. Thus,

$$\sin \frac{\pi}{12} = \sqrt{\frac{1 - \cos \frac{\pi}{6}}{2}}$$ ⟵ In Quadrant I, $\sin \frac{\alpha}{2} > 0$.

$$= \sqrt{\frac{1 - \frac{\sqrt{3}}{2}}{2}}$$

$$= \sqrt{\frac{\frac{2}{2} - \frac{\sqrt{3}}{2}}{2}}$$

$$= \sqrt{\frac{\frac{2 - \sqrt{3}}{2}}{2}}$$

$$= \sqrt{\frac{2 - \sqrt{3}}{4}}$$

$$= \frac{1}{2}\sqrt{2 - \sqrt{3}}$$

b. Since $\frac{\alpha}{2} = 255°$, the terminal side of $\frac{\alpha}{2}$ lies in Quadrant III and $\alpha = 510°$. Angle α is coterminal with an angle of 150° and the reference angle is $(180° - 150°)$, or 30°.

$$\cos 255° = -\sqrt{\frac{1 - \cos 30°}{2}}$$ ⟵ In Quadrant III, $\cos \frac{\alpha}{2} < 0$.

$$= -\sqrt{\frac{1 - \frac{\sqrt{3}}{2}}{2}}$$

$$= -\sqrt{\frac{\frac{2 - \sqrt{3}}{2}}{2}}$$

$$= -\sqrt{\frac{2 - \sqrt{3}}{4}}$$

$$= -\frac{1}{2}\sqrt{2 - \sqrt{3}}$$

Example 2

$P(-3, -4)$ is on the terminal side of an angle α in standard position, where $0 < \alpha < 2\pi$. Evaluate $\tan \frac{\alpha}{2}$. Simplify the answer.

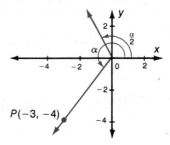

Solution: Since $x = -3$ and $y = -4$,

$$r = \sqrt{(-3)^2 + (-4)^2} = 5.$$

$$\therefore \cos \alpha = \frac{x}{r} = -\frac{3}{5}$$

Since α is in Quadrant III, $180° < \alpha < 270°$, and $90° < \frac{\alpha}{2} < 135°$. Thus, the terminal side of $\frac{\alpha}{2}$ lies in Quadrant II.

$$\tan \frac{\alpha}{2} = -\sqrt{\frac{1 - \cos \alpha}{1 + \cos \alpha}} \quad \longleftarrow \quad \text{In Quadrant II,} \\ \tan \frac{\alpha}{2} < 0.$$

$$= -\sqrt{\frac{1 - (-\frac{3}{5})}{1 + (-\frac{3}{5})}}$$

$$= -\sqrt{\frac{\frac{8}{5}}{\frac{2}{5}}} = -\sqrt{4}, \text{ or } -2$$

Classroom Exercises

If α is in the given quadrant, and if $0° < \alpha < 360°$, in what quadrant is $\frac{\alpha}{2}$?

1. Quadrant I   **2.** Quadrant II   **3.** Quadrant III   **4.** Quadrant IV

5. Given $\cos \alpha = \frac{3}{4}$, and $\frac{3\pi}{2} < \alpha < 2\pi$, evaluate $\cos \frac{\alpha}{2}$.

Written Exercises

a

Use the half-angle identities to evaluate each of the following. Write radicals in simplest form.

1. $\sin \frac{\pi}{8}$ **2.** $\cos \frac{\pi}{8}$ **3.** $\tan \frac{\pi}{8}$

4. $\sin 105°$ **5.** $\cos 105°$ **6.** $\tan 105°$

7. $\sin \frac{3\pi}{8}$ **8.** $\cos \frac{3\pi}{8}$ **9.** $\tan \frac{3\pi}{8}$

10. $\sin 165°$ **11.** $\cos 165°$ **12.** $\tan 165°$

13. $\sin \frac{5\pi}{24}$ **14.** $\cos \frac{3\pi}{16}$ **15.** $\tan \frac{\pi}{16}$

$P(-5, -12)$ is on the terminal side of an angle in standard position with measure α, where $0 < \alpha < 2\pi$. Evaluate each of the following. Write radicals in simplest form.

16. $\sin \alpha$ **17.** $\cos \alpha$ **18.** $\tan \alpha$

19. $\sin \frac{\alpha}{2}$ **20.** $\cos \frac{\alpha}{2}$ **21.** $\tan \frac{\alpha}{2}$

$P(3, -4)$ is on the terminal side of an angle in standard position with measure α, where $0 < \alpha < 2\pi$. Evaluate each of the following. Write radicals in simplest form.

22. $\cos \alpha$ **23.** $\sin \alpha$ **24.** $\tan \alpha$

25. $\cos \frac{\alpha}{2}$ **26.** $\sin \frac{\alpha}{2}$ **27.** $\tan \frac{\alpha}{2}$

28.–33. Repeat Exercises 16–21 for $2\pi < \alpha < 4\pi$.

34.–39. Repeat Exercises 22–27 for $2\pi < \alpha < 4\pi$.

In Exercises 40 and 41 begin with Identity **20,**

$$\tan \frac{\alpha}{2} = -\sqrt{\frac{1 - \cos \alpha}{1 + \cos \alpha}}$$

and derive the given identity. In each exercise explain why the $\pm$ sign is not needed.

40. $\tan \frac{\alpha}{2} = \dfrac{\sin \alpha}{1 + \cos \alpha}$ **41.** $\tan \frac{\alpha}{2} = \dfrac{1 - \cos \alpha}{\sin \alpha}$

Prove each identity.

42. $1 + \cos \alpha = 2 \cos^2 \frac{\alpha}{2}$ **43.** $\sin \alpha = 2 \sin \frac{\alpha}{2} \cos \frac{\alpha}{2}$

44. $\tan \frac{\alpha}{2} = \dfrac{\sec \alpha - 1}{\tan \alpha}$ **45.** $\sin \alpha = \dfrac{2 \tan \frac{\alpha}{2}}{1 + \tan^2 \frac{\alpha}{2}}$

46. $\sec \frac{\alpha}{2} = \pm \dfrac{\sqrt{2 + 2 \cos \alpha}}{1 + \cos \alpha}$ **47.** $\csc \frac{\alpha}{2} = \pm \dfrac{\sqrt{2 + 2 \cos \alpha}}{\sin \alpha}$

48. $\sin \alpha = \cot \frac{\alpha}{2}(1 - \cos \alpha)$ **49.** $\dfrac{1 - \tan \frac{\alpha}{2}}{1 + \tan \frac{\alpha}{2}} = \dfrac{1 - \sin \alpha}{\cos \alpha}$

50. $\tan \frac{\alpha}{2} + \cot \frac{\alpha}{2} = \sec \alpha (\cot \frac{\alpha}{2} - \tan \frac{\alpha}{2})$

51. Find the values of α for which $\cos \frac{\alpha}{2} = \frac{1}{2} \cos \alpha$

52. Find the values of α for which $\sin \frac{\alpha}{2} = \frac{1}{2} \sin \alpha$

Solve each system of equations for x and y.

1. $\begin{cases} x + y = 5 \\ x - y = 3 \end{cases}$ **2.** $\begin{cases} x + 2y = 0 \\ 2x - y = 5 \end{cases}$ **3.** $\begin{cases} 3x - 4y = 1 \\ 2x - 3y = -2 \end{cases}$

3-8 Sum/Product Identities

The <u>Sum/Product Identities</u> express sums or differences of two sine functions or two cosine functions as products. Conversely, the identities can also be used to express certain products as sums or differences.

Sum/Product Identities

21. $\sin(\alpha + \beta) + \sin(\alpha - \beta) = 2 \sin \alpha \cos \beta$

22. $\sin(\alpha + \beta) - \sin(\alpha - \beta) = 2 \cos \alpha \sin \beta$

23. $\cos(\alpha + \beta) + \cos(\alpha - \beta) = 2 \cos \alpha \cos \beta$

24. $\cos(\alpha + \beta) - \cos(\alpha - \beta) = -2 \sin \alpha \sin \beta$

The proofs of these identities use the Sum and Difference Identities for the sine and cosine.

Proof of Identity 21

$$\sin(\alpha + \beta) + \sin(\alpha - \beta) = (\sin \alpha \cos \beta + \cos \alpha \sin \beta) + (\sin \alpha \cos \beta - \cos \alpha \sin \beta)$$
$$= 2 \sin \alpha \cos \beta$$

The proofs of Identities **22–24** are similar. These proofs are asked for in the Exercises.

Example 1
Express $2 \cos 75° \cos 20°$ as a sum.

Solution: $2 \cos 75° \cos 20° = \cos(75° + 20°) + \cos(75° - 20°)$ ←——— Identity **23**
$$= \cos 95° + \cos 55°$$

Example 2 Express $\sin 76° + \sin 20°$ as a product.

Solution: First, find α and β such that $\alpha + \beta = 76°$ and $\alpha - \beta = 20°$.

$$\begin{cases} \alpha + \beta = 76° & (1) \\ \alpha - \beta = 20° & (2) \end{cases}$$ ← Add the left sides of the equations and the right sides of the equations.

$$2\alpha = 96°$$
$$\alpha = 48°$$
$$\beta = 28°$$ ← $48° + \beta = 76°$

Then, $\sin 76° + \sin 20° = 2 \sin 48° \cos 28°$ ← Identity **21**

Example 2 suggests an alternate form of the Sum/Product Identities.

Let $\alpha + \beta = s$ and $\alpha - \beta = t$.
Solve for α by addition as in Example 2. Then find β.

$$\alpha = \frac{s + t}{2} \qquad \beta = \frac{s - t}{2}$$

Now replace α with $\frac{s+t}{2}$ and β with $\frac{s-t}{2}$ in the Sum/Product Identities.

Sum/Product Identities (Alternate Form)

25. $\sin s + \sin t = 2 \sin \dfrac{s + t}{2} \cos \dfrac{s - t}{2}$

26. $\sin s - \sin t = 2 \cos \dfrac{s + t}{2} \sin \dfrac{s - t}{2}$

27. $\cos s + \cos t = 2 \cos \dfrac{s + t}{2} \cos \dfrac{s - t}{2}$

28. $\cos s - \cos t = -2 \sin \dfrac{s + t}{2} \sin \dfrac{s - t}{2}$

Examples 3 and 4 illustrate the use of this form.

Example 3 Express $\sin \dfrac{17\pi}{12} - \sin \dfrac{11\pi}{12}$ as a product.

Solution: Use Identity **26.**

$$\sin\frac{17\pi}{12} - \sin\frac{11\pi}{12} = 2\cos\frac{\frac{17\pi}{12}+\frac{11\pi}{12}}{2}\sin\frac{\frac{17\pi}{12}-\frac{11\pi}{12}}{2}$$

$$= 2\cos\frac{28\pi}{24}\sin\frac{6\pi}{24}$$

$$= 2\cos\frac{7\pi}{6}\sin\frac{\pi}{4}$$

Expressing sums and differences as products can simplify the proofs of identities.

Example 4

Prove: $\dfrac{\sin 5\theta + \sin 3\theta}{\cos 5\theta - \cos 3\theta} = \cot(-\theta)$

Proof: $\dfrac{\sin 5\theta + \sin 3\theta}{\cos 5\theta - \cos 3\theta} = \dfrac{2\sin 4\theta \cos\theta}{-2\sin 4\theta \sin\theta}$ ⟵ Identities **25** and **28**

$$= -\frac{\cos\theta}{\sin\theta}$$

$$= -\cot\theta, \text{ or } \cot(-\theta)$$

Classroom Exercises

1. Express $\sin\frac{\pi}{3}\cos\frac{\pi}{4}$ as a sum.

2. Express $\cos\frac{\pi}{3} - \cos\frac{\pi}{4}$ as a product.

Written Exercises

a

Express as a sum.

1. $2\cos 40°\cos 32°$

2. $2\sin 54°\sin 15°$

3. $2\sin\frac{\pi}{10}\cos\frac{\pi}{5}$

4. $2\cos 3\theta \sin 2\theta$

5. $2\sin 3x\cos x$

6. $2\cos 11t\sin 5t$

7. $2\cos 5x\cos 2x$

8. $-2\sin 7v\sin 5v$

9. $\sin 10x\sin 4x$

10. $\sin 15t\cos(-3t)$

11. $\cos 8x\sin 4x$

12. $\cos 3t\cos 5t$

13. $\cos 4x\cos 2x$

14. $\sin 3\theta\sin\theta$

15. $\cos 5\theta\ 7\theta$

Express as a product.

16. $\cos 51° - \cos 23°$

17. $\sin\frac{\pi}{8} + \sin\frac{\pi}{16}$

18. $\sin 131° - \sin 43°$

19. $\cos\frac{5\pi}{7} + \cos\frac{3\pi}{7}$

20. $\sin\frac{\pi}{3} + \sin\frac{\pi}{4}$

21. $\sin(x-\frac{\pi}{2}) - \sin(x+\frac{\pi}{2})$

22. $\cos\frac{1}{4} + \cos\frac{3}{4}$

23. $\cos(-3t) - \cos(5t)$

24. $\sin\frac{5\pi}{12} - \sin\frac{11\pi}{12}$

25. Prove Identity **22.**

26. Prove Identity **23.**

27. Prove Identity **24.**

Prove each identity.

28. $\dfrac{\cos 7t + \cos 5t}{\sin 7t - \sin 5t} = \dfrac{\csc t}{\sec t}$ (HINT: Use identities **27** and **26**.)

29. $\dfrac{\sin 3t + \sin t}{\sin 3t - \sin t} = \dfrac{2}{1 - \tan^2 t}$ (HINT: Use identities **25** and **26**.)

30. $\dfrac{\sin 4x + \sin 2x}{\cos 4x + \cos 2x} = \dfrac{1}{\cot 3x}$

31. $\dfrac{\sin 3x + \sin x}{\cos 3x - \cos x} = -\dfrac{\csc x}{\sec x}$

32. $\dfrac{\sin 3x - \sin x}{\cos 3x + \cos x} = \tan x$

33. $\dfrac{\cos 9x + \cos 5x}{\sin 9x - \sin 5x} = \dfrac{1 - \tan^2 x}{2 \tan x}$

34. $\dfrac{\sin \alpha - \sin \beta}{\sin \alpha + \sin \beta} = \dfrac{\tan \frac{\alpha - \beta}{2}}{\tan \frac{\alpha + \beta}{2}}$

35. $\dfrac{\sin 4x + \sin 2x}{\cos 4x + \cos 2x} = \tan 3x$

36. $\dfrac{\sin 5t - \sin 3t}{\sin 5t + \sin 3t} = \dfrac{\tan t}{\tan 4t}$

37. $\dfrac{\sin 2t - \sin 8t}{\cos 8t - \cos 2t} = \cot 5t$

Snell's Law

When a ray of light strikes a flat surface such as a pool of water, part of the light is reflected from the surface. The rest of the light enters the water but the direction of the ray's path in the water bends away from the direction of the original pathway. This bending occurs because the speed of light is different in air and in water. The figure below shows that three angles are formed by the broken path and the dashed line perpendicular to the pool's surface at the point where the light ray enters the water. These angles are the **angle of incidence** ϕ (phi), the **angle of reflection** θ, and the **angle of refraction** ϕ' (phi prime).

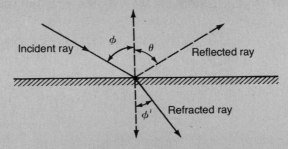

The measures of the angle of incidence and the angle of reflection are equal. That is, $\phi = \theta$. The relation between the measures of the angles of incidence and refraction is expressed by the following formula, known as **Snell's Law.**

$$\sin \phi = k \sin \phi'$$

The constant k is determined by the nature of the two substances (in this case air and water) through which the light ray travels. Values of k for several substances (where the light ray enters the substance from the air) are given below,

Substance	k
Water	1.33
Ethyl alcohol	1.36
Rock salt	1.54
Quartz	1.54
Glass	1.46–1.96
Diamond	2.42

Example: A ray of light strikes one of the faces of a diamond. The angle of incidence is 30°. Find, to the nearest degree, the angle of refraction ϕ'.

Solution: Use Snell's Law.

$$\sin \phi = k \sin \phi'$$

$\phi = 30°; k = 2.42$
(See the table)

$$\sin 30° = 2.42 \sin \phi'$$

$$\sin \phi' = \frac{\sin 30°}{2.42}$$

$$\sin \phi' = \frac{.5}{2.42} = .2066$$

From the table on pages 368– 372 or use a calculator.

$$\phi' = 12° \text{ (nearest degree)}$$

Light ray 30°

EXERCISES

1. A ray of light is shone on the surface of some ethyl alcohol. The angle of incidence is 60°. Find the angle of refraction.

2. A ray of light strikes a glass surface. The angle of incidence is 45° and the value of k is 1.6. Find the angle of reflection and angle of refraction.

3. A ray of light is reflected off the surface of a crystal of rock salt. The angle between the incident ray and the reflected ray is 50°. Find the angle of refraction.

4. A ray of light strikes the surface of a quartz crystal and is reflected in the quartz to form an angle of refraction of 24°. Find the angle of incidence.

5. A beam of light is directed onto the flat surface of a stationary, hard, transparent substance that is known to be either quartz or glass. The direction of the light beam is permitted to vary and, for each position of the light beam, a measurement is taken of the angle of incidence and the angle of refraction. These measurements are recorded in a table.

ϕ	15°	30°	40°	60°	80°
ϕ'	9.7°	16.1°	21.2°	28.6°	33.2°

Is the substance quartz or glass? (HINT: It is suspected that one of the measurements was recorded incorrectly.)

Trigonometric Identities **157**

3-9 Review of Identities

To prove an identity means to prove that the solution set consists of all values of the variable(s) for which the two sides of the equation are defined. Although there are no standardized methods for proving identities applying the strategies suggested in the chapter plus ingenuity and perseverance will help to produce the desired results.

Written Exercises

a

Prove each identity

1. $\sin^2 x(\csc^2 x - 1) = \cos^2 x$

2. $\csc^2 x(1 - \cos^2 x) = 1$

3. $\sec^2 x + \csc^2 x = \sec^2 x \csc^2 x$

4. $\tan x + \cot x = \sec x \csc x$

5. $\dfrac{\sec x}{\cot x + \tan x} = \sin x$

6. $\sec^2 x(1 - \sin^2 x) = 1$

7. $\dfrac{\cos x - \cos 2x}{\sin x + \sin 2x} = \tan \dfrac{x}{2}$

8. $\dfrac{\cos 2x - \cos 2y}{\cos x - \cos y} = 2\cos x + 2\cos y$

9. $\sin 2x \tan x = 2 \sin^2 x$

10. $\cos x \tan x \csc x = 1$

11. $\tan^2 x - \sin^2 x = \sin^2 x \tan^2 x$

12. $\csc^4 x - \cot^4 x = 1 + 2 \cot^2 x$

13. $(\tan x + \cot x)^2 = \sec^2 x + \csc^2 x$

14. $\sin x \tan x = \sec x - \cos x$

15. $\csc x \cot x \cos x + 1 = \csc^2 x$

16. $\sin x + \cot x \cos x = \csc x$

17. $\dfrac{\cos x + \sin x}{\sec x + \csc x} = \cos x \sin x$

18. $\dfrac{\cot x \cos x}{\cot x - \cos x} = \dfrac{\cos x}{1 - \sin x}$

19. $\dfrac{\sin x + \tan x}{\csc x + \cot x} = \sin x \tan x$

20. $\dfrac{\cot x + 1}{\sin x + \cos x} = \csc x$

21. $\sin^2 x \sec^2 x + \sin^2 x \csc^2 x = \sec^2 x$

22. $\cot^4 x + \cot^2 x = \csc^2 x(\csc^2 x - 1)$

23. $\tan^2 x - \cot^2 x = \sec^2 x - \csc^2 x$

24. $\dfrac{1}{1 + \sin x} + \dfrac{1}{1 - \sin x} = 2 \sec^2 x$

25. $\cot x - \sec x \csc x(1 - 2 \sin^2 x) = \tan x$

26. $\dfrac{1}{\csc x - \cot x} - \dfrac{1}{\csc x + \cot x} = 2\cot x$ **27.** $\dfrac{1 + \cot x}{\csc x} = \dfrac{1 + \tan x}{\sec x}$

28. $\dfrac{\cot^2 x - 1}{\csc^2 x} = \cos^2 x - \sin^2 x$ **29.** $\dfrac{2\tan x}{1 + \tan^2 x} = 2\sin x \cos x$

30. $\dfrac{\sin x}{1 - \cos x} = \dfrac{1 + \cos x}{\sin x}$ **31.** $\dfrac{1 - 2\cos^2 x}{\sin x \cos x} = \tan x - \cot x$

32. $\dfrac{\sin x \cos x}{\cos^2 x - \sin^2 x} = \dfrac{\tan x}{1 - \tan^2 x}$ **33.** $\dfrac{3\cos x + \cos 3x}{3\sin x - \sin 3x} = \cot^3 x$

34. $\dfrac{\sin 3x}{\sin x} - \dfrac{\cos 3x}{\cos x} = 2$ **35.** $\dfrac{\sin x}{1 + \sin x} = \tan \dfrac{x}{2}$

36. $\sin x + \sin 3x = 2\cos x \sin 2x$ **37.** $2\cos^2 \dfrac{x}{2} - \cos x = 1$

38. $\cos 4x = 8\cos^4 x - 8\cos^2 x + 1$ **39.** $\cos 4x = \dfrac{2 - \sec^2 x}{\sec^2 2x}$

40. $\cos 3x = 4\cos^3 x - 3\cos x$ **41.** $\sin 3x = 3\sin x - 4\sin^3 x$

42. $\tan 3x = \dfrac{3\tan x - \tan^3 x}{1 - 3\tan^2 x}$ **43.** $\csc^2 \dfrac{x}{2} - 1 = \dfrac{1 + \cos x}{1 - \cos x}$

44. $\sin^2 \left(\dfrac{x}{2}\right) = \dfrac{\tan x - \sin x}{2\tan x}$ **45.** $\tan x - \tan y = \sec x \sec y \sin(x - y)$

46. $\tan \dfrac{x}{2} = \dfrac{\tan x}{1 + \sec x}$ **47.** $\dfrac{\sin 5x - \sin 3x}{\cos 5x - \cos 3x} = -\cot 4x$

48. $\dfrac{\sin 6x - \sin 2x}{\sin 6x + \sin 2x} = \dfrac{\tan 2x}{\tan 4x}$ **49.** $\dfrac{\cos 2x - \cos 6x}{\cos 2x + \cos 6x} = \dfrac{\tan 4x}{\cot 2x}$

50. $\dfrac{\cos 2x - \cos x}{\cos 2x + \cos x} = \tan \dfrac{3x}{2} \tan \dfrac{x}{2}$ **51.** $\sin(x + y)\sin(x - y) = \sin^2 x - \sin^2 y$

52. $(a \sin x + b \cos x)^2 + (b \sin x - a \cos x)^2 = a^2 + b^2$

53. $\dfrac{2\sin x \cos x}{1 + \cos^2 x - \sin^2 x} = \tan x$

54. $\sin 4x - 4\sin x \cos^3 x - 4\cos x \sin^3 x$

55. $\dfrac{\sin x}{1 + \cos x} = \tan \dfrac{x}{2}$

56. $\dfrac{\sin x + \sin y}{\cos x + \cos y} = \tan \left(\dfrac{x + y}{2}\right)$

57. $\cos^2 3x - \sin^2 3x = \cos 6x$

58. $\cos^4 6x + 2\sin^2 6x \cos^2 6x + \sin^4 6x = 1$

59. $\dfrac{\sin 3x + \sin x}{\sin 6x - \sin 2x} = \cos x \sec 4x$

60. $\sin x + 2\sin 3x + \sin 5x = 4\cos^2 x \sin 3x$

61. Express $\sin(x + y + z)$ in terms of sines and cosines of x, y, and z.

Chapter Objectives and Review

Objective: To know the meanings of the mathematical terms in this chapter.

1. Be sure that you know the meanings of these mathematical terms.

Difference Identity
  for Cosine (p. 132)
  for Sine (p. 137)
  for Tangent (p. 141)
Double-Angle Identities (p. 145)
Fundamental Identities (p. 126)
Half-Angle Identities (p. 149)
Pythagorean Identities (p. 125)

Ratio Identities (p. 125)
Reciprocal Identities (p. 125)
Sum Identity
  for Cosine (p. 133)
  for Sine (p. 137)
  For Tangent (p. 141)
Sum/Product Identities (p. 153)
trigonometric identity (p. 124)

Objective: To prove the Fundamental Identities. (Section 3–1)

Use the definitions of the trigonometric functions to prove each identity.

2. $\tan \theta = \dfrac{\sin \theta}{\cos \theta}$

3. $\sin \theta = \dfrac{1}{\csc \theta}$

4. $\sin^2 \theta + \cos^2 \theta = 1$

5. $\cot \theta = \dfrac{1}{\tan \theta}$

Objective: To prove trigonometric identities using the Fundamental Identities. (Section 3–2)

Prove each identity.

6. $\cot \theta = \dfrac{\csc \theta}{\sec \theta}$

7. $\tan^2 \theta = \sec^2 \theta - \sin^2 \theta - \cos^2 \theta$

8. $\cos \theta \cdot \tan \theta = \dfrac{1}{\csc \theta}$

9. $\sin^2 \theta + \dfrac{1}{\sec^2 \theta} = \tan \theta \cot \theta$

10. $\dfrac{\sin \theta}{2 \csc \theta}\left(\tan^2 \theta + \sin^2 \theta + \dfrac{\sin^2 \theta}{\tan^2 \theta}\right) = \dfrac{1}{2} \tan^2 \theta$

11. $2 \sin^2 x - 1 = 1 - 2 \cos^2 x$

Objective: To apply the Sum and Difference Identities for cosine. (Section 3–3)

12. Evaluate $\cos 105°$ without using tables.

13. Evaluate $\cos 195°$ without using tables.

14. Prove: $\cos (3\pi + \theta) = -\cos \theta$

15. Prove: $\cos(\theta - \pi) = -\cos\theta$

16. If $\sin\alpha = \frac{4}{5}$ and $\cos\beta = \frac{12}{13}$, evaluate $\cos(\alpha - \beta)$. Assume that $0 < \alpha < \frac{\pi}{2}$ and $0 < \beta < \frac{\pi}{2}$.

Objective: To apply the Sum and Difference Identities for sine. (Section 3–4)

17. Evaluate $\sin 75°$ without using tables.

18. Evaluate $\sin 165°$ without using tables.

19. Prove: $\sin(180° - \theta) = \sin\theta$

20. Prove: $\sin(2\pi + \theta) = \sin\theta$

21. If $\sin\alpha = \frac{4}{5}$ and $\sec\beta = \frac{13}{12}$, evaluate $\sin(\alpha - \beta)$. Assume that $0 < \alpha < \frac{\pi}{2}$ and $0 < \beta < \frac{\pi}{2}$.

Objective: To apply the Sum and Difference Identities for tangent. (Section 3–5)

22. Evaluate $\tan 75°$ without using the tables.

23. Prove: $\tan(2\pi - \theta) = -\tan\theta$ **24.** Prove: $\tan(3\pi + \theta) = \tan\theta$

Objective: To apply the Double-Angle Identities. (Section 3–6)

If $\cos\alpha = \frac{4}{5}$ and α is in Quadrant I, evaluate each of the following.

25. $\sin 2\alpha$ **26.** $\cos 2\alpha$ **27.** $\tan 2\alpha$

28. Prove: $\csc 2\beta(1 - \cos 2\beta) = \tan\beta$

Objective: To apply the Half-Angle Identities. (Section 3–7)

29. Use the Half-Angle Identities to evaluate $\cos\frac{\pi}{12}$. Write radicals in simplest form.

30. The point $P(-3, 4)$ is on the terminal side of an angle α in standard position, where $0 < \alpha < \pi$. Evaluate $\tan\frac{\alpha}{2}$. Write radicals in simplest form.

31. Prove: $\sin\alpha = 2\sin\frac{\alpha}{2}\cos\frac{\alpha}{2}$

Objective: To apply the Sum/Product Identities. (Section 3–8)

32. Express $2\cos 55°\cos 21°$ as a sum.   **33.** Express $\sin\frac{\pi}{5} + \sin\frac{\pi}{10}$ as a product.

34. Express $\cos 53° - \cos 77°$ as a product.   **35.** Prove: $\dfrac{\sin 4\alpha + \sin 2\alpha}{\cos 4\alpha + \cos 2\alpha} = \dfrac{1}{\cot 3\alpha}$

Objective: To review trigonometric identities (Section 3–9)

Prove each identity.

36. $-\cos^2\theta = \dfrac{\cos^2 2\theta - 1}{4\sin^2\theta}$

37. $\left(\dfrac{1 + \tan\theta}{1 - \tan\theta}\right)^2 = \dfrac{1 + \sin 2\theta}{1 - \sin 2\theta}$

Chapter Test

Prove each identity.

1. $\cot = \dfrac{\cos \theta}{\sin \theta}$

2. $\dfrac{\tan \theta}{\sec \theta} = \sin \theta$

3. $\cot \theta = \cos \theta \cdot \csc \theta$

4. $\csc^2 \theta \tan^2 \theta - 1 = \tan^2 \theta$

5. $\dfrac{\tan \theta}{1 - \cos^2 \theta} = \sec \theta \cdot \csc \theta$

6. $\csc \theta + \cot \theta = \dfrac{1 + \cos \theta}{\sin \theta}$

7. $\dfrac{2 \tan \theta}{1 + \tan^2 \theta} = 2 \sin \theta \cos \theta$

8. $1 - \cos 2\beta = \sin 2\beta \cdot \tan \beta$

Evaluate without using tables.

9. $\cos 75°$

10. $\sin 105°$

11. $\sin \dfrac{\pi}{12}$

12. $\tan 195°$

13. Given that $\sin \alpha = \frac{3}{5}$ and $\cos \beta = \frac{5}{13}$, evaluate $\cos (\alpha + \beta)$ and $\sin (\alpha - \beta)$. Assume that $0 < \alpha < \frac{\pi}{2}$ and $0 < \beta < \frac{\pi}{2}$.

14. Prove the formula $\cos (270° + \theta) = \sin \theta$.

15. Given $\sin \alpha = .3584$ and $\cos \alpha = .9336$, evaluate $\sin 2\alpha$, $\cos 2\alpha$, and $\tan 2\alpha$.

If $\sin \alpha = \frac{8}{17}$ and α is in Quadrant I, evaluate each of the following.

16. $\cos 2\alpha$

17. $\tan \dfrac{\alpha}{2}$

18. Express $\cos \dfrac{\pi}{5} + \cos \dfrac{\pi}{10}$ as a product.

19. Express $2 \cos 75° \cos 20°$ as a sum.

20. Prove that $\sin (\alpha + \beta) - \sin (\alpha - \beta) = 2 \cos \alpha \sin \beta$.

Chapter 4
Trigonometry and Triangles

4-1 Law of Sines

Recall that solving a triangle means finding the measures of the unknown sides and angles of the triangle. In Chapter 1, right triangles were solved by using the Pythagorean Theorem and the definitions of the trigonometric functions. These techniques can be extended to the solution of any triangle.

Example 1

Find, to the nearest meter, the width of Perch Lake from point A to point B. The length of $\overline{AC}$, or b, equals 110 meters, and the measures of the angles of the triangle are as shown.

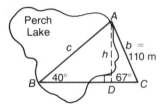

Solution: From point A, construct $\overline{AD}$ perpendicular to $\overline{BC}$.
In right triangle ADB,

$$\sin B = \frac{h}{c}, \text{ or } h = c \sin B.$$

In right triangle ADC,

$$\sin C = \frac{h}{b}, \text{ or } h = b \sin C.$$

Thus,
$$b \sin C = c \sin B$$

or
$$\frac{b}{\sin B} = \frac{c}{\sin C}.$$

Since the measures of $\angle B$, $\angle C$, and b are known, this equation can be used to calculate c, the distance from A to B.

$$\frac{110}{\sin 40°} = \frac{c}{\sin 67°}, \text{ or, } c = \frac{110 \cdot \sin 67°}{\sin 40°}$$

Thus,
$$c = \frac{110(.9205)}{.6428} \quad \longleftarrow \quad \text{Use the table or a calculator to evaluate } \sin \theta.$$

$$c = 158 \text{ meters} \quad \longleftarrow \quad \text{To the nearest meter}$$

Example 1 suggests the following theorem.

Theorem 4–1: Law of Sines

In any triangle, the ratio of the length of a side to the sine of the angle opposite that side is the same for each side-angle pair. That is, in any $\triangle ABC$,

$$\frac{a}{\sin A} = \frac{b}{\sin B} = \frac{c}{\sin C}.$$

$\frac{c}{\sin C} = \frac{b}{\sin B}$

To prove the Law of Sines, consider three cases.

Case I: Acute triangles

Case II: Obtuse triangles

Case III: Right triangles

Case I	**Case II**
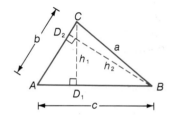	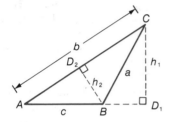

Proof: Each angle of $\triangle ABC$ is an acute angle. The measures of the altitudes from C and B are h_1 and h_2, respectively.

In right triangles ACD_1 and BCD_1, $\frac{h_1}{b} = \sin A$ and $\frac{h_1}{a} = \sin B$.

Thus, $\frac{b}{\sin B} = \frac{a}{\sin A}.$

In right triangles ABD_2 and CBD_2, $\frac{h_2}{c} = \sin A$ and $\frac{h_2}{a} = \sin C$.

Thus, $\frac{a}{\sin A} = \frac{c}{\sin C}.$

Therefore, $\frac{a}{\sin A} = \frac{b}{\sin B} = \frac{c}{\sin C}.$

In $\triangle ABC$, $\angle B$ is obtuse. The measures of the altitudes from C and B are h_1 and h_2, respectively.

In right triangles ACD_1 and BCD_1, $\frac{h_1}{b} = \sin A$ and $\frac{h_1}{a} = \sin (180° - B)$. Since $\sin (180° - B) = \sin B$, it follows that

$$\frac{b}{\sin B} = \frac{a}{\sin A}.$$

In right triangles ABD_2 and CBD_2, $\frac{h_2}{c} = \sin A$ and $\frac{h_2}{a} = \sin C$.

Thus, $\frac{a}{\sin A} = \frac{c}{\sin C}.$

Therefore, $\frac{a}{\sin A} = \frac{b}{\sin B} = \frac{c}{\sin C}.$

You are asked to prove Case III in the Exercises.

Example 2

In $\triangle ABC$, $b = 200$ cm, $C = 120°$, and $A = 34°$. Find BC and AB to the nearest centimeter.

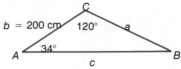

Solution: To apply the Law of Sines, you need to know the measure of the angle opposite the side whose length is given. Therefore, find the measure of $\angle B$.

$$B = 180° - (120° + 34°) = 26°$$

Now use the Law of Sines.

$$\frac{200}{\sin 26°} = \frac{a}{\sin 34°} \qquad \longleftarrow \qquad \frac{b}{\sin B} = \frac{a}{\sin A}$$

or $\qquad a = \frac{\sin 34° \cdot 200}{\sin 26°} \qquad \longleftarrow$ Use the table or a calculator.

$\qquad\qquad a = 255$ cm $\qquad \longleftarrow$ To the nearest centimeter

Similarly, $c = \dfrac{200 \cdot \sin 120°}{\sin 26°}$

$\qquad\qquad c = 395$ cm $\qquad \longleftarrow$ To the nearest centimeter

Classroom Exercises

In $\triangle ABC$, $B = 40°$, $C = 45°$ and $a = 10$. Give lengths to the nearest tenth.

1. Find A.   **2.** Find b.   **3.** Find c.

Written Exercises

In Exercises 1–10, solve $\triangle ABC$. Give lengths to the nearest tenth. Give measures of angles to the nearest 10 minutes.

1. $A = 71°$, $B = 42°$, $c = 15$

2. $A = 71°$, $a = 20$, $C = 62°$

3. $B = 41°$, $C = 130°$, $a = 10$

4. $C = 50°$, $B = 40°$, $c = 25$

5. $A = 90°$, $B = 30°$, $b = 12$

6. $a = 12$, $B = 110°$, $C = 35°$

7. $A = 41°$, $B = 75°$, $a = 13$

8. $B = 115°$, $A = 18°$, $a = 4$

9. $C = 37°10'$, $B = 42°20'$, $b = 32$

10. $B = 74°54'$, $C = 47°38'$, $a = 400$

11. Can you use the Law of Sines to solve $\triangle ABC$ if $A = 40°$, $b = 10$, $c = 20$? Explain.

12. Prove Theorem 4–1 for the case in which $C = 90°$.

13. Prove that for any triangle ABC, $\dfrac{a-b}{b} = \dfrac{\sin A - \sin B}{\sin B}$.

14. Use the Law of Sines to show that for any $\triangle ABC$, $a \geq b \sin A$.

In Exercises 15–18, find distances to the nearest unit.

15. Two lighthouses at points A and B are 40 kilometers apart. Each has visual contact with a freighter at point C. If $m \angle CAB = 20°30'$ and $m \angle CBA = 115°$, how far is the freighter from A? See the figure at the left below.

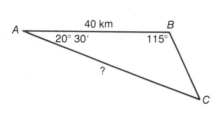
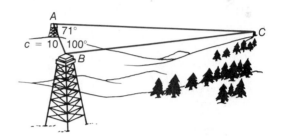

16. Two ranger stations located 10 kilometers apart receive a distress call from a camper. Electronic equipment allows them to determine that the camper is at an angle of 71° from the first station and 100° from the second, each angle having as one side the line segment connecting the stations. Which station is closer to the camper? How far away is it? See the figure at the right above.

17. What is the length of a side of a regular octagon if a diagonal is 15 centimeters long? See the figure at the left below.

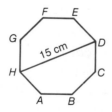
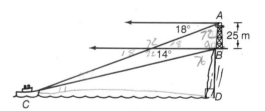

18. A tower 25 meters high stands at the top of a cliff. The lines of sight from a ship at point C to A and B make angles of 18° and 14°, respectively, with the horizontal. Find BD, the height of the cliff. See the figure at the right above.

Prove that each identity is true for any triangle ABC.

19. $\dfrac{a-b}{a+b} = \dfrac{\sin A - \sin B}{\sin A + \sin B}$

20. $\dfrac{a+b}{a-b} = \dfrac{\tan \frac{1}{2}(A+B)}{\tan \frac{1}{2}(A-B)}$

Review Capsule for Section 4-2

Solve each equation for the variable indicated.

Example: $3 = a^2 + b^2 - 2abd$; for d

Solution: $3 = a^2 + b^2 - 2abd$

$$2abd = a^2 + b^2 - 3 \quad \text{and} \quad d = \frac{a^2 + b^2 - 3}{2ab}$$

1. $7 = 4 + b^2 - 2bc$; for c

2. $a = b - 3cd$; for b

3. $x = 2 + b^2 + 3d$; for d

4. $x + y = a^2 + b^2$; for b

5. $a^2 = b^2 + c^2 - 2bcx$; for x

6. $c^2 = a^2 + b^2 - 2ab \cos C$; for $\cos C$

4-2 Law of Cosines

The Law of Sines <u>cannot</u> be used to solve a triangle when only two sides and the included angle are known or when only three sides are known. In these cases, you need the <u>Law of Cosines</u>.

Theorem 4-2: Law of Cosines

In any triangle, the square of a side is equal to the sum of the squares of the other sides minus twice the product of these sides and the cosine of the included angle. That is, in $\triangle ABC$,

$$a^2 = b^2 + c^2 - 2bc \cos A$$
$$b^2 = a^2 + c^2 - 2ac \cos B$$
$$c^2 = a^2 + b^2 - 2ab \cos C.$$

Proof: Given the lengths of the two sides, a and b, of $\triangle ABC$ and the measure of the included angle, θ, find the length of the third side, c.

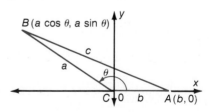

Place $\triangle ABC$ in the xy plane such that C is at the origin and side CA lies along the positive x axis. Since the length of side CA equals b, the coordinates of A are $(b, 0)$. Let (x, y) represent the coordinates of B. Then, since $\overrightarrow{CB}$ is the terminal side of an angle θ in standard position and B is a point on the terminal side of θ,

$$\cos \theta = \frac{x}{a} \qquad \text{and} \qquad \sin \theta = \frac{y}{a},$$

or $\qquad\qquad\qquad x = a \cos \theta \qquad \text{and} \qquad y = a \sin \theta.$

Thus, the coordinates of B are $(a \cos \theta, a \sin \theta)$.
Now, use the distance formula to find c.

$c = \sqrt{(b - a \cos \theta)^2 + (0 - a \sin \theta)^2}$ ◄——— Square both sides.

$c^2 = (b - a \cos \theta)^2 + (0 - a \sin \theta)^2$

$c^2 = b^2 - 2ab \cos \theta + a^2 \cos^2 \theta + a^2 \sin^2 \theta$

$c^2 = b^2 - 2ab \cos \theta + a^2(\cos^2 \theta + \sin^2 \theta)$ ◄——— $\cos^2 \theta + \sin^2 \theta = 1$

$c^2 = a^2 + b^2 - 2ab \cos \theta$, or   ◄——— Replace θ with C.

$c^2 = a^2 + b^2 - 2ab \cos C$

Although the proof of the Law of Cosines is shown for an angle θ with terminal side in Quadrant II, it applies to any angle θ where $0° < \theta < 180°$. By reorienting $\triangle ABC$ and following a procedure similar to the one shown above, you can prove the other two forms of the Law of Cosines.

Example 1 In $\triangle ABC$, $A = 40°$, $b = 10$, and $c = 20$. Find a to the nearest tenth.

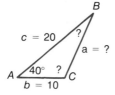

Solution: Since two sides and the included angle are given, use the Law of Cosines to find a.

$a^2 = b^2 + c^2 - 2bc \cos A$

$\quad = (10)^2 + (20)^2 - 2 \cdot 10 \cdot 20 \cdot \cos 40°$

$\quad = 100 + 400 - 400(.7660)$

$\quad = 500 - 306.4$

$\quad = 193.6$

$a = \sqrt{193.6}$, or about 13.9

Sometimes a triangle is solved by using first the Law of Cosines and then the Law of Sines.

Example 2

Find, to the nearest ten minutes, the measures of $\angle B$ and $\angle C$ in the triangle of Example 1.

Solution: Use the Law of Sines to find the measure of $\angle B$.

$$\sin B = \frac{b \sin A}{a}$$

$$= \frac{10 \sin 40°}{13.9} = \frac{10(.6428)}{13.9} = .4624$$

Since B is opposite side AC, the shortest side of $\triangle ABC$, it cannot be an obtuse angle.

$\therefore B$ is about $27°30'$. Finally, $C = 180° - (A + B)$

$$= 180° - (40° + 27°30') = 112°30'.$$

You can use the Law of Cosines to solve a triangle when the measures of three sides of the triangle are known.

Example 3

In $\triangle ABC$, $a = 14$, $b = 11$, and $c = 21$. Solve the triangle.

Solution:

$$a^2 = b^2 + c^2 - 2bc \cos A \quad \longleftarrow \quad \text{Solve for } \cos A.$$

$$\cos A = \frac{b^2 + c^2 - a^2}{2bc}$$

$$\cos A = \frac{11^2 + 21^2 - 14^2}{2 \cdot 11 \cdot 21} = .7922 \quad \longleftarrow \quad \begin{array}{l}\text{Since } \cos A \text{ is positive,} \\ A \text{ is in Quadrant I.}\end{array}$$

$$\therefore A = 37°40' \quad \longleftarrow \quad \begin{array}{l}\text{To the nearest ten} \\ \text{minutes}\end{array}$$

Next, find B. Use the Law of Sines.

$$\frac{\sin B}{b} = \frac{\sin A}{a} \quad \longleftarrow \quad \text{Solve for } \sin B.$$

$$\sin B = b\frac{\sin A}{a}$$

$$= 11\left(\frac{\sin 37°40'}{14}\right) = 11\left(\frac{.6111}{14}\right) = .4802$$

$\therefore B = 28°40'$, to the nearest ten minutes.

Finally, $C = 180° - (A + B) = 180° - (37°40' + 28°40')$

$\therefore C = 113°40'$, to the nearest ten minutes

Classroom Exercises

1. The sides of a triangle measure 8, 9, and 13. Without tables determine whether the largest angle is obtuse.

2. In $\triangle ABC$, $a = 6$, $b = 4$, and $c = 5$. Find the measures of the angles to the nearest degree.

Written Exercises

a

In Exercises 1–16, solve $\triangle ABC$. Give lengths of sides to the nearest tenth. Give measures of angles to the nearest ten minutes.

1. $a = 5$, $b = 8$, $C = 40°$
2. $b = 7$, $c = 10$, $A = 51°$
3. $a = 10$, $c = 15$, $B = 171°$
4. $b = 9$, $c = 11$, $A = 123°$
5. $a = 8$, $b = 6$, $C = 60°$
6. $a = 9$, $c = 5$, $B = 120°$
7. $a = 3$, $b = 7$, $c = 5$
8. $a = 1$, $b = 2$, $c = 2$
9. $a = 5$, $b = 4$, $c = 3$
10. $a = 10$, $b = 12$, $c = 14$
11. $a = 2$, $b = 3$, $c = 4$
12. $a = 4$, $b = 2$, $c = 5$
13. $a = 5$, $b = 17$, $C = 39°$
14. $a = 7$, $b = 8$, $c = 3$
15. $a = 12$, $b = 9$, $c = 16$
16. $A = 81°40'$, $b = 5$, $c = 6$

17. Why is the formula $c^2 = a^2 + b^2 - 2ab \cos C$ equivalent to the Pythagorean Theorem if $C = 90°$?

b

In Exercises 18–22 give lengths to the nearest unit. Give measures of angles to the nearest ten minutes.

18. A triangular lot has sides of 215 m, 185 m and 125 m. Find the measures of the angles at its corners.

19. From point C both ends A and B of a proposed railroad tunnel are visible. If $AC = 165$ m, $BC = 115$ m, and $C = 74°$, find AB, the length of the tunnel.

20. The distances from a boat B to two points A and C on the shore are known to be 100 m and 80 m, respectively, and $\angle ABC = 55°$. Find AC.

21. The radius of a circle is 20 cm, and two radii OX and OY form an angle which measures 115°. How long is the chord XY?

22. Two sides and a diagonal of a parallelogram measure 7, 9, and 15, respectively. Find the measures of the angles of the parallelogram.

C

23. Show that in $\triangle ABC$

$$1 + \cos C = \frac{(a+b+c)(a+b-c)}{2ab}.$$

24. Show that in $\triangle ABC$

$$1 - \cos C = \frac{(-a+b+c)(a-b+c)}{2ab}.$$

25. Prove that the sum of the squares of the diagonals of a parallelogram is equal to the sum of the squares of the lengths of the four sides.

26. In the figure at the right, points A and B are inaccessible, and AC, BC, AD, and BD also cannot be measured. If $CD = 100$ m, $\alpha = 33°$, $\beta = 42°$, γ (gamma) $= \alpha + \beta$, $\theta = 37°$, and ϕ (phi) $= 78°$, find AB to the nearest meter. (HINT: First use $\triangle BCD$ to find BC, and $\triangle ACD$ to find AC.)

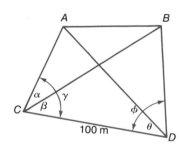

 ————————— *Law of Sines/Law of Cosines* —————

The Examples below show how to use a scientific calculator to solve for a, b, or c in the Law of Sines or the Law of Cosines.

Example 1 $c = \dfrac{110 \sin 67°}{\sin 40°}$; $c = $ _?_ (See Example 1, page 164.)

Solution:

  $\boxed{101.25553}$

$\boxed{4}\ \boxed{0}\ \boxed{\sin}\ \boxed{=}$ $\boxed{157.52565}$

Example 2 $a = \sqrt{10^2 + 20^2 - 2 \cdot 10 \cdot 20 \cos 40°}$; $a = $ _?_ (See Example 1, page 169.)

Solution:

$\boxed{1}\ \boxed{0}\ \boxed{x^2}\ \boxed{+}\ \boxed{2}\ \boxed{0}\ \boxed{x^2}\ \boxed{-}$ $\boxed{500.}$

$\boxed{2}\ \boxed{\times}\ \boxed{1}\ \boxed{0}\ \boxed{\times}\ \boxed{2}\ \boxed{0}\ \boxed{\times}$ $\boxed{400.}$

$\boxed{4}\ \boxed{0}\ \boxed{\cos}\ \boxed{=}\ \boxed{\sqrt{x}}$ $\boxed{13.913383}$

4-3 The Ambiguous Case

To apply either the Law of Sines or the Law of Cosines to solving a triangle, three parts of the triangle must be known.

Given	Model	Procedure
1. Three sides	$b = 8$, $a = 5$, $c = 7$ (triangle ABC)	Use the Law of Cosines.
2. Two sides and the included angle	$c = 4$, $61°$ at B, $a = 6$ (triangle ABC)	Use the Law of Cosines.
3. Two angles and any side	$70°$ at B, $a = 12$, $31°$ at A (triangle ABC)	Use the Law of Sines.
4. Two sides and an angle opposite one side	$b = 6$, $a = 9$, $62°$ at A (triangle ABC)	Use the Law of Sines (Ambiguous Case).

The last case in the table is called the **ambiguous case** because there may be no triangle, one triangle, or two triangles satisfying the given conditions.

The diagrams below illustrate the various possibilities, given *A*, *b*, and *a*. They are divided into two cases:

$$A < 90° \quad \text{and} \quad A \geq 90°.$$

Case I: $A < 90°$

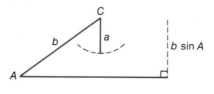

No solution: $a < b \sin A$

One solution: $a = b \sin A$

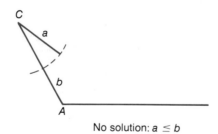

Two solutions: $a < b$
and $a > b \sin A$

One solution: $a \geq b$

Case II: $A \geq 90°$

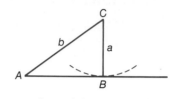

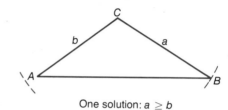

No solution: $a \leq b$

One solution: $a > b$

The Law of Sines can be used to solve a triangle for which the data are ambiguous. However, it is important to sketch and label the triangle first. Then determine the number of possible solutions.

Example 1 Find the number of solutions for each triangle.

a. $A = 30°$
$a = 6$
$b = 12$

$b \sin A$
$12 (\sin 30°)$
$12 (\frac{1}{2}) = 6$
$b \sin A = a$
1 solution

b. $A = 30°$
$a = 8$
$b = 12$

c. $A = 30°$
$a = 4$
$b = 12$

Solutions: Sketch each triangle.

a.

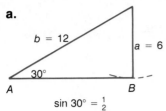

$\sin 30° = \frac{1}{2}$
$\therefore a = b \sin A$
One Solution

b.

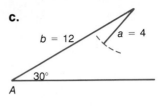

$a < b$ and $a > b \sin A$
Two Solutions

c.

$b = 12$ $a = 4$

$30°$

A

$a < b \sin A$
No Solutions

When there are two solutions, you must solve <u>both</u> triangles.

Example 2   In $\triangle ABC$, $A = 30°$, $a = 15$, and $b = 20$.

a. Find B to the nearest ten minutes.

b. Find c to the nearest tenth.

Solutions: a. Since $\sin 30° = \frac{1}{2}$ and $b \sin 30° = 10$,
$b \sin 30° < a < b$ ⟵ $10 < 15 < 20$
Thus, there are two solutions, corresponding to two triangles AB_1C and AB_2C.
First, solve $\triangle AB_1C$.

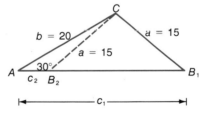

$$\frac{a}{\sin 30°} = \frac{b}{\sin B_1}$$

$$\frac{15}{\frac{1}{2}} = \frac{20}{\sin B_1}$$

$$\sin B_1 = \frac{10}{15} = \frac{2}{3}$$

B_1 is about $41°50'$.
$\therefore m \angle ACB_1$ is about $180° - (30° + 41°50')$, or $108°10'$.
b. Now find c_1, where $c_1 = AB_1$.

$$\frac{c_1}{\sin (108°10')} = \frac{15}{\sin 30°}$$

$$c_1 = \frac{\sin (108°10') \cdot 15}{\sin 30°}$$

$$= \sin (71°50') \cdot 30$$

$$= (.9502) \cdot 30$$

$$= 28.5 \quad ⟵ \quad \text{To the nearest tenth}$$

a. $\triangle B_1CB_2$ is isosceles. (Refer to the figure below.) Thus, its base angles have equal measures. In $\triangle AB_2C$,

$$m \angle AB_2C = 180° - m \angle CB_2B_1 \quad \longleftarrow \quad \begin{array}{l} m \angle CB_2B_1 = \\ m \angle AB_1C = 41°50' \end{array}$$
$$= 180° - 41°50'$$
$$= 138°10'.$$

Thus, $m \angle ACB_2 = 180° - (30° + 138°10')$
$$= 11°50'.$$

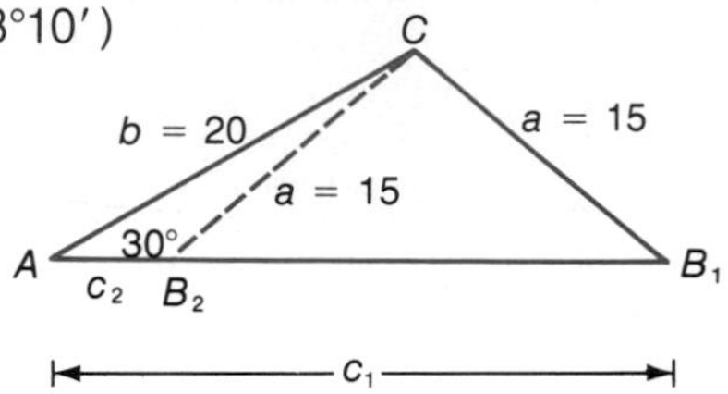

b. Now find c_2, where $c_2 = AB_2$.

$$\frac{c_2}{\sin ACB_2} = \frac{a}{\sin A}$$

$$c_2 = \frac{\sin(11°50') \cdot 15}{\sin 30°}$$

$$= (.2051)(30)$$

$$= 6.2 \quad \longleftarrow \quad \text{To the nearest tenth}$$

Example 3

In $\triangle ABC$, $C = 121°$, $c = 40$ and $b = 30$. Find B.

Solution: Since $C = 121°$, $C > 90°$. Also, $c > b$. Thus, there is one solution.

$$\frac{\sin B}{30} = \frac{\sin 121°}{40} \quad \longleftarrow \quad \sin 121° = \sin(180° - 59°) = \sin 59°$$

$$\sin B = \frac{30}{40} \sin 59°$$

$$\sin B = .6429 \quad \longleftarrow \quad \begin{array}{l} B \text{ equals either } 40° \text{ or } 140°. \\ \text{Since } C > 90°, B \neq 140°. \end{array}$$

$$B = 40°$$

Classroom Exercises

Find the number of solutions for the given data.

1. $A = 35°$, $a = 10$, $b = 20$

2. $A = 35°$, $a = 10$, $b = 17$

3. $A = 35°$, $a = 10$, $b = 9$

4. $A = 120°$, $a = 10$, $b = 10$

Written Exercises

2-32 evens
31-33

a

Find the number of solutions for the given data. Do not solve.

1. $A = 42°$, $a = 5$, $b = 10$

2. $A = 42°$, $a = 5$, $b = 7$

3. $A = 42°$, $a = 5$, $b = 4$

4. $A = 73°$, $a = 8$, $b = 9$

5. $A = 73°$, $a = 8$, $b = 8$ id   ?,?

6. $A = 73°$, $a = 8$, $b = 3$

7. $A = 93°$, $a = 4$, $b = 8$

8. $A = 93°$, $a = 4$, $b = 2$

9. $A = 173°$, $a = 9$, $b = 9.1$

10. $A = 173°$, $a = 9$, $b = 8.9$

11. $a = 2$, $b = 3$, $c = 6$

12. $C = 17°$, $a = 10$, $c = 11$

13. $B = 71°$, $a = 5$, $c = 275$

14. $A = 20°$, $a = 7$, $b = 29$

15. $A = 41°$, $B = 196°$, $a = 10$

16. $C = 20°$, $b = 10$, $c = 4$

17. $A = 60°$, $b = 2$, $a = \sqrt{3}$

18. $A = 90°$, $a = 20$, $b = 19$

19. $B = 140°$, $a = 3$, $b = 2$

20. $C = 120°$, $b = 14$, $c = 13$

21. $A = 105°$, $b = 13$, $C = 80°$

22. $a = 12$, $b = 6$, $c = 4$

23. $a = 12$, $c = 19$, $B = 71°$

24. $a = 6$, $b = 9$, $A = 27°$

25. $a = 4$, $b = 10$, $A = 50°$

26. $a = 17$, $b = 12$, $c = 26$

27. $A = 55°$, $B = 21°$, $c = 1$

28. $A = 30°$, $a = 4$, $b = 8$

Solve each triangle. If there are two solutions, give both of them. Express lengths to the nearest tenth and measures of angles to the nearest degree.

29. $a = 9$, $b = 7$, $C = 62°$

30. $A = 43°$, $B = 97°$, $a = 7$

31. $A = 41°$, $b = 13$, $a = 15$

32. $A = 41°$, $b = 13$, $a = 11$

33. $A = 30°$, $a = 14$, $b = 28$

34. $b = 10$, $c = 13$, $A = 120°$

35. $a = 4$, $b = 9$, $c = 10$

36. $A = 19°$, $C = 41°$, $c = 20$

37. $B = 32°$, $a = 10$, $b = 7$

38. $A = 115°$, $a = 5$, $b = 4$

b

39. The distance from the earth E to the sun S is approximately 1.5×10^8 kilometers. The distance from Venus to the sun is about 1.1×10^8 kilometers. The angle observed on the earth between Venus and the sun is 20°. Find the distance from the earth to Venus. (HINT: There are two possible positions for Venus, indicated by V_1 and V_2.)

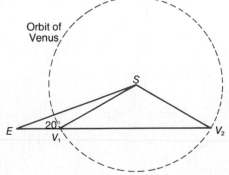

The Sun and Trigonometry

The amount of sunlight received by any given spot on the Earth's surface varies with the time of year. Let H represent the amount of light energy received from the sun on a one-centimeter-square patch when the sun is directly overhead. Then $H \sin \theta$ represents the amount of light energy received by a patch of the same area at a spot where the sun makes an angle of θ degrees with the horizon (see the figure at the left below).

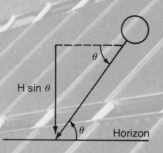

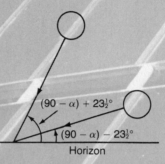

Since the Earth is slightly tilted with respect to the plane of its orbit about the sun, the extent of this tilt can be measured by the angle formed by this plane and the plane of the Earth's equator. The measure of this angle is about $23\frac{1}{2}°$. Thus, at a given point in the Northern Hemisphere, the angle of inclination of the sun at noon on the longest day of the year (around June 21) equals $(90 - \alpha) + 23\frac{1}{2}°$ where α is the angle of latitude (see pages 224–225). In the winter, the angle of inclination at noon on the shortest day of the year (around December 22) equals $(90 - \alpha) - 23\frac{1}{2}°$ where $\alpha \geq 23\frac{1}{2}°$ (see the figure at the right above).

see pages 224–225

EXERCISES

1. Find the amount of sunlight received at noon by a one-centimeter-square patch of earth at a latitude of 49° in the Northern Hemisphere on June 21 and on December 22.

2. Repeat Exercise 1 for a latitude of 60.5° in the Northern Hemisphere.

3. Repeat Exercise 1 for a latitude of 47.5° in the Northern Hemisphere.

4. For Exercise 1, write a ratio to compare the amount of light energy received on December 22 to that received on June 21. Express the ratio as a per cent.

4-4 Law of Tangents

The solution of triangles by the Law of Cosines involves a number of operations—squaring, adding, multiplying, and subtracting. The amount of computation can be decreased for the case of two sides and the included angle by using the <u>Law of Tangents</u>.

Theorem 4–3: Law of Tangents

In any triangle ABC,

$$\frac{a-b}{a+b} = \frac{\tan\left(\frac{A-B}{2}\right)}{\tan\left(\frac{A+B}{2}\right)} \qquad \frac{c-a}{c+a} = \frac{\tan\left(\frac{C-A}{2}\right)}{\tan\left(\frac{C+A}{2}\right)} \qquad \frac{b-c}{b+c} = \frac{\tan\left(\frac{B-C}{2}\right)}{\tan\left(\frac{B+C}{2}\right)}$$

The proof of the Law of Tangents uses the Law of Sines and the Sum/Product Identities (Section 3–8).

Proof: By the Law of Sines, $\dfrac{a}{\sin A} = \dfrac{b}{\sin B}$. Represent each side of this equation by k. Then

$$\frac{a}{\sin A} = k \text{ and } \frac{b}{\sin B} = k.$$

Thus, $a = k \sin A$	**1**	⟵ Multiplying by $\sin A$
and $b = k \sin B$	**2**	⟵ Multiplying by $\sin B$
$a - b = k(\sin A - \sin B)$	**3**	⟵ Subtracting the corresponding sides of equations **1** and **2**
$a + b = k(\sin A + \sin B)$	**4**	⟵ Adding the corresponding sides of equations **1** and **2**

$$\frac{a-b}{a+b} = \frac{\sin A - \sin B}{\sin A + \sin B} \qquad \longleftarrow \text{Dividing the corresponding sides of equations } \mathbf{3} \text{ and } \mathbf{4}$$

$$= \frac{2 \cos\left(\frac{A+B}{2}\right) \sin\left(\frac{A-B}{2}\right)}{2 \sin\left(\frac{A+B}{2}\right) \cos\left(\frac{A-B}{2}\right)} \qquad \longleftarrow \text{Sum/Product Identities}$$

$$= \cot\left(\frac{A+B}{2}\right) \tan\left(\frac{A-B}{2}\right)$$

$$= \frac{\tan\left(\frac{A-B}{2}\right)}{\tan\left(\frac{A+B}{2}\right)}$$

Similar proofs can be written for the other two forms of the Law of Tangents.

Example

In $\triangle ABC$, $a = 321$, $b = 234$, and $C = 71°$. Find A and B to the nearest 10 minutes. Find c to the nearest tenth.

Solution:

$$\frac{a - b}{a + b} = \frac{\tan \left(\frac{A - B}{2}\right)}{\tan \left(\frac{A + B}{2}\right)} \longleftarrow \quad A + B = 180° - C = 180° - 71°$$

$$\frac{321 - 234}{321 + 234} = \frac{\tan \left(\frac{A - B}{2}\right)}{\tan \left(\frac{109°}{2}\right)}$$

$$\frac{87}{555} = \frac{\tan \left(\frac{A - B}{2}\right)}{\tan 54°30'} \longleftarrow \quad \text{Solve for } \tan \left(\frac{A - B}{2}\right).$$

$$\tan \left(\frac{A - B}{2}\right) = \frac{87}{555} \cdot 1.4019 \longleftarrow \quad \tan 54°30' = 1.4019$$

$$= .2198$$

Thus, $\quad \dfrac{A - B}{2} = 12°20' \qquad \longleftarrow \quad$ To the nearest ten minutes

$$A - B = 24°40' \qquad \textbf{1}$$

But $\qquad A + B = 109° \qquad \textbf{2}$

Thus, $\qquad 2A = 133°40' \qquad \longleftarrow \quad$ Adding corresponding sides of equations **1** and **2**

$$A = 66°50' \qquad \longleftarrow \quad$$ To the nearest ten minutes

$$2B = 84°20' \qquad \longleftarrow \quad$$ Subtracting corresponding sides of equations **1** and **2**

$$B = 42°10' \qquad \longleftarrow \quad$$ To the nearest ten minutes

$$c = 321 \left(\frac{\sin 71°}{\sin 66°50'}\right) \qquad \longleftarrow \quad$$ Using the Law of Sines

$$c = 330.1 \qquad \longleftarrow \quad$$ To the nearest tenth

Classroom Exercises

In $\triangle ABC$, $b = 81$ meters, $c = 12$ meters, and $A = 21°$.

1. Find B.   **2.** Find C.   **3.** Find a.

Written Exercises

Use the Law of Tangents and the given data to solve $\triangle ABC$. Express lengths to the nearest tenth, and measures of angles to the nearest ten minutes.

1. $a = 64$, $b = 50$, $C = 72°$ **2.** $a = 672$, $b = 463$, $C = 59°$

3. $b = 18$, $c = 12$, $A = 123°$ **4.** $c = 381$, $a = 256$, $B = 106°$

5. $a = 218$, $b = 135$, $C = 117°$ **6.** $c = 157$, $a = 112$, $B = 42°$

7. $b = 38$, $c = 24$, $A = 12°$ **8.** $c = 20$, $a = 14$, $B = 32°$

9. $b = 8$, $c = 5$, $A = 83°$ **10.** $a = 11$, $c = 19$, $B = 83°$

11. The lengths of two sides of a triangle are 470 meters and 320 meters, and the measure of their included angle is 100°. Find the perimeter to the nearest meter.

12. The lengths of two sides of a parallelogram are 12 centimeters and 23 centimeters. The length of the longer diagonal is 30 centimeters. Find the length of the shorter diagonal to the nearest centimeter.

13. Two streets meet at an angle of 98°. At their intersection, a triangular lot has a frontage of 41.8 meters on one street and 57.3 meters on the other. Find the angles the third side makes with the two streets, and find the length of the third side. Give angles to the nearest degree and length to the nearest meter.

Review

In Exercises 1 and 2, use the Law of Sines to solve △ABC. Give measures of angles to the nearest degree and lengths to the nearest tenth. (Section 4–1)

1. $A = 47°$, $B = 70°$, $c = 5$ **2.** $B = 24°$, $C = 123°$, $a = 7$

In Exercises 3 and 4, use the Law of Cosines to solve △ABC. Give measures of angles to the nearest degree, and lengths to the nearest tenth. (Section 4–2)

3. $a = 5$, $b = 8$, $C = 120°$ **4.** $a = 24$, $c = 16$, $B = 38°$

In Exercises 5–8, find the number of solutions for the given data. Do not solve. (Section 4–3)

5. $A = 53°$, $a = 5$, $b = 6$ **6.** $A = 70°$, $a = 5$, $b = 6$

7. $A = 70°$, $a = 6$, $b = 5$ **8.** $A = 120°$, $a = 6$, $b = 5$

In Exercises 9 and 10, solve △ABC using the Law of Sines or the Law of Cosines, or both. If there are two solutions, give both solutions. Give measures of angles to the nearest degree, and lengths to the nearest tenth. (Section 4–3)

9. $B = 100°$, $b = 10$, $c = 6$ **10.** $A = 45°$, $a = 8$, $b = 10$

In Exercises 11 and 12, use the Law of Tangents to solve △ABC. Give measures of angles to the nearest degree and lengths to the nearest tenth. (Section 4–4)

11. $a = 10$, $b = 15$, $C = 55°$ **12.** $b = 23$, $c = 41$, $A = 130°$

4-5 The Area of a Triangle

Recall that if K represents the area of a triangle, then

$$K = \tfrac{1}{2}bh$$

where b is the measure of one side of the triangle and h is the altitude to that side. This formula may be used to obtain other formulas for the area of a triangle, given two sides and the sine of their included angle.

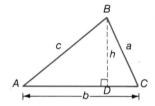 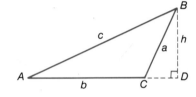

Given: b, a, C $(C < 90°)$ **Given:** b, a, C $(C > 90°)$

In each triangle above, $h = a \sin C$. Thus,

$$K = \tfrac{1}{2}ab \sin C.$$

Similarly it can be shown that

$$K = \tfrac{1}{2}bc \sin A$$

and *area SAS* $K = \tfrac{1}{2}ac \sin B.$

Example 1 In $\triangle ABC$, $B = 40°$, $a = 5$ cm, $c = 10$ cm. Find the area to the nearest square centimeter.

Solution: $K = \tfrac{1}{2}ac \sin B$

$K = \tfrac{1}{2} \cdot 5 \cdot 10 \sin 40°$

$ = \tfrac{1}{2} \cdot 5 \cdot 10 \cdot (.6428)$

$ = \tfrac{1}{2} \cdot 5 \cdot 6.428$

$ = 16.070$

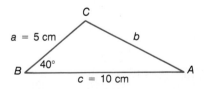

The area is about 16 cm². ⟵——— Read "cm²" as square centimeters.

When you know the measures of one side and at least two of the angles of a triangle, you can use the Law of Sines and the appropriate area formula to find the area of the triangle.

Example 2 In △ ABC, a = 10 m, A = 85°, and B = 60°. Find the area of △ ABC to the nearest square unit.

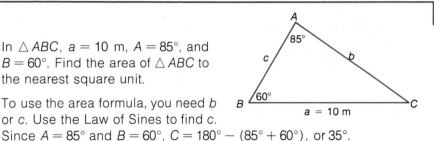

Solution: To use the area formula, you need b or c. Use the Law of Sines to find c.

Since A = 85° and B = 60°, C = 180° − (85° + 60°), or 35°.

Then, $\dfrac{a}{\sin A} = \dfrac{c}{\sin C}$ ⟵ Law of Sines

$\dfrac{10}{\sin 85°} = \dfrac{c}{\sin 35°}$ ⟵ Solve for c.

$c = \dfrac{10(.5736)}{(.9962)}$, or about 5.8 m

Now find the area. $K = \dfrac{1}{2} ac \sin 60°$

$= \dfrac{1}{2}(10)(5.8)(.8660)$

$= 25.11$, or about 25 m²

Classroom Exercises

1. In △ ABC, b = 281, c = 358 and A = 32°20′. Find K to the nearest square unit.

2. In △ ABC, A = 29°40′, B = 78°50′, and a = 69.7 m. Find K to the nearest square unit.

Written Exercises

In Exercises 1–10, find the area of triangle ABC to the nearest square unit.

1. a = 37 cm, b = 48 cm, C = 52°   **2.** b = 28 m, c = 42 m, A = 126°

3. a = 42, b = 50, C = 21°   **4.** a = 71 cm, b = 23 cm, C = 84°

5. a = 58 km, A = 35°, B = 67°   **6.** b = 36, C = 108°, B = 29°

7. a = 63 cm, C = 47°, A = 89°   **8.** a = 39 m, c = 63 m, B = 104°

9. b = 421 cm, A = 49°, B = 65°   **10.** b = 376, c = 538 km, A = 73°

Use the area formulas and the Law of Sines to prove each formula.

11. $K = \dfrac{c^2 \sin A \sin B}{2 \sin C}$

12. $K = \dfrac{a^2 \sin B \sin C}{2 \sin A}$

13. $K = \dfrac{b^2 \sin A \sin C}{2 \sin B}$

14. The diagonals of quadrilateral $EFGH$ intersect at X. $EX = GX = 52$ cm, $FX = HX = 64$ cm, and m $\angle GXH = 43°$. Find the area of $EFGH$ to the nearest square centimeter.

15. Find, to the nearest square meter, the area of quadrilateral $ABCD$. See the figure at the left below.

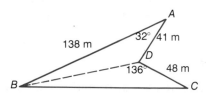
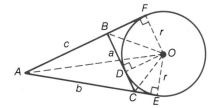

16. Given: $\triangle ABC$, circle O tangent to the sides of $\triangle ABC$ at D, E, and F. Find an expression for the radius r of circle O in terms of the area K, the semi-perimeter s, and the lengths a, b, and c of the sides of $\triangle ABC$. See the figure at the right above.

Prove that each of the following is true for any triangle ABC.

17. $a \cos B + b \cos A = c$

18. $a \cos B - b \cos A = \dfrac{a^2 - b^2}{c}$ (HINT: Use the equation in Exercise 17.)

19. $K = \dfrac{1}{2} a^2 \dfrac{\sin B \sin C}{\sin (B + C)}$

20. A plane region bounded by a central angle of a circle and the arc that the angle intercepts is a **sector of a circle.** Given a circle with radius r and a central angle of radian measure x, derive a formula for the area K of a sector in terms of r and x.

21. Find the area of the sector defined by a circle of radius 2 and a central angle of 45°.

22. A **segment of a circle** is a region bounded by an arc and the chord joining the endpoints of the arc. Derive a formula for the area, K, of segment ABP in terms of x, the radian measure of $\angle AOB$, and the radius r.
(HINT: $K = $ area of sector $AOB - $ area of $\triangle AOB$.)

23. Find the area of segment APB if $OB = 4$ and the measure of $\angle AOB$ is 120°.

4-6 Heron's Formula

When the lengths of three sides of a triangle are known, the area can be found in two ways.

1. Use the Law of Cosines to find the measure of one angle of the triangle. Then apply the appropriate formula from Section 4–5.
2. Use Heron's Formula. This usually involves less calculation.

Theorem 4-4   Heron's Formula

In triangle ABC with sides of lengths a, b, and c, the area K is given by the formula

$$K = \sqrt{s(s-a)(s-b)(s-c)}, \text{ where } s = \frac{a+b+c}{2}.$$

Proof: $K = \frac{1}{2}bc \sin A$ ⟵ Square both sides.

$K^2 = \frac{1}{4}b^2c^2 \sin^2 A$ ⟵ Replace $\sin^2 A$ with $1 - \cos^2 A$.

$K^2 = \frac{1}{4}b^2c^2(1 - \cos^2 A)$ ⟵ Factor $1 - \cos^2 A$.

$K^2 = \frac{1}{4}b^2c^2(1 + \cos A)(1 - \cos A)$

SSS

From the Law of Cosines, $\cos A = \dfrac{b^2 + c^2 - a^2}{2bc}$.

Then, $1 + \cos A = 1 + \dfrac{b^2 + c^2 - a^2}{2bc} = \dfrac{2bc + b^2 + c^2 - a^2}{2bc}$

$$= \frac{(b+c)^2 - a^2}{2bc} \quad \longleftarrow \text{ Factor.}$$

$$= \frac{(b+c+a)(b+c-a)}{2bc}.$$

Also, $1 - \cos A = 1 - \dfrac{b^2 + c^2 - a^2}{2bc} = \dfrac{2bc - b^2 - c^2 + a^2}{2bc}$

$$= \frac{a^2 - (b-c)^2}{2bc} \quad \longleftarrow \text{ Factor.}$$

$$= \frac{(a+b-c)(a-b+c)}{2bc}.$$

Thus, $K^2 = \frac{1}{4}b^2c^2 \cdot \dfrac{(b+c+a)(b+c-a)}{2bc} \cdot \dfrac{(a+b-c)(a-b+c)}{2bc}$

$$= \tfrac{1}{16}(a+b+c)(b+c-a)(a+b-c)(a+c-b).$$

Next, let $2s = a + b + c$. ⟵ The perimeter is $2s$.

or, $a + b + c = 2s$

Then, $b + c - a = 2(s - a)$, ⟵ Subtract $2a$ from both sides.

$a + b - c = 2(s - c)$, ⟵ Subtract $2c$ from both sides.

and $a + c - b = 2(s - b)$. ⟵ Subtract $2b$ from both sides.

Thus, $K^2 = \frac{1}{16} \cdot 2s \cdot 2(s - a) \cdot 2(s - b) \cdot 2(s - c)$

$K^2 = s(s - a)(s - b)(s - c)$. ⟵ Take the square root.

$\therefore K = \sqrt{s(s - a)(s - b)(s - c)}$ ⟵ s is the semi-perimeter.

Example 1 Triangle ABC has sides with lengths 5 cm, 8 cm, and 11 cm. Find the area to the nearest square centimeter.

Solution: $K = \sqrt{s(s - a)(s - b)(s - c)}$ ⟵ $s = \frac{5 + 8 + 11}{2} = 12$

$= \sqrt{12(12 - 5)(12 - 8)(12 - 11)}$

$= \sqrt{12(7)(4)(1)}$

$= 4\sqrt{21}$, or about 18 cm²

Heron's Formula is sometimes used to find the height of a triangle.

Example 2 In $\triangle ABC$, $a = 3$, $c = 5$, and $b = 6$. Find, to the nearest tenth, the height h from B to $\overline{AC}$.

Solution: Sketch the figure.
Then use Heron's Formula.

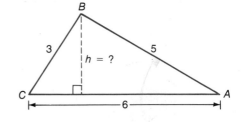

$K = \sqrt{s(s - a)(s - b)(s - c)}$

$K = \sqrt{7(4)(1)(2)}$

$= \sqrt{56}$

$= 2\sqrt{14}$

But $K = \frac{1}{2}(6)h$ ⟵ Area $= \frac{1}{2}bh$

$= 3h$.

$\therefore 3h = 2\sqrt{14}$ ⟵ Substitution

$h = \frac{2}{3}\sqrt{14}$, or about 2.5

Classroom Exercises

Find, to the nearest tenth of a square unit, the area of △ABC in each case.

1. $a = 2$, $b = 4$, $c = 4$

2. $a = 8$, $b = 8$, $c = 8$

Written Exercises

a

Use Heron's formula to find the area of each triangle ABC to the nearest square unit.

1. $a = 5$, $b = 6$, $c = 7$

2. $a = 21$, $b = 28$, $c = 11$

3. $a = 1$, $b = 2$, $c = 2$

4. $a = 5$, $b = 12$, $c = 13$

5. $a = 3$, $b = 3$, $c = 3$

6. $a = 8$, $b = 5$, $c = 5$

7. $a = 100$, $b = 50$, $c = 70$

8. $a = 35$, $b = 42$, $c = 37$

9. $a = 496$, $b = 564$, $c = 632$

10. $a = 212$, $b = 341$, $c = 285$

For Exercises 11–14, find, to the nearest tenth, the height to the longest side.

11. $a = 4$, $b = 5$, $c = 6$

12. $a = 3$, $b = 3$, $c = 5$

13. $a = 8$, $b = 8$, $c = 10$

14. $a = 5$, $b = 7$, $c = 9$

b

15. The length of each side of a rhombus is 10 centimeters. The length of one diagonal is 12 centimeters. Find the area to the nearest square centimeter.

16. Use Heron's formula to express the height, h, from B to $\overline{AC}$ in △ABC, in terms of a, b, and c.

17. A triangle has two equal sides, each with measure x. The length of the third side is 2. Use Heron's Formula to write a formula for the area of the triangle in terms of x.

18. The height of an isosceles triangle is 10 centimeters and the base is 8 centimeters. Use Heron's formula to find the length of each of the two congruent sides to the nearest tenth of a centimeter.

19. The area of an equilateral triangle is 10. Find, to the nearest tenth, the length of each side of the triangle. Use Heron's Formula.

20. Use Heron's Formula to find a formula for the area of an equilateral triangle, each of whose sides is of length a.

4-7 Vectors

The complete description of certain quantities, such as force, requires both a direction and a magnitude. Such quantities are called **vector quantities.**

Mathematically, a vector can be considered to be a <u>directed line</u> segment. The direction of a vector is given by its position in the plane. In the figure at the right, vectors AB and CD have the same direction. In the figure, vector AB, written $\overrightarrow{AB}$, has <u>initial point</u> $A(1, 3)$ and <u>terminal point</u> $B(4, 7)$. The length or **magnitude** of $\overrightarrow{AB}$ is symbolized as $|\overrightarrow{AB}|$. The distance formula is used to find the magnitude.

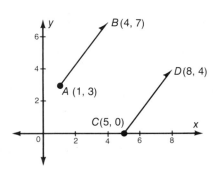

Example 1
Refer to the figure above to find the magnitude of each vector.

a. $\overrightarrow{AB}$ **b.** $\overrightarrow{CD}$

Solutions:

a. $|\overrightarrow{AB}| = \sqrt{(4-1)^2 + (7-3)^2}$ ← Distance Formula → **b.** $|\overrightarrow{CD}| = \sqrt{(8-5)^2 + (4-0)^2}$

$\qquad = \sqrt{9 + 16}$ $\qquad\qquad\qquad\qquad\qquad\qquad = \sqrt{9 + 16}$

$\qquad = \sqrt{25}$ $\qquad\qquad\qquad\qquad\qquad\qquad\;\; = \sqrt{25}$

$\qquad = 5$ $\qquad\qquad\qquad\qquad\qquad\qquad\quad\;\; = 5$

Example 1 shows that $|\overrightarrow{AB}| = |\overrightarrow{CD}|$. Vectors with the same magnitude and direction, such as $\overrightarrow{AB}$ and $\overrightarrow{CD}$, are called **equivalent vectors.**

In the figure at the right, $\overrightarrow{OB}$ is the **x-component** and $\overrightarrow{OC}$ is the **y-component** of $\overrightarrow{OA}$. The angle θ is called the **direction angle** of $\overrightarrow{OA}$. When you know the magnitude and direction angle of a vector you can find the magnitude of its x and y components.

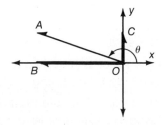

Example 2

In the figure at the right, $|\vec{OA}| = 100$ and $\theta = 160°$. Find each of the following to the nearest whole number.

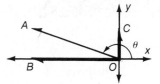

a. $|\vec{OB}|$ **b.** $|\vec{OC}|$

Solutions: a. Draw $\vec{BA}$. Since $\vec{BA}$ and $\vec{OC}$ have the same length and direction, they are equivalent. Thus, $m\angle ABO = 90°$. Since $m\angle AOB = 180° - 160° = 20°$,

$$\frac{|\vec{OB}|}{|\vec{OA}|} = \cos 20°$$

$$\begin{aligned} |\vec{OB}| &= |\vec{OA}| \cos 20° \\ &= 100 \cos 20° \\ &= 100(.9397) \\ &= 93.97, \text{ or about } 94 \end{aligned}$$

b. Similarly,

$$\frac{|\vec{BA}|}{|\vec{OA}|} = \sin 20°$$

$$\begin{aligned} |\vec{BA}| &= |\vec{OA}| \sin 20° \\ &= 100(.3420). \\ &= 34.20, \text{ or about } 34 \end{aligned}$$

Since $\vec{BA}$ and $\vec{OC}$ are equivalent, $|\vec{OC}|$ is about 34.

In Example 2, $\vec{OA}$ is called the **vector sum** of $\vec{OB}$ and $\vec{OC}$. In symbols, this is written

$$\vec{OA} = \vec{OB} + \vec{OC}.$$

Since $\vec{OC}$ and $\vec{BA}$ are equivalent and $\vec{OB}$ and $\vec{CA}$ are equivalent, the vector sum can also be written as follows.

$$\vec{OA} = \vec{OB} + \vec{BA}$$
$$\vec{OA} = \vec{CA} + \vec{OC}$$
$$\vec{OA} = \vec{CA} + \vec{BA}$$

The process of finding the vector sum when the x- and y-components are known is similar to finding the length of the hypotenuse of a right triangle.

Trigonometry and Triangles   **189**

Example 3

Refer to the figure at the right, to find the following.

a. $|\overrightarrow{OA}|$ to the nearest whole number

b. θ to the nearest 10°

Solutions: a. Draw $\overrightarrow{BA}$ equivalent to $\overrightarrow{OC}$. Then
$\overrightarrow{OA} = \overrightarrow{OB} + \overrightarrow{BA}$. Since $|\overrightarrow{OB}| = 5$,
and $|\overrightarrow{BA}| = |\overrightarrow{OC}| = 12$,

$$|\overrightarrow{OA}| = \sqrt{5^2 + 12^2}$$
$$= \sqrt{169}, \text{ or } 13.$$

b. To find θ, let $\alpha = m \angle BOA$.

Then $\cos \alpha = \dfrac{|\overrightarrow{OB}|}{|\overrightarrow{OA}|}$

$$\cos \alpha = \frac{5}{13} = .3846$$
$$\alpha = 67°20'.$$

Therefore, $\theta = 360° - 67°20'$, or $292°40'$.

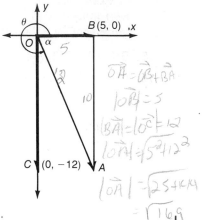

Handwritten annotations:
$\cos \alpha = \dfrac{|\overrightarrow{OB}|}{|\overrightarrow{OA}|}$
$\cos \alpha = \dfrac{5}{13}$
$\alpha = 67.38$
$\alpha = 67°20'$
$\alpha = 360 - 67°20' = 292°40'$

$\overrightarrow{OA} = \overrightarrow{OB} + \overrightarrow{BA}$
$|\overrightarrow{OB}| = 5$
$|\overrightarrow{BA}| = |\overrightarrow{OC}| = 12$
$|\overrightarrow{OA}| = \sqrt{5^2 + 12^2}$
$|\overrightarrow{OA}| = \sqrt{25 + 144}$
$= \sqrt{169}$
$= 13$

Classroom Exercises

Refer to the figure at the right to name each vector.

1. The x-component of $\overrightarrow{OA}$.

2. The y-component of $\overrightarrow{OA}$.

3. The vector sum $\overrightarrow{OB} + \overrightarrow{OC}$.

4. The vector equivalent to $\overrightarrow{OC}$.

5. The vector sum $\overrightarrow{OB} + \overrightarrow{BA}$.

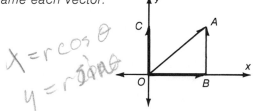

Handwritten: $x = r\cos\theta$ $y = r\sin\theta$

Written Exercises

In Exercises 1–8, find $|\overrightarrow{AB}|$ to the nearest tenth.

1. $A(4, 2)$; $B(3, 1)$

2. $A(4, 2)$; $B(3, -1)$

3. $A(5, -2)$; $B(1, -5)$

4. $A(1, -3)$; $B(2, -3)$

5. $A(-3, -7)$; $B(2, 1)$

6. $A(-3, 5)$; $B(-2, -5)$

7. $A(2, 5)$; $B(-7, -3)$

8. $A(1, 0)$; $B(-3, 5)$

In Exercises 9–12, $\vec{OA}$ has initial point $O(0, 0)$, and direction angle θ. Find $|\vec{OB}|$ and $|\vec{OC}|$, the x and y components of $\vec{OA}$, to the nearest whole number.

9. $|\vec{OA}| = 5$; $\theta = 120°$

10. $|\vec{OA}| = 3$; $\theta = 17°$ ^{60} $Ex\ 2$

11. $|\vec{OA}| = 8$; $\theta = 290°$

12. $|\vec{OA}| = 7$; $\theta = \frac{\pi}{3}$

In Exercises 13–16, B and C are the terminal points of $\vec{OB}$ and $\vec{OC}$, respectively, with initial point $O(0, 0)$. Find $|\vec{OA}|$, where $\vec{OA} = \vec{OB} + \vec{OC}$, to the nearest whole number. Find the direction angle θ to the nearest 10°.

13. $B(3, 0)$; $C(0, -4)$

14. $B(-6, 0)$; $C(0, 8)$ $Ex\ 3$

15. $B(-20, 0)$; $C(0, -15)$

16. $B(24, 0)$; $C(0, 18)$

In Exercises 17–20, A and B are the terminal points of $\vec{OA}$ and $\vec{OB}$, respectively, with initial point $O(0, 0)$. Find $|\vec{OA}|$, where $\vec{OA} = \vec{OB} + \vec{BA}$, to the nearest whole number. Find the measure of the direction angle θ to the nearest 10°.

17. $A(3, 4)$; $B(3, 0)$

18. $A(-12, -16)$; $B(-12, 0)$

19. $A(-10, 24)$; $B(0, 24)$

20. $A(8, -6)$; $B(0, -6)$

In Exercises 21 and 22, let $\vec{OA}$ with initial point $O(0, 0)$ have x-component $\vec{OB}$ and y-component $\vec{OC}$. Let α be the measure of the acute angle BOA.

21. Prove that $|\vec{OB}| = |\vec{OA}| \cos \alpha$.

22. Prove that $|\vec{OC}| = |\vec{OA}| \sin \alpha$.

Review

In Exercises 1–4, find the area of $\triangle ABC$ to the nearest square unit. (Section 4–5)

1. $a = 21$, $b = 18$, $C = 64°$

2. $a = 10$, $c = 12$, $B = 108°$

3. $b = 6$, $A = 80°$, $C = 30°$

4. $c = 13$, $A = 95°$, $B = 20°$

Use Heron's Formula to find the area of $\triangle ABC$ to the nearest square unit. (Section 4–6)

5. $a = 12$, $b = 14$, $c = 15$

6. $a = 25$, $b = 38$, $c = 50$

Find each of the following. Refer to the figure at the right (Section 4–7)

7. $|\vec{OA}|$ to the nearest tenth

8. θ to the nearest 10′

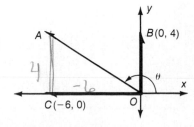

4-8 Applications: Displacement

The method of Example 3 in the previous section can be generalized to find the vector sum of any two vectors. For example, to find $\overrightarrow{OB} + \overrightarrow{OC}$ in the figure at the right, construct $\overrightarrow{BA}$ equivalent to $\overrightarrow{OC}$. Then

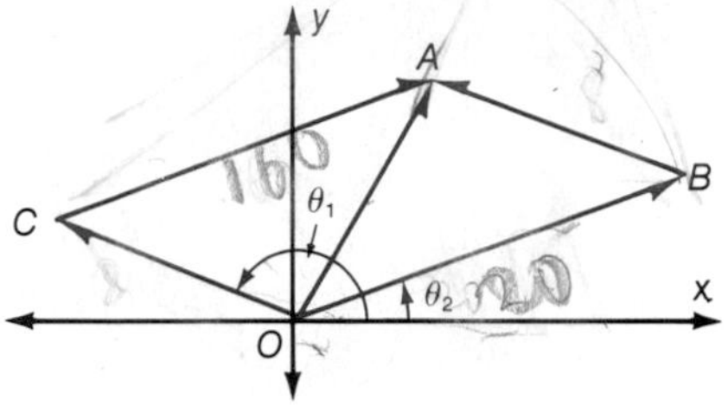

$$\overrightarrow{OA} = \overrightarrow{OB} + \overrightarrow{OC}.$$

The vector sum $\overrightarrow{OB} + \overrightarrow{OC}$ could also be obtained by constructing $\overrightarrow{CA}$ equivalent to $\overrightarrow{OB}$.

In the figure above, $\overrightarrow{OB}$ and $\overrightarrow{OC}$ are called **component vectors** of $\overrightarrow{OA}$. When you know the magnitude and direction angle of each component vector, you can find the magnitude and direction angle of the vector sum.

Example 1

In the figure above, $|\overrightarrow{OC}| = 10$, $|\overrightarrow{OB}| = 15$, $\theta_1 = 160°$, and $\theta_2 = 20°$. Find $|\overrightarrow{OA}|$ to the nearest tenth.

Solution: First find $|\overrightarrow{BA}|$ and m $\angle ABO$. Then use the Law of Cosines to find $|\overrightarrow{OA}|$. Since $\overrightarrow{OC}$ and $\overrightarrow{BA}$ are equivalent.

$$|\overrightarrow{BA}| = |\overrightarrow{OC}| = 10.$$

Also, the lines that contain $\overrightarrow{OC}$ and $\overrightarrow{BA}$ are parallel. Therefore,

$$\begin{aligned}
\text{m} \angle ABO &= 180° - \text{m} \angle COB \\
&= 180° - (\theta_1 - \theta_2) \\
&= 180° - (160° - 20°) \\
&= 180° - 140° \\
&= 40°.
\end{aligned}$$

$\therefore |\overrightarrow{OA}| = \sqrt{|\overrightarrow{OB}|^2 + |\overrightarrow{BA}|^2 - 2|\overrightarrow{OB}| \cdot |\overrightarrow{BA}| \cdot \cos 40°}$ ⟵   Law of Cosines: $b^2 = a^2 + c^2 - 2ac \cos B$

$= \sqrt{15^2 + 10^2 - 2(15)(10)(.7660)}$

$= \sqrt{95.2}$, or about 9.8.

In Example 1, $\vec{OA}$ is also the vector sum of $\vec{OB}$ and $\vec{BA}$, since $\vec{OC}$ and $\vec{BA}$ are equivalent. That is,

$$\vec{OA} = \vec{OB} + \vec{BA}.$$

Similarly, $\vec{OA} = \vec{OC} + \vec{CA}$, where $\vec{CA}$ is equivalent to $\vec{OB}$.

The method of Example 1 can be used to solve problems involving vectors. For example, when you travel from one place to another, you undergo a change of position or a **displacement.** Displacement is a vector quantity.

The magnitude of the displacement is the distance between the initial and terminal locations. In the figure at the right, $\vec{AB}$ and $\vec{BC}$ represent displacements from A to B and from B to C, respectively. The **net displacement** from A to C is given by $\vec{AC}$, the vector sum.

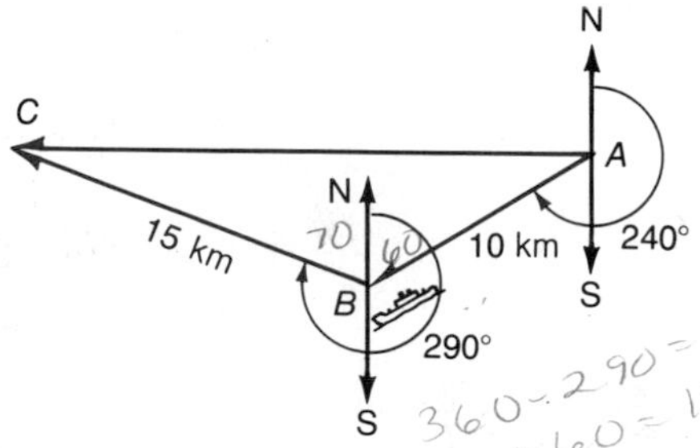

$$\vec{AC} = \vec{AB} + \vec{BC}.$$

The direction of a displacement can be given in many ways. For example, in navigation the course or bearing of a ship gives the direction of a displacement. The course is an angle measured clockwise from north.

Example 2

A ship sails 10 kilometers from point A to point B on course 240°. (Refer to the figure above.) At point B, the ship changes course to 290° and sails 15 kilometers to point C. Find, to the nearest kilometer, the magnitude of the net displacement from point A to point C.

Solution: The net displacement is represented by $\vec{AC} = \vec{AB} + \vec{BC}$. First, find m $\angle CBA$. Then use the Law of Cosines to find $|\vec{AC}|$.

$$\text{m} \angle CBA = \text{m} \angle CBN + \text{m} \angle ABN.$$

Since the north-south lines are parallel,

$$\text{m} \angle ABN = \text{m} \angle BAS = 60°.$$

Also, $\text{m} \angle CBN = 70°.$

Therefore, $\text{m} \angle CBA = 70° + 60° = 130°.$

Law of Cosines ⟶ $|\vec{AC}| = \sqrt{|\vec{AB}|^2 + |\vec{BC}|^2 - 2|\vec{AB}| \cdot |\vec{BC}| \cdot \cos(130°)}$

$$= \sqrt{10^2 + 15^2 - 2(10)(15)(-.6428)}$$

$$= \sqrt{517.84}, \text{ or about 23 km.}$$

Written Exercises

In Exercises 1–4, $\overrightarrow{OC}$ and $\overrightarrow{OB}$ are component vectors of $\overrightarrow{OA}$ and have initial point O(0, 0). Their direction angles are θ_1 and θ_2, respectively. Find $|\overrightarrow{OA}|$ to the nearest tenth.

1. $|\overrightarrow{OB}| = 5$, $|\overrightarrow{OC}| = 8$, $\theta_1 = 160°$, $\theta_2 = 20°$
2. $|\overrightarrow{OB}| = 7$, $|\overrightarrow{OC}| = 10$, $\theta_1 = 239°$, $\theta_2 = 110°$
3. $|\overrightarrow{OB}| = 9$, $|\overrightarrow{OC}| = 11$, $\theta_1 = 250°$, $\theta_2 = 193°$
4. $|\overrightarrow{OB}| = 10$, $|\overrightarrow{OC}| = 15$, $\theta_1 = 284°$, $\theta_2 = 275°$

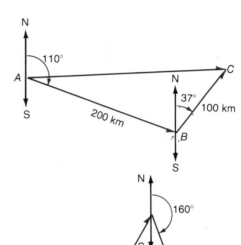

5. An airplane flies 200 kilometers on course 110° from point A to point B. At point B the pilot changes course to 37°. The plane then flies 100 kilometers to point C. Find, to the nearest kilometer, the magnitude of the net displacement from A to C.

6. A party of hikers walks 8 kilometers from camp on course 30°, then turns and walks 6 kilometers on course 160°. Find, to the nearest tenth of a kilometer, the magnitude of the net displacement from camp.

7. Find, to the nearest degree, the course to be followed by the hikers in Exercise 6 in order to return to camp by the most direct route.

8. A helicopter takes off on a straight-line path which makes a 65° angle with the horizontal. It reaches an altitude of 350 meters at point A. Find, to the nearest meter, the magnitude of the net displacement from the point of takeoff to point A.

9. A ship leaves port at point A and sails 61.1 kilometers on course 131°50′ to point B. Then it sails 76.5 kilometers on course 36°30′ to point C. Find, to the nearest 10 minutes, the bearing of point C from point A.

10. An aircraft carrier is traveling on course 204°20′ at 50 kilometers per hour. An airplane leaves the carrier at point A and flies on course 328°20′ to point C. The plane's displacement from A to C is 348 kilometers. What course, to the nearest 10 minutes, should the pilot set from point C in order to meet the carrier three hours after leaving it?

4-9 Applications: Force

It is not always necessary to use the coordinate plane to solve problems that involve vectors. In the figure at the right, $\vec{AD}$ and $\vec{AB}$ represent two forces exerted on the same object at point A. The lengths of the vectors represent the magnitudes of the forces. Angle DAB gives the directions of these forces. When two or more forces, called **component forces,** act on an object simultaneously, the **resultant force** is the single force which would produce the same effect. You can use vector addition to find the vector that represents the resultant force. In the figure above, $\vec{AC}$ represents the resultant force. It can be shown that $\vec{AC} = \vec{AD} + \vec{AB}$.

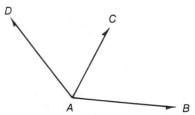

In the metric system the unit of force is the **newton,** abbreviated N. The weight of an object is the force of gravity exerted on the object. Therefore, weight is measured in newtons.

Example

An object weighing 433 newtons rests on a plank that makes an angle of 17° with the horizontal. In the figure, $\vec{AC}$ represents the weight, and $\vec{AD}$ and $\vec{AB}$ represent the components of the weight which act parallel and perpendicular, respectively, to the plank. Find $|\vec{AD}|$ and $|\vec{AB}|$ to the nearest newton.

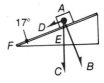

Solution: Since $\vec{AD}$ and $\vec{AB}$ are the components of $\vec{AC}$, $\vec{AC} = \vec{AD} + \vec{AB}$. Therefore, $\vec{AD}$ is equivalent to $\vec{BC}$. The force of gravity, $\vec{AC}$, acts perpendicular to the horizontal. Therefore, m $\angle FEA = 90°$. Thus,

$$\text{m} \angle FAE = 90° - 17° = 73° \quad \text{and} \quad \text{m} \angle CAB = 17°.$$

Therefore, $\quad \dfrac{|\vec{AB}|}{|\vec{AC}|} = \cos 17° \quad$ and $\quad \dfrac{|\vec{BC}|}{|\vec{AC}|} = \sin 17°.$

Thus, $|\vec{AB}| = |\vec{AC}| \cos 17° = 433(.9563)$ or about 414 N.

Similarly, $|\vec{BC}| = 433(.2924)$ or about 127 N.

Since $\vec{AD} = \vec{BC}$, $|\vec{AD}|$ is about 127 N.

Written Exercises

1-5

a

For Exercises 1–4, refer to the figure at the right below. A rope holds a block of ice in place. The plank on which the block of ice rests makes an angle θ with the horizontal.

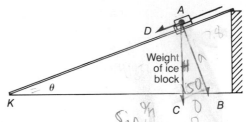

1. The tension in the rope is the component of the weight that acts parallel to the plank (neglect friction). Find the magnitude of the tension in the rope to the nearest newton, if θ = 17° and the ice weighs 432 newtons.

2. The force exerted on the plank by the ice is the component of the weight that acts perpendicular to the plank. If the block of ice weighs 415 newtons and θ = 23°, find the magnitude of the force exerted on the plank by the block of ice. Give the answer to the nearest newton.

3. Find θ to the nearest degree if the tension in the rope is 117.6 newtons and the magnitude of the force exerted on the plank by the block of ice is 49 newtons.

4. Find, to the nearest newton, the weight of the block of ice, if the tension in the rope is 58.8 newtons and the magnitude of the force exerted on the plank by the block of ice is 107.8 newtons.

5. A woman pushes with a force of 150 newtons on the handle of a lawn roller. (See the figure at the left below.) The handle makes a forty-five degree angle with the horizontal. Find, to the nearest newton, the magnitude of the horizontal and vertical components of this force.

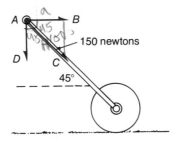

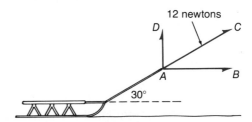

6. A sled is pulled over level snow by means of a rope which makes a thirty-degree angle with the horizontal. A twelve-newton force is exerted along the rope (see the figure at the right above). Find, to the nearest newton, the vertical and horizontal components of this force.

Refer to the figure below for Exercises 7–9. Scientists at NASA Langley Research Center designed an inclined plane apparatus to be used by astronauts in preparing for moon-gravity conditions. The apparatus is shown in the figure. An astronaut hangs on a freely moving cable and stands perpendicular to the inclined plane.

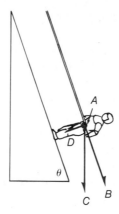

7. The astronaut weighs 784 newtons and $\theta = 70°$. Find, to the nearest newton, the magnitude of the component of the weight that is exerted perpendicular to the inclined plane.

8. Let w be the weight of the astronaut. Scientists must position the inclined plane so that the magnitude of the component exerted perpendicular to the inclined plane is $\frac{w}{6}$, since lunar gravity is $\frac{1}{6}$ earth gravity. Find, to the nearest degree, the value of θ that must be chosen.

9. What value of θ, to the nearest degree, would be required to simulate $\frac{1}{10}$ earth gravity?

10. A truck weighing 1.00×10^5 newtons is parked on a hill that rises three meters for each 100 meters of road. Find the magnitude of the force that tends to make the truck roll down the hill. Write the answer in scientific notation.

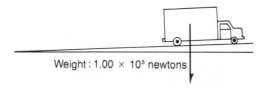

Weight: 1.00×10^5 newtons

11. A truck weighing 1.6×10^5 newtons is parked on a rising hill. The magnitude of the force that tends to make the truck roll downhill is 4000 newtons. Find the number of meters that the hill rises for each 100 meters of road.

12. A machine with a pulling force of 400 newtons is just able to pull an object weighing 650 newtons up an inclined ramp. Find, to the nearest degree, the angle at which the ramp is inclined.

4-10 Applications: Velocity

Velocity, like force, is a vector quantity. As with force, an observed velocity may be the **resultant** of two or more component velocities. That is to say, a velocity can be represented as the vector sum of vectors that represent component velocities.

The velocity of an airplane relative to the ground is the resultant of the wind velocity and of the velocity produced by the plane's engines. The magnitude of the velocity relative to the ground is called the **ground speed.** The magnitude and direction of the velocity produced by the engines are called the **air speed** and **heading,** respectively.

Example 1

Figure 1 at the left below shows the component velocities for an airplane with air speed 481.6 kilometers per hour on heading 85°12'. The wind velocity is 40 kilometers per hour in the direction 180°. Find, to the nearest unit, the ground speed of the airplane.

Solution: Draw $\overrightarrow{BC}$ equivalent to $\overrightarrow{AH}$. (See Figure 2.)

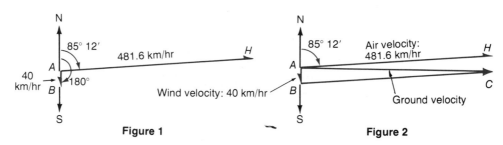

Figure 1   **Figure 2**

Then, $\overrightarrow{AC} = \overrightarrow{AB} + \overrightarrow{AH}$ represents the resultant ground velocity. Therefore, $|\overrightarrow{AC}|$ is the ground speed. Since $\overrightarrow{AH}$ and $\overrightarrow{BC}$ are equivalent, $|\overrightarrow{BC}| = |\overrightarrow{AH}| = 481.6$ and $m \angle ABC = m \angle NAH = 85°12'$. Therefore,

$$|\overrightarrow{AC}| = \sqrt{|\overrightarrow{AB}|^2 + |\overrightarrow{BC}|^2 - 2|\overrightarrow{AB}||\overrightarrow{BC}| \cos 85°12'}$$
$$= \sqrt{40^2 + (481.6)^2 - 2(40)(481.6)(.0837)} = 479.9.$$

Thus, the ground speed is about 480 km/hr.

Recall that the direction of motion of a ship or airplane is its <u>course</u>. The course of an airplane gives the direction of its actual velocity. In Example 2, m ∠ NAC is the course of the airplane.

Example 2 Find, to the nearest degree, the course of the airplane in Example 1.

Solution: The course is m ∠ NAC. Since m ∠ NAC + m ∠ CAB = m ∠ NAB = 180°,

$$m ∠ NAC = 180° - m ∠ CAB.$$

To find m ∠ CAB, use the Law of Sines.
Let m ∠ CAB = θ.

Then, $\dfrac{|\overrightarrow{BC}|}{\sin θ} = \dfrac{|\overrightarrow{AC}|}{\sin 85°12'}.$

Thus, $\sin θ = \dfrac{|\overrightarrow{BC}| \sin 85°12'}{|\overrightarrow{AC}|}$

$= \dfrac{481.6(.9965)}{480} = .9998.$

Thus, $θ = 89°$ ◄——— To the nearest degree.

Therefore the course is about 180° − 89°, or 91°.

Written Exercises

a

1. Find, to the nearest unit, the ground speed of an airplane with airspeed 480 km/hr and heading 90° if the wind velocity is 40 km/hr in the direction 180°.

2. Find, to the nearest degree, the course of the airplane in Exercise 1.

3. The velocity produced by the engines of a ship on a heading 157° has magnitude 35.2 km/hr. The water current has a velocity of 8 km/hr in the direction 213°. Find, to the nearest unit, the magnitude of the actual (observed) velocity of the ship.

4. Find the course of the ship in Exercise 3. Give your answer to the nearest degree.

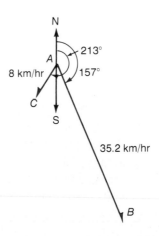

5. The angle between the heading and course of a ship or airplane is called the angle of drift. Find, to the nearest degree, the angle of drift of an airplane with air speed 240 km/hr on heading 90° in a 65 km/hr wind in the direction 0°. Refer to the figure at the left below.

6. Find, to the nearest unit, the ground speed of the airplane in Exercise 5.

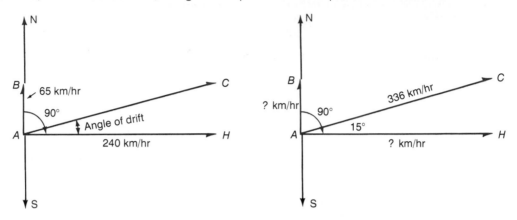

7. A wind in the direction 0° causes a 15° angle of drift for an airplane on heading 90°. Refer to the figure at the right above. The ground speed of the plane is 336 km/hr. Find the airspeed of the plane and the velocity of the wind. Give your answers to the nearest unit.

8. A river 3.2 kilometers wide flows due north at the rate of 2.4 km/hr. A small boat must travel directly across the river in 15 minutes. Refer to the figure at the right to find, to the nearest degree, the angle of drift, θ.

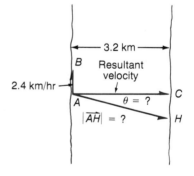

9. Find, to the nearest unit, the value of $|\overrightarrow{AH}|$, the magnitude of the velocity that must be maintained by the boat in Exercise 8. (HINT: Write the magnitude of the desired resultant velocity in km/hr.)

10. An airplane must fly at a ground speed of 450 km/hr on course 170° to be on schedule. The wind velocity is 25 km/hr in the direction 40°. Find the necessary heading to the nearest degree and the air speed to the nearest unit.

11. Point A is located on the western bank of a river, and point C is located downstream on the eastern bank. The river is of uniform width and flows due south at a rate of 2.4 km/hr. The line-of-sight distance from A to C is 762 meters. A boat crosses from A to C in 10 minutes. The boat's heading is 90°. Find, to the nearest unit, $|\overrightarrow{AH}|$, the magnitude of the velocity to be maintained by the boat.

12. Find, to the nearest degree, the angle of drift, θ, of the boat of Exercise 11.

BASIC: SOLVING TRIANGLES

An obvious use of the computer is the solution of triangles.

Problem: *Given the lengths of two sides of a triangle and the degree measure of the included angle, solve the triangle.*

The program below shows the steps for solving the triangle. The figure on the right is used as a reference in writing this program. A, B, and C represent the lengths of the sides. A1, B1, and C1 are the measures of the angles in degrees.

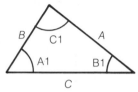

```
1Ø  READ B,C,A1
2Ø  LET R1 = A1*3.14159/18Ø          ◄──── R1 is the radian measure of A1.
3Ø  LET A = SQR(B↑2 + C↑2 - 2*B*C*COS(R1))
4Ø  IF C < B THEN 12Ø
5Ø  LET S = B*SIN(R1)/A               ◄──── S is the sine of B1.
6Ø  LET D = SQR(1 - S↑2)              ◄──── D is the cosine of B1.
7Ø  LET R2 = ATN(S/D)                 ◄──── R2 is the radian measure of B1.
8Ø  LET B1 = R2*18Ø/3.14159
9Ø  LET C1 = 18Ø - (A1 + B1)
1ØØ PRINT "SIDES ARE:";A;B;C
1Ø5 PRINT "ANGLES ARE:";A1;B1;C1
1Ø7 PRINT
11Ø GO TO 1Ø
12Ø LET S = C*SIN(R1)/A               ◄──── S is the sine of C1.
13Ø LET D = SQR(1 - S↑2)              ◄──── D is the cosine of C1.
14Ø LET R2 = ATN(S/D)                 ◄──── R2 is the radian measure of C1.
15Ø LET C1 = R2*18Ø/3.14159
16Ø LET B1 = 18Ø - (A1 + C1)
17Ø GO TO 1ØØ
18Ø DATA 5,8,35,6,9,49.33,7,5,152     ◄──── B=5, 6, and 7; C=8, 9, and 5
19Ø END                                      A1=35, 49.33, and 152
```

Analysis: Statement 2Ø: This changes degree measure to radian measure.

Statement 3Ø: This applies the Law of Cosines.

Statement 4Ø: This involves a decision.
   If C < B, *then* C1 *is an acute angle and statements* 12Ø *through* 15Ø *are used to compute* C1.
   If C > B, *then* B1 *is an acute angle and statements* 5Ø *through* 8Ø *are used to compute* B1.

Statements 5Ø and 12Ø: These apply the Law of Sines.

Statements 6Ø and 13Ø: These apply a Pythagorean Identity.

Statement 7Ø: This finds the radian measure of an angle that has a tangent with the value S/D. ATN is short for "arctangent."

Statement 1Ø7: This tells the computer to skip a line.

Output:

```
SIDES ARE:     4.84436    5    8
ANGLES ARE:      35     36.2994    1Ø8.7Ø1

SIDES ARE:     6.82761    6    9
ANGLES ARE:     49.33    41.8ØØ4     88.8696

SIDES ARE:    11.6536    7    5
ANGLES ARE:     152     16.3795    11.62Ø5

OUT OF DATA AT LINE 1Ø
```

Written Exercises

In each of Exercises 1–3, a formula for the area of a triangle is given. Write a program for the area based on the given formula.

1. $K = \frac{1}{2} ab \sin C$

2. $K = \frac{1}{2} \cdot \dfrac{c^2 \sin A \sin B}{\sin C}$

3. $K = \sqrt{s(s-a)(s-b)(s-c)}$, where $s = \dfrac{a+b+c}{2}$

4. Given three positive numbers, write a program to decide whether the numbers could be the lengths of three sides of a triangle.

For each of Exercises 5–8, write a program for solving a right triangle. In each exercise, you are given the measure of two parts of a right triangle, where angle measures are in degrees.

5. Hypotenuse and an acute angle

6. A leg and an acute angle

7. Two legs

8. Hypotenuse and a leg

For each of Exercises 9–12, write a program for solving a triangle. In each exercise you are given the measures of three parts, where angle measures are in degrees.

9. One side and two angles

10. Two angles and the included side

11. Three sides

12. Two sides and an angle

Chapter Objectives and Review

Objective: To know the meanings of the mathematical terms in this chapter.

1. Be sure that you know the meanings of these mathematical terms.

ambiguous case (p. 173)
component vectors (p. 192)
Heron's Formula (p. 185)
Law of Cosines (p. 168)
Law of Sines (p. 165)
Law of Tangents (p. 179)
solving a triangle (p. 164)

vector (p. 188)
  direction angle (p. 188)
  magnitude (p. 188)
  x-component (p. 188)
  y-component (p. 188)
vector quantity (p. 188)
vector sum (p. 189)

Objective: To use the Law of Sines to solve triangles. (Section 4–1)

In Exercises 2–5, solve $\triangle ABC$. Give measures of angles to the nearest degree, and lengths to the nearest tenth.

2. $A = 51°$, $C = 67°$, $a = 8$

3. $B = 17°$, $C = 112°$, $a = 12$

4. $A = 43°$, $a = 7$, $b = 4$

5. $a = 23$, $B = 81°$, $C = 87°$

Objective: To use the Law of Cosines to solve triangles. (Section 4–2)

In Exercises 6–9, solve $\triangle ABC$. Give measures of angles to the nearest degree, and lengths to the nearest tenth.

6. $a = 7$, $b = 6$, $C = 48°$

7. $a = 9$, $b = 11$, $c = 8$

8. $B = 130°$, $a = 6$, $c = 13$

9. $a = 15$, $b = 17$, $c = 20$

Objective: To find the number of solutions determined by the given data. (Section 4–3)

In Exercises 10–17, find the number of solutions for the given data.

10. $A = 25°$, $a = 7$, $b = 8$

11. $A = 63°$, $a = 5$, $b = 12$

12. $B = 51°$, $b = 7$, $c = 7$

13. $B = 110°$, $b = 17$, $a = 17$

14. $A = 47°$, $a = 17$, $b = 13$

15. $A = 135°$, $a = 6$, $b = 5$

16. $B = 81°$, $C = 100°$, $a = 5$

17. $a = 25$, $b = 25$, $c = 1$

Objective: To determine when to use the Law of Sines or the Law of Cosines, or both, to solve triangles. (Section 4–3)

In Exercises 18–21, solve △ABC using the Law of Sines or the Law of Cosines, or both. If there are two solutions, give both solutions. Give measures of angles to the nearest degree, and lengths to the nearest tenth.

18. $a = 20$, $b = 12$, $C = 56°$ **19.** $B = 49°$, $b = 16$, $c = 9$

20. $A = 40°$, $a = 9$, $b = 10$ **21.** $a = 9$, $b = 5$, $c = 10$

Objective: To use the Law of Tangents to solve triangles. (Section 4–4)

In Exercises 22–23, use the Law of Tangents to solve △ABC. Give measures of angles to the nearest degree and lengths to the nearest tenth.

22. $a = 5$, $b = 7$, $C = 64°$ **23.** $b = 25$, $c = 37$, $A = 106°$

Objective: To find the area of a triangle using formulas of the form $K = \frac{1}{2}ab \sin C$. (Section 4–5)

In Exercises 24–27, find the area of △ABC to the nearest square unit.

24. $a = 15$, $b = 20$, $C = 71°$ **25.** $b = 8$, $c = 26$, $A = 112°$

26. $b = 5$, $B = 20°$, $C = 93°$ **27.** $a = 17$, $A = 48°$, $B = 77°$

Objective: To find the area of a triangle using Heron's Formula. (Section 4–6)

Use Heron's Formula to find the area of △ABC to the nearest square unit.

28. $a = 10$, $b = 14$, $c = 16$ **29.** $a = 24$, $b = 20$, $c = 29$

30. $a = 30$, $b = 40$, $c = 25$ **31.** $a = 1$, $b = 1$, $c = 0.5$

Objective: To find the vector sum when the x and y components are known. (Section 4–7)

Find each of the following. Refer to the figure at the right.

32. $|\overrightarrow{OA}|$ to the nearest tenth

33. θ to the nearest 10 minutes

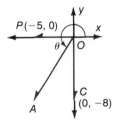

Objective: To find the vector sum when the magnitude and direction angle of each component vector are known. (Section 4–8)

34. In the figure below, $\vec{OA} = \vec{OC} + \vec{OB}$, $|\vec{OC}| = 12$, $|\vec{OB}| = 8$, $\theta_1 = 27°$, and $\theta_2 = 165°$. Find $|\vec{OA}|$ to the nearest tenth.

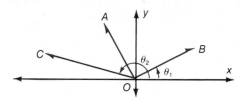

Objective: To apply the methods of vector addition in solving problems involving displacement, force, and velocity. (Sections 4–8, 4–9, and 4–10)

35. A ship leaves port at point A and sails 51 kilometers on course 264° to point B. Then it sails 78 kilometers on course 37° to point C. (See the figure at the left below.) Find, to the nearest kilometer, the magnitude of the net displacement from A to C.

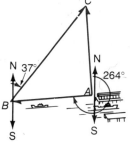
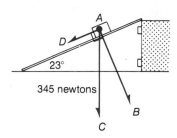

36. A crate weighing 345 newtons rests on a plank that makes a twenty-three-degree-angle with the horizontal. (See the figure at the right above.) Find, to the nearest newton, the magnitude of the component of the weight that acts parallel to the plank.

37. The figure below shows the component velocities for an airplane with airspeed 485 km/hr on heading 83°30'. The wind velocity is 38 km/hr in the direction 191°. Find, to the nearest unit, the ground speed of the airplane.

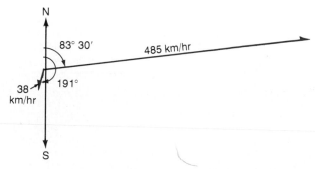

Chapter Test

For Exercises 1–3, use the Law of Sines or the Law of Cosines, or both. If there are two solutions, give both solutions. Give measures of angles to the nearest degree and lengths to the nearest unit.

1. In $\triangle ABC$, $A = 71°$, $B = 42°$, $c = 15$. Find a.

2. In $\triangle ABC$, $a = 5$, $b = 8$, $C = 40°$. Find c.

3. In $\triangle ABC$, $A = 30°$, $a = 15$, $b = 20$. Find c.

4. In $\triangle ABC$, $b = 7$, $c = 6$, $A = 48°$. Use the Law of Tangents to find B to the nearest degree.

Find the area of $\triangle ABC$ to the nearest square unit.

5. $B = 40°$, $a = 5$, $c = 10$

6. $a = 7$, $b = 8$, $c = 9$

In the figure at the right, $\overrightarrow{OB} = \overrightarrow{OE} + \overrightarrow{OD}$, $\overrightarrow{OA} = \overrightarrow{OB} + \overrightarrow{OC}$, $|\overrightarrow{OC}| = 4$, and $\theta_2 = 150°$. Use this figure to find each of the following.

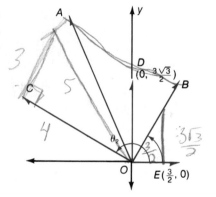

7. $|\overrightarrow{OB}|$

8. θ_1

9. $|\overrightarrow{OA}|$

10. A crate weighing 301 newtons rests on a plank that makes a 15°-angle with the horizontal. Find, to the nearest newton, the magnitude of the component of the weight that acts perpendicular to the plank.

Cumulative Review: Chapters 3 and 4

Write the letter of the response that best answers each question.

1. Find the equality that is <u>not</u> an identity. (Section 3–1)

   a. $1 = \csc^2\theta - \cot^2\theta$

   b. $\cot\theta = \dfrac{\cos\theta}{\sin\theta}$

   c. $\tan^2\theta - \sec^2\theta = 1$

   d. $\sin\theta = \dfrac{1}{\csc\theta}$

2. Find the equality that is an identity. (Section 3–2)

   a. $\sin^2\theta = \dfrac{\sec^2\theta - 1}{\sec\theta}$

   b. $\dfrac{\csc\theta}{\sin\theta} = \cot^2\theta + 1$

   c. $\sec\theta = \sin\theta \cdot \tan\theta$

   d. $\sin\theta \cdot \sec\theta = 1$

3. Evaluate $\cos 15°$ without tables. (Section 3–3)

   a. $\frac{\sqrt{2}}{2}(\sqrt{3}+1)$ **b.** $\frac{\sqrt{3}}{2}(\sqrt{2}-1)$ **c.** $\frac{\sqrt{3}}{4}(\sqrt{2}+1)$ **d.** $\frac{\sqrt{2}}{4}(\sqrt{3}+1)$

4. Given that $\sin 73° = .96$, $\cos 73° = .29$, $\sin 41° = .66$, and $\cos 41° = .75$, evaluate $\cos 32°$. (Section 3–3)

   a. $-.85$ **b.** $-.42$ **c.** $.42$ **d.** $.85$

5. Evaluate $\sin 75°$ without tables. (Section 3–4)

   a. $\frac{\sqrt{2}}{4}(\sqrt{3}-1)$ **b.** $\frac{\sqrt{2}}{4}(\sqrt{3}+1)$ **c.** $\frac{\sqrt{3}}{4}(\sqrt{2}-1)$ **d.** $\frac{\sqrt{3}}{4}(\sqrt{2}+1)$

6. Given that $\tan 24° = .45$ and $\tan 142° = -.78$, evaluate $\tan 118°$. (Section 3–5)

   a. -1.9 **b.** 1.9 **c.** $-.91$ **d.** $-.51$

7. Given that $\cos^2 1.3 = .0716$, evaluate $\cos 2.6$. (Section 3–6)

   a. $.8569$ **b.** $-.8568$ **c.** $.9284$ **d.** $-.9284$

8. Given that $\sin\alpha = .42$ and $\cos\alpha = .91$, evaluate $\sin 2\alpha$. (Section 3–6)

   a. $.76$ **b.** $-.76$ **c.** $.38$ **d.** $-.38$

9. Given that $\pi < \alpha < \frac{3\pi}{2}$, find an expression for $\sin\frac{\alpha}{2}$. (Section 3–7)

   a. $\sqrt{\frac{1+\cos\alpha}{2}}$ **b.** $-\sqrt{\frac{1+\cos\alpha}{2}}$ **c.** $\sqrt{\frac{1-\cos\alpha}{2}}$ **d.** $-\sqrt{\frac{1-\cos\alpha}{2}}$

10. Given that $\cos\alpha = -\frac{15}{17}$ and $\frac{\pi}{2} < \alpha < \pi$, evaluate $\tan\frac{\alpha}{2}$. (Section 3–7)

   a. 4 **b.** -4 **c.** $.25$ **d.** $-.25$

11. Express $\sin 40° - \sin 80°$ as a product. (Section 3–8)

   a. $2\sin 60° \cos 20°$

   b. $-2\cos 60° \sin 20°$

   c. $2\cos 60° \cos 20°$

   d. $2\sin 60° \sin 20°$

12. In $\triangle ABC$, $A = 32°$, $B = 71°$, and $a = 5.0$. Find b to the nearest tenth. (Section 4–1)

    a. 8.0   **b.** 8.7   **c.** 8.9   **d.** 9.2

13. In $\triangle ABC$, $a = 5.0$, $b = 2.0$, and $C = 80°$. Find c to the nearest tenth. (Section 4–2)

    a. 5.1   **b.** 5.4   **c.** 5.6   **d.** 5.8

14. In $\triangle ABC$, $a = 3$, $b = 4$, and $c = 6$. Find B to the nearest ten minutes. (Section 4–2)

    a. 35°50′   **b.** 36°   **c.** 36°20′   **d.** 36°40′

15. In $\triangle ABC$, $A = 71°$, $a = 8$, and $b = 8.2$. Find the number of solutions for the triangle. (Section 4–3)

    a. None   **b.** One   **c.** Two   **d.** Not enough information

16. For $\triangle ABC$, express $\dfrac{\tan\left(\frac{A-C}{2}\right)}{\tan\left(\frac{A+C}{2}\right)}$ in terms of sides a and c. (Section 4–4)

    a. $\dfrac{a+c}{a-c}$ **b.** $\dfrac{a-c}{a+c}$ **c.** $\dfrac{c-a}{c+a}$ **d.** $\dfrac{c+a}{c-a}$

17. In $\triangle ABC$, $a = 5.0$, $b = 10$, and $C = 47°$. Find the area of $\triangle ABC$ to the nearest square unit. (Section 4–5)

    a. 17   **b.** 18   **c.** 19   **d.** 20

18. In $\triangle ABC$, $a = 5$, $b = 5$, and $c = 6$. Find the area of the triangle to the nearest square unit. (Section 4–6)

    a. 16   **b.** 144   **c.** 10   **d.** 12

19. Vector OA has the origin as its initial point and 150° as its direction angle. $|\overrightarrow{OA}| = 8$. Find, to the nearest whole number, $\overrightarrow{OB}$, the magnitude of the x component of $\overrightarrow{OA}$. (Section 4–7)

    a. 4   **b.** 7   **c.** −4   **d.** −7

20. Vectors OB and OC are component vectors of $\overrightarrow{OA}$ and have the origin as their initial point. Their direction angles are 40° and 100° respectively and $|\overrightarrow{OB}| = 20$, $|\overrightarrow{OC}| = 10$. Find $|\overrightarrow{OA}|$ to the nearest unit. (Sections 4–7, 4–8, 4–9, and 4–10)

    a. 17   **b.** 26   **c.** 8   **d.** None of these

Chapter 5
Inverse Functions/ Equations

5-1 Inverse Relations

Recall that every relation has an **inverse relation** which is formed by interchanging the elements of each ordered pair in the relation. This inverse relation is often simply called the **inverse.**

$$\text{Relation:} \quad \{(-4, 3), (-2, 3), (1, 2), (3, 4)\}$$
$$\text{Inverse:} \quad \{(3, -4), (3, -2), (2, 1), (4, 3)\}$$

By interchanging the elements of each ordered pair, the domain of a relation becomes the range of its inverse, and the range of the relation becomes the domain of its inverse.

Relation: $\{(-4, 3), (-2, 3), (1, 2), (3, 4)\}$	Inverse: $\{(3, -4), (3, -2), (2, 1), (4, 3)\}$
Domain: $\{-4, -2, 1, 3\}$	Domain: $\{2, 3, 4\}$
Range: $\{2, 3, 4\}$	Range: $\{-4, -2, 1, 3\}$

Note that the relation on the left is a function because each first element is paired with a unique second element. However, its inverse is <u>not</u> a function because the ordered pairs $(3, -4)$ and $(3, -2)$ have the same first element, 3, paired with different second elements, -4 and -2.

If a function is defined by an equation such as $y = 2x + 3$, you can write its inverse by first interchanging the variables. Then, if possible, solve for the independent variable.

Example 1

Find the inverse of $y = 2x + 3$. Then graph $y = 2x + 3$ and its inverse on the same coordinate plane.

Solution: 1. Interchange x and y in $y = 2x + 3$.

$$x = 2y + 3$$

2. Now solve this equation for y.

$$y = \tfrac{1}{2}x - \tfrac{3}{2}$$

3. Next, graph the function and its inverse.

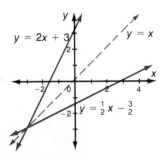

Note that the graph of $y = \frac{1}{2}x - \frac{3}{2}$ is a "mirror image" or <u>reflection</u> of the graph of $y = 2x + 3$ with respect to the line $y = x$. That is, a point on the graph of $y = 2x + 3$, such as (1, 5) and the corresponding point (5, 1) on the graph of its inverse, $y = \frac{1}{2}x - \frac{3}{2}$, are equidistant from the line $y = x$.

Example 2 shows that although it may be possible to write a rule for the inverse of a function, it may not always be possible to solve for y.

Example 2 Graph $y = \cos x$ and its inverse in the same coordinate plane.

Solution: First, interchange x and y in $y = \cos x$.

$$x = \cos y$$

The rule $x = \cos y$ <u>cannot</u> be solved for y at this time. However, the graph can be drawn. Begin by sketching the graph of $y = \cos x$.

To graph $x = \cos y$, interchange the coordinates of the points on the graph of $y = \cos x$. For example, the point $(2\pi, 1)$ on the graph of $y = \cos x$ becomes the point $(1, 2\pi)$ on the graph of $x = \cos y$.

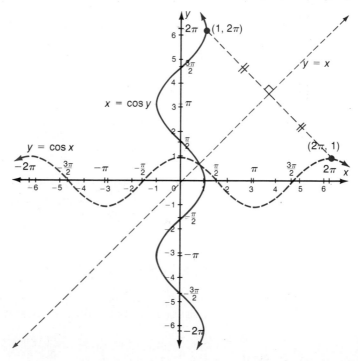

Note that the graph of $x = \cos y$ is the reflection of the graph of $y = \cos x$ with respect to the line $y = x$.

In Example 2, $x = \cos y$ cannot be solved algebraically for y. Thus, the inverse of $y = \cos x$ is a new relation to which we must assign a name and meaning.

$y = \mathbf{arc\ cos\ x}$ means y is a number whose cosine is x.

For example, $y = \text{arc cos } \frac{\sqrt{2}}{2}$ means y is a number whose cosine is $\frac{\sqrt{2}}{2}$ or $\cos y = \frac{\sqrt{2}}{2}$. As the graph of $x = \cos y$ shows, there are an infinite number of values of y for which $\cos y = \frac{\sqrt{2}}{2}$. Thus, when $x = \frac{\sqrt{2}}{2}$,

$$y = \cdots -\frac{9\pi}{4}, -\frac{7\pi}{4}, -\frac{\pi}{4}, \frac{\pi}{4}, \frac{7\pi}{4}, \frac{9\pi}{4}, \cdots .$$

It is often convenient to use **set-builder notation** to describe a general solution. The general solution is called the **solution set**. For example, the general solution above can be written as

$$\{y: \quad y = \tfrac{\pi}{4} + 2n\pi \quad \underline{\text{or}} \quad y = -\tfrac{\pi}{4} + 2n\pi, n \in \mathscr{I}\}.$$

This is read:

"The set of all y such that $y = \frac{\pi}{4} + 2n\pi$ <u>or</u> $y = -\frac{\pi}{4} + 2n\pi$, where n is an integer."

Clearly, $y = \text{arc cos } x$ does <u>not</u> define a function.

The inverse of the sine function can be assigned a name in a similar manner.

$y = \mathbf{arc\ sin\ x}$ means y is a number whose sine is x.

Example 3

Evaluate each expression in terms of radians.

a. arc cos $\left(-\frac{\sqrt{2}}{2}\right)$ **b.** arc sin $\frac{\sqrt{2}}{2}$

Solutions: **a.** Let $y = \text{arc cos } \left(-\frac{\sqrt{2}}{2}\right)$. Then $\cos y = -\frac{\sqrt{2}}{2}$.

Since the cosine function is negative in Quadrants II and III,

$$y = \frac{3\pi}{4}, \frac{3\pi}{4} \pm 2\pi, \frac{3\pi}{4} \pm 4\pi, \cdots \quad \underline{\text{or}} \quad y = -\frac{3\pi}{4}, -\frac{3\pi}{4} \pm 2\pi, -\frac{3\pi}{4} \pm 4\pi, \cdots .$$

Thus, arc cos $\left(-\frac{\sqrt{2}}{2}\right) = \{y: \quad y = \frac{3\pi}{4} + 2n\pi \quad \text{or} \quad y = -\frac{3\pi}{4} + 2n\pi, n \in \mathscr{I}\}$

b. Let $y = \text{arc sin } \frac{\sqrt{2}}{2}$. Then $\sin y = \frac{\sqrt{2}}{2}$.

Since the sine function is positive in Quadrants I and II,

$$y = \frac{\pi}{4}, \frac{\pi}{4} \pm 2\pi, \frac{\pi}{4} \pm 4\pi, \cdots \quad \underline{\text{or}} \quad y = \frac{3\pi}{4}, \frac{3\pi}{4} \pm 2\pi, \frac{3\pi}{4} \pm 4\pi, \cdots .$$

Thus, arc sin $\frac{\sqrt{2}}{2} = \{y: \quad y = \frac{\pi}{4} + 2n\pi \quad \underline{\text{or}} \quad y = \frac{3\pi}{4} + 2n\pi, n \in \mathscr{I}\}$.

Classroom Exercises

Let R be the relation {(0, 1), (2, 1)}.

1. Is R a function?

2. Is the inverse of R a function?

Write an equation for the inverse of each function.

3. $y = 2x$ **4.** $y = 4$ **5.** $y = -3x$ **6.** $y = x - 1$

Find two solutions for each equation, where $-\pi \le y \le \pi$.

7. $y = \text{arc sin} \frac{1}{2}$ **8.** $y = \text{arc cos}(-1)$ **9.** $y = \text{arc sin } 0$ **10.** $y = \text{arc cos} \frac{\sqrt{3}}{2}$

Written Exercises

all odd

Let R be the relation {(0, 1), (1, 4), (2, 4)}.

1. Find the domain of R.

2. Find the range of R.

3. Find the inverse of R.

4. Is R a function?

5. Is the inverse of R a function?

6. Find the domain of the inverse of R.

Graph each function and its inverse on the same coordinate plane.

7. $y = x + 1$ **8.** $y = 2x$ **9.** $y = 3x - 1$ **10.** $y = 2$

11. $y = 1 - x$ **12.** $y = x^2$ **13.** $y = x^3$ **14.** $y = |x|$

15. Tell which functions in Exercises 7–14 have inverses that are <u>not</u> functions.

Write an equation for the inverse of each function.

16. $y = 3x - 1$ **17.** $y = -1$ **18.** $y = \cos x$ **19.** $y = \sin x$

20. Graph $y = \sin x$ and its inverse on the same coordinate plane.

Evaluate each expression in terms of radians.

21. arc sin 1   **22.** arc cos 0   **23.** arc cos 1   **24.** arc sin $\frac{1}{2}$

25. arc sin (-1) **26.** arc cos $\frac{1}{2}$ **27.** arc sin $(-\frac{1}{2})$ **28.** arc sin $\frac{\sqrt{3}}{2}$

29. arc cos $\frac{\sqrt{3}}{2}$ **30.** arc cos (-1) **31.** arc cos $(-\frac{1}{2})$ **32.** arc sin $(-\frac{\sqrt{2}}{2})$

Inverse Functions/Equations   **213**

5-2 Inverse Sine and Cosine Functions

Since equations such as $y = \text{arc sin } \frac{\sqrt{2}}{2}$ and $y = \text{arc cos } (-\frac{\sqrt{2}}{2})$ have more than one solution, the inverses of the sine and cosine functions are not functions. However, by restricting the range of the corresponding inverse relation, it is possible to define an inverse sine function and an inverse cosine function.

Example 1 Restrict the range of $y = \text{arc sin } x$ so that the result is a function.

Solution: First sketch the graph of $y = \sin x$. Then, to graph $y = \text{arc sin } x$, interchange the coordinates of the points on the graph of $y = \sin x$.

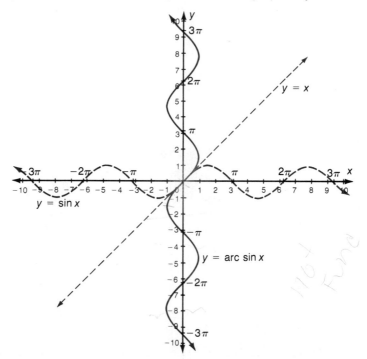

Note that the graph of $y = \text{arc sin } x$ is a reflection of the graph of $y = \sin x$ with respect to the line $y = x$.

There are several ways to restrict the range of $y = \text{arc sin } x$ in order to obtain a function.

Three ways are shown below.

Function 1	**Function 2**	**Function 3**
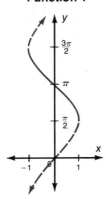	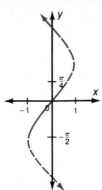	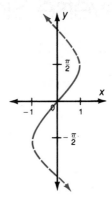
Range: $\frac{\pi}{2} \leq y \leq \frac{3\pi}{2}$	Range: $-\frac{\pi}{2} \leq y \leq \frac{\pi}{4}$	Range: $-\frac{\pi}{2} \leq y \leq \frac{\pi}{2}$
Domain: $-1 \leq x \leq 1$	Domain: $-1 \leq x \leq \frac{\sqrt{2}}{2}$	Domain: $-1 \leq x \leq 1$

Function 3 in Example 1 above is the one that we shall use to define an inverse sine function because it is the most convenient. Note that its range is restricted to $-\frac{\pi}{2} \leq y \leq \frac{\pi}{2}$.

To distinguish this "new" function from the relation $y = \text{arc } \sin x$, the "A" in arc sin x is capitalized.

Definition

$y = \text{Arc } \sin x$ means $y = \text{arc } \sin x$ <u>and</u> $-\frac{\pi}{2} \leq y \leq \frac{\pi}{2}$.

The numbers in the restricted range $-\frac{\pi}{2} \leq y \leq \frac{\pi}{2}$ are called the **principal values** of the inverse sine relation. The range of the relation $y = \text{arc } \cos x$ can also be restricted so that an inverse cosine function can be defined. (The graph of $y = \text{arc } \cos x$ was drawn in Section 5–1 on page 211.) Note that the range for $y = \text{Arc } \cos x$ is <u>not</u> the same as the range for $y = \text{Arc } \sin x$.

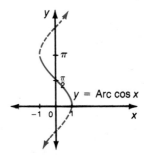

Definition

$y = \text{Arc } \cos x$ means $y = \text{arc } \cos x$ <u>and</u> $0 \leq y \leq \pi$.

Example 2 Evaluate each expression.

a. Arc sin $\left(-\frac{1}{2}\right)$ in terms of degrees

b. Arc cos $\left(-\frac{1}{2}\right)$ in terms of radians

Solutions: **a.** Let $y =$ Arc sin $\left(-\frac{1}{2}\right)$. Then $\sin y = -\frac{1}{2}$ and $-90° \le y \le 90°$.
For $-90° \le y \le 90°$, $\sin y$ is negative only in Quadrant IV.
Thus, $y = -30°$, and

$$\text{Arc sin}\left(-\frac{1}{2}\right) = -30°.$$

b. Let $y =$ Arc cos $\left(-\frac{1}{2}\right)$. Then $\cos y = -\frac{1}{2}$ and $0 \le y \le \pi$.
For $0 \le y \le \pi$, $\cos y$ is negative only in Quadrant II.
Thus, $y = \frac{2\pi}{3}$, and

$$\text{Arc cos}\left(-\frac{1}{2}\right) = \frac{2\pi}{3}.$$

To evaluate an expression such as $\sin\left(\text{Arc cos}\,\frac{1}{2}\right)$, first find the value of the expression in parentheses. Then evaluate the resulting expression.

Example 3 Evaluate each expression.

a. $\sin\left(\text{Arc cos}\,\frac{1}{2}\right)$

b. $\sin\left(\text{Arc sin}\,\frac{\sqrt{2}}{2}\right)$

Solutions: Begin by evaluating the function in parentheses.

a. Let $y =$ Arc cos $\frac{1}{2}$. Then
$\cos y = \frac{1}{2}$ and $y = 60°$.
Therefore,
$\sin\left(\text{Arc cos}\,\frac{1}{2}\right) = \sin 60°$,

and $\sin 60° = \frac{\sqrt{3}}{2}$. Thus,

$$\sin\left(\text{Arc cos}\,\frac{1}{2}\right) = \frac{\sqrt{3}}{2}.$$

b. Let $y =$ Arc sin $\frac{\sqrt{2}}{2}$. Then
$\sin y = \frac{\sqrt{2}}{2}$ and $y = \frac{\pi}{4}$.
Therefore,
$\sin\left(\text{Arc sin}\,\frac{\sqrt{2}}{2}\right) = \sin\frac{\pi}{4}$,

and $\sin\frac{\pi}{4} = \frac{\sqrt{2}}{2}$. Thus,

$$\sin\left(\text{Arc sin}\,\frac{\sqrt{2}}{2}\right) = \frac{\sqrt{2}}{2}.$$

In Example **3b** note that $\sin\left(\text{Arc sin}\,\frac{\sqrt{2}}{2}\right) = \frac{\sqrt{2}}{2}$. This suggests the following property.

$$\textbf{sin (Arc sin } x\textbf{)} = x$$

However, it does not follow that Arc sin (sin x) = x, as the following illustrates.

$$y = \text{Arc sin } (\sin \tfrac{5\pi}{4}) \qquad\qquad y = \text{Arc sin } (\sin \tfrac{\pi}{4})$$

$$y = \text{Arc sin } (-\tfrac{\sqrt{2}}{2}) \qquad\qquad y = \text{Arc sin } \tfrac{\sqrt{2}}{2}$$

$$y = -\tfrac{\pi}{4} \qquad\qquad\qquad\qquad y = \tfrac{\pi}{4}$$

$$\therefore \text{Arc sin } (\sin \tfrac{5\pi}{4}) \neq \tfrac{5\pi}{4} \qquad \therefore \text{Arc sin } (\sin \tfrac{\pi}{4}) = \tfrac{\pi}{4}$$

Thus, for Arc sin (sin x) = x to be true, the values of x must be restricted to $-\tfrac{\pi}{2} \le x \le \tfrac{\pi}{2}$. Note that $\tfrac{\pi}{4}$ is in this interval and $\tfrac{5\pi}{4}$ is not.

Similar properties exist for the cosine and inverse cosine functions.

NOTE: The superscript, −1, is sometimes used to denote the inverse. Thus,

$$\textbf{Cos}^{-1}\,\textbf{x} \text{ means Arc cos } x,$$

$$\textbf{cos}^{-1}\,\textbf{x} \text{ means arc cos } x,$$

and so on. If this notation is used, do not confuse $\cos^{-1} x$ with the reciprocal, $\frac{1}{\cos x}$. The superscript −1 in $\cos^{-1} x$ is <u>not</u> an exponent.

Classroom Exercises _____

Evaluate each expression in terms of degrees.

1. Arc cos $\tfrac{\sqrt{2}}{2}$

2. Arc sin $\tfrac{\sqrt{2}}{2}$

Evaluate each expression.

3. cos (Arc sin $\tfrac{1}{2}$)

4. sin (Arc sin $\tfrac{1}{2}$)

Written Exercises

Give the domain and range of each function.

1. $y = $ Arc sin x

2. $y = $ Arc cos x

Find the domain of each function for the given restriction on the range.

3. $y = $ arc sin x, where $0 \le y \le \tfrac{\pi}{2}$

4. $y = $ arc sin x, where $-\tfrac{\pi}{2} < y < 0$

5. $y = $ arc cos x, where $0 < y < \tfrac{\pi}{2}$

6. $y = $ arc cos x, where $\tfrac{\pi}{2} \le y \le \pi$

Graph each of the following.

7. $y = \text{Arc sin } x$ **8.** $y = \text{Arc cos } x$

9. $y = \text{arc sin } x$, where $-\pi \le y \le \pi$ **10.** $y = \text{arc cos } x$, where $0 \le y \le 2\pi$

Evaluate each expression in terms of degrees.

11. $\text{Arc sin} \frac{1}{2}$ **12.** $\text{Arc cos} \frac{1}{2}$ **13.** $\text{Arc sin} \frac{\sqrt{2}}{2}$

14. $\text{Arc cos} \frac{\sqrt{2}}{2}$ **15.** $\text{Arc cos} (-\frac{\sqrt{2}}{2})$ **16.** $\text{Arc sin} (-\frac{\sqrt{2}}{2})$

Evaluate each expression in terms of radians.

17. $\text{Arc cos } 1$ **18.** $\text{Arc cos } 0$ **19.** $\text{Arc sin } 0$

20. $\text{Arc sin } 1$ **21.** $\text{Arc sin } (-1)$ **22.** $\text{Arc cos } (-1)$

23. $\text{Arc sin} \frac{\sqrt{3}}{2}$ **24.** $\text{Arc cos} (-\frac{\sqrt{3}}{2})$ **25.** $\text{Arc sin} (-\frac{\sqrt{3}}{2})$

26. $\text{Arc cos} \frac{\sqrt{3}}{2}$ **27.** $\text{Arc sin} (-\frac{1}{2})$ **28.** $\text{Arc cos} (-\frac{1}{2})$

Evaluate each expression.

29. $\sin (\text{Arc sin} \frac{1}{2})$ **30.** $\cos (\text{Arc sin} \frac{1}{2})$ **31.** $\cos (\text{Arc cos} \frac{\sqrt{3}}{2})$

32. $\cos (\text{Arc sin} \frac{\sqrt{3}}{2})$ **33.** $\sin (\text{Arc cos} \frac{\sqrt{3}}{2})$ **34.** $\sin (\text{Arc cos} \frac{1}{2})$

35. $\sin (\text{Arc cos } 0)$ **36.** $\cos (\text{Arc sin } 1)$ **37.** $\sin (\text{Arc cos} \frac{\sqrt{2}}{2})$

38. $\cos (\text{Arc cos } 1)$ **39.** $\cos (\text{Arc sin } 0)$ **40.** $\sin (\text{Arc cos } 0)$

41. $\sin [\text{Arc cos } (-1)]$ **42.** $\cos [\text{Arc sin } (-\frac{\sqrt{2}}{2})]$ **43.** $\sin [\text{Arc cos } (-\frac{1}{2})]$

44. $\sin [\text{Arc sin } (-\frac{1}{2})]$ **45.** $\cos [\text{Arc cos } (-\frac{\sqrt{3}}{2})]$ **46.** $\cos [\text{Arc sin } (-1)]$

Example: Evaluate $\cos [\text{Arc sin } (-\frac{5}{13})]$.

Let $y = \text{Arc sin } (-\frac{5}{13})$. Then $\sin y = -\frac{5}{13}$ and $-\frac{\pi}{2} \le y \le \frac{\pi}{2}$.
For $-\frac{\pi}{2} \le y \le \frac{\pi}{2}$, $\sin y$ is negative only in Quadrant IV.
Thus, y is an angle in Quadrant IV. From Identity **6** on page 125,

$$\sin^2 y + \cos^2 y = 1$$
$$(-\tfrac{5}{13})^2 + \cos^2 y = 1$$
$$\cos^2 y = \tfrac{144}{169}$$
$$\cos y = \tfrac{12}{13} \text{ or } \cos y = -\tfrac{12}{13}.$$

Since cosine is positive in Quadrant IV,

$$\cos [\text{Arc sin } (-\tfrac{5}{13})] = \cos y = \tfrac{12}{13}.$$

47. $\sin (\text{Arc cos} \frac{3}{5})$ **48.** $\cos (\text{Arc sin} \frac{4}{5})$ **49.** $\sin [\text{Arc sin } (-\frac{12}{13})]$

50. $\csc (\text{Arc sin} \frac{2}{3})$ **51.** $\sec (\text{Arc cos} \frac{3}{4})$ **52.** $\csc (\text{Arc cos} \frac{3}{5})$

53. $\tan (\text{Arc sin} \frac{\sqrt{2}}{2})$ **54.** $\cot [\text{Arc cos } (-\frac{\sqrt{2}}{2})]$ **55.** $\csc [\text{Arc sin } (-\frac{4}{5})]$

56. $\sin\left(\text{Arc cos}\frac{2}{3}\right)$ **57.** $\cos\left(\text{Arc sin}\frac{1}{4}\right)$ **58.** $\tan\left[\text{Arc sin}\left(-\frac{\sqrt{3}}{2}\right)\right]$

59. $\sin\left[\left(\text{Arc cos}\frac{1}{2}\right)+\frac{\pi}{2}\right]$ **60.** $\cos\left[\left(\text{Arc sin}\frac{\sqrt{2}}{2}\right)+\pi\right]$ **61.** $\sec\left(\text{Arc cos}\,a\right)$

62. Find a range other than $-\frac{\pi}{2}\le y\le\frac{\pi}{2}$ that could have been used to define the Arc sin function.

63. Find a range other than $0\le y\le\pi$ that could have been used to define the Arc cos function.

Show that the following expressions have no meaning.

64. Arc sin $\sqrt{2}$ **65.** Arc cos $(-\sqrt{3})$ **66.** $\tan\left(\text{Arc sin}\,1\right)$

C   **67.** Find all real numbers x that satisfy the equation

$$\text{Arc cos}\,(\cos x) = x.$$

Sketch the inverse of each function. Tell whether the inverse is a function.

68.

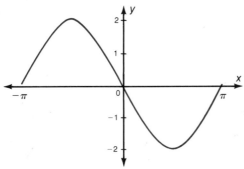

69.

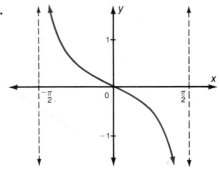

70. Find the value of $\sin\left(\text{Arc sin}\frac{1}{4}+\text{Arc cos}\frac{1}{2}\right)$.

(HINT: Let $x_1 = \text{Arc sin}\frac{1}{4}$ and $x_2 = \text{Arc cos}\frac{1}{2}$. Then find $\sin\left(x_1 + x_2\right)$.)

Review Capsule for Section 5-3 _____

Give the range of each trigonometric function. (For further review, see pages 108–111.)

1. $y = \tan x$ on the interval $-\frac{\pi}{4} < x < \frac{\pi}{4}$

2. $y = \tan x$ on the interval $-\frac{\pi}{2} < x < \frac{\pi}{2}$

3. $y = \sec x$ on the interval $0 \le x \le \frac{\pi}{2}$

4. $y = \sec x$ on the interval $0 \le x \le \pi, x \ne \frac{\pi}{2}$

5-3 Inverses of Tan, Cot, Sec, and Csc

As with the sine and cosine functions, the inverse relations of the tan, cot, sec, and csc functions are written as <u>arc tan</u>, <u>arc cot</u>, <u>arc sec</u>, and <u>arc csc</u>, respectively.

y = arc tan x means y is a real number whose tan is x, or $x = \tan y$.

y = arc cot x means y is a real number whose cot is x, or $x = \cot y$.

y = arc sec x means y is a real number whose sec is x, or $x = \sec y$.

y = arc csc x means y is a real number whose csc is x, or $x = \csc y$.

By restricting the range of each inverse relation to selected principal values, inverse functions can be defined.

Definitions:

y = Arc tan x	means	$y = \text{arc tan } x$ <u>and</u> $-\frac{\pi}{2} < y < \frac{\pi}{2}$.
y = Arc cot x	means	$y = \text{arc cot } x$ <u>and</u> $0 < y < \pi$.
y = Arc sec x	means	$y = \text{arc sec } x$ <u>and</u> $0 \le y \le \pi$, $y \ne \frac{\pi}{2}$.
y = Arc csc x	means	$y = \text{arc csc } x$ <u>and</u> $-\frac{\pi}{2} \le y \le \frac{\pi}{2}$, $y \ne 0$.

The graphs of $y = \text{Arc tan } x$ and $y = \text{Arc cot } x$ are shown below. The graphs of $y = \text{Arc sec } x$ and $y = \text{Arc csc } x$ are shown on page 221.

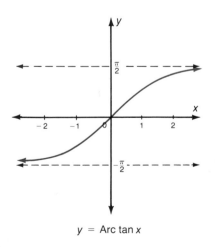

$y = \text{Arc tan } x$ $y = \text{Arc cot } x$

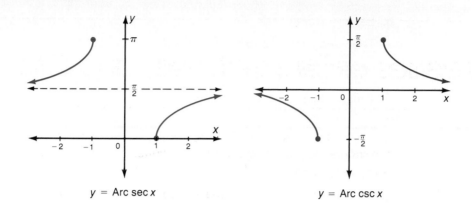

$y = \text{Arc sec } x$ $y = \text{Arc csc } x$

Example 1

Evaluate each expression in terms of radians.

a. arc tan $\frac{\sqrt{3}}{3}$ **b.** Arc sec $\sqrt{2}$

Solutions: a. Let $y = $ arc tan $\frac{\sqrt{3}}{3}$. Then tan $y = \frac{\sqrt{3}}{3}$.
The tangent function is positive in Quadrants I and III.
Thus, for tan $y = \frac{\sqrt{3}}{3}$, $y = \frac{\pi}{6}$, $\frac{\pi}{6} \pm \pi$, $\frac{\pi}{6} \pm 2\pi$, $\frac{\pi}{6} \pm 3\pi$, $\cdots$

or, $\{y: \ y = \frac{\pi}{6} + n\pi, n \in \mathscr{I}\}$.

b. Let $y = $ Arc sec $\sqrt{2}$. Then sec $y = \sqrt{2}$ and $0 \le y \le \pi$.
For this range, $0 \le y \le \pi$, sec y is positive only in Quadrant I.
Thus, $y = \frac{\pi}{4}$, and Arc sec $\sqrt{2} = \frac{\pi}{4}$.

As with the Arc sin and Arc cos functions, the restricted range (principal values) can be expressed in degrees.

Example 2

Evaluate each expression.

a. tan (Arc cot $\frac{\sqrt{3}}{3}$)   **b.** sec (Arc sec $\frac{2\sqrt{3}}{3}$)

Solutions: Begin by evaluating the function in parentheses.

a. Let $y = $ Arc cot $\frac{\sqrt{3}}{3}$.
Then cot $y = \frac{\sqrt{3}}{3}$ and $y = 60°$. Therefore,
tan (Arc cot $\frac{\sqrt{3}}{3}$) = tan 60° and tan 60° = $\sqrt{3}$.

Thus, tan $\left(\text{Arc cot } \frac{\sqrt{3}}{3}\right) = \sqrt{3}$.

Inverse Functions/Equations   **221**

b. Let $y = \text{Arc sec } \frac{2\sqrt{3}}{3}$.

Then $\sec y = \frac{2\sqrt{3}}{3}$ and $y = \frac{\pi}{6}$. Therefore,

$\sec (\text{Arc sec } \frac{2\sqrt{3}}{3}) = \sec \frac{\pi}{6}$ and $\sec \frac{\pi}{6} = \frac{2\sqrt{3}}{3}$.

Thus, $\sec \left(\text{Arc sec } \frac{2\sqrt{3}}{3} \right) = \frac{2\sqrt{3}}{3}$.

Recall that $\sin (\text{Arc sin } x) = x$. Example **2b** suggests a similar property for the secant and inverse secant functions.

$$\textbf{sec (Arc sec } x\textbf{)} = x$$

Similar properties can be stated for the tan and Arc tan functions, cot and Arc cot functions, and the csc and Arc csc functions.

Classroom Exercises

Evaluate each expression in terms of degrees.

1. arc sec (-1) **2.** arc cot (-1) **3.** Arc csc $\frac{2}{\sqrt{3}}$

Evaluate each expression in terms of radians.

4. arc tan $\frac{1}{\sqrt{3}}$ **5.** Arc sec 2   **6.** Arc cot $(-\sqrt{3})$

7. Evaluate tan (Arc csc $\frac{25}{7}$).

Written Exercises

Evaluate each expression in terms of degrees.

1. arc tan 1   **2.** arc tan $\sqrt{3}$ **3.** arc cot $\sqrt{3}$

4. arc sec (-2) **5.** arc sec $\frac{2\sqrt{3}}{3}$ **6.** arc sec 1

Evaluate each expression in terms of radians.

7. arc cot 0   **8.** arc tan (-1) **9.** arc csc (-2)

10. Arc cot (-1) **11.** Arc csc $\frac{2\sqrt{3}}{3}$ **12.** Arc sec (-2)

13. Arc tan $(-\frac{\sqrt{3}}{3})$ **14.** Arc tan $(-\sqrt{3})$ **15.** Arc csc 1

Evaluate each expression.

16. tan (Arc tan 4)　　　　　　　　　　**17.** sin [Arc tan (−1)]

18. sin [Arc cot (−$\frac{3}{4}$)]　　　　　　　**19.** cos [Arc tan (−$\frac{12}{5}$)]

20. sec (Arc csc 4)　　　　　　　　　**21.** csc [Arc csc (−3)]

22. sec [Arc csc (−3)]　　　　　　　**23.** sin (Arc sec 4)

24. sin (Arc tan 2)　　　　　　　　　**25.** csc [Arc cot (−$\frac{3}{2}$)]

26. sin [2 Arc tan (−$\frac{8}{15}$)]　　　　　**27.** tan [2 Arc cot ($\frac{24}{7}$)]

28. tan [$\frac{1}{2}$ Arc tan ($\frac{3}{5}$)]　　　　　　**29.** tan [$\frac{1}{2}$ Arc sec (−$\frac{13}{5}$)]

30. Prove:　Arc cot $x = \frac{\pi}{2} -$ Arc tan x　　**31.** Prove:　Arc tan $\frac{1}{2} +$ Arc tan $\frac{1}{3} = \frac{\pi}{4}$.

Find all the real numbers for which each equation holds.

32. Arc tan (tan x) = x　　　　　　**33.** Arc cot (cot x) = x

34. Arc sec (sec x) = x　　　　　　**35.** Arc csc (csc x) = x

36. Arc sec x = Arc cos $\frac{1}{x}$　　　　　**37.** Arc csc x = Arc sin $\frac{1}{x}$

C

Prove each of the following.

38. tan [Arc tan a + Arc tan 1] = $\frac{1 + a}{1 - a}$　　**39.** tan [Arc tan b − Arc tan a] = $\frac{b - a}{1 + ab}$

40. Arc sin $\frac{3}{5}$ + Arc cos $\frac{5}{13}$ = Arc sin $\frac{63}{65}$　　**41.** 2 Arc tan $\frac{1}{3}$ + Arc tan $\frac{1}{7} = \frac{\pi}{4}$

42. Arc tan $\frac{1}{5}$ − Arc tan (−5) = $\frac{\pi}{2}$　　**43.** Arc cos x + Arc sin $x = \frac{\pi}{2}$

 ———————————— *Inverse Functions* ————————————

The Examples below show how to use a scientific calculator to find an angle measure in terms of degrees or radians when the trigonometric function of that angle is known.

Example 1　Evaluate arc tan 1 in terms of degrees.

Solution:　[1] [INV] [tan]

$\boxed{45.}$

Example 2　Evaluate arc sin $\frac{\sqrt{3}}{2}$ in terms of radians.

Solution:　[3] [$\sqrt{x}$] [÷] [2] [=] [INV] [sin]

$\boxed{1.0471976 \atop \text{RAD}}$

Example 3　Evaluate sin (arc cos $\frac{1}{2}$).

Solution:　[1] [÷] [2] [=] [INV] [cos] [sin]

$\boxed{0.8660254}$

Navigation

A navigator of a ship or aircraft can use trigonometry to find the shortest path between any two points on the earth's surface. Since, for large distances, the earth may be regarded as a sphere, the shortest path between two points is the length of an arc of a <u>great circle</u>.

A **great circle** of a sphere is the circle formed by the intersection of the sphere and a plane passing through its center. The **equator** is the great circle formed by the plane passing through the center of the earth and perpendicular to the earth's North-South axis. A circle formed by a plane parallel to the equator is called a **parallel of latitude,** or simply a **parallel.** Half of a great circle passing through the North and South poles is called a **meridian of longitude,** or a **meridian.** By international agreement, the **prime meridian** is the meridian which passes through the city of Greenwich, England.

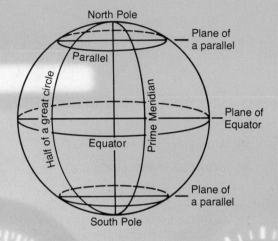

To find the shortest path between two points on the earth's surface, a navigator must know these data.

1. The **latitude** or angular measure of each point's parallel of latitude north or south of the equator.

2. The **longitude** or angular measure of each point's meridian east or west of the prime meridian.

In the figure at the right, α_1 represents the latitude of P_1 and β_1 represents its longitude. Thus, for $\alpha_1 = 60.5°$ and $\beta_1 = 48.2°$, the location of P_1 is represented as follows.

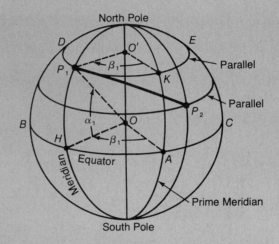

Latitude: ←——   **60.5° north of the**
60.5° N   **equator**

Longitude: ←——   **48.2° west of the**
48.2° W   **prime meridian**

It can be shown that for any two points P_1 and P_2 in the Northern Hemisphere, the arc measure θ (in radians) of $\overset{\frown}{P_1P_2}$ can be found by the formula

$$\theta = \text{Arc cos } (\sin \alpha_1 \sin \alpha_2 + \cos \alpha_1 \cos \alpha_2 \cos \phi)$$

where α_1 and α_2 are the latitudes of points P_1 and P_2 respectively and ϕ is the change in longitude from P_1 to P_2.

Since $s = R\theta$ (see page 67), the length of the path from P_1 to P_2 along a great-circle route can be given by

$$s = R\theta = R \text{ Arc cos } (\sin \alpha_1 \sin \alpha_2 + \cos \alpha_1 \cos \alpha_2 \cos \phi)$$

where R is the radius of the earth. Although in the above formula θ must be in radians, α_1, α_2, and ϕ may be expressed in degree measure.

E X E R C I S E S

1. Find the shortest flight path for a plane flying from Seattle (latitude: 47.5° N; longitude: 122.5° W) to New York (latitude: 40.5° N; longitude: 74.0° W). The radius of the earth is approximately 3959 miles. Round your answer to the nearest ten miles. (HINT: Since Seattle and New York are both west of the prime meridian, $\phi = 122.5° - 74°$).

2. Find the shortest flight path for a plane flying from Boston (latitude: 42.5° N; longitude: 71.0° W) to Paris (latitude: 49.0° N; longitude: 2.5° E). Round your answer to the nearest ten miles. (HINT: Since Boston and Paris are on opposite sides of the prime meridian, $\phi = 71.0° + 2.5°$).

5-4 Trigonometric Equations

A **conditional equation** is an equation that is true for some, but not all, permissible replacements of the variable. Thus, the trigonometric equation

$$\sin x = \frac{1}{2}$$

is a conditional equation because it is true only for the following values of x.

$$\cdots -\frac{11\pi}{6}, -\frac{7\pi}{6}, \frac{\pi}{6}, \frac{5\pi}{6}, \frac{13\pi}{6}, \cdots$$

Two basic techniques that are often useful in solving trigonometric equations are factoring and applying known identities. Example 1 shows how both of these techniques are used in solving a trigonometric equation.

Example 1 Solve: $\sin 2x = 4 \sin x$

Solution: First use an identity to obtain all expressions in terms of x.

$$\sin 2x = 4 \sin x \qquad \longleftarrow \text{ Use a Double-angle Identity.}$$
$$2 \sin x \cos x = 4 \sin x \qquad \longleftarrow \begin{array}{l}\text{Obtain an equation with}\\ \text{one side equal to zero.}\end{array}$$
$$\sin x \cos x - 2 \sin x = 0 \qquad \longleftarrow \text{ Factor.}$$
$$\sin x (\cos x - 2) = 0 \qquad \longleftarrow \begin{array}{l}\text{If } ab = 0, \text{ then}\\ a = 0 \ \underline{\text{or}} \ b = 0.\end{array}$$
$$\sin x = 0 \ \ \underline{\text{or}} \ \ \cos x - 2 = 0$$

If $\sin x = 0$, then $x = n\pi$, $n \in \mathscr{I}$.

If $\cos x - 2 = 0$, then $\cos x = 2$. But the range of $y = \cos x$ is $\{y: \ -1 \le y \le 1, \ y \in R\}$. Therefore, the equation $\cos x = 2$ has no solution.

Thus, the solution set is

$$\{x: \ x = n\pi, \ n \in \mathscr{I}\}, \ \ \underline{\text{or}} \ \ \{x: \ x = n \cdot 180°, \ n \in \mathscr{I}\}.$$

It is often necessary to express trigonometric functions in an equation in terms of one trigonometric function only.

Example 2

Solve: $\sin^2 x - \cos^2 x = 0$

Solution:

$\sin^2 x - \cos^2 x = 0$ ⟵ Express in terms of the sine function only.

$\sin^2 x - (1 - \sin^2 x) = 0$ ⟵ $\cos^2 x = 1 - \sin^2 x$

$2 \sin^2 x - 1 = 0$

$\sin^2 x = \dfrac{1}{2}$

$\sin x = \dfrac{\sqrt{2}}{2}$ or $\sin x = -\dfrac{\sqrt{2}}{2}$

If $\sin x = \frac{\sqrt{2}}{2}$, then $x = 45° + n \cdot 360°$ or $x = 135° + n \cdot 360°$, $n \in \mathscr{I}$.

If $\sin x = -\frac{\sqrt{2}}{2}$, then $x = 225° + n \cdot 360°$ or $x = 315° + n \cdot 360°$, $n \in \mathscr{I}$.

The angle measures 45°, 135°, 225°, and 315° differ by 90°. Thus, the solution set is

$$\{x \quad x = 45° + n \cdot 90°, n \in \mathscr{I}\}, \quad \text{or} \quad \left\{x\colon \; x = \dfrac{\pi}{4} + \dfrac{n\pi}{2}, n \in \mathscr{I}\right\}.$$

Sometimes you are asked to find the solution set over a restricted domain, such as $0 \le x < 2\pi$, or $0 \le x < 360°$.

Example 3

Solve: $2 \sin^2 x - \cos x - 1 = 0$, where $0 \le x < 2\pi$.

Solution:

$2 \sin^2 x - \cos x - 1 = 0$ ⟵ Replace $\sin^2 x$ with $1 - \cos^2 x$.

$2(1 - \cos^2 x) - \cos x - 1 = 0$

$2 - 2 \cos^2 x - \cos x - 1 = 0$

$-2 \cos^2 x - \cos x + 1 = 0$

$2 \cos^2 x + \cos x - 1 = 0$

$(2 \cos x - 1)(\cos x + 1) = 0$

$\cos x = \dfrac{1}{2}$ or $\cos x = -1$ ⟵ Find x for the interval $0 \le x < 2\pi$.

For $\cos x = \frac{1}{2}$, $x = \frac{\pi}{3}$ or $x = \frac{5\pi}{3}$; for $\cos x = -1$, $x = \pi$.

Thus, the solution set is

$$\left\{\dfrac{\pi}{3}, \pi, \dfrac{5\pi}{3}\right\} \quad \text{or} \quad \{60°, 180°, 300°\}.$$

Whenever you square both sides of an equation or multiply both sides by a variable that could equal zero, it is important to <u>check</u> all solutions. The reason for this is that such procedures may result in solutions that do not satisfy the original equation. Example 4 illustrates this.

Example 4 Solve: $\dfrac{1 - \cos x}{\sin x} = 1$, where $0 \le x < 2\pi$.

Solution:

$$\dfrac{1 - \cos x}{\sin x} = 1 \qquad \longleftarrow \text{ Multiply both sides by } \sin x.$$

$$1 - \cos x = \sin x \qquad \longleftarrow \begin{array}{l}\text{Square both sides in order to} \\ \text{obtain a Pythagorean identity.}\end{array}$$

$$(1 - \cos x)^2 = (\sin x)^2$$

$$1 - 2 \cos x + \cos^2 x = \sin^2 x \qquad \longleftarrow \text{ Express in terms of } \cos x \text{ only.}$$

$$1 - 2 \cos x + \cos^2 x = 1 - \cos^2 x$$

$$2 \cos^2 x - 2 \cos x = 0$$

$$2 \cos x (\cos x - 1) = 0$$

$$2 \cos x = 0 \quad \underline{\text{or}} \quad \cos x - 1 = 0$$

$$\cos x = 0 \quad \underline{\text{or}} \qquad \cos x = 1$$

If $\cos x = 0$ and $0 \le x < 2\pi$, then $x = \frac{\pi}{2}$ <u>or</u> $x = \frac{3\pi}{2}$.

If $\cos x = 1$ and $0 \le x < 2\pi$, then $x = 0$.

Thus,

$$x = \frac{\pi}{2} \quad \underline{\text{or}} \quad x = \frac{3\pi}{2} \quad \underline{\text{or}} \quad x = 0$$

and the three <u>possible</u> values for x in the interval $0 \le x < 2\pi$ are 0, $\frac{\pi}{2}$, and $\frac{3\pi}{2}$.

Check: $x = 0$: $\quad \dfrac{1 - \cos 0}{\sin 0} = \dfrac{1 - 1}{0} = \dfrac{0}{0} \qquad \longleftarrow \begin{array}{l}\frac{0}{0} \text{ is undefined, so } x = 0 \\ \text{is } \underline{\text{not}} \text{ a solution.}\end{array}$

$x = \dfrac{\pi}{2}$: $\quad \dfrac{1 - \cos \frac{\pi}{2}}{\sin \frac{\pi}{2}} = \dfrac{1 - 0}{1} = 1 \qquad \longleftarrow \begin{array}{l}\text{The original equation} \\ \text{is true for } x = \frac{\pi}{2}.\end{array}$

$x = \dfrac{3\pi}{2}$: $\quad \dfrac{1 - \cos \frac{3\pi}{2}}{\sin \frac{3\pi}{2}} = \dfrac{1 - 0}{-1} = -1 \qquad \longleftarrow \begin{array}{l}\text{The original equation} \\ \text{is false for } x = \frac{3\pi}{2}.\end{array}$

Thus, the solution set is $\left\{\dfrac{\pi}{2}\right\}$, or $\{90°\}$.

Classroom Exercises

Solve for x.

1. $2 \sin x - \sqrt{3} = 0$

2. $\tan x - 1 = 0 \; (0 \le x < 2\pi)$

3. $\tan x \sec x = \tan x \; (0 \le x < 2\pi)$

4. $2 \sin x - \sqrt{2} = 0 \; (-\frac{\pi}{2} \le x \le \frac{\pi}{2})$

5. $\tan^2 x - 1 = 0 \; (0 \le x \le \pi)$

6. $2 \cos^2 x = \cos x \; (-\frac{\pi}{2} \le x \le \frac{\pi}{2})$

Written Exercises

1-16
odd

a

Solve each equation.

1. $2 \sin x + \sqrt{3} = 0$

2. $\sqrt{3} \cot x + 1 = 0$

3. $\sqrt{2} \cos x - 1 = 0$

4. $4 \tan x - 4 = 0$

5. $\sqrt{3} \sec x + 2 = 0$

6. $2 \cos^2 x = \cos x$

Solve each equation on the interval $0 \le x < 2\pi$.

7. $4 \sin^2 x = 1$

8. $3 \sin^2 x - \cos^2 x = 0$

9. $2 \tan^2 x - 3 \sec x + 3 = 0$

10. $\sqrt{3} \csc^2 x + 2 \csc x = 0$

11. $\cos 2x + \sin x = 1$

12. $\sin 2x + \cos x = 0$

13. $4 \tan x + \sin 2x = 0$

14. $\sin 2x = 2 \sin x$

15. $\sin x = \cos x$

16. $\dfrac{\sin x}{1 + \cos x} = 1$

b

17. $\tan 2x \cot x - 3 = 0$

18. $\cos^2 x + \cos 2x = \frac{5}{4}$

19. $\cos^2 x - \cos 2x = -\frac{3}{4}$

20. $\tan x + \tan 2x = 0$

21. $\sin 2x \sin x + \cos 2x \cos x = 1$

22. $\cos x = \cos 2x$

23. $\sin \left(\frac{\pi}{4} + x\right) - \sin \left(\frac{\pi}{4} - x\right) = \frac{\sqrt{2}}{2}$

24. $\cos \left(\frac{\pi}{4} + x\right) + \cos \left(\frac{\pi}{4} - x\right) = 1$

25. $\sin 2x \cos x - \cos 2x \sin x = -\frac{\sqrt{3}}{2}$

26. $|\sin x| = \frac{1}{2}$

27. $\cos 2x + 3 \cos x - 1 = 0$

28. $\cos 2x + \cos x = 0$

29. $\sin^2 x - \cos^2 x - \cos x - 1 = 0$

30. $\cos 3x + \cos x = 0$

31. $\sin 3x + \sin x = 0$

32. $\cot x \tan 2x = 3$

c

33. $6 \cos^2 x + 5 \cos x + 1 = 0$

34. $\tan^3 2x = \tan 2x$

35. $2 \tan x - 2 \cot x = -3$

36. $2 \sin^3 x - \sin x = 0$

37. $\sin 2x + \cos 3x = 0$

38. $4 \sin^4 x + \sin^2 x = 3$

For Exercises 39–41, use the Sum/Product Identities.

39. $\sin x + \sin 2x + \sin 3x = 0$

40. $\cos x - \cos 3x - \cos 5x = 0$

41. $\sin x + \sin 2x - \sin 4x = 0$

____ Review _____

Find the inverse of each function. Then graph the given function and its inverse on the same coordinate plane. (Section 5–1)

1. $y = x + 2$ **2.** $y = -\sin x$ **3.** $y = \frac{1}{3}x$

Evaluate each expression in terms of degrees. (Section 5–1)

4. arc sin $\left(-\frac{\sqrt{2}}{2}\right)$ **5.** arc sin 1 **6.** arc cos $\frac{\sqrt{3}}{2}$

Evaluate each expression. Give angle measures in radians. (Section 5–2)

7. Arc cos 1 **8.** cos (Arc sin $\frac{\sqrt{2}}{2}$) **9.** sin [Arc cos $\left(-\frac{3}{5}\right)$]

Evaluate each expression. (Section 5–3)

10. arc tan $\sqrt{3}$ **11.** Arc sec 1 **12.** csc (Arc csc 4)

Solve each equation. (Section 5–4)

13. $4 \sin x - \sqrt{12} = 0$ **14.** $\sqrt{3} \csc x - 2 = 0$ **15.** $\tan^2 x = \tan x$

16. $2 \cos^2 x + \sin x - 1 = 0$

Review Capsule for Section 5-5 _____

Change each degree measure to radian measure in terms of π.

1. 180° **2.** 270° **3.** −720° **4.** 240°

5. 300° **6.** 120° **7.** −60° **8.** 150°

Change each radian measure to degree measure.

9. $-\frac{\pi}{3}$ **10.** $\frac{2\pi}{5}$ **11.** $\frac{5\pi}{6}$ **12.** -4π

13. $\frac{7\pi}{12}$ **14.** $\frac{7\pi}{2}$ **15.** $-\frac{3\pi}{4}$ **16.** $\frac{9\pi}{4}$

5-5 Polar Coordinates

The position of any point P in the plane is determined by an ordered pair of numbers (x, y). These are called **rectangular,** or **Cartesian, coordinates.** (See Figure 1.) Another way of determining the position of a point is to give its polar coordinates.

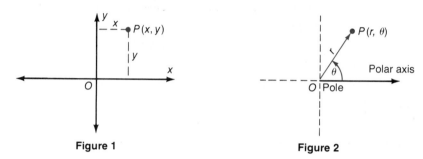

Figure 1   Figure 2

In Figure 2, the numbers r and θ in the ordered pair (r, θ) are called the **polar coordinates** of P. Vertex O of angle θ is called the **pole** and the ray corresponding to the positive x axis is called the **polar axis.** In a polar coordinate system, angle measures may be designated in either radian measure or degree measure.

A point in a polar coordinate system may be named by many different pairs of polar coordinates.

Example 1 Point P has polar coordinates $(2, 210°)$. Graph point P. Then find three other polar coordinate pairs that represent P.

Solution: 1. Draw an angle of $210°$ in standard position.

2. Locate point P two units away from the pole O on the terminal side of the angle.

3. Find other polar coordinates for P by adding or subtracting multiples of $360°$ to $210°$. Thus, some other coordinates for P are $P(2, 570°)$, $P(2, 930°)$, $P(2, -150°)$.

Example 1 suggests the following.

$$P(r, \theta) = P(r, \theta + n \cdot 360°), \text{ where } n \in \mathscr{I}$$

or $$P(r, \theta) = P(r, \theta + 2n\pi), \text{ where } n \in \mathscr{I}$$

There is still another way to identify the same point. In polar coordinates, r is not restricted to positive numbers. When r is negative, you first draw the angle θ. Then mark the point $|r|$ units from the pole on the ray opposite to the terminal side of θ. Recall that the ray opposite to the terminal side of θ is the terminal side of $\theta + 180°$. Thus, in the figure at the right $(-2, 30°)$ is the same point as $(2, 210°)$.

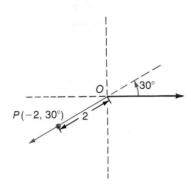

Example 2 Graph each point.

a. $P(-2, 40°)$ **b.** $Q(-2, -170°)$

Solutions: 1. Draw θ in standard position.

2. Measure $|r|$ units from the pole on the ray opposite to the terminal side of θ.

a.

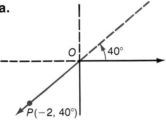

b.

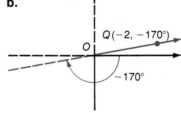

In general, if r is any real number and $\overrightarrow{OQ}$ is the terminal side of an angle with measure θ in standard position, then point P with polar coordinates (r, θ) is located as follows.

1. If $r > 0$, P is on $\overrightarrow{OQ}$.

2. If $r < 0$, P is on the ray opposite $\overrightarrow{OQ}$.

3. If $r = 0$, P is the pole, O.

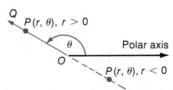

Written Exercises

a

In each of Exercises 1–4, the numbers of the given ordered pair are the polar coordinates of point P. Graph point P. Then find two other polar coordinate pairs that represent P.

1. $(2, 45°)$ **2.** $(3, 180°)$ **3.** $(4, \frac{5\pi}{4})$ **4.** $(\frac{1}{2}, 135°)$

Graph each point. Then, for each point, write polar coordinates (r, θ), where $r > 0$.

5. $(-5, \frac{\pi}{6})$ **6.** $(-\frac{1}{2}, \frac{3\pi}{2})$ **7.** $(-2, 0)$ **8.** $(-2, 390°)$

Change the given rectangular coordinates to polar coordinates.

9. $(1, 1)$ **10.** $(0, 4)$ **11.** $(-9, 0)$ **12.** $(-1, 1)$
13. $(1, -\sqrt{3})$ **14.** $(-1, -\sqrt{3})$ **15.** $(-2, \sqrt{12})$ **16.** $(0, -1)$

Change the given polar coordinates to rectangular coordinates.

17. $(4, \frac{3\pi}{4})$ **18.** $(1, \frac{7\pi}{6})$ **19.** $(5, 270°)$ **20.** $(-2, 90°)$
21. $(-1, \frac{7\pi}{6})$ **22.** $(3, -\frac{2\pi}{3})$ **23.** $(-9, \frac{13\pi}{3})$ **24.** $(-\frac{3}{2}, -\frac{\pi}{4})$

Express each equation in polar coordinate form.

25. $y = -4$ **26.** $2x - y = 3$
27. $y = x$ **28.** $x^2 + y^2 + 2y = 0$
29. $xy = 16$ **30.** $y^2 = 4x$
31. $x^2 + y^2 = 16$ **32.** $x^2 + y^2 + 4x = 0$

Express each equation in rectangular coordinate form.

33. $r = 2 \cos \theta$ **34.** $r = 3 \sin \theta$
35. $r = 3$ **36.** $r = -4$
37. $r = 2 \sin \theta + 2 \cos \theta$ **38.** $r \sin \theta = 6$

b

39. $r = 4 \sec \theta$ **40.** $r^2 - 3r + 2 = 0$
41. $\theta = 0°$ **42.** $\theta = \frac{\pi}{4}$

c

43. Prove that if a point (x_1, y_1) has polar coordinates (r, θ) and if (x_2, y_2) has polar coordinates $(r, -\theta)$, then $x_1 = x_2$ and $y_1 = -y_2$.

5-6 Graphs of Polar Equations

Equations involving polar coordinates are called **polar equations.** The set of points (r, θ) that satisfy a polar equation is called a **polar graph.**

Example 1 Graph $r = 5$.

Solution: The polar graph of $r = 5$ is the set of points with polar coordinates $(5, \theta)$. Thus, the graph is a circle with center at the pole and radius 5. (Note that θ can have any value, either in radian measure or degree measure).

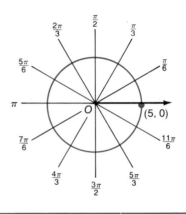

Example 2 Graph $\theta = 140°$.

Solution: The graph of $\theta = 140°$ is the set of points with coordinates $(r, 140°)$. Thus, the graph is a line through the pole. (Note that r can have any real number value.)

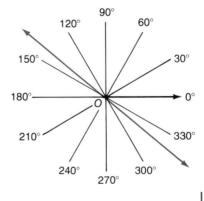

It is usually necessary to construct a table of values before drawing a polar graph.

Example 3 Graph $r = 3 \cos \theta$.

Solution: First, make a table of values. The numbers to the right of the table provide a way of indicating corresponding points on the graph.

θ	$\cos \theta$	r	
0	1	3	1
$\frac{\pi}{6}$	$\frac{\sqrt{3}}{2}$	$\frac{3\sqrt{3}}{2}$	2
$\frac{\pi}{3}$	$\frac{1}{2}$	$\frac{3}{2}$	3
$\frac{\pi}{2}$	0	0	4
$\frac{2\pi}{3}$	$-\frac{1}{2}$	$-\frac{3}{2}$	5
$\frac{5\pi}{6}$	$-\frac{\sqrt{3}}{2}$	$-\frac{3\sqrt{3}}{2}$	6
π	-1	-3	7
$\frac{7\pi}{6}$	$-\frac{\sqrt{3}}{2}$	$-\frac{3\sqrt{3}}{2}$	8
$\frac{4\pi}{3}$	$-\frac{1}{2}$	$-\frac{3}{2}$	9
$\frac{3\pi}{2}$	0	0	10
$\frac{5\pi}{3}$	$\frac{1}{2}$	$\frac{3}{2}$	11

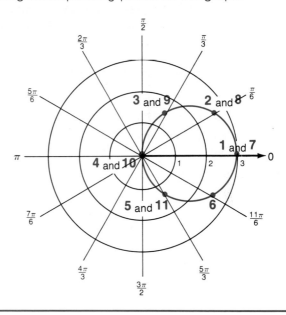

The graph in Example 3 appears to be a circle. To check this, change the equation to rectangular coordinates to determine if it is of the form

$$(x - h)^2 + (y - k)^2 = a^2.$$

$r = 3 \cos \theta$	← Multiply each side by r.
$r^2 = 3r \cos \theta$	← $r^2 = x^2 + y^2$ and $r \cos \theta = x$.
$x^2 + y^2 = 3x$	
$x^2 - 3x + y^2 = 0$	← Complete the square on x.
$x^2 - 3x + \left(\frac{3}{2}\right)^2 + y^2 = \left(\frac{3}{2}\right)^2$	← Express $x^2 - 3x + \left(\frac{3}{2}\right)^2$ as
$\left(x - \frac{3}{2}\right)^2 + y^2 = \left(\frac{3}{2}\right)^2$	the square of a binomial.

Since $\left(x - \frac{3}{2}\right)^2 + y^2 = \left(\frac{3}{2}\right)^2$ is the equation of a circle with center at $\left(\frac{3}{2}, 0\right)$ and radius $\frac{3}{2}$, the polar graph is a circle.

When making a table for a polar graph, be sure that the values in the table extend two or three points beyond where the points begin to repeat. The reason for this is shown in Examples 4 and 5.

Example 4 Graph $r = 1 + \sin \theta$.

Solution: Make a table of values. Here, degree measure is used.
The graph is shown at the right below. The points that are shown correspond to the values in the table.

θ	$\sin \theta$	$r \; (= 1 + \sin \theta)$	
0°	0	1	
30°	.5	1.5	
60°	.866	1.866	
90°	1	2	
120°	.866	1.866	
150°	.5	1.5	
180°	0	1	
210°	−.5	.5	
240°	−.866	.134	
270°	−1	0	
300°	−.866	.134	
330°	−.5	.5	
360°	0	1	← Repeated point
390°	.5	1.5	← Repeated point
420°	.866	1.866	← Repeated point

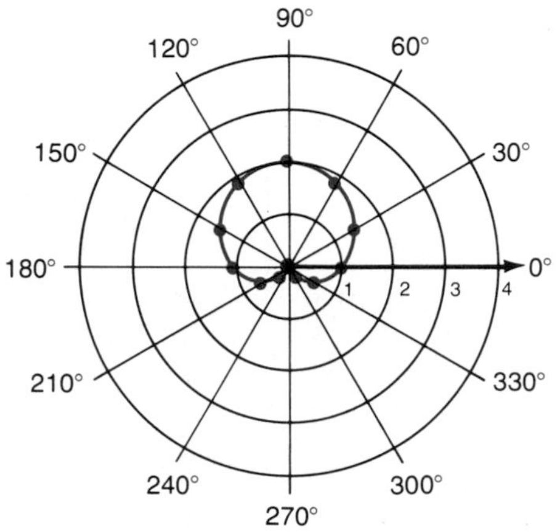

The graph in Example 4 is called a <u>cardioid</u>. It is important to plot more than just one "repeated point" to be sure that the entire curve has been graphed. Example 5 shows that a "repeated point" of a polar graph may be followed by a point that is <u>not</u> repeated.

Example 5 Graph $r = \sin 2\theta$.

Solution: Four successive portions are identified at the right of the table and also on the graph. Arrows show the order in which the points are plotted.

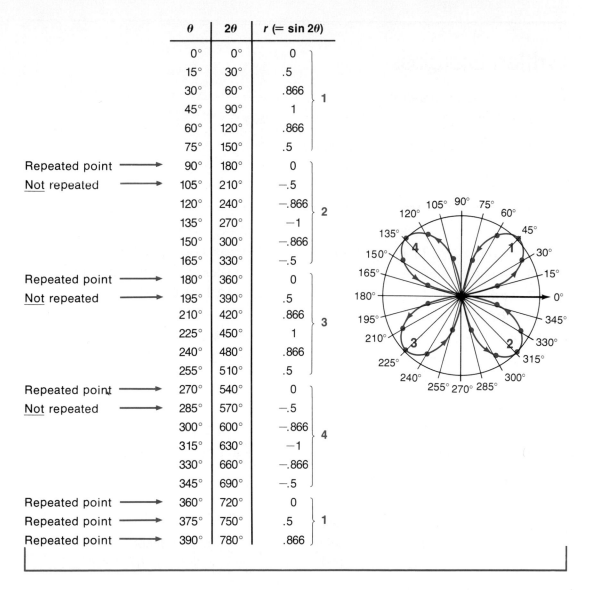

θ	2θ	$r\ (= \sin 2\theta)$	
0°	0°	0	
15°	30°	.5	
30°	60°	.866	1
45°	90°	1	
60°	120°	.866	
75°	150°	.5	
Repeated point → 90°	180°	0	
Not repeated → 105°	210°	−.5	
120°	240°	−.866	2
135°	270°	−1	
150°	300°	−.866	
165°	330°	−.5	
Repeated point → 180°	360°	0	
Not repeated → 195°	390°	.5	
210°	420°	.866	3
225°	450°	1	
240°	480°	.866	
255°	510°	.5	
Repeated point → 270°	540°	0	
Not repeated → 285°	570°	−.5	
300°	600°	−.866	4
315°	630°	−1	
330°	660°	−.866	
345°	690°	−.5	
Repeated point → 360°	720°	0	
Repeated point → 375°	750°	.5	1
Repeated point → 390°	780°	.866	

The graph of Example 5 is called a four-leaved rose.

Classroom Exercises

Graph each equation. Then describe the graph.

1. $r = 2$ **2.** $\theta = 45°$ **3.** $r = 6 \sin \theta$ **4.** $r \cos \theta = 2$

Written Exercises

Graph each equation.

1. $r = 1$

2. $r = -3$

3. $\theta = 75°$

4. $\theta = -\frac{\pi}{3}$

5. $r = \cos\theta$

6. $r = 2\sin\theta$

7. $r = -3\cos\theta$

8. $r = -4\sin\theta$

9. $r = 1 + \cos\theta$ (cardioid)

10. $r = \sin\theta - 1$ (cardioid)

11. $r = \cos\theta - 1$ (cardioid)

12. $r = 2\sin 3\theta$ (3-leaved rose)

13. $r = 3\cos 3\theta$ (3-leaved rose)

14. $r = \cos 2\theta$ (4-leaved rose)

15. $r = \sin 5\theta$ (5-leaved rose)

16. $r = \cos 4\theta$ (8-leaved rose)

17. $r = 2 + \cos\theta$ (limaçon)

18. $r = 2 + \sin\theta$ (limaçon)

19. $r = 2 + 3\cos\theta$ (limaçon)

20. $r = 1 - 2\sin\theta$ (limaçon)

21. $r^2 = 4\sin 2\theta$ (lemniscate)

22. $r^2 = 4\cos 2\theta$ (lemniscate)

23. $r = 2\theta$ (spiral of Archimedes)

24. $r\theta = 3$ (hyperbolic spiral)

25. $r = \dfrac{4}{1 - \cos\theta}$ (parabola)

26. $r = \dfrac{2}{1 - \frac{1}{2}\cos\theta}$ (ellipse)

27. $r = \dfrac{16}{1 - 4\cos\theta}$

28. $r = \dfrac{2}{3 - 2\cos\theta}$

29. A circle has radius a and the polar coordinates of its center are (r_1, θ_1). Show that a polar equation of the circle is

$$a^2 = r^2 + r_1^{\,2} - 2rr_1\cos(\theta - \theta_1).$$

(HINT: Use the Law of Cosines.)

30. A perpendicular is drawn from the pole to a line that does not include the pole. The polar coordinates of the intersection point are (r_1, θ_1). Show that a polar equation of the line is $r = r_1\sec(\theta - \theta_1)$.

31. Graph $r = \sin\theta + 1$ and $r = \sin\theta - 1$ on the same polar coordinate system.

32. Show that there is no ordered pair (r_1, θ_1) that satisfies the equation $r = \sin\theta + 1$ and $r = \sin\theta - 1$ simultaneously. Explain why this fact is consistent with the result of Exercise 31.

33. Graph $r = \sin\theta - 1$ and $r = \cos\theta + 1$ on the same polar coordinate system. Determine which intersection points correspond to simultaneous solutions of $r = \sin\theta - 1$ and $r = \cos\theta + 1$.

Chapter Objectives and Review

Objective: To know the meanings of the mathematical terms in this chapter.

1. Be sure that you know the meanings of these mathematical terms.

arc cosecant (p. 220)
Arc cosecant (p. 220)
arc cosine (p. 212)
Arc cosine (p. 215)
arc cotangent (p. 220)
Arc cotangent (p. 220)
arc secant (p. 220)
Arc secant (p. 220)
arc sine (p. 212)
Arc sine (p. 215)
arc tangent (p. 220)
Arc tangent (p. 220)
conditional equation (p. 226)
inverse relation (p. 210)
polar axis (p. 231)
polar coordinates (p. 231)

polar equation (p. 236)
polar graph (p. 236)
pole (p. 231)
principal values
   of arc cosecant (p. 220)
   of arc cosine (p. 215)
   of arc cotangent (p. 220)
   of arc secant (p. 220)
   of arc sine (p. 215)
   of arc tangent (p. 220)
rectangular coordinates (p. 231)
reflection (p. 211)
set-builder notation (p. 212)
solution set (p. 212)
trigonometric equation (p. 226)

Objective: To find the inverse of a given function. (Section 5–1)

Find the inverse of each function. Then graph the given function and its inverse on the same coordinate plane.

2. $y = 2x - 4$ **3.** $y = \sin x$ **4.** $y = \frac{1}{2}x$ **5.** $y = \cos x$

Objective: To evaluate expressions involving arc sine and arc cosine. (Section 5–1)

Evaluate each expression in terms of degrees.

6. arc $\sin\left(-\frac{\sqrt{3}}{2}\right)$ **7.** arc $\sin 0$ **8.** arc $\cos\frac{\sqrt{2}}{2}$ **9.** arc $\cos\frac{1}{2}$

Objective: To evaluate expressions involving Arc sine and Arc cosine. (Section 5–2)

Evaluate each expression. Give angle measures in radians.

10. Arc $\sin\frac{1}{2}$ **11.** Arc $\cos\frac{\sqrt{2}}{2}$ **12.** $\sin\left(\text{Arc } \sin\frac{\sqrt{3}}{2}\right)$

13. $\cos\left(\text{Arc } \cos 1\right)$ **14.** $\sin\left[\text{Arc } \cos\left(-\frac{1}{2}\right)\right]$ **15.** $\cos\left[\text{Arc } \sin(-1)\right]$

Objective: To evaluate expressions involving the inverses of the tangent, cotangent, secant, and cosecant functions. (Sections 5–3)

Evaluate each expression in terms of degrees.

16. arc csc (-2) **17.** Arc cot $(-\sqrt{3})$

Objective: To solve trigonometric equations. (Section 5–4)

Solve each equation.

18. $2 \cos x + \sqrt{3} = 0$ **19.** $2 \sin^2 x = \sin x$

20. Solve: $3 \cos^2 x - \sin^2 x = 0$, where $0 \le x < 2\pi$.

Objective: To graph a point given its polar coordinates. (Section 5–5)

In each of Exercises 21–24, the numbers in the given ordered pair are the polar coordinates of point P. Graph point P.

21. $(2, 30°)$ **22.** $(-3, 150°)$ **23.** $(5, -\frac{3\pi}{4})$ **24.** $(-1, -\frac{5\pi}{6})$

Objective: To express rectangular coordinates as polar coordinates and vice versa. (Section 5–5)

Change the given polar coordinates to rectangular coordinates and change the given rectangular coordinates to polar coordinates.

25. Polar coordinates $(3, \frac{5\pi}{6})$ **26.** Rectangular coordinates $(-3, -3)$

27. Rectangular coordinates $(5, -\sqrt{75})$ **28.** Polar coordinates $(-2, -30°)$

Express each equation in the form indicated.

29. $x = 5$; polar coordinate form

30. $r = 2 \cos \theta$; rectangular coordinate form

31. $x^2 = 4y$; polar coordinate form

32. $r \cos \theta = 6$; rectangular coordinate form

Objective: To graph polar equations. (Section 5–6)

Graph each equation.

33. $r = 3$ **34.** $\theta = \frac{2\pi}{3}$

35. $r = -3 \sin \theta$ **36.** $r = 3 \cos 2\theta$

1. Find the inverse of the function $y = 4x - 12$.

2. Graph $y = \cos x$ and its inverse on the same coordinate plane.

Evaluate each expression.

3. arc sin $\left(-\frac{1}{2}\right)$

4. sin [Arc cos (-1)]

5. arc sec (-2)

6. cot $\left(\text{Arc tan} \frac{2}{5}\right)$

Solve.

7. $\sqrt{3} \cot x + 1 = 0$

8. $\tan^2 x - \frac{1}{2} \sec^2 x = 0$

Change the given rectangular coordinates to polar coordinates.

9. $(-1, 1)$

10. $(-2, \sqrt{12})$

Change the given polar coordinates to rectangular coordinates.

11. $(-2, 90°)$

12. $\left(-\frac{3}{2}, -\frac{\pi}{4}\right)$

13. Graph the point with polar coordinates $\left(-3, -\frac{2\pi}{3}\right)$.

14. Express the equation $y^2 = 9x$ in polar coordinate form.

15. Graph: $r = -2 \sin 2\theta$

Cumulative Review: Chapters 1-5

Write the letter of the response that best answers each question.

1. Find the distance between $P(2, -3)$ and $Q(1, 4)$. (Section 1–1)

    a. $\sqrt{10}$ **b.** $\sqrt{2}$ **c.** $5\sqrt{2}$ **d.** $\sqrt{58}$

2. Find the measure of an angle that is coterminal under two clockwise rotations with an angle that is in standard position and that has a measure of 45°. (Section 1–2)

    a. 405°   **b.** −405°   **c.** 765°   **d.** −675°

Point P(x, y) is on the terminal side of an angle with measure θ in standard position. P is located r units from the origin. Find the correct ratio for each given trigonometric function.

3. $\cos \theta$ (Section 1–3)

    a. $\frac{y}{r}$ **b.** $\frac{x}{y}$ **c.** $\frac{x}{r}$ **d.** $\frac{y}{x}$

4. $\cot \theta$ (Section 1–4)

    a. $\frac{x}{y}$ **b.** $\frac{y}{x}$ **c.** $\frac{y}{r}$ **d.** $\frac{x}{r}$

5. Find the value of $\cos 60°$. (Section 1–5)

    a. $\frac{\sqrt{3}}{2}$ **b.** $\frac{1}{2}$ **c.** 1   **d.** 0

6. Use the table on pages 368–372 to evaluate $\tan 243°30'$. (Section 1–6)

    a. 2.006   **b.** −1.052   **c.** .4986   **d.** −.8949

7. Refer to the figure at the right. Then find θ to the nearest 10 minutes. (Sections 1–7 and 1–8)

    a. 32°   **b.** 32°10′

    c. 32°15′   **d.** 32°20′

8. The rafters of a roof meet the horizontal plate to form an angle of 45°. Find the pitch. (Sections 1–9, 1–10, 1–11, and 1–12)

    a. 1   **b.** 2   **c.** $\frac{1}{2}$ **d.** $\frac{1}{4}$

9. Change 47° to radian measure. (Section 2–1)

    a. 94π **b.** 47π **c.** $47 \cdot 180\pi$ **d.** $\frac{47\pi}{180}$

10. A wheel is spinning with an angular velocity of 100 rad/sec. Find the distance in centimeters traveled during $\frac{1}{10}$ second by a point on the wheel that is 10 cm from the wheel's center. (Section 2–3)

    a. 100   **b.** 10   **c.** 10π **d.** 100π

11. Find the equality that is <u>not</u> an identity. (Section 2–4)

a. $\sin(\pi + \theta) = \sin\theta$ **b.** $\cos(\pi + \theta) = -\cos\theta$

c. $\sin(\pi - \theta) = \sin\theta$.   **d.** $\cos(\pi - \theta) = -\cos\theta$

12. Tell how the graph of $y = 3 + \sin x$ is related to the graph of $y = \sin x$. (Section 2–5)

a. 3 units below $y = \sin x$ **b.** 3 units right of $y = \sin x$

c. 3 units above $y = \sin x$ **d.** 3 units left of $y = \sin x$

13. Find the period of $y = -\frac{2}{3}\cos(4x - \frac{\pi}{2})$. (Sections 2–6, 2–7, and 2–8)

a. $\frac{\pi}{8}$ **b.** $\frac{2}{3}$ **c.** $\frac{\pi}{2}$ **d.** π

14. Find the function that is undefined at π and has period π. (Section 2–10)

a. $y = \tan x$ **b.** $y = \cot x$ **c.** $y = \sec x$ **d.** $y = \csc x$

15. Find the equality that is an identity. (Sections 3–1 and 3–2)

a. $\tan^2\theta = 1 - \cot^2\theta$ **b.** $\tan\theta = \cos\theta - 1$

c. $\sin^2\theta \cdot \tan^2\theta = 1 - \sin^2\theta$ **d.** $1 + \tan^2\theta = \dfrac{1}{1 - \sin^2\theta}$

16. Evaluate $\cos\frac{13\pi}{12}$ without tables. (Section 3–3)

a. $-\frac{1}{4}(\sqrt{2} + \sqrt{6})$ **b.** $\frac{1}{4}(\sqrt{2} + \sqrt{6})$ **c.** $-\frac{1}{4}(\sqrt{2} - \sqrt{6})$ **d.** $\frac{1}{4}(\sqrt{2} - \sqrt{6})$

17. Evaluate $\sin 15°$ without tables. (Section 3–4)

a. $\frac{1}{4}(\sqrt{6} + \sqrt{2})$ **b.** $-\frac{1}{4}(\sqrt{6} + \sqrt{2})$ **c.** $\frac{1}{4}(\sqrt{6} - \sqrt{2})$ **d.** $-\frac{1}{4}(\sqrt{6} - \sqrt{2})$

18. Evaluate $\tan 345°$ without using tables. (Section 3–5)

a. $2 - \sqrt{3}$ **b.** $-2 - \sqrt{3}$ **c.** $-2 + \sqrt{3}$ **d.** $2 + \sqrt{3}$

19. If $\cos\alpha = \frac{3}{5}$ and α is in Quadrant I, evaluate $\tan 2\alpha$. (Section 3–6)

a. $\frac{12}{7}$ **b.** $-\frac{7}{12}$ **c.** $-\frac{7}{24}$ **d.** $-\frac{24}{7}$

20. Find the equality that is <u>not</u> an identity. (Section 3–7)

a. $\sin\frac{\alpha}{2} = \pm\sqrt{\dfrac{1 + \cos\alpha}{2}}$ **b.** $\tan\frac{\alpha}{2} = \pm\sqrt{\dfrac{1 - \cos\alpha}{1 + \cos\alpha}}$

c. $\tan\frac{\alpha}{2} = \dfrac{\sin\alpha}{1 + \cos\alpha}$ **d.** $\tan\frac{\alpha}{2} = \dfrac{1 - \cos\alpha}{\sin\alpha}$

21. In $\triangle ABC$, $A = 24°$, $C = 100°$, and $b = 100$ cm. Find a to the nearest centimeter. (Section 4–1)

a. 49 cm   **b.** 50 cm   **c.** 119 cm   **d.** 118 cm

22. In $\triangle ABC$, $a = 10$, $b = 20$, and $C = 47°$. Find c to the nearest whole number. (Section 4–2)

a. 19   **b.** 15   **c.** 20   **d.** 360

23. In $\triangle ABC$, $A = 30°$, $a = 7$, and $b = 15$. Find the number of solutions for the triangle. (Section 4–3)

    a. 2   **b.** 1   **c.** 0   **d.** Not enough information

24. In $\triangle ABC$, $a = 250$, $b = 150$, and $C = 60°$. Find $\tan\left(\frac{A-B}{2}\right)$. (Section 4–4)

    a. $\frac{1}{12}\sqrt{3}$ **b.** $-\frac{1}{12}\sqrt{3}$ **c.** $\frac{1}{4}\sqrt{3}$ **d.** $-\frac{1}{4}\sqrt{3}$

25. In $\triangle ABC$, $a = 5.1$, $b = 7.2$, and $C = 63°$. Find the area of $\triangle ABC$ to the nearest square unit. (Section 4–5)

    a. 32.7   **b.** 33   **c.** 8   **d.** 16

26. In $\triangle ABC$, $a = 5$, $b = 7$, and $c = 10$. Find the area of the triangle to the nearest square unit. (Section 4–6)

    a. 15   **b.** 16   **c.** 225   **d.** 264

27. Vectors OF and OW are component vectors of $\overrightarrow{OR}$ and have the origin as their initial point. Their direction angles are $150°$ and $270°$, respectively, and $|\overrightarrow{OF}| = 10$, $|\overrightarrow{OW}| = 4$. Find $|\overrightarrow{OR}|$ to the nearest unit. (Sections 4–7, 4–8, 4–9, and 4–10)

    a. 12   **b.** 9   **c.** 7   **d.** 14

28. Name the inverse of $y = \sin x$. (Section 5–1)

    a. $y = \dfrac{x}{\sin x}$ **b.** $y = \dfrac{1}{\sin x}$

    c. $y = \text{arc } \sin x$ **d.** $y = \cos x$

29. Evaluate $\cos\left(\text{Arc } \cos \frac{1}{4}\right)$. (Section 5–2)

    a. $75.5°$ **b.** $14.5°$ **c.** $.25$ **d.** $.97$

30. Evaluate $\tan[\text{Arc } \cot(-1)]$. (Section 5–3)

    a. -1 **b.** 1   **c.** 0   **d.** Undefined

31. Solve $\cot x + \sqrt{3} = 0$ on the interval $0 \leq x < 2\pi$. (Section 5–4)

    a. $\left\{\frac{5\pi}{6}\right\}$ **b.** $\left\{\frac{5\pi}{6}, \frac{11\pi}{6}\right\}$ **c.** $\left\{\frac{2\pi}{3}\right\}$ **d.** $\left\{\frac{2\pi}{3}, \frac{5\pi}{3}\right\}$

32. Name the pair of polar coordinates that refers to the same point as $(-2, 40°)$. (Section 5–5)

    a. $(2, 220°)$ **b.** $(2, -320°)$ **c.** $(-2, 220°)$ **d.** $(-2, -40°)$

33. Describe the graph of the polar equation $r = 16$. (Section 5–6)

    a. A circle with radius 4   **b.** A circle with radius 16

    c. A line through the origin   **d.** A cardioid

Chapter 6
Circular Functions and Applications

6-1 The Wrapping Function

A circle whose center is the origin and whose radius is one is a **unit circle.** A point $P(x, y)$ is on the unit circle if and only if the coordinates of P satisfy the equation of the circle; that is, if and only if

$$x^2 + y^2 = 1.$$

If the point $P(x, y)$ is on the unit circle, then $P_1(-x, y)$, $P_2(-x, -y)$, and $P_3(x, -y)$ are also on the circle.

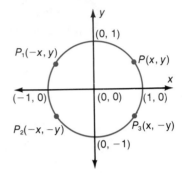

Recall that the formula for the circumference of a circle is $C = 2\pi r$. Since $r = 1$ in the unit circle, the circumference is $C = 2\pi$. Thus, a string wrapped around the unit circle exactly once would be 2π units long.

Consider a vertical number line that is tangent to the unit circle at $A(1, 0)$. The scale on the number line is the same as the scale on the x– and y–axes. Imagine "wrapping" the positive part of the number line around the unit circle in a counterclockwise direction. The point $\frac{\pi}{2}$ on the number line would coincide with the point $B(0, 1)$ on the unit circle, because arc AB is one-fourth of the circle and

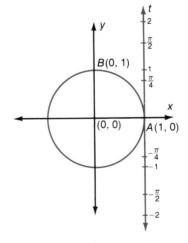

$$\frac{1}{4} \cdot 2\pi = \frac{\pi}{2}.$$

Similar reasoning would result in the following table of pairings, where each point with coordinate t on the number line is paired with a point with coordinates (x, y) on the unit circle.

t	0	$\frac{\pi}{2}$	π	$\frac{3\pi}{2}$	2π	$\frac{5\pi}{2}$	3π	$\frac{7\pi}{2}$	4π
(x, y)	(1, 0)	(0, 1)	(−1, 0)	(0, −1)	(1, 0)	(0, 1)	(−1, 0)	(0, 1)	(1, 0)

Table 1

If the negative part of the number line is wrapped around the unit circle in a clockwise direction, then the pairings shown in Table 2 are obtained.

t	$-\frac{\pi}{2}$	$-\pi$	$-\frac{3\pi}{2}$	-2π	$-\frac{5\pi}{2}$	-3π	$-\frac{7\pi}{2}$	-4π
(x, y)	$(0, -1)$	$(-1, 0)$	$(0, 1)$	$(1, 0)$	$(0, -1)$	$(-1, 0)$	$(0, 1)$	$(1, 0)$

Table 2

Continuing this wrapping, you will find that each point with coordinate t on the number line maps onto exactly one point with coordinates (x, y) on the unit circle. Thus, the wrapping process defines a function that is called the **wrapping function** W. W maps the set of real numbers onto the points of the unit circle.

Therefore, if t is a number on the vertical number line tangent to the unit circle at $A(1, 0)$ and if (x, y) is a point on the circle, then

$$W(t) = (x, y).$$

Thus, $W(0) = (1, 0)$, $W(\frac{\pi}{2}) = (0, 1)$, $W(\pi) = (-1, 0)$ and so on.

Example   Given that $W(2) = (-0.4161, 0.9093)$, find the following.

a. $W(\pi + 2)$ **b.** $W(-2)$ **c.** $W(\pi - 2)$

Solutions: **a.** Since $W(\pi + 2)$ is half way around the circle from $W(2)$, it is on the end of a diameter with $W(2)$ as one endpoint. Thus,

$$W(\pi + 2) = (0.4161, -0.9093).$$

b. $W(-2)$ is the same distance from $A(1, 0)$ as $W(2)$ but in a clockwise direction. Thus,

$$W(-2) = (-0.4161, -0.9093).$$

c. Think of $\pi - 2$ as $-2 + \pi$. Then $W(\pi - 2)$ is halfway around the circle from $W(-2)$. This means that it is on the end of a diameter with $W(-2)$ as one endpoint. Thus,

$$W(\pi - 2) = (0.4161, 0.9093).$$

The example suggests the following properties of the wrapping function W.

If $W(t) = (x, y)$, then

1. $W(-t) = (x, -y)$
2. $W(\pi - t) = (-x, y)$
3. $W(\pi + t) = (-x, -y)$

As Table 1 shows, real numbers greater than 2π are mapped onto points that are already paired with the real numbers t, where $0 \leq t \leq 2\pi$. If (x, y) is paired with t, then (x, y) is also paired with $t + 2\pi$, $t + 4\pi$, $t + 6\pi$, and so on.

Table 2 shows a similar relationship for $t \leq 0$. That is, if (x, y) is paired with t, then (x, y) is also paired with $t - 2\pi$, $t - 4\pi$, $t - 6\pi$, and so on. Thus, W is a periodic function with period 2π, or

4. $W(t + 2n\pi) = (x, y)$, where n is an integer.

Classroom Exercises

Tell whether each point is on a unit circle.

1. $(-0.6, 0.4)$ **2.** $(0.8, -0.6)$ **3.** $(0.5, 0.5)$ **4.** $\left(\frac{\sqrt{3}}{2}, \frac{1}{2}\right)$

5. What is the period of the wrapping function W?

Given: $W(t) = \left(\frac{\sqrt{2}}{2}, \frac{\sqrt{2}}{2}\right)$. *Find each of the following.*

6. $W(-t)$ **7.** $W(\pi - t)$ **8.** $W(\pi + t)$ **9.** $W(2\pi + t)$

Written Exercises

For each value of t in Exercises 1–8, find the quadrant in which $W(t) = (x, y)$ lies.

1. 1 **2.** -1 **3.** 6 **4.** -3

5. $-\dfrac{7\pi}{6} + \pi$ **6.** $\dfrac{1004\pi}{3}$ **7.** $\dfrac{15\pi}{2}$ **8.** -22

9. Given: $W(\frac{1}{2}) = (0.8776, 0.4794)$. Find two positive numbers other than $\frac{1}{2}$ that map onto this point.

10. Given: $W(-\frac{1}{2}) = (0.8776, -0.4794)$. Find two negative numbers other than $-\frac{1}{2}$ that map onto this point.

11. Given: $W(2.1) = (-0.5048, 0.8632)$. Find a negative number and a positive number other than 2.1 that map onto this point.

In Exercises 12–17, W(t) is given. Find $W(-t)$, $W(\pi - t)$, $W(\pi + t)$, and $W(2\pi + t)$.

12. $W(t) = (\frac{12}{13}, \frac{5}{13})$

13. $W(t) = (-\frac{4}{5}, \frac{3}{5})$

14. $W(t) = (-\frac{1}{2}, -\frac{\sqrt{3}}{2})$

15. $W(t) = (\frac{2}{3}, \frac{\sqrt{5}}{3})$

16. $W(t) = (-\frac{\sqrt{42}}{7}, \frac{\sqrt{7}}{7})$

17. $W(t) = (\frac{\sqrt{13}}{\sqrt{14}}, -\frac{1}{\sqrt{14}})$

In Exercises 18–21, $W(-\pi) = (-1, 0)$. Find each of the following.

18. $W(\pi)$ **19.** $W(-3\pi)$ **20.** $W(3\pi)$ **21.** $W(21\pi)$

The wrapping function $W(t) = (x, y)$ is periodic with period 2π. Describe whether x increases (becomes more positive) or decreases (becomes more negative) as t increases over the given interval.

22. $0 \le t \le \frac{\pi}{2}$ **23.** $\frac{\pi}{2} < t \le \pi$ **24.** $\pi < t \le \frac{3\pi}{2}$ **25.** $\frac{3\pi}{2} < t \le 2\pi$

26–29. Describe whether y increases or decreases under the same conditions given in Exercises 22–25.

30. Given: $(0.3135, y)$ is a point on the unit circle. Find y.

31. Given: $(x, 0.1358)$ is a point on the unit circle. Find x.

32. Explain why it is impossible for $W(t)$ to equal $(1, 1)$.

The wrapping function S is defined by wrapping a number line around a square as shown at the right.

Evaluate S(t) for the given values of t.

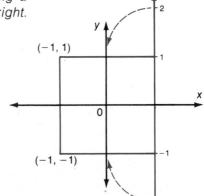

33. 0   **34.** 1

35. 2   **36.** 3

37. 4   **38.** 5

39. 6   **40.** 7

41. 8   **42.** 9

43. 10   **44.** −3

45. Is S periodic? If so, what is the period?

6-2 Circular Functions

Since the wrapping function W maps each real number t onto an ordered pair of real numbers (x, y) such that $x^2 + y^2 = 1$, two new functions can be defined. The first maps t onto the first coordinate x; the second maps t onto the second coordinate y.

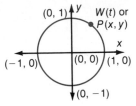

Definitions: Given $W(t) = (x, y)$, the cosine function, $\cos t$, maps t onto x, and the sine function, $\sin t$, maps t onto y. In symbols,

$$\cos t = x \quad \text{and} \quad \sin t = y.$$

These functions are called **circular functions,** because their values are coordinates of points on the unit circle. The domain for $\cos t$ and $\sin t$ is the set of real numbers and the range is the set of real numbers between -1 and 1 inclusive. Since $(x, y) = (\cos t, \sin t)$ and $x^2 + y^2 = 1$, it readily follows that $(\cos t)^2 + (\sin t)^2 = 1$, or

$$\cos^2 t + \sin^2 t = 1.$$

Example 1

Given $\sin t = -\frac{5}{13}$, use $\cos^2 t + \sin^2 t = 1$ to find $\cos t$.

Solution: $x^2 + \left(-\dfrac{5}{13}\right)^2 = 1$ ⟵ Replace y with $-\dfrac{5}{13}$.

$$x^2 = 1 - \frac{25}{169} = \frac{144}{169}$$

Thus, $x = \pm\sqrt{\dfrac{144}{169}} = \pm\dfrac{12}{13}.$

Since $\sin t < 0$, $P(x, y)$ is in Quadrant III or IV.

$\therefore x = -\dfrac{12}{13}$ ⟵ If $P(x, y)$ is in Quadrant III

<u>or</u> $x = \dfrac{12}{13}$ ⟵ If $P(x, y)$ is in Quadrant IV

You can use Tables 1 and 2 on pages 248 and 249 to find the value of the sine and cosine functions when $P(x, y)$ is on an axis.

Example 2 Use Table 1 on page 248 to find the following.

a. $\sin \frac{\pi}{2}$ **b.** $\cos \frac{5\pi}{2}$

Solutions: a. Since $\frac{\pi}{2}$ maps onto $(0, 1)$ and $\sin t = y$, $\sin \frac{\pi}{2} = 1$.

b. Since $\frac{5\pi}{2}$ maps onto $(0, 1)$ and $\cos t = x$, $\cos \frac{5\pi}{2} = 0$.

Since $W(t) = (\cos t, \sin t)$ and $(x, y) = (\cos t, \sin t)$, the four properties of W given in Section 6–1 can be used to derive properties of the sine and cosine functions.

Circular Functions: Sine and Cosine Properties

1. $W(-t) = (x, -y) = (\cos t, -\sin t)$.
$\therefore \cos(-t) = \cos t$ and $\sin(-t) = -\sin t$
2. $W(\pi - t) = (-x, y) = (-\cos t, \sin t)$
$\therefore \cos(\pi - t) = -\cos t$ and $\sin(\pi - t) = \sin t$
3. $W(\pi + t) = (-x, -y) = (-\cos t, -\sin t)$
$\therefore \cos(\pi + t) = -\cos t$ and $\sin(\pi + t) = -\sin t$
4. $W(t + 2n\pi) = (x, y) = (\cos t, \sin t), n \in \mathscr{I}$
$\therefore \cos(t + 2n\pi) = \cos t, n \in \mathscr{I}$ and $\sin(t + 2n\pi) = \sin t, n \in \mathscr{I}$

The sine and cosine functions are used to define four other circular functions—the <u>tangent</u> (tan), <u>cotangent</u> (cot), <u>secant</u> (sec), and <u>cosecant</u> (csc).

Definitions: For any real number t,

$$\tan t = \frac{\sin t}{\cos t}, \cos t \neq 0 \qquad \sec t = \frac{1}{\cos t}, \cos t \neq 0$$

$$\cot t = \frac{\cos t}{\sin t}, \sin t \neq 0 \qquad \csc t = \frac{1}{\sin t}, \sin t \neq 0$$

Properties 1–4 on page 250 can be used to derive properties for these circular functions.

Example 3 Use properties 1–4 on page 250 and the definitions on page 253 to express each of the following as a function of t alone.

a. $\tan (-t)$ **b.** $\tan (\pi - t)$

Solutions: a. $\tan (-t) = \dfrac{\sin (-t)}{\cos (-t)}$ ⟵   Definition of $\tan t$

$\qquad\qquad\qquad = \dfrac{-\sin t}{\cos t}$ ⟵   Property **1**

$\qquad\therefore \tan (-t) = -\dfrac{\sin t}{\cos t}$, or $\tan (-t) = -\tan t$

b. $\tan (\pi - t) = \dfrac{\sin (\pi - t)}{\cos (\pi - t)}$

$\qquad\qquad\qquad = \dfrac{\sin t}{-\cos t}$ ⟵   Property **2**

$\qquad\therefore \tan (\pi - t) = -\dfrac{\sin t}{\cos t}$, or $\tan (\pi - t) = -\tan t$

As you might suspect, <u>all</u> of the properties of the trigonometric functions apply to the circular functions. Therefore, you may wonder why it is necessary to examine the circular functions. Circular functions are functions of <u>real numbers</u> with no need to think of angles. Thus, circular functions can be applied to the solution of problems from science, engineering, and mathematics that involve real numbers that do not explicitly represent angles at all. Some of these applications are considered in the following four sections.

Classroom Exercises

Use $\sin^2 t + \cos^2 t = 1$ *to find the value of* $\sin t$ *or* $\cos t$, *given the following information.*

1. $\cos t = \frac{3}{5}$,   Quadrant I

2. $\cos t = \frac{3}{5}$, Quadrant IV

3. $\sin t = -\frac{2\sqrt{2}}{3}$,   Quadrant III

4. $\sin t = -\frac{2\sqrt{2}}{3}$, Quadrant IV

Use Properties 1–4 on page 250 and the definitions on page 253 to express each of the following as a function of t alone.

5. $\sec (\pi + t)$.   **6.** $\csc (\pi - t)$

Written Exercises

In Exercises 1–18, complete each statement.

1. A unit circle is a circle with center at _____ and with radius _____.
2. A point $P(x, y)$ is on the unit circle if _____.
3. The functions $\cos t = x$ and $\sin t = y$ are called circular functions because _____.
4. The domain of the circular functions $\cos t = x$ and $\sin t = y$ is _____.
5. The range of the circular functions $\cos t = x$ and $\sin t = y$ is _____.
6. The wrapping function W pairs the real number $\frac{3\pi}{2}$ with the ordered pair _____ on the unit circle.
7. For the wrapping function W, if $W(t) = (x, y)$, then $W(-t) = $ _____.
8. For the wrapping function W and for any integer n, $W(t + 2n\pi) = $ _____.
9. Since $W(t) = W(t + 2\ \pi)$, $\cos t = \cos$ _____ and $\sin t = \sin$ _____.
10. The wrapping function W is periodic with period _____.
11. The circular functions $\cos t = x$ and $\sin t = y$ are periodic with period _____.
12. If $W(1) = (0.5403, 0.8415)$, $W(\pi - 1) = $ _____.
13. If $W(1) = (0.5403, 0.8415)$, $W(\pi + 1) = $ _____.
14. If $\cos 1 = 0.5403$, $\cos(\pi - 1) = $ _____.
15. If $\sin 1 = .8415$, $\sin(\pi - 1) = $ _____.
16. If $\cos 1 = .5403$, $\cos(\pi + 1) = $ _____.
17. If $\sin 1 = .8415$, $\sin(\pi + 1) = $ _____.
18. The properties of the trigonometric functions are also the properties of the _____ functions.

Use $\sin^2 t + \cos^2 t = 1$ to find the value of $\sin t$ or $\cos t$ given the following information.

19. $\sin t = -\frac{3}{5}$, Quadrant IV
20. $\sin t = \frac{12}{13}$, Quadrant II
21. $\sin t = -\frac{4}{5}$, Quadrant III
22. $\cos t = \frac{12}{13}$, Quadrant IV

Use properties 1–4 on page 250 and the definitions on page 253 to express each of the following as a function of t alone.

23. $\tan(\pi + t)$
24. $\cot(-t)$
25. $\cot(\pi - t)$
26. $\cot(\pi + t)$
27. $\sec(-t)$
28. $\sec(\pi - t)$
29. $\csc(-t)$
30. $\tan(3\pi + t)$

b

Refer to the figure at the right for Exercises 31–33.

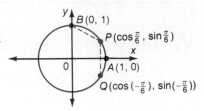

31. Express the coordinates of Q in terms of $\cos \frac{\pi}{6}$ and $\sin \frac{\pi}{6}$.

32. Explain why $BP = PQ$.

33. Find the values of $\cos \frac{\pi}{6}$ and $\sin \frac{\pi}{6}$. (HINT: use the distance formula and $\cos^2 t + \sin^2 t = 1$.)

Find the value of each function.

34. $\sin \frac{7\pi}{6}$ **35.** $\cos \frac{7\pi}{6}$ **36.** $\sin \frac{5\pi}{6}$ **37.** $\cos \frac{5\pi}{6}$

38. $\sin -\frac{\pi}{6}$ **39.** $\cos -\frac{\pi}{6}$ **40.** $\tan \frac{7\pi}{6}$ **41.** $\sec -\frac{\pi}{6}$

Review Capsule for Section 6-3 _____

Match each function with its graph. (For further review, see page 82 and pages 88–104.)

1. $y = -2 \sin x$ **2.** $y = \cos 2x$ **3.** $y = 2 \cos x$ **4.** $y = 2 \sin 2(x + \pi)$

a.

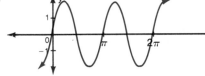

b.

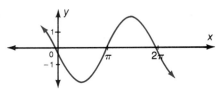

c.

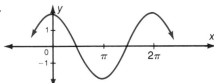

d.

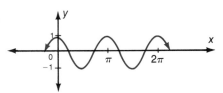

Graph the following functions for $-2\pi \le x \le 4\pi$. (For further review, see pages 80–95.)

5. $y = \sin x$ **6.** $y = \cos x$ **7.** $y = 2 \sin x$ **8.** $y = \sin 2x$

Find the period of each function. (For further review, see pages 93–95 and pages 99–101.)

9. $y = \cos x$ **10.** $y = 2 \sin x$ **11.** $y = \sin 2x$ **12.** $y = \cos \frac{1}{2}x$

13. $y = \sin (x - \frac{\pi}{2})$ **14.** $y = \cos 2(x + \frac{\pi}{3})$

15. $y = \cos (2x + \frac{\pi}{3})$ **16.** $y = -\sin (\frac{1}{2}x - \pi)$

6-3 Periodic Motion

In Chapter 2 you learned that functions of the form

$$y = A \sin B(x - C) \quad \text{and} \quad y = A \cos B(x - C)$$

are <u>periodic functions</u>, that is, functions that repeat their values at regular intervals. (See page 83 for the formal definition.) If t (for "time") replaces x as the independent variable, then the above formulas may represent **periodic motion,** such as the oscillation of a coiled spring.

One complete execution of an event that is repeated over and over again is called a **cycle.** The graph below shows several cycles for the function $y = \sin t$.

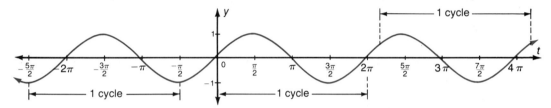

In Section 2–8 the period of the function $y = A \sin B(t - C)$ was shown to be $\frac{2\pi}{|B|}$. On the graph, the period p is the distance along the t axis that is required to complete one cycle. For periodic motion, the period p is the duration of time required for one cycle. The **frequency** f is the number of cycles per unit of time.

Example 1

One complete rotation of the earth about its axis is one cycle. Find the frequency f in cycles per hour of the periodic rotation of the earth about its axis.

Solution: Since the earth completes one cycle per day and there are 24 hours in a day,

$$f = \frac{1 \text{ cycle}}{1 \text{ day}}$$

$$= \frac{1 \text{ cycle}}{24 \text{ hours}}$$

$$= \tfrac{1}{24} \text{ or about .042 cycles per hour.}$$

Since 24 hours is the period p for the rotation of the earth about its axis, Example 1 suggests the following formula.

$$f = \frac{1}{p} \hspace{6em} \textbf{1}$$

Formula **1** is used to define the frequency f for trigonometric and circular functions.

Example 2 Find the frequency of each function.

a. $y = -5 \sin(2t + \pi)$ **b.** $y = \tan 3\pi t$

Solutions: a. Write $2t + \pi$ as $2(t + \frac{\pi}{2})$.

$y = -5 \sin 2(t + \frac{\pi}{2})$

$p = \dfrac{2\pi}{|B|}$ ⟵ See page 94.

$p = \dfrac{2\pi}{2}$, or π

Thus, $f = \dfrac{1}{p}$. ⟵ From Formula 1

$f = \dfrac{1}{\pi}$ cycles per unit

b. $p = \dfrac{\pi}{|B|}$ ⟵ See page 111.

$p = \dfrac{\pi}{3\pi}$

$p = \dfrac{1}{3}$

⟶ $f = \dfrac{1}{p}$

$f = \dfrac{1}{\frac{1}{3}}$, or 3 cycles per unit

Example 2 suggests the following formula for the frequency f of functions of the form $y = A \sin B(t - C)$ or $y = A \cos B(t - C)$.

$$f = \frac{|B|}{2\pi} \hspace{6em} \textbf{2}$$

Example 3 Find a function of the form $y = \sin Bt$ with a frequency of two cycles per unit.

Solution: Since $f = 2$, $\dfrac{|B|}{2\pi} = 2$.   ⟵ From Formula 2

Therefore, $|B| = 2 \cdot 2\pi = 4\pi$.   ⟵ $B = 4\pi$ <u>or</u> $B = -4\pi$.

Thus, $y = \sin 4\pi t$ and $y = \sin(-4\pi t)$ are functions that are of the required form.

Classroom Exercises

Find the frequency of each function.

1. $y = \sin\frac{1}{2}t$

2. $y = \sec 4t$

3. $y = \cos 2t$

4. $y = \tan(6t + \pi)$

5. $y = \csc 4t$

6. $y = \cot(\frac{1}{2}t + \pi)$

Written Exercises

a

In Exercises 1–4, find the frequency in cycles per unit for each function.

1.

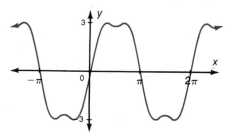

2.

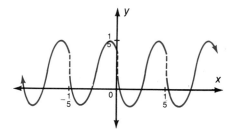

3.

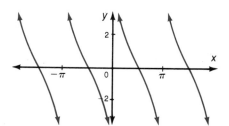

4.

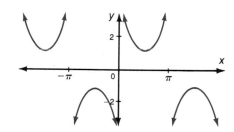

Find the frequency f of each function.

5. $y = 5 \cos 3t$

6. $y = 2 \sec(-6\pi t)$

7. $y = \sin(4t + \pi)$

8. $y = -3 \cot(2t + \frac{\pi}{2})$

Find a function of the form $y = \sin Bt$ with the given frequency, where f is in cycles per unit.

9. $f = \frac{1}{2}$

10. $f = \frac{1}{2\pi}$

11. $f = \frac{1}{\pi}$

12. $f = 4$

Find a function of the form $y = \cos Bt$ with the given frequency, where f is in cycles per unit.

13. $f = \frac{2}{\pi}$

14. $f = 1$

15. $f = \frac{1}{4}$

16. $f = 3$

17. Rotary motion is a special type of periodic motion. For an object in rotary motion, each complete revolution is one cycle. Find the frequency f in cycles per second of an object in rotary motion with angular velocity $\omega = 32\pi$ rad/min.

Find a function that is of the form $y = A \sin B(t + C)$ and that has the given characteristics.

18. Amplitude: 5; frequency: $\frac{1}{2\pi}$; phase shift: $\frac{\pi}{3}$ units to the right

19. Amplitude: $\frac{2}{3}$; frequency: $\frac{4}{\pi}$; phase shift: $\frac{\pi}{8}$ units to the left

6-4 Applications: Simple Harmonic Motion

In Figure 1 below, an object is attached to a spring, and it is hanging at rest (position 0). When the spring is stretched to the position in Figure 2 (position $-A$) and then released, the object will oscillate, or travel up and down repeatedly. The maximum displacement of the object above and below its rest position 0 will be A units. The object is in simple periodic motion, or **simple harmonic motion,** as long as no outside force, such as friction, acts to change the motion. If pictures could be taken of the "bouncing object" at equal intervals of time, Figure 3 would result.

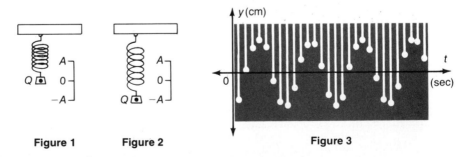

Figure 1   **Figure 2**   **Figure 3**

The **sinusoidal** (sine-like) shape of the graph suggests that the position of the object is a function of time.

It can be shown that the following two formulas are models for any object in simple harmonic motion.

$$y = A \sin B(t - C), \ t \geq 0 \qquad y = A \cos B(t - C), \ t \geq 0$$

These equations give the position, y, of an object as a function of the time, t (in seconds), where $|A|$ is the amplitude (maximum displacement of the object).

Example 1 In the figure at the right, a spherical ball oscillates in simple harmonic motion. The sphere is shown at its low point (minimum) in the cycle. At this point, $t = 0$. The sphere completes $\frac{1}{2}$ cycle in one second. Graph the vertical position of the sphere as a function of time.

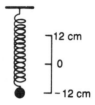

Solution: The sphere oscillates between a minimum of -12 and a maximum of 12. Thus, the amplitude of the graph is 12.

The frequency is $\frac{1}{2}$ cycle per second. Thus, the period is given by

$$p = \frac{1}{f} \qquad \longleftarrow \text{ See page 258.}$$

$$= \frac{1}{\frac{1}{2}}$$

$$= 2$$

The period is 2 seconds.
Now you can use this information to sketch the graph.

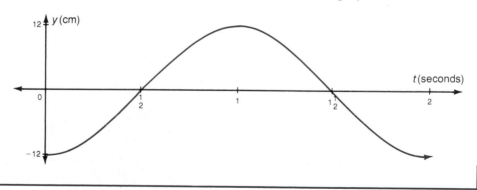

Using one of the formulas for harmonic motion, you can find an equation of the graph of Example 1. This is shown in Example 2 on the next page.

Example 2 Find an equation of the graph in Example 1.

Solution: To find an equation of the graph choose either of the equations of simple harmonic motion, say

$$y = A \sin B(t - C).$$

The amplitude is given by

$$\text{amplitude} = |A|.$$

So, from the graph, either $A = 12$ or $A = -12$. Choose one of these two possibilities:

$$A = 12 \quad \longleftarrow \quad \text{If } -12 \text{ is chosen for } A, \text{ a different but correct equation will result.}$$

Thus, $\qquad y = 12 \sin B(t - C).$ **1**

Next, let $t = 0$. Then, from the graph, $y = -12$ and equation **1** becomes:

$$-12 = 12 \sin B(0 - C)$$

or $\qquad -12 = 12 \sin (-BC) \quad \longleftarrow \quad \sin(-x) = -\sin x$

$$-12 = -12 \sin BC$$

$$1 = \sin BC$$

$$\text{Arc} \sin 1 = \text{Arc} \sin (\sin BC) \quad \longleftarrow \quad \begin{array}{l}\text{Arc} \sin (\sin x) = x \text{ only}\\ \text{if } x \text{ is a principal value.}\end{array}$$

$$\frac{\pi}{2} = BC \qquad \qquad \mathbf{2}$$

Next, let $t = 1$. From the graph, $y = 12$ and equation **1** can be used again to give another equation involving B and C.

$$12 = 12 \sin B(1 - C)$$

$$1 = \sin (B - BC)$$

$$\text{Arc} \sin 1 = B - BC \quad \longleftarrow \quad \begin{array}{l}B - BC \text{ must be a}\\ \text{principal value.}\end{array}$$

$$\frac{\pi}{2} = B - BC \qquad \qquad \mathbf{3}$$

Add the corresponding sides of equations **2** and **3**. The result is

$$\pi = B.$$

Substitute this value in equation **2**. The result is

$$C = \frac{1}{2}.$$

Thus, $y = 12 \sin \pi (t - \frac{1}{2}).$ $\qquad \longleftarrow \quad A = 12; \ B = \pi; \ C = \frac{1}{2}$

There are several other equations such as

$$y = -12 \cos \pi t$$

that describe the motion in Example 1. You are asked to verify some of these alternate solutions in the exercises.

Written Exercises

a

In the figure at the right, a spherical ball oscillates in simple harmonic motion between the two extreme points shown. In Exercises 1–6, use the figure and the given information to graph the vertical position of the sphere as a function of time.

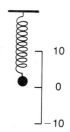

1. $t = 0$ when $y = 0$. The sphere completes one cycle in one second.
2. $t = 0$ when $y = 0$. The sphere completes $\frac{1}{4}$ cycle in one second.
3. $t = 0$ when $y = -10$. The sphere completes $\frac{1}{4}$ cycle in one second.
4. $t = 0$ when $y = -10$. The sphere completes 2 cycles in one second.
5. $t = 0$ when $y = 10$. The sphere completes $\frac{1}{2}$ cycle in one second.
6. $t = 0$ when $y = 10$. The sphere completes $\frac{1}{4}$ cycle in one second.

7–12. Use the graphs of Exercises 1–6 to find an equation for each graph. Use $y = A \sin B(t - C)$.

b

The figure at the right illustrates the simple harmonic motion of a sphere oscillating between two points. In Exercises 13 and 14, use the figure and the given information to graph the vertical position of the sphere as a function of time. Then use the graph to find an equation of this motion. Use $y = A \sin B(t - C)$.

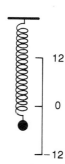

13. $t = 0$ when $y = -5$. Frequency: $\frac{1}{4}$ cycle per second
14. $t = 0$ when $y = 5$. Frequency: $\frac{1}{2}$ cycle per second

15–20. For the graphs of Exercise 1–6, find an equation of simple harmonic motion based upon $y = A \cos B(t - C)$ rather than $y = A \sin B(t - C)$.

21. Sketch the graph of $y = 12 \cos \pi(t - 1)$ for $0 < t < 2$.
22. Sketch the graph of $y = -12 \cos \pi t$ for $0 < t < 2$.

C **23.** Use the results of Exercises 21 and 22 to verify that $y = 12 \cos \pi (t - 1)$ and $y = -12 \cos \pi t$ are equations that describe the simple harmonic motion of the object in Example 1. If an object oscillates in simple harmonic motion with equation $y = A \sin B(t - C)$, then the velocity of the object as a function of time can be shown to be

$$v = AB \cos B(t - C).$$

In Exercises 24–29, find the velocity of the oscillating sphere under the conditions of Exercise 1 for the times given.

24. $t = 0$ second   **25.** $t = \frac{1}{4}$ second

26. $t = \frac{1}{2}$ second   **27.** $t = \frac{3}{4}$ second

28. $t = 1$ second   **29.** $t = 2$ second

In the figure at the right, a particle moves from point S(0, −2) at time t = 0 to point P along the right half of the circle. The particle casts a shadow, represented by point Q, on the y-axis. When the particle reaches point T(0, 2), it continues to move down the left half of the circle until it reaches S and then begins the entire cycle over again. The angular velocity of the particle is 4π rad/sec. The particle's shadow oscillates between S and T in simple harmonic motion with equation $y = A \sin B(t - C)$. Use these facts to answer Exercises 30–36.

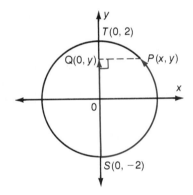

30. Find the amplitude, $|A|$.

31. Find the time required for the particle to complete one cycle. (HINT: $\theta = \omega t$; see page 71. Note that $\theta = 0$ when $t = 0$.)

32. Find the frequency of the simple harmonic motion in cycles/second.

33. Find the period, $\dfrac{2\pi}{|B|}$.

34. Find the phase shift, C.

35. Write an equation for the vertical position, y, of the particle's shadow as a function of time, t.

36. Show that if A is the radius of the circle of Exercises 30–35 and ω is the angular velocity of the rotating particle, then

$$y = A \sin \left(\omega t - \tfrac{\pi}{2} \right)$$

is an equation for the position of the particle's shadow on the y axis as a function of time.

6-5 Applications: Electricity

In this figure, copper wire coiled around an armature is rotating through a magnetic field. When it is rotating at a constant angular velocity ω, an electromotive force, or **emf,** is created. The magnitude E of an emf can be expressed as a function of time t (in seconds) by the formula,

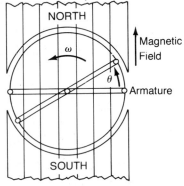

$$E = E_m \sin \omega t, \; t \geq 0,$$

where E_m is the maximum emf. The unit of measure for E is the volt. This formula is an instance of the formula $y = A \sin B(t - C)$, with E_m and ω used, respectively, instead of A and B. Here, E_m and ω are positive numbers and the phase shift, C, is taken to be 0.

The number of rotations, or cycles, that the coil completes in one second is the frequency of the alternating emf. (See Section 6–3.) The frequency is given by

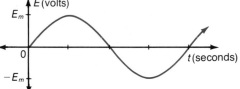

$$f = \frac{\omega}{2\pi} \text{ cycles per second.}$$

A graph of emf as a function of time is shown at the right.

Example 1 A wire coil rotates counterclockwise through a magnetic field and is described by the equation $E = 100 \sin 250\pi t$.

a. Find the frequency of the rotation of the coil.
b. Find E when $t = .001$ seconds.

Solutions: a. In the equation, $E = 100 \sin 250\pi t$, $\omega = 250\pi$.

$$f = \frac{\omega}{2\pi}$$

$$= \frac{250\pi}{2\pi} = 125 \text{ cycles per second}$$

Circular Functions and Applications **265**

b. When $t = .001$ seconds:

$$E = 100 \sin 250\pi (.001)$$
$$= 100 \sin .25\pi$$
$$= 100 \cdot \frac{\sqrt{2}}{2}$$
$$= 50 \sqrt{2}, \text{ or about 70.7 volts}$$

Example 2

A wire coil rotates through a magnetic field 60 times each second. The maximum value (E_m) of the alternating emf is 110 volts. Write an equation of the form $E = E_m \sin \omega t$.

Solution: Since the frequency is 60 cycles per second,

$$f = \frac{\omega}{2\pi}$$

$$60 = \frac{\omega}{2\pi}$$

$120\pi = \omega$, or $\omega = 120\pi$ radians per second.

Since $E_m = 110$, $E = 110 \sin 120\pi t$.

Example 3 illustrates that the emf can change rapidly in a very small interval of time.

Example 3

Use the equation of Example 2 to find the value of E at the times given.

a. $\frac{1}{2}$ sec

b. $\frac{181}{360}$ sec

Solutions: a. When $t = \frac{1}{2}$:

$$E = 110 \sin \left(120\pi \cdot \tfrac{1}{2}\right)$$
$$= 110 \sin 60\pi$$
$$= 110 \sin (0 + 2\pi \cdot 30) \quad \longleftarrow \quad \sin (t + 2\pi n) = \sin t$$
$$= 110 \sin 0$$
$$= 110(0), \text{ or 0 volts}$$

b. When $t = \dfrac{181}{360}$: (This is only $\dfrac{1}{360}$ second later than in **a.**)

$$E = 110 \sin \left(120\pi \cdot \dfrac{181}{360}\right)$$
$$= 110 \sin \dfrac{181\pi}{3}$$
$$= 110 \sin \left(\dfrac{\pi}{3} + 60\pi\right) \quad \longleftarrow \quad 60\pi = 30 \cdot 2\pi$$
$$= 110 \sin \dfrac{\pi}{3}$$
$$= 110 \cdot \dfrac{\sqrt{3}}{2}$$
$$= 55\sqrt{3}, \text{ or about 95 volts}$$

Written Exercises

a

In Exercises 1 and 2, an equation for an alternating emf is given. Find the frequency of the rotation of the coil.

1. $E = 120 \sin 90\pi t$

2. $E = 100 \sin 150\pi t$

3. In Exercise 1, find the value of E when $t = .1$ sec.

4. In Exercise 2, find the value of E when $t = .01$ sec.

In Exercises 5–8, write an equation for an alternating emf of the form $E = E_m \sin \omega t$ with the given E_m and frequency.

5. $E_m = 220$ volts
Frequency = 60 cycles per second

6. $E_m = 110$ volts
Frequency = 50 cycles per second

7. $E_m = 100$ volts
Frequency = 62 cycles per second

8. $E_m = 120$ volts
Frequency = 58 cycles per second

In Exercises 9–12, use the equation of Exercise 5 to find the emf at the times given.

9. $t = \dfrac{1}{60}$ sec

10. $t = \dfrac{1}{360}$ sec

11. $t = 1$ sec

12. $t = \dfrac{361}{720}$ sec

b

Graph the following functions. (Choose scales for the t– and E–axes that are different from each other.)

13. $E = 100 \sin 250\pi t$

14. $E = 110 \sin 120\pi t$

15. $E = 120 \sin 90\pi t$

16. $E = 100 \sin 150\pi t$

C **17.** The flow of electric current I is measured in **amperes** (or amps). For a simple generator, this can be given as a function of the form

$$I = I_m \sin \omega t.$$

Write an equation for a 60-cycle current with a maximum of 15 amps.

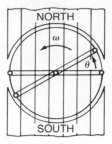

6-6 Applications: Sound

Vibrations, such as those created by the strings of a guitar, produce <u>sound</u>. Sound, in turn, can be converted to a graphical image on the television-like screen of an <u>oscilloscope</u>.

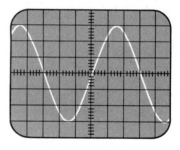

If the graphical image has a sinusoidal shape, then the sound is called a **simple sound,** and can be represented by an equation of the form

$$y = A \sin Bt, t \geq 0 \qquad \text{or} \qquad y = A \cos Bt, t \geq 0,$$

where $|A|$ is the amplitude and t is the time in seconds. The <u>frequency of the vibrations</u> is given by the formula

$$f = \frac{|B|}{2\pi},$$

where B is the coefficient of t in the equation $y = A \sin Bt$ or $y = A \cos Bt$. As in earlier applications, the phase shift is taken to be zero.

An example of a simple sound is "middle C" on the musical scale. Middle C has a frequency of 264 cycles per second.

Example 1 Write an equation of the form $y = A \sin Bt$ for the oscilloscope image of middle C. Assume that the amplitude is 1 unit.

Solution: Since the amplitude, $|A|$, is 1, $A = 1$ or $A = -1$. For convenience, choose $A = 1$.

The frequency is 264 cycles per second. Thus,

$$f = \frac{|B|}{2\pi}$$

or, $|B| = 2\pi f.$

Thus, $|B| = 2\pi (264)$

$= 528\pi,$

and so

$B = 528\pi$ or $B = -528\pi.$

Thus, either

$y = \sin 528\pi t,$

or $y = -\sin 528\pi t$ ◀———— Recall: $\sin(-t) = -\sin t$

is an equation for the oscilloscope graph of "middle C."

Example 2 Graph the equations of the oscilloscope image of Example 1.

Solutions:

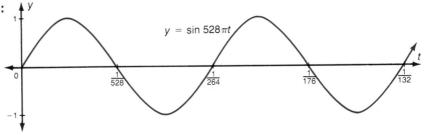

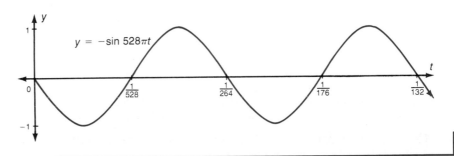

Example 2 illustrates the fact that the exact form of an equation of simple harmonic motion depends upon where you begin clocking the motion, in other words, at what point in the cycle you choose to have $t = 0$.

Circular Functions and Applications   **269**

Written Exercises

In Exercises 1–4, write an equation of the form $y = A \sin Bt$ for the oscilloscope image for a simple sound with the given pitch and frequency. Assume that the amplitude is one unit. (HINT: In each case two equations can be found. Find both equations.)

1. c'' (one octave above middle C)
   Frequency: 528 cycles per second

2. c''' (two octaves above middle C)
   Frequency: 1056 cycles per second

3. c (one octave below middle C)
   Frequency: 132 cycles per second

4. C (two octaves below middle C)
   Frequency: 66 cycles per second

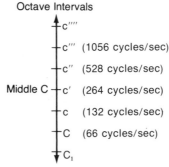

In Exercises 5–8, graph the given equations.

5. The two equations of Exercise 1

6. The two equations of Exercise 2

7. The two equations of Exercise 3

8. The two equations of Exercise 4

In Exercises 9–11, sketch the second graph of Example 2 but with the y axis located as indicated.

9. $\frac{1}{2}$ cycle to the right of its original position

10. $\frac{1}{4}$ cycle to the right of its original position

11. $\frac{1}{4}$ cycle to the left of its original position

In Exercises 12–14, write an equation for each of the following.

12. The graph of Exercise 9

13. The graph of Exercise 10 (HINT: Use $y = A \cos Bt$.)

14. The graph of Exercise 11 (HINT: Use $y = A \cos Bt$.)

15. Acoustical engineers often study the pressure variations in a sound wave. In a certain sound wave, the pressure p is given by

$$p = 10 \sin 200\pi \left(t - \frac{x}{1000}\right) \text{ dynes/cm}^2$$

where t is in seconds, and the distance x from the source of the sound is in centimeters. Sketch the graph of p as a function of x at time $t = \frac{1}{200}$.

*The strings of a musical instrument such as a guitar can vibrate only at certain frequencies. Since the ends of the string cannot move, plucking the string produces a traveling wave that is reflected from the ends to produce a **standing wave.** (See the figure below.)*

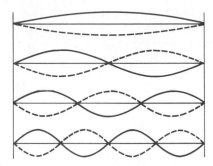

The possible frequencies in cycles per second at which a string can vibrate are given by

$$\frac{V}{2L}, \; 2\frac{V}{2L}, \; 3\frac{V}{2L}, \; \cdots$$

where L is the length of the string and V is the velocity of waves in the string.

Use this information for Exercises 16 and 17.

16. The velocity of waves in a certain string is 480 m/sec. The string is 60 centimeters long. Find the smallest frequency at which it can vibrate.

17. A string usually vibrates at many of its possible frequencies simultaneously. The sounds corresponding to the frequencies

$$n\frac{V}{2L}, \; \text{where } n > 1$$

are called **overtones.**

For the string in Exercise 16, find the value of *n* for the highest overtone that could be heard by a person who can hear frequencies up to 10,000 cycles per second.

AM/FM Radio

In order to transmit the sounds of a program, a radio station <u>modulates</u> (varies) either the <u>amplitude</u> or <u>frequency</u> of its **carrier wave.** The graph of the carrier wave is a sine curve whose frequency (see page 258) is that assigned to the station.

One way of using the carrier wave to transmit program sounds is by imposing a periodic change on the amplitude of the wave. This is done by forcing the amplitude to vary with time while keeping the period and frequency constant (**amplitude modulation,** or **AM**). AM stations broadcast on frequencies between 535,000 and 1,605,000 cycles per second.

The carrier wave can also be used to transmit program sounds by forcing the frequency to vary with the time while keeping the amplitude constant (**frequency modulation,** or **FM**). FM stations broadcast on frequencies between 88,000,000 and 108,000,000 cycles per second.

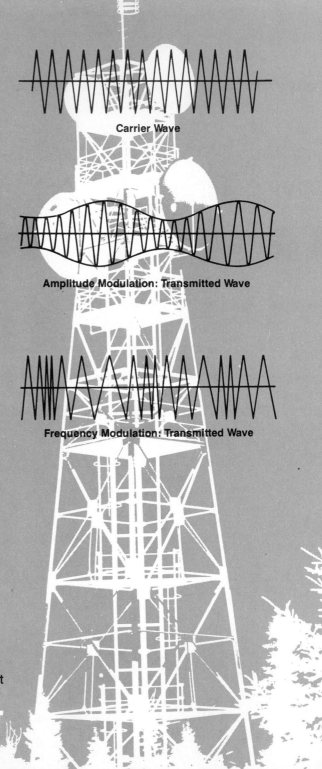

Carrier Wave

Amplitude Modulation: Transmitted Wave

Frequency Modulation: Transmitted Wave

EXERCISES

1. What is the difference between amplitude modulation and frequency modulation?

   The equation of the carrier wave of a radio station is of the form $y = A_0 \sin Bt$ *where* $B > 0$. *The equation* $B = 2\pi f$ *relates B to the frequency f of the carrier wave (see page 258).*

2. Write the equation of the carrier wave in terms of *f* and *t*.

3. Write the equation of the carrier wave of a station that operates at a frequency of 1,000,000 cycles per second.

Chapter Objectives and Review

Objective: To know the meanings of the mathematical terms in this chapter.

1. Be sure that you know the meanings of these mathematical terms.

circular function (p. 252)
cycle (p. 257)
frequency (p. 257)
periodic motion (p. 257)

simple harmonic motion (p. 260)
sinusoidal (p. 260)
unit circle (p. 248)
wrapping function (p. 249)

Objective: To use the wrapping function, W, to find points on the unit circle. (Section 6–1)

In Exercises 2–5, $W(\frac{\pi}{4}) = (\frac{\sqrt{2}}{2}, \frac{\sqrt{2}}{2})$. Find each of the following.

2. $W(\pi + \frac{\pi}{4})$ 3. $W(\pi - \frac{\pi}{4})$ 4. $W(2\pi + \frac{\pi}{4})$ 5. $W(\frac{121\pi}{4})$

Objective: To apply the properties of circular functions. (Section 6–2)

6. If $\cos(.48) = .8870$, $\cos(\pi + .48) = $ _____ .

7. If $\sin(.48) = .4617$, $\sin(\pi - .48) = $ _____ .

Use properties 1–4 on page 250 and the definitions on page 253 to express each of the following as a function of t alone.

8. $\csc(\pi + t)$ 9. $\tan(3\pi - t)$

Objective: To find the frequency of trigonometric and circular functions. (Section 6–3)

Find the frequency f of each function.

10. $y = -\cos 2t$ 11. $y = 3\tan(t + \pi)$

12. $y = 3\sin(6t - \frac{\pi}{2})$ 13. $y = \csc(\pi t - 2)$

Objective: To find a function of the form $y = \sin Bt$ or $y = \cos Bt$ that has a given frequency. (Section 6–3)

14. Find a function of the form $y = \sin Bt$ with a frequency of $\frac{2}{\pi}$ cycles per unit.

15. Find a function of the form $y = \cos Bt$ with a frequency of 5 cycles per unit.

Objective: To apply equations of simple harmonic motion to problems involving oscillation, electricity, and sound. (Sections 6–4, 6–5, and 6–6)

16. A spherical ball oscillates in simple harmonic motion between a minimum of -6 and a maximum of 6. At the minimum, $t = 0$. The sphere completes $\frac{1}{8}$ cycle in one second. Graph the vertical position of the sphere as a function of time.

17. Find an equation of the graph of Exercise 16.

18. The frequency of an alternating emf is 80 cycles per second and the maximum emf is 200 volts. Write an equation for the emf of the form $E = E_m \sin \omega t$.

19. Use the equation of Exercise 18 to find the emf when $t = \frac{1}{60}$ seconds.

20. Write an equation of the form $y = A \sin Bt$ for the oscilloscope image for a simple sound that has a frequency of 440 cycles per second. Assume that the amplitude is one unit. (This is the note a' – "A above middle C.")

Chapter Test

In Exercises 1 and 2, $W(t)$ is given. Find $W(-t)$, $W(\pi-t)$, $W(\pi+t)$, and $W(2\pi+t)$.

1. $W(t) = (-\frac{3}{5}, \frac{4}{5})$

2. $W(t) = (\frac{\sqrt{15}}{4}, \frac{1}{4})$

3. If $\sin(.48) = .4618$, $\sin(\pi + .48) =$ _____.

4. If $\cos(.48) = .8870$, $\cos(\pi - .48) =$ _____.

5. Use properties 1–4 on page 250 and the definitions on page 253 to express $\cot(3\pi + t)$ as a function of t alone.

Find the frequency f of each function.

6. $y = \sin 3\pi t$

7. $y = \cos(12t - \pi)$

8. Find a function of the form $y = \sin Bt$ with a frequency of $\frac{3}{2\pi}$ cycles per unit.

9. A spherical ball oscillates in simple harmonic motion between a minimum of -20 and maximum of 20. At the maximum, $t = 0$. The sphere completes $\frac{1}{3}$ cycle in one second. Graph the vertical position of the sphere as a function of time.

10. Find an equation of the graph in Exercise 9.

Chapter 7
Complex Numbers

7-1 Addition and Subtraction

By the definition of square root, if $x^2 = 5$ then $x = \sqrt{5}$ or $x = -\sqrt{5}$. Similarly, if $x^2 = -1$, then $x = \sqrt{-1}$ or $x = -\sqrt{-1}$. Since there is no real number whose square is -1, mathematicians invented the numbers i and $-i$.

Definition: $i = \sqrt{-1}$

With this definition, it follows that

$$i^2 = (\sqrt{-1})^2 = -1$$

and

$$(-i)^2 = (-\sqrt{-1})^2 = -1$$

If a is a real number and $-a < 0$, then $\sqrt{-a}$ is a **pure imaginary number.**

$$\sqrt{-a} = i\sqrt{a} = \sqrt{a} \cdot i$$

A complex number is the sum of a real number and a pure imaginary number.

Definition: A **complex number** is a number of the form $a + bi$, where a and b are real numbers and $i = \sqrt{-1}$.

The form $a + bi$ is called the **standard form** of a complex number. If $a = 0$ and $b \neq 0$, then $a + bi$ equals bi, a pure imaginary number. If $b = 0$, then $a + bi$ equals a, a real number. Thus, the **set of complex numbers** C is the union of the set of real numbers and the set of imaginary numbers.

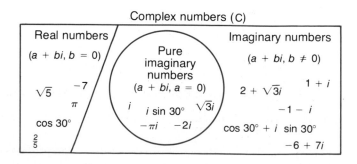

Complex numbers (C)

Real numbers ($a + bi$, $b = 0$) : $\sqrt{5}$, -7, π, $\cos 30°$, $\frac{2}{5}$

Pure imaginary numbers ($a + bi$, $a = 0$) : i, $i \sin 30°$, $\sqrt{3}i$, $-\pi i$, $-2i$

Imaginary numbers ($a + bi$, $b \neq 0$) : $2 + \sqrt{3}i$, $1 + i$, $-1 - i$, $\cos 30° + i \sin 30°$, $-6 + 7i$

Equality of complex numbers is defined in terms of the equality of real numbers.

Definition: Two complex numbers $a + bi$ and $c + di$ are equal if and only if $a = c$ and $b = d$.

For example, if $x + yi = \sqrt{3} + 2i$ then $x = \sqrt{3}$ and $y = 2$.

The familiar properties of the operations for real numbers apply to the complex numbers. The definition for the addition of complex numbers uses the commutative, associative, and distributive properties of real numbers.

Definition: Addition of Complex Numbers

For all real numbers a, b, c, and d,

$$(a + bi) + (c + di) = (a + c) + (b + d)i.$$

To add two complex numbers, find the sum of the real parts and the sum of the imaginary parts.

Example 1 Add: **a.** $(2 + 3i) + (6 - 5i)$ **b.** $(3 - 5i) + (-3 + 5i)$

Solutions:
a. $(2 + 3i) + (6 - 5i) = (2 + 6) + (3 - 5)i$
$= 8 + (-2)i = 8 - 2i$

b. $(3 - 5i) + (-3 + 5i) = (3 + (-3)) + (-5 + 5)i$
$= (3 - 3) + (-5 + 5)i = 0 + 0i = 0$

As Example **1b** suggests, every complex number $a + bi$ has a unique <u>additive inverse</u>, $-(a + bi)$ or $(-a - bi)$. Subtraction of complex numbers is defined in terms of additive inverses.

Definition: Subtraction of Complex Numbers

To subtract a complex number $a + bi$, add its additive inverse, $-a - bi$.

Example 2 Subtract: $(5 - 2i) - (4 + 6i)$

Solution:
$$
\begin{aligned}
(5 - 2i) - (4 + 6i) &= (5 - 2i) + (-4 - 6i) \\
&= (5 + (-4)) + (-2 + (-6))i \\
&= (5 - 4) + (-2 - 6)i \\
&= 1 - 8i
\end{aligned}
$$

Classroom Exercises

Add.

1. $(-2 + 3i) + (5 - 6i)$

2. $(6 + 5i) + (-6 - 6i)$

Write the additive inverse.

3. $-3 + 2i$

4. $8 - 4i$

Subtract.

5. $(4 + 2i) - (5 - 6i)$

6. $(9 - 3i) - (-7 + 4i)$

Written Exercises

a

In Exercises 1–9, use the given information and the definition of equality for complex numbers to give the values of a and b.

1. $a + bi = 4 - 7i$ **2.** $a + bi = 16$ **3.** $a + bi = 3i$

4. $a + bi = 0$ **5.** $a + bi = \sqrt{3} + i$ **6.** $a + bi = 5 - 1$

7. $a + bi = 1 + i\sqrt{3}$ **8.** $a + bi = 1 + \sqrt{3} + i$ **9.** $a + bi = (1 + \sqrt{3})i$

Add.

10. $(4 + 5i) + (2 + 7i)$ **11.** $(1 + 8i) + (11 + i)$

12. $(-3 + 2i) + (4 - i)$ **13.** $(-1 - 4i) + (-2 + 3i)$

14. $-8 + (6 - 3i)$ **15.** $(-3 + 4i) + 3i$

16. $(\sqrt{2} + 3i) + (1 - \sqrt{2} - \frac{5}{2}i)$ **17.** $(4 - i\sqrt{3}) + (-4 + i\sqrt{3})$

Write the additive inverse.

18. $2 + 4i$ **19.** $-3 + 6i$ **20.** $-1 - i$ **21.** $5 - 4i$

22. $\sqrt{2} + i\sqrt{3}$ **23.** $7i$ **24.** 4 **25.** 0

Subtract.

26. $(8 + 3i) - (2 + i)$ **27.** $(3 + 4i) - (13 - 5i)$

28. $(-1 + 2i) - (-4 - 3i)$ **29.** $(5 - i) - (-2 - 6i)$

30. $(2 - i\sqrt{2}) - (-2 - i\sqrt{2})$ **31.** $(1 - \sqrt{2} + i) - (-\sqrt{2} + 2i)$

b

In Exercises 32–35, perform the indicated operations. Then use the definitions of equality for complex numbers to give the values of a and b.

32. $a + bi = 9 + (6 - 2i)$

33. $a + bi = (2 - 3i) + (2 + 3i)$

34. $a + bi = (4 + 2i) - (5 - 6i) + (3 + 2i)$

35. $a + bi = 7i - (6 - 3\sqrt{2}\,i) + (1 - 2i) - 4i$

c

36. Prove or disprove the following statement: *The sum of two pure imaginary numbers is a pure imaginary number.*

37. The complex number $c + di$ is called the **additive identity element** if the equation

$$(a + bi) + (c + di) = (a + bi)$$

holds for any complex number $a + bi$. Find $c + di$.

For Exercises 38–40, assume the following properties of real numbers.

a. Closure Property for Addition:
    If x and y are real numbers, then $x + y$ is a real number.

b. Commutative Property for Addition:
    If x and y are real numbers, then $x + y = y + x$.

c. Associative Property for Addition:
    If x, y, and z are real numbers, then $(x + y) + z = x + (y + z)$.

38. Prove that the set of complex numbers is closed with respect to addition.

39. Prove that addition of complex numbers is commutative.

40. Prove that addition of complex numbers is associative. (HINT: Show that $[(a + bi) + (c + di)] + (e + fi) = (a + bi) + [(c + di) + (e + fi)]$ by simplifying each side independently.)

7-2 Multiplication and Division

Multiplication of complex numbers is defined as follows.

Definition: Multiplication of Complex Numbers

For all real numbers a, b, c, and d,

$$(a + bi)(c + di) = (ac - bd) + (ad + bc)i.$$

When multiplying complex numbers, it is generally easier to follow the method of Example 1 than to apply the definition. To multiply complex numbers, use the "foil method" for multiplying binomials.

Example 1   Multiply: $(5 - 3i)(4 - 2i)$

Solution: $(5 - 3i)(4 - 2i) = 5(4) + (5)(-2i) + (-3i)(4) + (-3i)(-2i)$

$\qquad\qquad\qquad\qquad = 20 - 10i - 12i + 6i^2 \quad \longleftarrow \quad i^2 = -1$

$\qquad\qquad\qquad\qquad = 20 - 22i - 6$

$\qquad\qquad\qquad\qquad = 14 - 22i$

Complex numbers of the form $a + bi$ and $a - bi$ are called **conjugates** of each other.

Example 2   Multiply: $(2 + 5i)(2 - 5i)$

Solution: $(2 + 5i)(2 - 5i) = 4 - 10i + 10i - 25i^2 \quad \longleftarrow \quad -10i + 10i = 0$

$\qquad\qquad\qquad\qquad = 4 - 25(-1) = 4 + 25 = 29$

As Example 2 suggests, the product of two conjugates is a real number. In standard form, the answer would be written as $29 + 0i$.

As with real numbers, division of complex numbers is defined in terms of multiplying by a reciprocal.

Definition: Division of Complex Numbers

For any real numbers a, b, c, and d, where $c \neq 0$ or $d \neq 0$,

$$(a + bi) \div (c + di) = (a + bi)\left(\frac{1}{c + di}\right).$$

To multiply by a reciprocal, you express the reciprocal in another form by using the conjugate and the multiplication property of one. This is illustrated in Example 3.

Example 3 Divide: $(2 - 4i) \div (8 - 6i)$

Solution: Apply the definition. Write in fraction form. Then multiply the numerator and denominator by the conjugate of the denominator.

$$(2 - 4i) \div (8 - 6i) = (2 - 4i)\left(\frac{1}{8 - 6i}\right)$$

$$= \frac{2 - 4i}{8 - 6i} \cdot \frac{8 + 6i}{8 + 6i}$$

$$= \frac{(2 - 4i)(8 + 6i)}{64 + 36}$$

$$= \frac{16 + 12i - 32i - 24i^2}{100}$$

$$= \frac{16 + 24 - 20i}{100} = \frac{40 - 20i}{100}$$

$$= \frac{2}{5} - \frac{1}{5}i \longleftarrow \text{ Standard form}$$

Thus, as Example 3 illustrates, to divide $a + bi$ by $c + di$, multiply $\frac{a + bi}{c + di}$ by $\frac{c - di}{c - di}$.

It can be shown that the complex numbers satisfy the properties of a field. You were asked to verify some of these properties in the Exercises in the previous section. You will be asked to verify other properties in the Exercises of this section.

Classroom Exercises _____

Multiply.

1. $(3 + 5i)(6 - i)$

2. $(-2 + 3i)(-5 - 2i)$

Write the conjugate of each number.

3. $3 + 4i$

4. $-3 + 2i$

5. $1 - 2i$

6. $-2 - i$

Divide.

7. $(1 - i) \div (3 + 4i)$

8. $(3 + 5i) \div (1 - 2i)$

Written Exercises

a

Multiply.

1. $(2 + i)(3 + i)$

2. $(1 + 2i)(1 + 3i)$

3. $(4 - i)(2 - 2i)$

4. $(-2 + 3i)(4 - 5i)$

5. $(-5 - 3i)(4 + i)$

6. $(-3 + 2i)(-5 - 6i)$

7. $(6 + 3i)(6 - 3i)$

8. $(1 - 4i)(1 + 4i)$

9. $(2i)(3 - 6i)$

10. $(5)(-4 + i)$

11. $(-5 - 7i)(0)$

12. $(3 + 4i)^2$

13. $(2 - i)^2$

14. $(-5 - 6i)^2$

15. $(\frac{1}{2} + 3i)(1 - \frac{1}{2}i)$

16. $(\frac{3}{2} - .2i)^2$

For Exercises 17–24, multiply the given complex number by its conjugate.

17. $7 + 2i$

18. $3 - 5i$

19. $-2 + i$

20. $-5 - 4i$

21. 12

22. $3i$

23. $\sqrt{2} + 1 - i$

24. $-\pi i$

Divide. Express each answer in standard form.

25. $(3 + i) \div (2 + 2i)$

26. $(3 + 4i) \div (-2 + 2i)$

27. $(4 + 3i) \div (1 + 2i)$

28. $(2 - 3i) \div (1 + i)$

29. $(-1 + 5i) \div (6 - 5i)$

30. $(2 - 5i) \div (6 + 4i)$

31. $(4 + 2i) \div (1 + i)$

32. $(6 - 5i) \div (2 + 3i)$

33. $2 \div (1 - i)$

34. $-3 \div (3 + i)$

35. $i \div (3 + 2i)$

36. $-2i \div (-1 + 3i)$

37. $(-3 - 4i) \div i$

38. $(7 + 2i) \div (-3i)$

Express each complex number in standard form.

39. $\dfrac{3 - 5i}{5}$

40. $\dfrac{1 - 2i}{3 - 4i}$

41. $\dfrac{2(1 + 5i) - 3(2 - i)}{i}$

42. $\dfrac{\frac{1}{2}(2 - i) - \frac{2}{3}(-2 + 3i)}{2 - i}$

43. Complete the table.

x	1	2	3	4	5	6	7	8	9	10	11	12
i^x	i	-1	$-i$	1	?	?	?	?	?	?	?	?

44. Assume that $i^{ab} = (i^a)^b$ where a and b are whole numbers. Prove that $i^{4x} = 1$, where x is any whole number.

45. Let x be any whole number. Prove that the function $f(x) = i^x$ is periodic with period 4. (HINT: Prove that $f(x + 4) = f(x)$, where x is any whole number.)

46. Prove that the sum of a complex number and its conjugate is a real number.

47. Show that the **multiplicative inverse** of any complex number $a + bi$, where $a \neq 0$ or $b \neq 0$, is

$$\frac{a}{a^2 + b^2} - \frac{b}{a^2 + b^2}i.$$

48. Show that $1 + 0i$ is the **multiplicative identity** element for the set of complex numbers.

For Exercises 49–52, assume the following properties of real numbers.

a. Closure Property for Multiplication:
If x and y are real numbers, then $x \cdot y$ is a real number.

b. Commutative Property for Multiplication:
If x and y are real numbers, then $x \cdot y = y \cdot x$.

c. Associative Property for Multiplication:
If x, y, and z are real numbers, then $(x \cdot y) \cdot z = x \cdot (y \cdot z)$.

d. Distributive Property:
If x, y, and z are real numbers, then $x \cdot (y + z) = x \cdot y + x \cdot z$.

49. Prove that the set of complex numbers is closed with respect to multiplication.

50. Prove that multiplication of complex numbers is commutative.

51. Prove that multiplication of complex numbers is associative. (HINT: Show that $[(a + bi) \cdot (c + di)] \cdot (e + fi) = (a + bi) \cdot [(c + di) \cdot (e + fi)]$ by simplifying each side independently.)

52. Prove that the complex numbers satisfy the Distributive Property.

7-3 Complex Numbers in Polar Form

Every complex number $a + bi$ can also be written as an ordered pair (a, b), where a is the **real part** and b is the **imaginary part** (the coefficient of i).

Standard Form	$2 + 4i$	$2 - 4i$	$0 + 0i$	$0 + \sqrt{3}i$	$\pi + 0i$
Ordered Pair Form	$(2, 4)$	$(2, -4)$	$(0, 0)$	$(0, \sqrt{3})$	$(\pi, 0)$

Thus, each complex number can be associated with a point on the coordinate plane called the **complex number plane.** The **real axis** represents complex numbers of the form $(a, 0)$. The **imaginary axis** represents complex numbers of the form $(0, b)$. In this plane, a complex number can be represented by a <u>point</u> or by a <u>vector</u> drawn from the origin to the point.

A vector representing $a + bi$ or (a, b) has both length <u>and</u> direction. The symbol $|a + bi|$, read the <u>absolute value</u> of $a + bi$, represents the length. You find the length or <u>magnitude</u> by applying the distance formula.

length: $(a, b) = |a + bi| = \sqrt{a^2 + b^2}$

The direction of a vector is indicated by the angle determined by the positive real axis and the vector. The measure, θ, of this angle is called the **argument** of the complex number.

argument: $(a + bi) = \theta$

Any of the trigonometric functions can be used to find θ. However, since a and b are known and $\tan \theta = \dfrac{b}{a}$, θ can be found by applying one of the following.

For $a > 0$, $\theta = \text{Arc} \tan \dfrac{b}{a}$. For $a = 0$ and $b > 0$, $\theta = \dfrac{\pi}{2}$.

For $a < 0$, $\theta = \pi + \text{Arc} \tan \dfrac{b}{a}$. For $a = 0$ and $b < 0$, $\theta = \dfrac{3\pi}{2}$.

For $a = b = 0$, θ can have any real value.

The letter z is often used to represent a complex number.

Example 1 Let $z = -2\sqrt{3} + 2i$.

a. Find $|z|$.　　　　　**b.** Find θ.　　　　　**c.** Graph z.

Solutions: a. $|z| = |-2\sqrt{3} + 2i|$

$$= \sqrt{(-2\sqrt{3})^2 + 2^2}$$

$$= \sqrt{16}, \text{ or } 4$$

b. Since $a = -2\sqrt{3}$, $a < 0$.

Thus, $\theta = \pi + \text{Arc} \tan \left(\dfrac{2}{-2\sqrt{3}} \right)$

$$= \pi + \text{Arc} \tan \left(-\dfrac{1}{\sqrt{3}} \right)$$

$$= \pi + \left(-\dfrac{\pi}{6} \right) = \dfrac{5\pi}{6}, \text{ or } 150°$$

c.

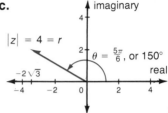

Let $a + bi$ be a complex number, with $r = |a + bi| \neq 0$ and with argument θ. Then

$$\sin \theta = \frac{b}{r} \text{ and } \cos \theta = \frac{a}{r}.$$

Thus, $b = r \sin \theta$ and $a = r \cos \theta$. Hence, $a + bi = r \cos \theta + ir \sin \theta$ or

$a + bi = r(\cos \theta + i \sin \theta)$.

The right side of this equation is called the **polar** or **trigonometric form** of the complex number $a + bi$, where r is the **modulus** and θ is the argument. Note that the modulus is equal to the length of the vector. When $r = 0$, θ can have any real value, since $r = 0$ if and only if $a = b = 0$.

The expression "$\cos \theta + i \sin \theta$" is sometimes abbreviated as **cis θ,** where c represents cos, i represents i, and s represents sin. Thus, $a + bi = r$ cis θ.

Example 2 Express $3 - \sqrt{3}i$ in polar form.

Solution:　$\theta = \text{Arc} \tan \left(-\dfrac{\sqrt{3}}{3} \right) = -30°$

$$r = \sqrt{3^2 + (-\sqrt{3})^2} = \sqrt{12} = 2\sqrt{3}$$

Thus, $3 - \sqrt{3}i = 2\sqrt{3}[\cos (-30°) + i \sin (-30°)]$　　←　$\sin (-30°) = \sin 330°$

or　　$3 - \sqrt{3}i = 2\sqrt{3}(\cos 330° + i \sin 330°)$

Classroom Exercises

Express in ordered pair form.

1. $2 + 6i$.

2. $-3i$

3. -1

4. $3 + \sqrt{-4}$

Express in standard form.

5. $(4, 3)$

6. $(3, 0)$

7. $(0, 4)$

8. $(-\sqrt{2}, -1)$

Written Exercises

a

For each complex number z, find |z| and θ. Then graph z.

1. $z = -1 + i$

2. $z = -1 - i$

3. $z = \frac{\sqrt{3}}{2} + \frac{1}{2}i$

4. $z = \sqrt{3} - i$

5. $z = \sqrt{2} + i\sqrt{2}$

6. $z = 2 - 2i$

7. $z = -3i$

8. $z = 2$

9. $z = -3 + 3i\sqrt{3}$

10. $z = -1$

11. $z = 3 + 3i$

12. $z = 4i$

13. $z = 1 + i$

14. $z = 3i$

15. $z = -\sqrt{3} - i$

16. $z = -2$

Use the figure below for Exercises 17–24. Express the complex number repre-sented by the given vector in standard form. Then find its modulus, r, and argument, θ.

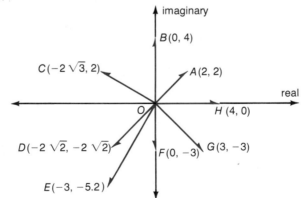

17. $\overrightarrow{OA}$

18. $\overrightarrow{OB}$

19. $\overrightarrow{OC}$

20. $\overrightarrow{OD}$

21. $\overrightarrow{OE}$

22. $\overrightarrow{OF}$

23. $\overrightarrow{OG}$

24. $\overrightarrow{OH}$

25. Which of the vectors in Exercises 17–24 represent real numbers?

26. Which of the vectors in Exercises 17–24 represent pure imaginary numbers?

Express in polar form.

27. $-3 + 3i$ **28.** $2\sqrt{2} - 2i\sqrt{2}$ **29.** $2\sqrt{3} + 2i$ **30.** $-\sqrt{3} - i$

31. $1 - i\sqrt{3}$ **32.** $5 + 5i$ **33.** $-\sqrt{3} - i\sqrt{3}$ **34.** $4i$

35. -3 **36.** $\sqrt{3} - i$ **37.** $-3 + i\sqrt{3}$ **38.** 2

39. $2 + 2i\sqrt{3}$ **40.** $-3i$ **41.** $-1 + i\sqrt{3}$ **42.** $-2 - 2i\sqrt{3}$

b

Express in standard form.

43. $3(\cos 60° + i \sin 60°)$ **44.** $\sqrt{2}[\cos (-45°) + i \sin (-45°)]$

45. $\cos (-\frac{\pi}{6}) + i \sin (-\frac{\pi}{6})$ **46.** $2(\cos \frac{3\pi}{4} + i \sin \frac{3\pi}{4})$

47. $4(\cos \frac{\pi}{2} + i \sin \frac{\pi}{2})$ **48.** $2\sqrt{3}(\cos \pi + i \sin \pi)$

49. $2 \text{ cis } 30°$ **50.** $\text{cis } (-\frac{\pi}{4})$

Complete each statement.

51. If $a + bi$ is a real number, then $|a + bi| = $? .

52. If $a + bi$ is a pure imaginary number, then $|a + bi| = $? .

53. If $|a + bi| = 13$ and $a = -5$, then $b = $? or ? .

Graph the given complex number and its conjugate on the same coordinate plane.

54. $4 + 3i$ **55.** -4 **56.** $2i$ **57.** $3 - 3i$

c

58. Prove that the modulus of any complex number equals the modulus of its conjugate.

59. Prove that if θ is the argument of a complex number, then $-\theta$ is the argument of its conjugate.

Review

Perform the indicated operations. (Sections 7–1 and 7–2)

1. $(6 - 3i) + (9 + i)$ **2.** $(4 - 6i) - (6 - 5i)$

3. $(18 - 4i) - (-7 + i)$ **4.** $(-1 - i) + (4 - 15i)$

5. $(8 - 5i)(8 + 5i)$ **6.** $(12 - 7i) \div (1 + i)$

7. $(-9 - 11i) \div (6 + 5i)$ **8.** $(-15 + 2i)(-4 - 3i)$

For each complex number z, find $|z|$ and θ. Then write z in polar form. (Section 7–3)

9. $z = 2 + 2i$ **10.** $z = \sqrt{2} - i\sqrt{6}$

11. $z = 3i$ **12.** $z = -1 + i$

The Push-Button Telephone

On a push-button telephone, each button causes a different sound to be produced. Whenever a button is pushed, an electronic device, called an **oscillator,** is turned on. The oscillator produces electrical impulses that are converted to sound. Each sound is a signal, or code, for the number on the button.

The sound produced when a button is pushed is actually a combination of two tones, or <u>simple sounds.</u> A **simple sound** is one which an oscilloscope translates to a curve of the form

$$y = A \sin Bt,$$

where B equals 2π times the frequency (see Section 6–3). The figure at the right shows that each button identifies one low frequency and one high frequency. For example, the two tones produced when the 3-button is pushed would have the oscilloscope graphs shown below. For these graphs, the amplitudes A_1 and A_2 are assumed to be equal to a common value, A.

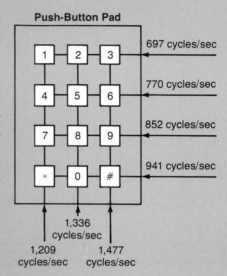

Push-Button Pad

697 cycles/sec
770 cycles/sec
852 cycles/sec
941 cycles/sec

1,336 cycles/sec
1,209 cycles/sec
1,477 cycles/sec

Low frequency for 3:
697 cycles per second

Function: $y_1 = A_1 \sin (697)(2\pi)t$, or
$\qquad y_1 = A \sin 1394\pi t$

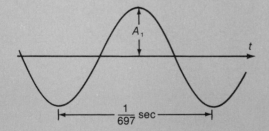

$\frac{1}{697}$ sec

High frequency for 3:
1477 cycles per second

Function: $y_2 = A_2 \sin (1477)(2\pi)t$, or
$\qquad y_2 = A \sin 2954\pi t$

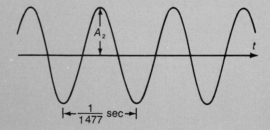

$\frac{1}{1477}$ sec

EXERCISES

1. Use the method of addition of ordinates (see Section 2–9) and the graphs shown above to sketch the graph of $y = A(\sin 1394\pi t + \sin 2954\pi t)$. This is the graph of the sound that results from combining the two simple sounds. (HINT: Let $A = 1$.)
2. Sketch the oscilloscope graphs for the low frequency and high frequency tones produced when the 6-button is pushed.
3. Sketch the graph for the combinations of the two simple sounds in Exercise 2.

7-4 Products and Quotients in Polar Form

Addition and subtraction of complex numbers in standard form are simple operations. However, multiplication and division are complicated. When complex numbers are expressed in polar form, multiplication and division are simplified.

Theorem 7-1:   Multiplication in Polar Form

If z_1 and z_2 are complex numbers, where $z_1 = r_1(\cos \theta_1 + i \sin \theta_1)$, and $z_2 = r_2(\cos \theta_2 + i \sin \theta_2)$, then

$$z_1 \cdot z_2 = r_1 \cdot r_2[\cos (\theta_1 + \theta_2) + i \sin (\theta_1 + \theta_2)].$$

Proof:

$$z_1 \cdot z_2 = r_1(\cos \theta_1 + i \sin \theta_1) \cdot r_2(\cos \theta_2 + i \sin \theta_2)$$
$$= r_1 \cdot r_2 \cdot (\cos \theta_1 + i \sin \theta_1) \cdot (\cos \theta_2 + i \sin \theta_2)$$
$$= r_1 \cdot r_2(\cos \theta_1 \cos \theta_2 + i \cos \theta_1 \sin \theta_2 + i \sin \theta_1 \cos \theta_2 + i^2 \sin \theta_1 \sin \theta_2)$$
$$= r_1 \cdot r_2(\cos \theta_1 \cos \theta_2 + i^2 \sin \theta_1 \sin \theta_2 + i \cos \theta_1 \sin \theta_2 + i \sin \theta_1 \cos \theta_2)$$
$$= r_1 \cdot r_2[\cos \theta_1 \cos \theta_2 - \sin \theta_1 \sin \theta_2 + i(\cos \theta_1 \sin \theta_2 + \sin \theta_1 \cos \theta_2)]$$

But
$$\cos (\theta_1 + \theta_2) = \cos \theta_1 \cos \theta_2 - \sin \theta_1 \sin \theta_2, \qquad \text{and}$$
$$\sin (\theta_1 + \theta_2) = \cos \theta_1 \sin \theta_2 + \sin \theta_1 \cos \theta_2.$$

Thus,
$$z_1 \cdot z_2 = r_1 \cdot r_2[\cos (\theta_1 + \theta_2) + i \sin (\theta_1 + \theta_2)].$$

Example 1   Find $z_1 \cdot z_2$ when $z_1 = 2(\cos 40° + i \sin 40°)$ and $z_2 = 4(\cos 60° + i \sin 60°)$. Express the product in standard form.

Solution:
$$z_1 \cdot z_2 = 2(\cos 40° + i \sin 40°) \cdot 4(\cos 60° + i \sin 60°)$$
$$= 2 \cdot 4[\cos (40° + 60°) + i \sin (40° + 60°)]$$
$$= 8(\cos 100° + i \sin 100°)$$
$$= 8(-\cos 80° + i \sin 80°) \qquad \longleftarrow \begin{array}{l} \cos (180° - \theta) = -\cos \theta \\ \sin (180° - \theta) = \sin \theta \end{array}$$
$$= 8(-.1736 + .9848i)$$
$$= -1.3888 + 7.8784i \qquad \longleftarrow \text{ Standard form}$$

The procedure for division is similar.

Theorem 7-2: Division in Polar Form

If $z_1 = r_1(\cos \theta_1 + i \sin \theta_1)$ and $z_2 = r_2(\cos \theta_2 + i \sin \theta_2)$, $z_2 \neq 0$,

then

$$\frac{z_1}{z_2} = \frac{r_1}{r_2}[\cos(\theta_1 - \theta_2) + i \sin(\theta_1 - \theta_2)].$$

Proof: The symbol $\bar{z}$ represents the conjugate of z. As when dividing complex numbers in standard form, you can multiply the numerator and denominator by the conjugate of the denominator. That is,

$$\frac{z_1}{z_2} = \frac{z_1 \cdot \bar{z}_2}{z_2 \cdot \bar{z}_2}.$$

It can be shown that if $z = r(\cos \theta + i \sin \theta)$, then

$$\bar{z} = r[\cos(-\theta) + i \sin(-\theta)].$$

Thus, $\dfrac{z_1}{z_2} = \dfrac{r_1[\cos \theta_1 + i \sin \theta_1]}{r_2[\cos \theta_2 + i \sin \theta_2]} \cdot \dfrac{r_2[\cos(-\theta_2) + i \sin(-\theta_2)]}{r_2[\cos(-\theta_2) + i \sin(-\theta_2)]}$

$$= \frac{r_1 r_2[\cos(\theta_1 - \theta_2) + i \sin(\theta_1 - \theta_2)]}{r_2 r_2[\cos(\theta_2 - \theta_2) + i \sin(\theta_2 - \theta_2)]} \quad \longleftarrow \text{Theorem 7-1}$$

$$= \frac{r_1[\cos(\theta_1 - \theta_2) + i \sin(\theta_1 - \theta_2)]}{r_2(1 + 0i)}$$

$$= \frac{r_1}{r_2}[\cos(\theta_1 - \theta_2) + i \sin(\theta_1 - \theta_2)].$$

Example 2 Find $z_1 \div z_2$ when $z_1 = 2(\cos 40° + i \sin 40°)$ and $z_2 = 4(\cos 60° + i \sin 60°)$. Express the quotient in standard form.

Solution: $\dfrac{z_1}{z_2} = \dfrac{2(\cos 40° + i \sin 40°)}{4(\cos 60° + i \sin 60°)}$

$$= \frac{2}{4}[\cos 40° - 60°) + i \sin(40° - 60°)]$$

$$= \frac{1}{2}[\cos(-20°) + i \sin(-20°)] \quad \longleftarrow \begin{array}{l} \cos(-\theta) = \cos\theta; \\ \sin(-\theta) = -\sin\theta \end{array}$$

$$= \frac{1}{2}(\cos 20° - i \sin 20°)$$

$$= \frac{1}{2}(.9397 - .3420i)$$

$$= .4698 - .1710i \quad \longleftarrow \text{Standard form}$$

Classroom Exercises

Let $z_1 \cdot z_2 = r(\cos \theta + i \sin \theta)$. Find r and θ.

1. $z_1 = 2(\cos 35° + i \sin 35°)$; $z_2 = 5(\cos 13° + i \sin 13°)$

2. $z_1 = \frac{1}{2}(\cos 48° + i \sin 48°)$; $z_2 = 8(\cos 22° + i \sin 22°)$

Let $z_1 \div z_2 = r(\cos \theta + i \sin \theta)$. Find r and θ.

3. $z_1 = 8(\cos 75° + i \sin 75°)$; $z_2 = 4(\cos 25° + i \sin 25°)$

4. $z_1 = 27(\cos 40° + i \sin 40°)$; $z_2 = 3(\cos 10° + i \sin 10°)$

Written Exercises

a

Find $z_1 \cdot z_2$. Express the product in standard form.

1. $z_1 = 4(\cos 30° + i \sin 30°)$; $z_2 = 6(\cos 60° + i \sin 60°)$

2. $z_1 = 4(\cos \frac{\pi}{6} + i \sin \frac{\pi}{6})$; $z_2 = 2(\cos \frac{\pi}{3} + i \sin \frac{\pi}{3})$

3. $z_1 = 5(\cos 90° + i \sin 90°)$; $z_2 = 2(\cos 150° + i \sin 150°)$

4. $z_1 = 3(\cos 105° + i \sin 105°)$; $z_2 = 11(\cos 15° + i \sin 15°)$

5. $z_1 = 7(\cos 18° + i \sin 18°)$; $z_2 = 8(\cos 25° + i \sin 25°)$

Find $z_1 \div z_2$. Express the quotient in standard form.

6. $z_1 = 6(\cos 135° + i \sin 135°)$; $z_2 = 2(\cos 120° + i \sin 120°)$

7. $z_1 = 6(\cos \frac{3\pi}{4} + i \sin \frac{3\pi}{4})$; $z_2 = 2(\cos \frac{\pi}{4} + i \sin \frac{\pi}{4})$

8. $z_1 = 3(\cos 180° + i \sin 180°)$; $z_2 = \cos 270° + i \sin 270°$

9. $z_1 = 2(\cos 25° + i \sin 25°)$; $z_2 = 3(\cos 70° + i \sin 70°)$

10. $z_1 = 6(\cos 80° + i \sin 80°)$; $z_2 = 2(\cos 35° + i \sin 35°)$

Find $z_1 \cdot z_2$ and $\frac{z_1}{z_2}$. Express the answers in standard form.

11. $z_1 = 3(\cos 80° + i \sin 80°)$; $z_2 = \frac{1}{2}(\cos 40° + i \sin 40°)$

12. $z_1 = 5(\cos \frac{2\pi}{3} + i \sin \frac{2\pi}{3})$; $z_2 = 4(\cos \frac{3\pi}{4} + i \sin \frac{3\pi}{4})$

13. $z_1 = 2(\cos 135° + i \sin 135°)$; $z_2 = \frac{2}{3}(\cos 150° + i \sin 150°)$

14. $z_1 = 3(\cos \frac{\pi}{2} + i \sin \frac{\pi}{2})$; $z_2 = \cos(-\frac{\pi}{2}) + i \sin(-\frac{\pi}{2})$

15. $z_1 = 1 - i$; $z_2 = -i$

16. $z_1 = 5i$; $z_2 = 2 - 3i$

C

For Exercises 17–20, let $z = r(\cos \theta + i \sin \theta)$. Let $\bar{z}$ be the conjugate of z.

17. Prove that

$$\frac{1}{z} = \frac{1}{r}(\cos \theta - i \sin \theta) = \frac{1}{r^2} \cdot \bar{z}, \quad \text{where } z \neq 0.$$

18. Prove that $z^2 = r^2(\cos 2\theta + i \sin 2\theta)$.

19. Prove that $z \div \bar{z} = \cos 2\theta + i \sin 2\theta$.

20. Prove that $\bar{z} \div z = \cos 2\theta - i \sin 2\theta$.

21. Prove that the quotient of two complex numbers with the same argument is a real number.

7-5 De Moivre's Theorem

The second power of a complex number in polar form, $z = r(\cos \theta + i \sin \theta)$, is obtained by multiplying the number by itself.

$z \cdot z = z^2 = r^2[\cos (\theta + \theta) + i \sin (\theta + \theta)]$ ◄——— Theorem 7–1

$\quad = r^2(\cos 2\theta + i \sin 2\theta)$

Note that the coefficient of θ is the same as the exponent of r. Compare the coefficient of θ and the exponent of r in the third power of z.

$z^3 = z^2 \cdot z = [r^2(\cos 2\theta + i \sin 2\theta)][r(\cos \theta + i \sin \theta)]$

$\quad = r^3[\cos (2\theta + \theta) + i \sin (2\theta + \theta)]$

$\quad = r^3(\cos 3\theta + i \sin 3\theta)$

These two examples suggest a theorem named after the mathematician Abraham De Moivre (1667–1754). The proof of this theorem is asked for in the exercises.

Theorem 7–3: De Moivre's Theorem

If $z = r(\cos \theta + i \sin \theta)$, then

$$z^n = r^n(\cos n\theta + i \sin n\theta),$$

where n is a positive integer.

Example 1

Evaluate $(1 + i)^8$. Express the answer in standard form.

Solution: First express $1 + i$ in polar form.

$$\theta = \text{Arc tan}\frac{1}{1} \quad \longleftarrow \quad \text{Since } a > 0 \text{ (see pages 284–285)}$$
$$= 45°$$
$$r = \sqrt{1^2 + 1^2} = \sqrt{2}$$

Thus, $\quad 1 + i = \sqrt{2}(\cos 45° + i \sin 45°)$

and $\quad (1 + i)^8 = (\sqrt{2})^8[(\cos (8 \cdot 45°) + i \sin (8 \cdot 45°)]$
$$= 16(\cos 360° + i \sin 360°)$$
$$= 16 = 16 + 0i. \quad \longleftarrow \quad \text{Standard form}$$

Theorem 7–3 can also be shown to be true when n is a negative integer provided $z \neq 0$. This proof is asked for in the Exercises.

Example 2

Evaluate $[2(\cos \frac{\pi}{6} + i \sin \frac{\pi}{6})]^{-3}$. Express the answer in standard form.

Solution: $[2(\cos \frac{\pi}{6} + i \sin \frac{\pi}{6})]^{-3} = 2^{-3}[\cos (-3 \cdot \frac{\pi}{6}) + i \sin (-3 \cdot \frac{\pi}{6})]$
$$= \frac{1}{8}[\cos (-\frac{\pi}{2}) + i \sin (-\frac{\pi}{2})]$$
$$= \frac{1}{8}[0 - i]$$
$$= -\frac{1}{8}i = 0 - \frac{1}{8}i \quad \longleftarrow \quad \text{Standard form}$$

De Moivre's Theorem may also be used to derive certain identities.

Example 3

Derive identities in terms of $\cos \theta$ and $\sin \theta$.

    a. $\cos 3\theta$ **b.** $\sin 3\theta$

Solutions: a. Let $a = \cos \theta$ and $b = \sin \theta$. Then use De Moivre's Theorem:

$$\cos 3\theta + i \sin 3\theta = (\cos \theta + i \sin \theta)^3$$
$$= (a + bi)^3 \quad \longleftarrow \quad \text{Expand the right side.}$$
$$= a^3 + 3ia^2b + 3ab^2i^2 + b^3i^3 \quad \longleftarrow \quad i^2 = -1; \ i^3 = -i$$
$$= a^3 - 3ab^2 + i(3a^2b - b^3)$$
$$\therefore \cos 3\theta + i \sin 3\theta = \cos^3 \theta - 3 \cos \theta \sin^2 \theta + i(3 \cos^2 \theta \sin \theta - \sin^3 \theta)$$

By the definition of equality of complex numbers, the real parts of the left and right sides of this last equation are equal, as are the imaginary parts of both sides. This leads to the following.

$$\cos 3\theta = \cos^3 \theta - 3 \cos \theta \sin^2 \theta \quad \longleftarrow \quad \sin^2 \theta = 1 - \cos^2 \theta$$
$$= \cos^3 \theta - 3 \cos \theta (1 - \cos^2 \theta)$$
$$= \cos^3 \theta - 3 \cos \theta + 3 \cos^3 \theta$$
$$= 4 \cos^3 \theta - 3 \cos \theta$$

b. $\sin 3\theta = 3 \cos^2 \theta \sin \theta - \sin^3 \theta \quad \longleftarrow \quad \cos^2 \theta = 1 - \sin^2 \theta$
$$= 3(1 - \sin^2 \theta) \sin \theta - \sin^3 \theta$$
$$= 3 \sin \theta - 3 \sin^3 \theta - \sin^3 \theta$$
$$= 3 \sin \theta - 4 \sin^3 \theta$$

A similar procedure may be used to derive identities for $\sin n\theta$ and $\cos n\theta$, where n is any integer.

Classroom Exercises

Evaluate.

1. $(-2 - 2i)^5$

2. $(-i)^7$

3. $[2(\cos 30° + i \sin 30°)]^2$

Written Exercises

a

Evaluate each of the following. Express answers in standard form.

1. $[2(\cos \frac{\pi}{3} + i \sin \frac{\pi}{3})]^3$

2. $[3(\cos 120° + i \sin 120°)]^3$

3. $[3(\cos 30° + i \sin 30°)]^3$

4. $[\frac{1}{2}(\cos \frac{\pi}{4} + i \sin \frac{\pi}{4})]^5$

5. $[2(\cos \frac{\pi}{12} + i \sin \frac{\pi}{12})]^{-6}$

6. $[2(\cos 30° + i \sin 30°)]^{-4}$

7. $(4[\cos (-15°) + i \sin (-15°)])^3$

8. $[1(\cos \frac{\pi}{4} - i \sin \frac{\pi}{4})]^5$

9. $(\sqrt{3} + i)^6$ **10.** $(-\sqrt{3} + i)^3$ **11.** $(1 - i)^6$ **12.** $(1 + i)^6$

13. $(1 + i)^{-4}$ **14.** $(i)^{-4}$ **15.** $(1 - \sqrt{3}i)^5$ **16.** $(-1 + \sqrt{3}i)^{-5}$

17. $(-i)^{-5}$ **18.** $(-\sqrt{3} - i)^3$ **19.** $(2 - 2\sqrt{3}i)^2$ **20.** $(2i)^{-5}$

21. Derive identities for $\sin 2\theta$ and $\cos 2\theta$ in terms of $\sin \theta$ and $\cos \theta$.

22. Derive identities for $\sin 4\theta$ and $\cos 4\theta$ in terms of $\sin \theta$ and $\cos \theta$.

23. Derive identities for $\sin 5\theta$ and $\cos 5\theta$ in terms of $\sin \theta$ and $\cos \theta$.

Graph the equations in Exercises 24 and 25 on the same set of coordinate axes for the values n = 1, 2, 3, 4, 5, 6.

24. $z^n = \cos n90° + i \sin n90°$ 25. $z^n = \cos n30° + i \sin n30°$

26. What patterns do you notice in Exercises 24 and 25?

27. Prove De Moivre's Theorem. (HINT: Use mathematical induction. That is, show that if P_k represents the sentence,

$$z^k = r^k(\cos k\theta + i \sin k\theta), \text{ then}$$

    a. P_1 is true, and
    b. P_k implies P_{k+1}.

Then by the induction hypothesis, the theorem will be true for all positive integers.)

28. Use De Moivre's Theorem and the definition $z^{-n} = \dfrac{1}{z^n}$ where $z \neq 0$, to prove that if $z = r(\cos \theta + i \sin \theta)$ and n is a positive integer, then

$$z^{-n} = r^{-n}(\cos n\theta - i \sin n\theta).$$

Review Capsule for Section 7-6 _____

In Exercises 1–4, find the described roots. Simplify all radicals.

Example: The one real cube root of −16.   **Solution:** $\sqrt[3]{-16} = \sqrt[3]{(-2)^3 \cdot 2}$
$$= -2\sqrt[3]{2}$$

1. The two real square roots of 169.   2. The two real fourth roots of 32.

3. The one real cube root of −27.   4. The one real cube root of 81.

In Exercises 5–8, find the real numbers described. Simplify all radicals.

Example: A real number z such that $z^4 = 256$. ←——— Take the square root of each side.

    Solution: $z^2 = 16$ <u>or</u> $z^2 = -16$ ←——— If $z^2 = -16$, then z is not a real number.
            Since $z^2 = 16$, $z = 4$ <u>or</u> $z = -4$.

5. A real number z such that $z^2 = 52$.   6. A real number z such that $z^3 = -125$.

7. A real number z such that $z^4 = 162$.   8. A real number z such that $z^5 = -64$.

7-6 Roots of Complex Numbers

In algebra you found that two real values of x satisfy $x^2 = 4$, but only one real value of x satisfies $x^3 = 8$. However, there are three complex numbers that satisfy $x^3 = 8$. Theorem 7–4, which follows from De Moivre's theorem, can be used to find these three numbers. In Theorem 7–4, 360° could be replaced by 2π.

Theorem 7–4:   Complex Roots Theorem

If n is a positive integer and $z \neq 0$, then $z^n = r(\cos \theta + i \sin \theta)$ has n roots which are given by

$$z = \sqrt[n]{r}\left[\cos\left(\frac{\theta}{n} + \frac{k \cdot 360°}{n}\right) + i \sin\left(\frac{\theta}{n} + \frac{k \cdot 360°}{n}\right)\right],$$

where $k = 0, 1, 2, \cdots, n-1$.

Proof: First show that for any integer k,

$$\sqrt[n]{r}\left[\cos\left(\frac{\theta}{n} + \frac{k \cdot 360°}{n}\right) + i \sin\left(\frac{\theta}{n} + \frac{k \cdot 360°}{n}\right)\right]$$

satisfies the equation $z^n = r(\cos \theta + i \sin \theta)$. That is, if

$$z = \sqrt[n]{r}\left[\cos\left(\frac{\theta}{n} + \frac{k \cdot 360°}{n}\right) + i \sin\left(\frac{\theta}{n} + \frac{k \cdot 360°}{n}\right)\right],$$

then $\quad z^n = \left(\sqrt[n]{r}\left[\cos\left(\frac{\theta}{n} + \frac{k \cdot 360°}{n}\right) + i \sin\left(\frac{\theta}{n} + \frac{k \cdot 360°}{n}\right)\right]\right)^n$.

De Moivre's $\longrightarrow \quad = (\sqrt[n]{r})^n\left[\cos n\left(\frac{\theta}{n} + \frac{k \cdot 360°}{n}\right) + i \sin n\left(\frac{\theta}{n} + \frac{k \cdot 360°}{n}\right)\right]$
Theorem

$\qquad\qquad = r[\cos(\theta + k \cdot 360°) + i \sin(\theta + k \cdot 360°)]$

$\qquad\qquad = r(\cos \theta + i \sin \theta)$.

Also, since 360° is the period of $\sin \theta$ and $\cos \theta$, the value of

$$\sqrt[n]{r}\left[\cos\left(\frac{\theta}{n} + \frac{k \cdot 360°}{n}\right) + i \sin\left(\frac{\theta}{n} + \frac{k \cdot 360°}{n}\right)\right]$$

for an integer k where $k < 0$ or $k > n-1$ will be the same as one of the n distinct values given when $k = 0, 1, 2, \cdots, n-1$.

Example 1

Find the three cube roots of 8. Express the roots in polar form and in standard form.

Solution: $8 = 8 + 0i = 8(\cos 0° + i \sin 0°)$ ⟵— Express in polar form.

Let $z^3 = 8 = 8(\cos 0° + i \sin 0°)$. Therefore, by Theorem 7-4,

$$z = \sqrt[3]{8}\left[\cos\left(\frac{0°}{3} + \frac{k \cdot 360°}{3}\right) + i \sin\left(\frac{0°}{3} + \frac{k \cdot 360°}{3}\right)\right]$$

$$z = 2[\cos(k \cdot 120°) + i \sin(k \cdot 120°)].$$

For $k = 0$:

$$z_0 = 2[\cos 0° + i \sin 0°] = 2 + 0i, \text{ or } 2$$

For $k = 1$:

$$z_1 = 2[\cos 120° + i \sin 120°] = 2\left(-\tfrac{1}{2} + \tfrac{\sqrt{3}}{2}i\right) = -1 + \sqrt{3}i$$

For $k = 2$:

$$z_2 = 2[\cos 240° + i \sin 240°] = 2\left(-\tfrac{1}{2} - \tfrac{\sqrt{3}}{2}i\right) = -1 - \sqrt{3}i$$

Example 2

Find the four fourth roots of $-\tfrac{1}{2} + \tfrac{\sqrt{3}}{2}i$. Express the roots in polar form and in standard form. Graph the roots.

Solution: To find a fourth root of $-\tfrac{1}{2} + \tfrac{\sqrt{3}}{2}i$ means to find a number z such that $z^4 = -\tfrac{1}{2} + \tfrac{\sqrt{3}}{2}i$. Thus, if

$$z^4 = -\tfrac{1}{2} + \tfrac{\sqrt{3}}{2}i = (\cos 120° + i \sin 120°), \quad ⟵— \text{Polar form}$$

then $z - \sqrt[4]{1}\left[\cos\left(\frac{120°}{4} + \frac{k \cdot 360°}{4}\right) + i \sin\left(\frac{120°}{4} + \frac{k \cdot 360°}{4}\right)\right]$ ⟵— Theorem 7-4

$$z = \cos(30° + k \cdot 90°) + i \sin(30° + k \cdot 90°).$$

For $k = 0$:

$$z_0 = \cos 30° + i \sin 30° = \tfrac{\sqrt{3}}{2} + \tfrac{1}{2}i$$

For $k = 1$:

$$z_1 = \cos 120° + i \sin 120° = -\tfrac{1}{2} + \tfrac{\sqrt{3}}{2}i$$

For $k = 2$:

$$z_2 = \cos 210° + i \sin 210° = -\tfrac{\sqrt{3}}{2} - \tfrac{1}{2}i$$

For $k = 3$:

$$z_3 = \cos 300° + i \sin 300° = \tfrac{1}{2} - \tfrac{\sqrt{3}}{2}i$$

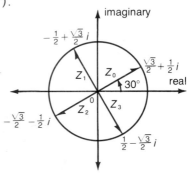

Examples 1 and 2 illustrate a general situation. The n roots of z are distributed regularly around a circle whose radius is r. If there are n roots, then the measure of the angle between two consecutive roots is

$$\frac{360°}{n} \quad \text{or} \quad \frac{2\pi}{n}.$$

Classroom Exercises

1. Find the three cube roots of 1.

2. Find the three cube roots of 27.

Written Exercises

(1—23 odds)

a Find the indicated roots. Express the roots in polar form and in standard form. Graph the roots.

1. The two square roots of $4i$

2. The two square roots of $-i$

3. The three cube roots of $8i$

4. The three cube roots of $-27i$

5. The three cube roots of -27

6. The three cube roots of -1

7. The four fourth roots of -81

8. The six sixth roots of 1

9. The four fourth roots of $-8 + 8i\sqrt{3}$

10. The two square roots of $8 + 8i\sqrt{3}$

b Solve each equation for z. Express the roots in polar form and in standard form.

11. $z^4 = 1$

12. $z^3 = i$

13. $z^6 = -1$

14. $z^3 = -1 + \sqrt{3}i$

15. $z^{12} = 1$

16. $z^2 = -1$

17. $z^5 = 1 - \sqrt{3}i$

18. $z^4 = 16$

19. $z^3 + 8 = 0$

20. $z^4 + 2 - 2i\sqrt{3} = 0$

21. $z^3 = i^2$

22. $z^3 = (1 + i)^2$

23. $z^3 = (1 + \sqrt{3}i)^2$ (Leave answers in polar form.)

24. Show that the four fourth roots of 1 are 1, ω, ω^2 and ω^3, where $\omega = \cos\frac{\pi}{2} + i\sin\frac{\pi}{2}$.

BASIC: COMPLEX NUMBERS

BASIC cannot handle complex numbers as such. To perform operations with a complex number $a + bi$ you can program the computer to perform the appropriate operations on the real numbers a and b.

Problem: Given a complex number $a + bi$, find the results when it is raised to the powers 1, 2, 3, and 4.

In the program below, the computer converts $A + Bi$ to polar form $R(\cos T + i \sin T)$ and then uses De Moivre's Theorem.

```
1Ø   READ A,B                    8Ø   LET T = Ø
2Ø   LET R = SQR(A↑2 + B↑2)      9Ø   GO TO 16Ø
3Ø   IF R = Ø THEN 8Ø           1ØØ   LET T = 3.14159 + ATN(B/A)
4Ø   IF A = Ø THEN 12Ø          11Ø   GO TO 16Ø
5Ø   IF A < Ø THEN 1ØØ          12Ø   IF B > Ø THEN 15Ø
6Ø   LET T = ATN(B/A)           13Ø   LET T = 3*3.14159/2
7Ø   GO TO 16Ø                  14Ø   GO TO 16Ø
        15Ø   LET T = 3.14159/2
        16Ø   PRINT "POWER    REAL & IMAGINARY  PARTS"
        17Ø   PRINT 1;TAB(8);A;TAB(2Ø);B
        18Ø   LET N = 2
        19Ø   PRINT N;TAB(8);R↑N*COS(N*T);TAB(2Ø);R↑N*SIN(N*T)
        2ØØ   LET N = N + 1
        21Ø   IF N < 5 THEN 19Ø
        22Ø   PRINT
        23Ø   GO TO 1Ø
        24Ø   DATA 1,1,Ø,2,-3,Ø,-1,1   ◄——— A = 1,0,−3, and −1
        25Ø   END                              B = 1,2,0, and 1
```

Analysis: Statement 2Ø: This computes and stores the value of R.

Statements 3Ø, 4Ø, 5Ø, and 12Ø: These statements determine which formula should be used for computing the value of T. Thus, a value for T is computed and stored by one of the following statements: 6Ø, 8Ø, 1ØØ, 13Ø, 15Ø.

Statement 17Ø: This prints the first power, which is simply $A + Bi$.

Statements 18Ø through 21Ø: These are used to print the powers of $A + Bi$ for N = 2 through N = 4.

Here is the output from the program just shown. In the output, "E" indicates a power of ten. Thus, 9.3628E—8 means 9.3628×10^{-8}.

```
Output:  POWER      REAL  &  IMAGINARY PARTS
           1          1                1
           2          Ø                2
           3         -2                2.
           4         -4                9.36268E-8

         POWER      REAL  &  IMAGINARY PARTS
           1          Ø                2
           2         -4                1.Ø7671E-5
           3     -3.25821E-5    -8
           4         16              -8.61366E-5

         POWER      REAL  &  IMAGINARY PARTS
           1         -3                Ø
           2          9              -4.84518E-5
           3        -27               2.174Ø1E-4
           4         81              -8.72133E-4

         POWER      REAL  &  IMAGINARY PARTS
           1         -1                1
           2     2.98Ø23E-8      -2
           3         -2.              -2.
           4         -4.             -9.36268E-8
```

Written Exercises

Write a program for each exercise.

1. Given the complex numbers A + B*i* and C + D*i*, find their sum.

2. Given the complex numbers A + B*i* and C + D*i*, find their product.

3. Given the complex numbers A + B*i* and C + D*i*, find the difference of the first minus the second.

4. Given the complex numbers A + B*i* and C + D*i*, find the quotient (if it exists) of the first divided by the second.

5. Given a complex number in polar form, convert it to standard form.

6. Given a complex number A + B*i* and a positive integer N, use De Moivre's Theorem to compute the N Nth roots of A + B*i*.

Chapter Objectives and Review

Objective: To know the meanings of the mathematical terms in this chapter.

1. Be sure that you know the meanings of these mathematical terms.

additive inverse (p. 277)
cis θ (p. 285)
complex number (p. 276)
   argument (p. 284)
   conjugate (p. 280)
   imaginary part (p. 284)
   length (p. 284)
   modulus (p. 285)
   ordered pair form (p. 284)
   polar form (p. 285)
   real part (p. 284)

standard form (p. 276)
trignometric form (p. 285)
complex number plane (p. 284)
   imaginary axis (p. 284)
   real axis (p. 284)
De Moivre's Theorem (p. 292)
i (p. 276)
pure imaginary number (p. 276)
reciprocal (p. 281)
vector (p. 284)
   magnitude (p. 284)

Objective: To add and subtract complex numbers. (Section 7–1)

Perform the indicated operations.

2. $(8 - 7i) + (5 + 2i)$

3. $(9 + 5i) - (2 - 3i)$

4. $(3 - i) - (-4 + 6i)$

5. $(-12 - 4i) + (-8 + i)$

Objective: To multiply and divide complex numbers. (Section 7–2)

Perform the indicated operations.

6. $(3 - 5i)(4 + 7i)$

7. $(8 + 3i) \div (2 - 4i)$

8. $(-6 - 7i) \div (3 + 2i)$

9. $(-12 + 2i)(-12 - 2i)$

Objective: To express complex numbers in polar form. (Section 7–3)

For each complex number z, find $|z|$ and θ. Then express z in polar form.

10. $z = 3 - 3i$ **11.** $z = -\sqrt{3} + i$ **12.** $z = -\sqrt{2} - \sqrt{2}i$ **13.** $z = \frac{1}{2} - \frac{\sqrt{3}}{2}i$

Objective: To multiply complex numbers expressed in polar form. (Section 7–4)

Find $z_1 \cdot z_2$. Express the product in standard form.

14. $z_1 = 3(\cos 165° + i \sin 165°); z_2 = 8(\cos 45° + i \sin 45°)$

15. $z_1 = 5(\cos \frac{5\pi}{6} + i \sin \frac{5\pi}{6}); z_2 = 9(\cos \frac{7\pi}{6} + i \sin \frac{7\pi}{6})$

Objective: To divide complex numbers expressed in polar form. (Section 7–4)

Find $z_1 \div z_2$. Express the quotient in standard form.

16. $z_1 = 36\left(\cos\frac{9\pi}{8} + i\sin\frac{9\pi}{8}\right)$; $z_2 = 27\left(\cos\frac{3\pi}{8} + i\sin\frac{3\pi}{8}\right)$

17. $z_1 = 2(\cos 327° + i\sin 327°)$; $z_2 = 18(\cos 147° + i\sin 147°)$

Objective: To use De Moivre's Theorem to evaluate powers of complex numbers. (Section 7–5)

Evaluate each expression. Express answers in standard form.

18. $[2(\cos 60° + i\sin 60°)]^4$

19. $\left[3\left(\cos\frac{\pi}{6} + i\sin\frac{\pi}{6}\right)\right]^{-2}$

20. $[4(\cos 15° + i\sin 15°)]^{-3}$

21. $(1 + i)^5$

Objective: To find roots of complex numbers. (Section 7–6)

Find the indicated roots. Express the answers in polar form and in standard form. Graph the roots.

22. The three cube roots of $64i$

23. The four fourth roots of -16

Solve each equation for z.

24. $z^4 = -1$

25. $z^3 = 1 - \sqrt{3}i$

Chapter Test

Perform the indicated operations.

1. $(6 - i) - (5 - 3i)$

2. $(15 + 5i) \div (1 - 2i)$

3. $(9 + 7i)(-3 + i)$

4. $(3 + 8i) + (3 - 8i)$

5. Express $1 - \sqrt{3}i$ in polar form.

6. Find $z_1 \cdot z_2$ where $z_1 = \frac{1}{3}\left(\cos\frac{9\pi}{8} + i\sin\frac{9\pi}{8}\right)$ and $z_2 = -18\left(\cos\frac{3\pi}{8} + i\sin\frac{3\pi}{8}\right)$. Express the product in standard form.

7. Find $z_1 \div z_2$ where $z_1 = -3(\cos 25° + i\sin 25°)$ and $z_2 = 6(\cos 70° + i\sin 70°)$. Express the quotient in standard form.

8. Use De Moivre's Theorem to evaluate $(1 - i)^8$. Express the answer in standard form.

9. Graph the four fourth roots of $-81i$.

10. Solve: $z^4 = -2 + 2\sqrt{3}i$

Chapter 8
Sequences and Series

8-1 Sequences

The list of numbers below follows a pattern.

$$1, 4, 9, 16, 25, \cdots, a_n, \cdots$$

Each number in the list is called a **term.** The symbol a_n (read "a sub n") designates the nth term. The subscript n in a_n is called the **index** of the term. For the list above,

$$a_1 = 1, \; a_2 = 4, \; a_3 = 9, \; a_4 = 16, \; \text{and} \; a_5 = 25.$$

You can find a rule for some patterns. To find a rule for the pattern above, pair each term with its index.

Example 1

a. Find a rule for 1, 4, 9, 16, 25, $\cdots$.

b. Use the rule to write the next three terms.

Solutions: a.

1 2 3 4 5 $\cdots$ n $\cdots$ $\longleftarrow$ Positive Integers

1 4 9 16 25 $\cdots$ a_n $\cdots$ $\longleftarrow$ Terms

<u>Rule</u>: Each term is the square of the positive integer with which it is paired. That is, $a_n = n^2$.

b. $a_6 = 6^2 = 36;\quad a_7 = 7^2 = 49;\quad a_8 = 8^2 = 64$

The pairings in Example 1 can be written as a set of ordered pairs.

$$\{(1, 1), (2, 4), (3, 9), (4, 16), (5, 25), \cdots, (n, n^2), \cdots\}$$

or $\qquad \{(n, n^2)\}$, where n is a positive integer.

Note that $\{(n, n^2)\}$ is a function, since for each positive integer n, n^2 is unique. A function such as this is called a <u>sequence</u>. In Example 1, n^2 is called the <u>general term</u> of the sequence.

Definitions: A **sequence** is a function whose domain is the set of positive integers. The numbers in the range of the function are the **terms of the sequence.** When the rule for the terms of a sequence is written in symbols, it is called the **general term** of the sequence.

For convenience, we often refer to the <u>terms of a sequence</u> as a sequence, since the domain is always the set of positive integers. Thus, the sequence with general term n^2 is denoted as

$$\{1, 4, 9, 16, \cdots, n^2, \cdots\} \quad \text{or} \quad \{n^2\}.$$

Example 2

Write the first four terms of each sequence.

a. $\{a_n\} = \{-n^3 + \frac{13}{2}n^2 - \frac{21}{2}n + 6\}$ **b.** $\{b_n\} = \{\frac{1}{2}n^2 + \frac{1}{2}n\}$

Solutions: a. $a_1 = -(1)^3 + \frac{13}{2}(1)^2 - \frac{21}{2}(1) + 6 = 1$ **b.** $b_1 = \frac{1}{2}(1)^2 + \frac{1}{2}(1) = 1$

$a_2 = -(2)^3 + \frac{13}{2}(2)^2 - \frac{21}{2}(2) + 6 = 3$ $b_2 = \frac{1}{2}(2)^2 + \frac{1}{2}(2) = 3$

$a_3 = -(3)^3 + \frac{13}{2}(3)^2 - \frac{21}{2}(3) + 6 = 6$ $b_3 = \frac{1}{2}(3)^2 + \frac{1}{2}(3) = 6$

$a_4 = -(4)^3 + \frac{13}{2}(4)^2 - \frac{21}{2}(4) + 6 = 4$ $b_4 = \frac{1}{2}(4)^2 + \frac{1}{2}(4) = 10$

Classroom Exercises

Write the rule for each sequence.

1. 4, 7, 10, 13, $\cdots$ **2.** 1, 3, 5, 7, $\cdots$ **3.** 0, -1, -2, -3, $\cdots$

Write the first three terms of each sequence.

4. $\{a_n\} = \{2n\}$ **5.** $\{a_n\} = \{\frac{1}{3}n\}$ **6.** $\{a_n\} = \{3^{n-3}\}$

Written Exercises

a

Write the rule for each sequence.

1. 2, 5, 8, 11, $\cdots$ **2.** -2, -1, 0, 1, $\cdots$ **3.** 1, $\frac{1}{2}$, $\frac{1}{3}$, $\frac{1}{4}$, $\cdots$

4. 1, $\frac{1}{3}$, $\frac{1}{9}$, $\frac{1}{27}$, $\cdots$ **5.** -3, 2, 7, 12, $\cdots$ **6.** 2.5, 3, 3.5, 4, $\cdots$

Write the first five terms of each sequence.

7. $\{a_n\} = \{n\}$ **8.** $\{b_n\} = \{1^{n-1}\}$ **9.** $\{b_n\} = \{n^3\}$

10. $\{a_n\} = \{\frac{1}{n}\}$ **11.** $\{b_n\} = \{-\frac{1}{n}\}$ **12.** $\{a_n\} = \{5\left(\frac{1}{2}\right)^{n-1}\}$

13. $\{a_n\} = \{(-1)^{n-1}\}$ **14.** $\{a_n\} = \left\{\left(-\dfrac{1}{2}\right)^{n-1}\right\}$ **15.** $\{b_n\} = \left\{\left(\dfrac{1}{n}\right)^2\right\}$

16. $\{a_n\} = \{n(n-1)\}$ **17.** $\{b_n\} = \left\{\dfrac{n+2}{2n+3}\right\}$ **18.** $\{a_n\} = \left\{\cos^2\dfrac{n\pi}{2}\right\}$

Use the rule to find the indicated term.

19. $a_n = \dfrac{1}{n}$; 7th term   **20.** $a_n = -\dfrac{1}{n}$; 7th term   **21.** $a_n = 2n$; 5th term

22. $a_n = \dfrac{2}{n-1}$; 8th term   **23.** $a_n = n^2$; 9th term   **24.** $a_n = 3^{1-n}$; 5th term

25. $a_n = n^2(1+n)$; 6th term **26.** $a_n = (-1)^{n+1}$; 7th term **27.** $a_n = (-1)^{2n-1}$; 8th term

28. $a_n = \left(\dfrac{1}{n}\right)^2$; 6th term   **29.** $a_n = \left(\dfrac{1}{3}\right)^n$; 5th term   **30.** $a_n = \left(\dfrac{-2}{3}\right)^n$; 4th term

Write the first four terms of each sequence. (Recall: $n! = 1 \cdot 2 \cdot 3 \cdots n$; $0! = 1$.)

31. $\left\{\dfrac{n}{2}(n+1)\right\}$ **32.** $\left\{\dfrac{2n-1}{3n+2}\right\}$ **33.** $\{n^2(n-1)\}$

34. $\{3^{1-n}\}$ **35.** $\left\{\dfrac{2n+1}{2n-1}\right\}$ **36.** $\{i^n\}$

37. $\left\{\dfrac{1}{(n-1)!}\right\}$ **38.** $\left\{\dfrac{1-(-1)^n}{n^2}\right\}$ **39.** $\{\cos^n 60°\}$

40. $\left\{1+\dfrac{1}{n}\right\}$ **41.** $\left\{1+\dfrac{3n+4}{2n-3}\right\}$ **42.** $\left\{\dfrac{1}{n^2+3}\right\}$

43. $\left\{(-1)^n \sin\dfrac{n\pi}{4}\right\}$ **44.** $\left\{\dfrac{\sqrt{n+1}}{n}\right\}$ **45.** $\left\{\dfrac{(-1)^{n-1}}{n!}\right\}$

46. $\left\{\dfrac{(2x)^n}{(2n-1)^3}\right\}$ **47.** $\left\{\dfrac{\cos nx}{x^2+n^2}\right\}$ **48.** $\left\{\dfrac{(-1)^{n+1}x^{2n-1}}{(2n-1)!}\right\}$

49. $\left\{\dfrac{(-1)^{n+1}x^{2n-2}}{(2n-2)!}\right\}$ **50.** $\left\{\dfrac{x^{n-1}}{(n-1)!}\right\}$ **51.** $\left\{\dfrac{i^{n-1}x^{n-1}}{(n-1)!}\right\}$

A sequence is said to be defined **recursively** when the first term and a relationship between each term and its successor are given. Each sequence below is defined recursively. Write the first six terms of each sequence.

52. $a_1 = -2$; $a_{n+1} = a_n + 5$ **53.** $a_1 = 4$; $a_{n+1} = (-1)^n a_n$

54. Write the first 8 terms of the **Fibonacci sequence** which is defined recursively below.

$$a_1 = 1; \; a_2 = 1; \; a_{n+1} = a_n + a_{n-1} \;\; \text{for } n \geq 2$$

55. Prove that the first k terms of the sequence $\{a_n\}$ are identical to the first k terms of the sequence $\{b_n\} = \{a_n + (n-1)(n-2)(n-3)\cdots(n-k)\}$.

56. Prove that the $(k+1)$st terms of the sequences in Exercise 55 are <u>not</u> equal.

8-2 Arithmetic and Geometric Sequences

If each term of a sequence is obtained by adding a constant d to the preceding term, the sequence is called an **arithmetic sequence.** The constant d that is added to each term is called the **common difference.**

Example 1 The first term of an arithmetic sequence is a_1 and the common difference is d. Write the next three terms.

Solution: $a_2 = a_1 + d$ $\qquad$ $a_3 = a_2 + d$ $\qquad\qquad$ $a_4 = a_3 + d$

$\qquad\qquad\qquad\qquad\qquad = (a_1 + d) + d \qquad\qquad = (a_1 + 2d) + d$

$\qquad\qquad\qquad\qquad\qquad = a_1 + 2d \qquad\qquad\qquad = a_1 + 3d$

In general if $\{a_n\}$ represents an arithmetic sequence with first term a_1 and common difference d, then

$$\{a_n\} = \{a_1 + (n-1) \cdot d\}.$$

Example 2 Write the 25th term of the arithmetic sequence $\{a_n\}$ with $a_1 = 1$ and $a_2 = 4$.

Solution: Since $\{a_n\}$ is an arithmetic sequence and $a_1 = 1$,

$$a_{25} = 1 + (25 - 1) \cdot d.$$

Also, since $a_2 = a_1 + d$ and $a_2 = 4$, $4 = 1 + d$.
Therefore, $d = 3$.
Then, $\qquad\qquad\qquad a_{25} = 1 + (25 - 1) \cdot 3$

$\qquad\qquad\qquad\qquad\quad = 1 + (24) \cdot 3$, or 73.

If each term of a sequence is obtained by multiplying the preceding term by some fixed number r, the sequence is called a **geometric sequence.** The constant r is called the **common ratio.**

Example 3

The first term of a geometric sequence is a_1 and the common ratio is r. Find the next three terms.

Solution:

$$a_2 = a_1 \cdot r \qquad\qquad \begin{aligned} a_3 &= a_2 \cdot r \\ &= (a_1 \cdot r) \cdot r \\ &= a_1 \cdot r^2 \end{aligned} \qquad\qquad \begin{aligned} a_4 &= a_3 \cdot r \\ &= (a_1 \cdot r^2) \cdot r \\ &= a_1 \cdot r^3 \end{aligned}$$

In general, if $\{a_n\}$ represents a geometric sequence with first term a_1 and common ratio r, then

$$\{a_n\} = \{a_1 \cdot r^{(n-1)}\}.$$

Example 4

Write the 8th term of the geometric sequence $\{a_n\}$ with $a_1 = 3$ and $a_2 = 6$.

Solution: Since $\{a_n\}$ is a geometric sequence and $a_1 = 3$,

$$a_8 = 3 \cdot r^{(8-1)}.$$

Also, since $a_2 = a_1 \cdot r$ and $a_2 = 6$, $6 = 3 \cdot r$.   Therefore, $r = 2$.

Then,

$$\begin{aligned} a_8 &= 3 \cdot (2)^{8-1} \\ &= 3 \cdot 128, \text{ or } 384. \end{aligned}$$

Classroom Exercises

In Exercises 1–3, a is the first term of an arithmetic sequence and d is the common difference. Write the next four terms.

1. $a = 1$; $d = 5$
2. $a = -30$; $d = -2$
3. $a = 25b$; $d = -5b$

Use the given values of a_1 and a_2 to write the indicated terms of the arithmetic sequence $\{a_n\}$.

4. $a_1 = -3$, $a_2 = 0$; 24th term
5. $a_1 = \frac{1}{2}$, $a_2 = 2$; 15th term

In Exercises 6–8, a is the first term of a geometric sequence and r is the common ratio. Write the next four terms.

6. $a = 2$; $r = -2$
7. $a = \frac{1}{4}$; $r = 2$
8. $a = \sqrt{2}$; $r = \sqrt{2}$

Use the given values of a_1 and a_2 to write the indicated terms of the geometric sequence $\{a_n\}$.

9. $a_1 = -1$, $a_2 = -2$; 8th term

10. $a_1 = 96$, $a_2 = 24$; 6th term

Written Exercises

a

For Exercises 1–6, a is the first term of an arithmetic sequence and d is the common difference. Write the next four terms.

1. $a = 2$; $d = 3$

2. $a = -5$; $d = 3$

3. $a = -3$; $d = -2$

4. $a = 7$; $d = -5$

5. $a = 100$; $d = -50$

6. $a = \frac{1}{2}$; $d = -\frac{3}{4}$

For Exercises 7–12, use the given values of a_1 and a_2 to write the indicated term of the arithmetic sequence $\{a_n\}$.

7. $a_1 = 4$, $a_2 = 5$; 20th term

8. $a_1 = 18$; $a_2 = 14$; 24th term

9. $a_1 = -3$, $a_2 = -3\frac{1}{2}$; 11th term

10. $a_1 = 2\sqrt{3}$, $a_2 = \sqrt{3}$; 9th term

11. $a_1 = \frac{1}{2}$, $a_2 = 2$; 25th term

12. $a_1 = x$, $a_2 = x + 2$; 12th term

For Exercises 13–18, a is the first term of a geometric sequence with common ratio r. Write the next four terms.

13. $a = 2$; $r = 3$

14. $a = -5$; $r = \frac{1}{3}$

15. $a = 1$; $r = -\frac{1}{2}$

16. $a = 7$; $r = \frac{2}{5}$

17. $a = 5$; $r = -\frac{2}{3}$

18. $a = -5$; $r = -2$

For Exercises 19–24, use the given values of a_1 and a_2 to write the indicated term of the geometric sequence $\{a_n\}$.

19. $a_1 = 2$, $a_2 = 6$; 8th term

20. $a_1 = \frac{1}{2}$, $a_2 = 1$; 10th term

21. $a_1 = -27$, $a_2 = -18$; 14th term

22. $a_1 = \frac{1}{4}$, $a_2 = \frac{1}{2}$; 12th term

23. $a_1 = 9$, $a_2 = -6$; 5th term

24. $a_1 = .7$, $a_2 = -2.8$; 6th term

b

For Exercises 25–28, write the general term of the arithmetic sequence having the first three terms as given.

25. $\{-6, -2, 2, \cdots\}$

26. $\{2x, x, 0, \cdots\}$

27. $\{x + 1, 2x, 3x - 1, \cdots\}$

28. $\{-x^2, 0, x^2, \cdots\}$

For Exercises 29–32, write the general term of the geometric sequence with the first three terms as given.

29. $\{\sqrt{2}, 2, 2\sqrt{2}, \cdots\}$

30. $\{1, -1, 1, \cdots\}$

31. $\{96, -24, 6, \cdots\}$

32. $\{i, -1, -i, \cdots\}$

For Exercises 33–36, write the first four terms of the geometric sequence with the given term and common ratio.

33. $a_2 = -18$; $r = -\frac{3}{2}$

34. $a_5 = (1.03)^4$; $r = 1.03$

35. $a_3 = y^{12}$; $r = y^3$

36. $a_2 = 4x^8$; $r = -2x^2$

37. The third term of an arithmetic sequence is 5 and the eighth term is 20. Find the first and second terms.

38. The third term of a geometric sequence is 9 and the seventh term is 81. Find the fifth term.

39. The seventh term of an arithmetic sequence is 6 and the thirteenth term is −18. Find the first five terms.

40. Find the twentieth term of the arithmetic sequence where the fourth term is 9 and the seventh term is 10.

41. Find the common difference for the arithmetic sequence where the first term is 38 and the seventy-fifth term is 1.

42. The arithmetic sequence $\{a_n\}$ has $a_1 = 100$, $d = -2$, and $a_j = -14$. Find j.

43. The geometric sequence $\{\frac{1}{4}, \frac{1}{2}, 1, \cdots\}$ has $a_j = 128$. Find j.

44. Find the value of y so that $y + 1$, $y - 1$, $y - 2$ are the first three terms of a geometric sequence

45. The arithmetic sequence $\{5, 2, -1, -4, \cdots\}$ has $a_j = -85$. Find j.

46. How many even integers are there between 16 and 250? (HINT: $a_1 = 18$, $a_j = 248$, $d = 2$.)

47. How many odd integers are there between 35 and 261?

48. Find the value of y so that $y + 1$, $3y - 2$, and $4y + 5$ are the first three terms of an arithmetic sequence.

49. Write the general term for the sequence of positive integers.

50. Write the general term for the sequence of positive even integers.

51. Write the general term for the sequence of positive odd integers.

*The terms between any two given terms of an arithmetic sequence are called **arithmetic means.***

52. Find three arithmetic means between 6 and 11. (HINT: Let $a_1 = 6$ and $a_5 = 11$.)

53. Find the arithmetic mean between x and y.

*The terms between any two given terms of a geometric sequence are called **geometric means.***

54. Find one positive geometric mean between 3 and 48.

55. Find a geometric mean between x and y.

56. The fundamental frequencies of the notes of the musical scale are consecutive terms of a geometric sequence. The frequency of the 13th note is double that of the first. Find the common ratio.

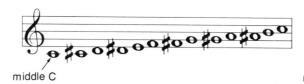

middle C

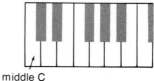

middle C

57. The fundamental frequency of middle C is 264 cycles per second. Find the fundamental frequency of A, the ninth note above middle C.

58. The figure at the right contains a large square with a side of length 12 and a succession of smaller squares formed by connecting midpoints of consecutive sides. The areas of the squares form a geometric sequence. Find the common ratio and the area of the red square.

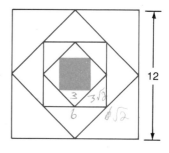

C **59.** Write a recursive definition for $\{a_n = a_1 + (n-1)d\}$. (See Exercises 52–54 on page 306.)

60. Write a general term for a sequence whose first four terms are 2, 4, 8, and 16, but whose fifth term is <u>not</u> 32.

61. Prove that if a is the general term of a geometric sequence then $b_n = (a_n)^2$ is also the general term of a geometric sequence.

62. Prove that if $\{a_n\}$ and $\{b_n\}$ are arithmetic sequences then $\{c_n\} = \{a_n + b_n\}$ is also an arithmetic sequence.

63. Prove that for a constant c and an arithmetic sequence $\{a_n\}$, the sequence $\{b_n\} = \{c \cdot a_n\}$ is arithmetic.

Review Capsule for Section 8-3 _____

Solve for n.

Example: $\dfrac{2}{n+1} < \dfrac{1}{100}$　　　　**Solution:** $\dfrac{2}{n+1} < \dfrac{1}{100}$

$$2 \cdot 100 < n+1$$
$$199 < n$$

1. $\dfrac{7}{2n} < \dfrac{1}{1000}$　　**2.** $\dfrac{1}{n+3} < \dfrac{1}{10000}$　　**3.** $\dfrac{n+2}{n+3} < \dfrac{1}{100}$　　**4.** $\dfrac{4}{n^2} < \dfrac{1}{100}$

8-3 Limit of a Sequence

For a sequence $\{a_n\}$ it sometimes happens that as n increases, the terms of the sequence tend to cluster around some fixed number.

Example 1 Graph the first eight terms of the sequence below.

$$\{a_n\} = \left\{2, 1\tfrac{1}{2}, 1\tfrac{1}{3}, \cdots, 1 + \frac{1}{n}, \cdots\right\}$$

Solution:

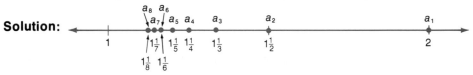

In Example 1, the terms of the sequence appear to "close in" on 1 as n increases. In fact, if you choose any positive number, no matter how small, you can find a natural number m such that a_m and all succeeding terms differ from 1 by less than the number chosen.

Example 2 For the sequence $\{a_n\} = \{1 + \frac{1}{n}\}$, find a natural number m such that a_m and all succeeding terms differ from 1 by less than $\frac{1}{1000}$.

Solution: The difference between 1 and a_m can be expressed as $|1 - a_m|$. To find m so that the difference between 1 and a_m is less than $\frac{1}{1000}$, solve the following inequality for m.

$$\left|1 - \left(1 + \frac{1}{m}\right)\right| < \frac{1}{1000} \quad \longleftarrow \quad \text{Since } a_m = 1 + \frac{1}{m}$$

$$\left|-\frac{1}{m}\right| < \frac{1}{1000}$$

$$\frac{1}{m} < \frac{1}{1000} \quad \longleftarrow \quad \text{Since } m > 0$$

$$1000 < m$$

Thus, a_{1001} and all succeeding terms will differ from 1 by less than $\frac{1}{1000}$.

The sequence in Example 2 is said to <u>converge</u> to the <u>limit</u> 1.

Definitions: For a sequence $\{a_n\}$ and a real number L, if, corresponding to any positive number d (no matter how small), a natural number m can be found such that a_m and all succeeding terms differ from L by less than d, that is,

$$|L - a_n| < d \text{ for all } n \ge m,$$

then the sequence **converges** to the **limit** L. In symbols,

$$\lim_{n \to \infty} a_n = L$$

and $\{a_n\}$ is called a **convergent sequence.** A sequence that does not converge to a limit is called a **divergent sequence.**

Each term, a_n, of a convergent sequence can be used to estimate the limit. In general, the estimate gets better as n increases.

Example 3

a. Guess the number to which the terms of $\left\{\dfrac{n+2}{n+3}\right\}$ converge.

b. Use the definition to prove that this number is the limit.

Solutions: a. Replace n with $1, 2, 3, \cdots$.

$$\left\{\frac{n+2}{n+3}\right\} = \left\{\frac{3}{4}, \frac{4}{5}, \frac{5}{6}, \frac{6}{7}, \frac{7}{8}, \cdots, \frac{n+2}{n+3}, \cdots\right\}$$

By inspection, the terms appear to converge to 1.

b. To prove that 1 is the limit, show that for any positive number d, there is a natural number m such that $\left|1 - \dfrac{n+2}{n+3}\right|$ is less than d for all $n \ge m$. Since 1 can be written as $\dfrac{n+3}{n+3}$ and n is positive,

$$\left|1 - \frac{n+2}{n+3}\right| = \left|\frac{n+3}{n+3} - \frac{n+2}{n+3}\right| = \left|\frac{1}{n+3}\right| = \frac{1}{n+3}.$$

Therefore $\left|1 - \dfrac{n+2}{n+3}\right| < d$ is equivalent to $\dfrac{1}{n+3} < d$, which is true whenever $\dfrac{1}{d} < n + 3$ or $\dfrac{1}{d} - 3 < n$. Thus, for any positive number d, you can choose m to be any natural number greater than $\dfrac{1}{d} - 3$. Then $\left|1 - \dfrac{n+2}{n+3}\right|$ will be less than d whenever $n \ge m$. Therefore,

$$\lim_{n \to \infty} \frac{n+2}{n+3} = 1.$$

Classroom Exercises

Simplify each expression.

1. $1 - \dfrac{n+3}{n+4}$

2. $1 - \left(1 - \dfrac{1}{n}\right)$

Find the tenth term of each sequence.

3. $\left\{\dfrac{1}{n^2}\right\}$

4. $\{(-1)^n\}$

Written Exercises

a   Graph the first 10 terms of each sequence.

1. $\left\{\dfrac{1}{n}\right\}$

2. $\left\{\dfrac{3}{n}\right\}$

3. $\left\{\dfrac{-5}{n}\right\}$

4. $\left\{\dfrac{-10}{n}\right\}$

5. $\left\{1 - \dfrac{1}{n}\right\}$

6. $\left\{2 - \dfrac{1}{n}\right\}$

7. $\left\{5 + \dfrac{3}{n}\right\}$

8. $\left\{3 - \dfrac{10}{n}\right\}$

9. $\left\{\dfrac{4}{n^2}\right\}$

10. $\left\{\dfrac{4}{3n}\right\}$

11. $\left\{\dfrac{7}{2n}\right\}$

12. $\left\{\dfrac{2}{n+1}\right\}$

13. $\{n\}$

14. $\left\{n + \dfrac{1}{n}\right\}$

15. $\{n^2\}$

16. $\left\{n - \dfrac{1}{n}\right\}$

17. $\{(-1)^n\}$

18. $\{3(-1)^n\}$

19. $\left\{(-1)^n \dfrac{1}{n}\right\}$

20. $\{(-1)^n \, n^2\}$

21. $\{1 + (-1)^n\}$

22. $\left\{\dfrac{n}{n+1}\right\}$

23. $\left\{\dfrac{n}{n+3}\right\}$

24. $\left\{\dfrac{n+3}{n+4}\right\}$

For each sequence $\{a_n\}$ in Exercises 25–32, find a natural number m such that a_m and all succeeding terms differ from L by less than $\dfrac{1}{10,000}$.

25. $\{a_n\} = \left\{1 - \dfrac{1}{n}\right\}$; $L = 1$

26. $\{a_n\} = \left\{2 - \dfrac{1}{n}\right\}$; $L = 2$

27. $\{a_n\} = \left\{5 + \dfrac{3}{n}\right\}$; $L = 5$

28. $\{a_n\} = \left\{3 - \dfrac{10}{n}\right\}$; $L = 3$

29. $\{a_n\} = \left\{\dfrac{1}{n}\right\}$; $L = 0$

30. $\{a_n\} = \left\{\dfrac{-5}{n}\right\}$; $L = 0$

31. $\{a_n\} = \left\{\dfrac{2}{n+1}\right\}$; $L = 0$

32. $\{a_n\} = \left\{\dfrac{7}{2n}\right\}$; $L = 0$

For each of the following sequences, graph the first six terms. Then guess the limit L to which the terms converge.

33. $\left\{\left(\dfrac{1}{2}\right)^n\right\}$

34. $\left\{\left(-\dfrac{1}{2}\right)^n\right\}$

35. $\left\{\dfrac{2^n}{3^n}\right\}$

36. $\left\{\left(-\dfrac{2}{3}\right)^n\right\}$

37. $\left\{4\left(\dfrac{1}{2}\right)^n\right\}$

38. $\left\{5 - \left(\dfrac{1}{2}\right)^n\right\}$

39. $\left\{3 + \dfrac{(-1)^n}{3^n}\right\}$

40. $\left\{-1\left(\dfrac{1}{2}\right)^n\right\}$

For Exercises 41–56, use the definition of a limit to prove that the sequence $\{a_n\}$ converges to the limit L.

41. $\{a_n\} = \left\{\dfrac{n+1}{n+3}\right\}$; $L = 1$

42. $\{a_n\} = \left\{\dfrac{n}{n+3}\right\}$; $L = 1$

43. $\{a_n\} = \left\{\dfrac{n}{n+1}\right\}$; $L = 1$

44. $\{a_n\} = \left\{\dfrac{n+3}{n+4}\right\}$; $L = 1$

45. $\{a_n\} = \left\{1 + \dfrac{1}{n}\right\}$; $L = 1$

46. $\{a_n\} = \left\{1 - \dfrac{1}{n}\right\}$; $L = 1$

47. $\{a_n\} = \left\{2 - \dfrac{1}{n}\right\}$; $L = 2$

48. $\{a_n\} = \left\{5 + \dfrac{3}{n}\right\}$; $L = 5$

49. $\{a_n\} = \left\{\dfrac{1}{n}\right\}$; $L = 0$

50. $\{a_n\} = \left\{\dfrac{-500}{n}\right\}$; $L = 0$

51. $\{a_n\} = \left\{\dfrac{4}{n^2}\right\}$; $L = 0$

52. $\{a_n\} = \left\{\dfrac{n^2}{n^2+4}\right\}$; $L = 1$

53. $\{a_n\} = \left\{\dfrac{(-1)^n}{n+1}\right\}$; $L = 0$

54. $\{a_n\} = \left\{\dfrac{(-1)^n}{n^2}\right\}$; $L = 0$

55. $\{a_n\} = \left\{\dfrac{1}{10^n}\right\}$; $L = 0$

56. $\{a_n\} = \left\{\left(-\dfrac{1}{3}\right)^n\right\}$; $L = 0$

For Exercises 57–60, guess the limit to which the terms of the given sequence converges. Then prove that your guess is correct.

57. $\left\{2 + \dfrac{1}{n}\right\}$

58. $\left\{-\dfrac{1}{n}\right\}$

59. $\left\{\dfrac{1}{n+1}\right\}$

60. $\left\{\dfrac{1-n}{n+4}\right\}$

8-4 Finding the Limit of a Sequence

Statements **1–3** below are theorems that can be proved using the definition of limit.

1. $\lim\limits_{n \to \infty} c = c$ for any real number c.

2. $\lim\limits_{n \to \infty} \dfrac{c}{n} = 0$ for any real number c.

3. $\lim\limits_{n \to \infty} \left(\dfrac{b}{c}\right)^n = 0$ for any pair of real numbers b and c such that $c \neq 0$ and $\left|\dfrac{b}{c}\right| < 1$.

Statements **4–7** are also theorems that can be proved using the definition of limit provided $\{a_n\}$ and $\{b_n\}$ converge.

4. $\lim\limits_{n \to \infty} (a_n \pm b_n) = \lim\limits_{n \to \infty} a_n \pm \lim\limits_{n \to \infty} b_n$

5. $\lim\limits_{n \to \infty} a_n \cdot b_n = (\lim\limits_{n \to \infty} a_n) \cdot (\lim\limits_{n \to \infty} b_n)$

6. $\lim\limits_{n \to \infty} \dfrac{a_n}{b_n} = \dfrac{\lim\limits_{n \to \infty} a_n}{\lim\limits_{n \to \infty} b_n}$, where $b_n \neq 0$ and $\lim\limits_{n \to \infty} b_n \neq 0$

7. $\lim\limits_{n \to \infty} c \cdot a_n = c \cdot \lim\limits_{n \to \infty} a_n$ for any real number c.

These theorems can be used to find the limit of some sequences.

Example 1   Find $\lim\limits_{n \to \infty} 3\left(\dfrac{1}{2}\right)^{n-1}$, if it exists.

Solution: Rewrite $3\left(\dfrac{1}{2}\right)^{n-1}$ as $3 \cdot \left(\dfrac{1}{2}\right)^{-1} \cdot \left(\dfrac{1}{2}\right)^{n}$.

$\lim\limits_{n \to \infty} [3 \cdot (\tfrac{1}{2})^{n-1}] = \lim\limits_{n \to \infty} [3 \cdot (\tfrac{1}{2})^{-1} \cdot (\tfrac{1}{2})^{n}]$

$\qquad\qquad\qquad = 3 \cdot (\tfrac{1}{2})^{-1} \cdot \lim\limits_{n \to \infty} (\tfrac{1}{2})^{n}$ ⟵ Limit Theorem 7

$\qquad\qquad\qquad = 3 \cdot (\tfrac{1}{2})^{-1} \cdot 0 = 0$ ⟵ Limit Theorem 3

In Example 1, $\{3(\tfrac{1}{2})^{n-1}\}$ is a geometric sequence with $a_1 = 3$ and $r = \tfrac{1}{2}$. Example 1 suggests the following statement, part of which you are asked to prove in the exercises.

The geometric sequence $\{a_1 \cdot r^{n-1}\}$ <u>converges</u> whenever $|r| < 1$. In this case,

$$\lim_{n \to \infty} a_1 r^{n-1} = 0.$$

The sequence diverges whenever $|r| > 1$.

Example 2 Find $\lim_{n \to \infty} [\tfrac{1}{8} \cdot (-2)^{n-1}]$, if it exists.

Solution: Since $\{\tfrac{1}{8} \cdot (-2)^{n-1}\}$ is a geometric sequence and $|r| = |-2|$ is greater than 1, the sequence diverges. Thus, no limit exists.

Example 3 Find $\lim_{n \to \infty} \left(1 + \dfrac{3n + 4}{2n - 3}\right)$, if it exists.

Solution: Rewrite $\dfrac{3n + 4}{2n - 3}$ as $\dfrac{3 + \dfrac{4}{n}}{2 - \dfrac{3}{n}}$ by dividing the numerator and denominator

by n. Then, by Limit Theorem **2**,

$$\lim_{n \to \infty} \left(\dfrac{4}{n}\right) = 0 \text{ and } \lim_{n \to \infty} \left(\dfrac{3}{n}\right) = 0.$$

$$\therefore \quad \lim_{n \to \infty} \left(\dfrac{3 + \dfrac{4}{n}}{2 - \dfrac{3}{n}}\right) = \dfrac{\lim\limits_{n \to \infty} 3 + \lim\limits_{n \to \infty} \left(\dfrac{4}{n}\right)}{\lim\limits_{n \to \infty} 2 - \lim\limits_{n \to \infty} \left(\dfrac{3}{n}\right)} \quad \longleftarrow \text{ Limit Theorems 6 and 4}$$

$$= \dfrac{3 + 0}{2 - 0} \quad \longleftarrow \text{ Limit Theorems 1 and 2}$$

Thus, $\lim_{n \to \infty} \left(\dfrac{3n + 4}{2n - 3}\right) = \dfrac{3}{2}$, and

$$\lim_{n \to \infty} \left(1 + \dfrac{3n + 4}{2n - 3}\right) = \lim_{n \to \infty} 1 + \lim_{n \to \infty} \left(\dfrac{3n + 4}{2n - 3}\right) \quad \longleftarrow \text{ Limit Theorem 4}$$

$$= 1 + \dfrac{3}{2} \quad \longleftarrow \text{ Limit Theorem 1}$$

$$= 2\tfrac{1}{2}$$

Example 4 Find $\lim\limits_{n\to\infty}\left(\dfrac{1}{n^2+3}\right)$, if it exists.

Solution: Rewrite $\dfrac{1}{n^2+3}$ as $\dfrac{\frac{1}{n^2}}{1+\frac{3}{n^2}}=\dfrac{\frac{1}{n}\cdot\frac{1}{n}}{1+\frac{3}{n}\cdot\frac{1}{n}}$. Then, by Limit Theorem **2**,

$$\lim_{n\to\infty}\frac{1}{n}=0 \quad\text{and}\quad \lim_{n\to\infty}\frac{3}{n}=0.$$

Thus, $\lim\limits_{n\to\infty}\left(\dfrac{1}{n^2+3}\right)=\lim\limits_{n\to\infty}\left(\dfrac{\frac{1}{n^2}}{1+\frac{3}{n^2}}\right)=\lim\limits_{n\to\infty}\left(\dfrac{\frac{1}{n}\cdot\frac{1}{n}}{1+\frac{3}{n}\cdot\frac{1}{n}}\right)$

$$=\frac{\left(\lim\limits_{n\to\infty}\frac{1}{n}\right)\left(\lim\limits_{n\to\infty}\frac{1}{n}\right)}{\lim\limits_{n\to\infty}1+\left(\lim\limits_{n\to\infty}\frac{3}{n}\right)\left(\lim\limits_{n\to\infty}\frac{1}{n}\right)}=\frac{0\cdot 0}{1+0\cdot 0}=0$$

Classroom Exercises

Identify the limit theorem used to find each limit.

1. $\lim\limits_{n\to\infty}\left(\dfrac{1}{3}\right)^n=0$
2. $\lim\limits_{n\to\infty}\left(-\dfrac{3}{5}\right)^n=0$
3. $\lim\limits_{n\to\infty}\dfrac{2}{n}=0$
4. $\lim\limits_{n\to\infty}\dfrac{2000}{n}=0$

Find each limit.

5. $\lim\limits_{n\to\infty}\dfrac{80}{n}$
6. $\lim\limits_{n\to\infty}\left(-\dfrac{1}{8}\right)^n$
7. $\lim\limits_{n\to\infty}5\left(-\dfrac{1}{8}\right)^n$
8. $\lim\limits_{n\to\infty}\left(\dfrac{166}{167}\right)^n$

Written Exercises

a

For Exercises 1–29, find the limit if it exists; otherwise write <u>no limit exists</u>.

1. $\lim\limits_{n\to\infty}\dfrac{-1}{n}$
2. $\lim\limits_{n\to\infty}\left(\dfrac{2}{5}\right)^n$
3. $\lim\limits_{n\to\infty}3\left(-\dfrac{1}{2}\right)^n$
4. $\lim\limits_{n\to\infty}3\left(-\dfrac{1}{2}\right)^{n-1}$

5. $\lim\limits_{n\to\infty}\left(\dfrac{5}{2}\right)^n$
6. $\lim\limits_{n\to\infty}\left(-\dfrac{1}{2}\right)(3)^{n-1}$
7. $\lim\limits_{n\to\infty}4\left(\dfrac{3}{5}\right)^{n-1}$
8. $\lim\limits_{n\to\infty}\dfrac{3}{n}$

9. $\lim\limits_{n\to\infty}-\dfrac{5}{n}$
10. $\lim\limits_{n\to\infty}\left(2+\dfrac{3}{n}\right)$
11. $\lim\limits_{n\to\infty}\left(12-\dfrac{4}{n}\right)$
12. $\lim\limits_{n\to\infty}\left(\dfrac{1+\frac{1}{n}}{2+\frac{3}{n}}\right)$

13. $\lim\limits_{n \to \infty} \left[\frac{2}{n} + 3\left(\frac{1}{2}\right)^{n-1} \right]$

14. $\lim\limits_{n \to \infty} \left[2 + \frac{3}{n} \right]\left[1 - \left(\frac{1}{2}\right)^{n} \right]$

15. $\lim\limits_{n \to \infty} \left(\frac{3n - 4}{2n + 3} \right)$

16. $\lim\limits_{n \to \infty} \left(2 + \frac{3n - 4}{2n + 3} \right)$

17. $\lim\limits_{n \to \infty} \left(\frac{3n}{n + 1} \right)$

18. $\lim\limits_{n \to \infty} \left(\frac{5n - 1}{5n + 2} \right)$

19. $\lim\limits_{n \to \infty} \left(\frac{1}{2}\right)^{n-1} \left(\frac{3n}{n + 1} \right)$

20. $\lim\limits_{n \to \infty} \left(\frac{2}{n} - \frac{5n - 1}{5n + 2} \right)$

21. $\lim\limits_{n \to \infty} \left(\frac{1}{n^2 - 3} \right)$

22. $\lim\limits_{n \to \infty} \left(\frac{3}{n^2 + 2} \right)$

23. $\lim\limits_{n \to \infty} \left(5 + \frac{n^2 - 3}{n^2 + 2} \right)$

24. $\lim\limits_{n \to \infty} \left(\frac{-3}{5}\right)^{n-1} \left(\frac{3}{n^2 + 2} \right)$

25. $\lim\limits_{n \to \infty} \left(\frac{3n^2 - 5n}{5n^2 + 2n - 6} \right)$

26. $\lim\limits_{n \to \infty} \left(\frac{n(n + 2)}{n + 1} - \frac{n^3}{n^2 + 1} \right)$

27. $\lim\limits_{n \to \infty} \left(\frac{4n^2 + 3n}{2n - 1} + \frac{4n^5 - 2n^2}{3n^7 + n^3 - 10} \right)$

28. $\lim\limits_{n \to \infty} \left(\frac{3n^2}{n^3 - 3} - \frac{2n - 4n^2}{n^2 - 2} \right)$

29. $\lim\limits_{n \to \infty} \left[\left(\frac{1}{3}\right)^{n} - \frac{1}{n} + 4\left(\frac{1}{2}\right)^{n} \right]$

C

30. Use the definition of limit to prove Limit Theorem **2**.

31. Use the limit theorems to prove that for a geometric sequence $\{a_1 \cdot r^{n-1}\}$

$$\lim\limits_{n \to \infty} a_1 \cdot r^{n-1} = 0 \text{ whenever } |r| < 1.$$

_____ Review _____

Write the first five terms of the sequence. (Section 8–1)

1. $\{a_n\} = \{-n\}$

2. $\{a_n\} = \{(\frac{1}{2})^n\}$

3. Write the 8th term of the geometric sequence $\{a_n\}$ with $a_1 = 3$ and $a_2 = 5$. (Section 8–2)

For each sequence $\{a_n\}$, find a natural number m such that a_m and all succeeding terms differ from L by less than $\frac{1}{100}$. (Section 8–3)

4. $\{a_n\} = \left\{ 3 + \frac{1}{n} \right\};\ L = 3$

5. $\{a_n\} = \left\{ \frac{3}{n + 2} \right\};\ L = 0$

Use the definition of limit to prove that the sequence $\{a_n\}$ converges to the limit L. (Section 8–3)

6. $\{a_n\} = \left\{ \frac{n}{n + 2} \right\};\ L = 1$

7. $\{a_n\} = \left\{ \frac{6}{n} \right\};\ L = 0$

Find the limit of each sequence. (Section 8–4)

8. $\lim\limits_{n \to \infty} 4\left(-\frac{1}{4}\right)^n$

9. $\lim\limits_{n \to \infty} \left(1 + \frac{2}{n} \right)$

Fibonacci Sequences

The sequence

$$1, 1, 2, 3, 5, 8, 13, 21, 34, \cdots$$

is called a **Fibonacci sequence.** Each term, except for the first and second, is the sum of the two preceding terms. The general term is defined by the rule $a_n = a_{n-1} + a_{n-2}$ where $n \geq 3$. (See Exercise 54 on page 306 for the recursive definition of the Fibonacci sequence.)

A sequence of "Fibonacci fractions" is obtained by dividing each term of the Fibonacci sequence (after the first) by the preceding term.

$$\frac{1}{1}, \frac{2}{1}, \frac{3}{2}, \frac{5}{3}, \frac{8}{5}, \frac{13}{8}, \frac{21}{13}, \frac{34}{21}, \cdots$$

This sequence has a limit called the **Golden Ratio.** It is represented by the Greek letter ϕ (phi) and is equal to $\frac{1+\sqrt{5}}{2}$, or about 1.618.

In the figure at the left below, rectangle $ABCD$ has $AB:BC = \phi:1$. This is called a **Golden Rectangle.** The square $AEFD$ with $AE = BC$ is cut off from this rectangle. The remaining rectangle $EBCF$ is another Golden Rectangle. From this rectangle, the square $EBGH$, with $EB = BG$, is cut off leaving another Golden Rectangle $FCGH$. This process, when continued indefinitely, produces a "sequence" of Golden Rectangles. The point 0, which is the "limit" of this sequence, is the pole of the equiangular spiral which passes through points $D, E, G, J, \ldots$ This curve is the graph of a polar equation of the form

$$r = ae^{\theta \cos \alpha}$$

where a and α are constants. The outer shell of the chambered nautilus is such a curve.

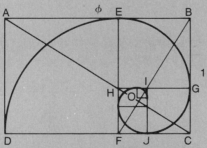

Chambered Nautilus

EXERCISES

1. Use $\sqrt{5} = 2.236067977$ to determine a good approximation for the Golden Ratio.
2. Evaluate a_6 in the sequence of Fibonacci fractions. How close is this to the approximation in Exercise 1?
3. Evaluate a_{12} in the sequence of Fibonacci fractions. How close is this to the approximation in Exercise 1?
4. Evaluate a_{20} in the sequence of Fibonacci fractions. How close is this to the approximation in Exercise 1?

8-5 Series

The terms of a sequence can be combined to form a new sequence.

Example 1 Use the terms of the sequence
$$\{a_n\} = \{1, \tfrac{1}{2}, \tfrac{1}{4}, \tfrac{1}{8}, \tfrac{1}{16}, \cdots , (\tfrac{1}{2})^{n-1}, \cdots \}$$
and the rule
$$S_n = a_1 + a_2 + a_3 + \cdots + a_n$$
to write the first five terms of the sequence $\{S_n\}$.

Solution: $S_1 = a_1 = 1$

$S_2 = a_1 + a_2 = S_1 + a_2 = 1 + \tfrac{1}{2} = 1\tfrac{1}{2}$

$S_3 = a_1 + a_2 + a_3 = S_2 + a_3 = 1\tfrac{1}{2} + \tfrac{1}{4} = 1\tfrac{3}{4}$

$S_4 = a_1 + a_2 + a_3 + a_4 = S_3 + a_4 = 1\tfrac{3}{4} + \tfrac{1}{8} = 1\tfrac{7}{8}$

$S_5 = a_1 + a_2 + a_3 + a_4 + a_5 = S_4 + a_5 = 1\tfrac{7}{8} + \tfrac{1}{16} = 1\tfrac{15}{16}$

In Example 1, $\{S_n\}$ is called a <u>sequence of partial sums</u>. The Greek letter $\sum$ (sigma) is used to represent a sum. For example, $\displaystyle\sum_{j=1}^{5} a_j$ is read "the sum of a sub j from $j=1$ to $j=5$." Thus,

$$S_5 = \sum_{j=1}^{5} a_j = a_1 + a_2 + a_3 + a_4 + a_5.$$

Similarly, the expression $a_1 + a_2 + a_3 + \cdots + a_n + \cdots$ is represented by $\displaystyle\sum_{j=1}^{\infty} a_j$. The letter j is called an **index.**

Definitions: $\displaystyle\sum_{j=1}^{\infty} a_j$ is an <u>infinite</u> <u>series</u> or simply a **series** for the sequence $\{a_n\}$. $S_n = \displaystyle\sum_{j=1}^{n} a_j$ is the ***n*th partial sum** of the series.

$\{S_n\} = \left\{ \displaystyle\sum_{j=1}^{n} a_j \right\}$ is the **sequence of partial sums** of the series.

The series for an arithmetic sequence is called an **arithmetic series.** The series for a geometric sequence is called a **geometric series.**

Example 2 Identify each series as arithmetic or geometric. Then write each series in expanded form.

a. $\displaystyle\sum_{j=1}^{\infty} \left[3 \cdot \left(\frac{1}{2}\right)^{(j-1)} \right]$ **b.** $\displaystyle\sum_{j=1}^{\infty} [2 + (j-1) \cdot 3]$

Solutions: a. Geometric, since $\left\{ 3 \cdot \left(\frac{1}{2}\right)^{(n-1)} \right\}$ is a geometric sequence.

$$\sum_{j=1}^{\infty} \left[3 \cdot \left(\frac{1}{2}\right)^{(j-1)} \right] = 3 + \frac{3}{2} + \frac{3}{4} + \frac{3}{8} + \cdots + \left[3 \cdot \left(\frac{1}{2}\right)^{(n-1)} \right] + \cdots$$

b. Arithmetic, since $\{2 + (n-1) \cdot 3\}$ is an arithmetic sequence.

$$\sum_{j=1}^{\infty} [2 + (j-1) \cdot 3] = 2 + 5 + 8 + 11 + \cdots + [2 + (n-1) \cdot 3] + \cdots$$

The following Example shows a technique for evaluating the partial sums of an arithmetic series.

Example 3 Evaluate the eleventh partial sum of the arithmetic series for the sequence $\{5 + (n-1) \cdot 4\}$.

Solution: $S_{11} = \displaystyle\sum_{j=1}^{11} [5 + (j-1) \cdot 4]$ ⟵ Definition of nth partial sum

$S_{11} = \ \ 5 + \ \ 9 + 13 + 17 + 21 + 25 + 29 + 33 + 37 + 41 + 45$ ⟵ Reverse the order of the terms.
$+\ S_{11} = 45 + 41 + 37 + 33 + 29 + 25 + 21 + 17 + 13 + \ \ 9 + \ \ 5$
$\overline{2S_{11} = 50 + 50 + 50 + 50 + 50 + 50 + 50 + 50 + 50 + 50 + 50}$ ⟵ Sum

$2S_{11} = 11 \cdot 50$

$\therefore S_{11} = \dfrac{11 \cdot 50}{2} = 275$

Example 3 suggests the following general term for the sequence of partial sums of an arithmetic series. Recall that the general term for an arithmetic sequence is $a_n = a_1 + (n-1) \cdot d$.

$$\boldsymbol{S_n = \frac{n \cdot (a_1 + a_n)}{2}} \quad \text{or} \quad \boldsymbol{S_n = \frac{n \cdot [2a_1 + (n-1) \cdot d]}{2}}$$

You can also find the general term for the sequence of partial sums of a geometric series. Recall that the general term for a geometric sequence is $a_1 r^{n-1}$.

$$S_n = \sum_{j=1}^{n} a_1 r^{(j-1)} \quad \longleftarrow \quad \text{Definition of } s_n$$

$$S_n = a_1 + a_1 r + a_1 r^2 + \cdots + a_1 r^{n-1} \quad \longleftarrow \quad \text{Multiply by } -r$$

$$\underline{-r \cdot S_n = \qquad -a_1 r - a_1 r^2 - \cdots - a_1 r^{n-1} - a_1 r^n}$$

$$S_n - r S_n = a_1 \qquad\qquad\qquad\qquad\qquad - a_1 r^n \quad \longleftarrow \quad \text{Sum}$$

$$S_n(1 - r) = a_1 - a_1 r^n$$

$$S_n = \frac{a_1 - a_1 r^n}{1 - r} \quad \text{or} \quad S_n = \frac{a_1}{1 - r}(1 - r^n), \ (r \neq 1)$$

Example 4
Find the eighth partial sum of the series for each sequence.

a. $\{1 + (n - 1) \cdot 3\}$ **b.** $\{3 \cdot (2)^{(n-1)}\}$

Solutions: **a.** This is an arithmetic sequence with $a_1 = 1$ and $d = 3$.

$$S_n = \frac{n \cdot [2 \cdot a_1 + (n - 1) \cdot d]}{2} \quad \longleftarrow \quad \begin{array}{l}\text{Substitute } n = 8,\\ a_1 = 1, \text{ and } d = 3.\end{array}$$

$$S_8 = \frac{8 \cdot [2 \cdot 1 + (8 - 1) \cdot 3]}{2} = 92$$

b. This is a geometric sequence with $a_1 = 3$ and $r = 2$.

$$S_n = \frac{a_1(1 - r^n)}{1 - r} \quad \longleftarrow \quad \begin{array}{l}\text{Substitute } n = 8, \ a = 3,\\ \text{and } r = 2.\end{array}$$

$$S_8 = \frac{3 \cdot (1 - 2^8)}{1 - 2} = 765$$

Classroom Exercises

Identify each series as arithmetic (A) or geometric (G).

1. $\displaystyle\sum_{j=1}^{\infty} \left[2 \cdot \left(\frac{1}{3}\right)^{j-1}\right]$ **2.** $\displaystyle\sum_{j=1}^{\infty} [3 + (j - 1) \cdot 2]$

Write each partial sum in expanded form.

3. $\displaystyle\sum_{j=1}^{4} \frac{1}{j}$ **4.** $\displaystyle\sum_{j=1}^{5} \left(1 + \frac{1}{j}\right)$

Written Exercises

Use the terms of the sequence $\{a_n\}$ and the rule $S_n = a_1 + a_2 + \cdots + a_n$ to write the first five terms of the sequence $\{S_n\}$.

1. $\{a_n\} = \{(-1)^n\}$

2. $\{a_n\} = \{1^n\}$

3. $\{a_n\} = \{n^2\}$

4. $\{a_n\} = \{n - 1\}$

5. $\{a_n\} = \{1 + (n-1) \cdot 2\}$

6. $\{a_n\} = \{2 + (n-1) \cdot 2\}$

7. $\{a_n\} = \{2n\}$

8. $\{a_n\} = \left\{1 + \dfrac{1}{n}\right\}$

9. $\{a_n\} = \left\{2 - \dfrac{1}{n}\right\}$

10. $\{a_n\} = \left\{n - \dfrac{1}{n}\right\}$

Identify each series as arithmetic (A) or geometric (G). Then write the series in expanded form.

11. $\displaystyle\sum_{j=1}^{\infty} [2 + (j-1) \cdot 4]$

12. $\displaystyle\sum_{j=1}^{\infty} [3 + (j-1)]$

13. $\displaystyle\sum_{j=1}^{\infty} [2 \cdot (3)^{j-1}]$

14. $\displaystyle\sum_{j=1}^{\infty} [100 - 50(j-1)]$

15. $\displaystyle\sum_{j=1}^{\infty} [-3 - 2(j-1)]$

16. $\displaystyle\sum_{j=1}^{\infty} [3 \cdot (-1)^{j-1}]$

17. $\displaystyle\sum_{j=1}^{\infty} \left[4 \cdot \left(-\dfrac{1}{1}\right)^{j-1}\right]$

18. $\displaystyle\sum_{j=1}^{\infty} \left[-2 \cdot \left(\dfrac{3}{4}\right)^{j-1}\right]$

Evaluate the indicated partial sum of the arithmetic series for the given arithmetic sequence.

19. $\{4 + (n-1) \cdot 5\}$; 10th

20. $\{4 - 5 \cdot (n-1)\}$; 10th

21. $\{7 - 5 \cdot (n-1)\}$; 8th

22. $\{100 - 50 \cdot (n-1)\}$; 12th

23. $\{1 + (n-1)\}$; 25th

24. $\{1 + (n-1) \cdot 2\}$; 20th

Find the indicated partial sum of the series for each sequence.

25. $\left\{2 \cdot \left(\dfrac{1}{2}\right)^{n-1}\right\}$; 8th

26. $\left\{2 \cdot \left(-\dfrac{1}{2}\right)^{n-1}\right\}$; 8th

27. $\left\{3 \cdot \left(\dfrac{1}{2}\right)^{n-1}\right\}$; 9th

28. $\{2 + (n-1) \cdot 2\}$; 20th

29. $\{2 - 2 \cdot (n-1)\}$; 20th

30. $\{4 \cdot (-1)^{n-1}\}$; 9th

31. $\{7 \cdot (-1)^{n-1}\}$; 8th

32. $\{-3 \cdot (-1)^{n-1}\}$; 10th

Write the general term for the sequence of partial sums of the series for each of the following sequences.

33. $\{-3 - 2 \cdot (n-1)\}$

34. $\{-5 + (n-1) \cdot 3\}$

35. $\left\{2 \cdot \left(\frac{1}{3}\right)^{n-1}\right\}$

36. $\left\{4 \cdot \left(-\frac{1}{2}\right)^{n-1}\right\}$

37. $\{-50 + (n-1) \cdot 100\}$

38. $\{1 + (n-1) \cdot 2\}$

39. Show that the sum of the first n positive integers is equal to $\frac{n(n+1)}{2}$. (HINT: See Exercise 49 on page 310.)

40. Show that the sum of the first n positive even integers is equal to $n(n+1)$. (HINT: See Exercise 50 on page 310.)

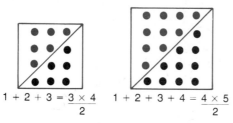

$1 + 2 + 3 = \dfrac{3 \times 4}{2}$ $1 + 2 + 3 + 4 = \dfrac{4 \times 5}{2}$

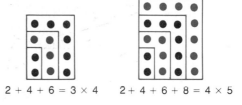

$2 + 4 + 6 = 3 \times 4$ $2 + 4 + 6 + 8 = 4 \times 5$

41. Show that the sum of the first n positive odd integers is equal to n^2. (HINT: See Exercise 51 on page 310.)

42. Find the sum of the even integers between 115 and 203.

43. Find the sum of the odd integers between 80 and 108.

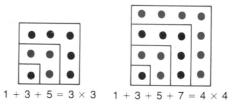

$1 + 3 + 5 = 3 \times 3$ $1 + 3 + 5 + 7 = 4 \times 4$

44. Find a formula for the nth partial sum of the series for the geometric sequence with first term $a_1 = 1$ and common ratio $r = 1$.

45. Show that if S_n is the nth partial sum of the series for $\{a_n\}$, then $a_n = S_n - S_{n-1}$.

46. Write the first five terms of the sequence $\{a_n\}$ where the general term for the sequence of partial sums is $S_n = \dfrac{n}{n+1}$. $\left(\text{HINT: } S_{n-1} = \dfrac{n-1}{n}.\right)$

47. For the constant c and the sequence $\{a_n\}$, prove that $\displaystyle\sum_{j=1}^{n} c \cdot a_j = c \cdot \sum_{j=1}^{n} a_j$.

That is, prove that if the series for $\{a_n\}$ has nth partial sum S_n, then the series for $\{c \cdot a_n\}$ has nth partial sum $c \cdot S_n$.

48. Given the sequences $\{a_n\}$ and $\{b_n\}$, prove that $\displaystyle\sum_{j=1}^{n} (a_j + b_j) = \sum_{j=1}^{n} a_j + \sum_{j=1}^{n} b_j$.

That is, prove that if the series for $\{a_n\}$ and $\{b_n\}$ have nth partial sums S_n and S_n' respectively, then the series for $\{a_n + b_n\}$ has nth partial sum $S_n + S_n'$.

8-6 The Sum of a Series

In Section 8–4, you found that some sequences converge to a limit. Therefore, the sequence of partial sums of a series may converge to a limit.

Example 1 Find the limit, if it exists, of the sequence $\{S_n\}$ of partial sums of the following geometric series.

$$\sum_{j=1}^{\infty}\left[3\left(\frac{1}{2}\right)^{j-1}\right] = 3 + \frac{3}{2} + \frac{3}{4} + \cdots + 3\left(\frac{1}{2}\right)^{n-1} + \cdots$$

Solution: This is the series for the geometric sequence with $a_1 = 3$ and $r = \frac{1}{2}$.

Therefore, $S_n = \dfrac{a}{1-r}(1 - r^n)$

$$= \frac{3}{1 - \frac{1}{2}}\left[1 - \left(\frac{1}{2}\right)^n\right]$$

$$= 6\left[1 - \left(\frac{1}{2}\right)^n\right].$$

Thus, $\displaystyle\lim_{n\to\infty} S_n = \lim_{n\to\infty} 6\left[1 - \left(\frac{1}{2}\right)^n\right]$

$$= 6 \cdot \lim_{n\to\infty}\left[1 - \left(\frac{1}{2}\right)^n\right] \qquad \longleftarrow \text{Limit Theorem 7}$$

$$= 6\left[\lim_{n\to\infty} 1 - \lim_{n\to\infty}\left(\frac{1}{2}\right)^n\right] \qquad \longleftarrow \text{Limit Theorem 4}$$

$$= 6\left[1 - \lim_{n\to\infty}\left(\frac{1}{2}\right)^n\right] \qquad \longleftarrow \text{Limit Theorem 1}$$

$$= 6(1 - 0) \qquad \longleftarrow \text{Limit Theorem 3}$$

$$= 6$$

In Example 1, the limit, 6, is called the <u>sum of the series</u>. That is,

$$6 = 3 + \frac{3}{2} + \frac{3}{4} + \cdots + 3\left(\frac{1}{2}\right)^{n-1} + \cdots$$

Example 1 suggests the following statement which you are asked to prove in the Exercises.

The sequence $\{S_n\} = \left\{\displaystyle\sum_{j=1}^{n} a_1 \cdot r^{j-1}\right\}$ of partial sums of the geometric series $\displaystyle\sum_{j=1}^{\infty} a_1 \cdot r^{j-1}$ converges whenever $|r| < 1$. In this case

$$\lim_{n\to\infty} S_n = \frac{a_1}{1-r}.$$

The sequence diverges whenever $|r| \geq 1$ and $a_1 \neq 0$.

Definition: When the sequence of partial sums of the series $\displaystyle\sum_{j=1}^{\infty} a_j$ converges to a limit S, the <u>series</u> <u>converges</u>. S is called the **sum of the series.** In symbols,

$$\sum_{j=1}^{\infty} a_j = S, \quad \text{or} \quad S = a_1 + a_2 + \cdots + a_n + \cdots.$$

When the sequence of partial sums diverges, the <u>series</u> <u>diverges</u>.

Therefore, every geometric series with $|r| < 1$ converges. In symbols,

$$\sum_{j=1}^{\infty} a_1 \cdot r^{j-1} = \frac{a_1}{1-r} \text{ whenever } |r| < 1.$$

Example 2

Find the sum S, if it exists, of $\displaystyle\sum_{j=1}^{\infty} \left[5 \cdot \left(-\frac{1}{2}\right)^{j-1}\right].$

Solution: Since $|r| = \left|-\frac{1}{2}\right|$ is less than 1, you have the following.

$$S = \lim_{n\to\infty} S_n = \frac{a}{1-r} = \frac{5}{1 - \left(-\frac{1}{2}\right)} = \frac{10}{3}, \text{ or } 3.\overline{3}$$

Various methods can be used to test for the convergence of a series. One such method is the **ratio test.**

Ratio Test: If for $\displaystyle\sum_{j=1}^{\infty} a_j = a_1 + a_2 + \cdots + a_n + \cdots$ the sequence $\left\{\left|\dfrac{a_{n+1}}{a_n}\right|\right\}$ converges and $\lim_{n\to\infty} \left|\dfrac{a_{n+1}}{a_n}\right| = L$, then

1. the series converges if $L < 1$;
2. the series diverges if $L > 1$;
3. the test fails if $L = 1$.

Example 3

Use the ratio test to determine whether the following series converges. Recall that $n!$ means $1 \cdot 2 \cdot 3 \cdot \cdots n$ and that $0! = 1$.

$$\sum_{j=1}^{\infty} \frac{1}{(j-1)!} = 1 + 1 + \frac{1}{2} + \frac{1}{6} + \frac{1}{24} + \cdots + \frac{1}{(n-1)!} + \cdots$$

Solution: Since $a_n = \dfrac{1}{(n-1)!}$, $a_{n+1} = \dfrac{1}{(n-1+1)!} = \dfrac{1}{n!}$. Therefore,

$$\left| \frac{a_{n+1}}{a_n} \right| = \left| \frac{\frac{1}{n!}}{\frac{1}{(n-1)!}} \right| = \left| \frac{1}{n!} \cdot \frac{(n-1)!}{1} \right|$$

$$= \left| \frac{1}{n} \right|.$$

Thus, $\displaystyle\lim_{n \to \infty} \left| \frac{a_{n+1}}{a_n} \right| = \lim_{n \to \infty} \left| \frac{1}{n} \right|$

$$= \lim_{n \to \infty} \frac{1}{n} \quad \longleftarrow \text{ Since } n \text{ is a positive integer.}$$

$$= 0. \quad \longleftarrow \text{ Limit Theorem 2}$$

Since $L = 0$ and 0 is less than 1, the series converges.

Example 3 shows that a series may be known to converge although its sum is not known. Since the sum of a series is the limit of the sequence of partial sums, the sum of the series can be approximated by estimating the limit of the sequence of partial sums.

Example 4

Find, to two decimal places, an approximate value for the sum of the series in Example 3.

$$\sum_{j=1}^{\infty} \frac{1}{(j-1)!} = 1 + 1 + \frac{1}{2} + \frac{1}{6} + \frac{1}{24} + \frac{1}{120} + \frac{1}{720} \cdots + \frac{1}{(n-1)!} + \cdots$$

Solution: It is sufficient to find the sum of the first five terms of this series since each term after the fifth is very small and, therefore, will add very little to a two-decimal place approximation.

$$S_5 = 1 + 1 + \frac{1}{2} + \frac{1}{6} + \frac{1}{24} = 2.708\overline{3}$$

$$\therefore \sum_{j=1}^{\infty} \frac{1}{(j-1)!} \approx 2.71 \quad \longleftarrow \text{ To two decimal places}$$

Classroom Exercises

Determine whether the given geometric series converges or diverges.

1. $\displaystyle\sum_{j=1}^{\infty} 7\left(\frac{2}{5}\right)^{j-1}$

2. $\displaystyle\sum_{j=1}^{\infty} \left(\frac{16}{17}\right)^{j-1}$

3. $\displaystyle\sum_{j=1}^{\infty} \frac{1}{2}\left(\frac{5}{3}\right)^{j-1}$

4. $\displaystyle\sum_{j=1}^{\infty} 5(-1)^{j-1}$

5. $\displaystyle\sum_{j=1}^{\infty} (1.001)^{j-1}$

6. $\displaystyle\sum_{j=1}^{\infty} 500\left(-\frac{1}{2}\right)^{j-1}$

Written Exercises

a

Use the limit theorems on page 316 to find the limit of the sequence $\{S_n\}$ of partial sums of the given geometric series.

1. $\displaystyle\sum_{j-1}^{\infty} 2\left(\frac{1}{3}\right)^{j-1}$

2. $\displaystyle\sum_{j-1}^{\infty} 200\left(\frac{1}{3}\right)^{j-1}$

3. $\displaystyle\sum_{j=1}^{\infty} 4\left(\frac{2}{3}\right)^{j-1}$

4. $\displaystyle\sum_{j=1}^{\infty} 2540\left(-\frac{9}{10}\right)^{j-1}$

5. $\displaystyle\sum_{j=1}^{\infty} 100\left(-\frac{1}{5}\right)^{j-1}$

6. $\displaystyle\sum_{j=1}^{\infty} \frac{5}{3}\left(\frac{1}{2}\right)^{j-1}$

Use the formula on page 327 to find the sum S, if it exists, of the given geometric series. Otherwise, write <u>diverges</u>.

7. $\displaystyle\sum_{j=1}^{\infty} 5\left(-\frac{2}{7}\right)^{j-1}$

8. $\displaystyle\sum_{j=1}^{\infty} 15\left(\frac{5}{3}\right)^{j-1}$

9. $\displaystyle\sum_{j=1}^{\infty} \frac{1}{2}(-1)^{j-1}$

10. $\displaystyle\sum_{j=1}^{\infty} -1\left(-\frac{1}{2}\right)^{j-1}$

11. $\displaystyle\sum_{j=1}^{\infty} 500\left(-\frac{3}{5}\right)^{j-1}$

12. $\displaystyle\sum_{j=1}^{\infty} 2(1.9)^{j-1}$

13. $\displaystyle\sum_{j=1}^{\infty} 9(.5)^{j-1}$

14. $\displaystyle\sum_{j=1}^{\infty} 100(.99)^{j-1}$

15. $\displaystyle\sum_{j=1}^{\infty} 12(-1.2)^{j-1}$

Use the ratio test to determine whether each of the following series converges. For each exercise write <u>converges</u>, <u>diverges</u>, or <u>test fails</u>.

16. $\displaystyle\sum_{j=1}^{\infty} \frac{1}{j!}$

17. $\displaystyle\sum_{j=1}^{\infty} \frac{1}{j}$

18. $\displaystyle\sum_{j=1}^{\infty} (-1)^{j}$

19. $\displaystyle\sum_{j=1}^{\infty} \frac{1}{10^{j}}$

20. $\displaystyle\sum_{j=1}^{\infty} \frac{1}{10j}$

21. $\displaystyle\sum_{j=1}^{\infty} \frac{10^{j}}{j!}$

22. $\displaystyle\sum_{j=1}^{\infty} \frac{j}{j+2}$

23. $\displaystyle\sum_{j=1}^{\infty} \frac{2^{j-1}}{j(j+1)}$

24. $\displaystyle\sum_{j=1}^{\infty} 5\left(\frac{1}{2}\right)^{j}$

25. $\displaystyle\sum_{j=1}^{\infty} j$

26. $\displaystyle\sum_{j=1}^{\infty} \frac{(-1)^{j-1}}{j}$

27. $\displaystyle\sum_{j=1}^{\infty} \frac{(-1)^{j-1}}{(2j-1)!}$

Each of the following series converges. Use the first five terms to find, to two decimal places, an approximate value for the sum of each series.

28. $\displaystyle\sum_{j=1}^{\infty} \frac{1}{2^j + 1}$ **29.** $\displaystyle\sum_{j=1}^{\infty} \frac{\cos(j \cdot \frac{\pi}{2})}{j^2}$ **30.** $\displaystyle\sum_{j=1}^{\infty} \frac{(-1)^{j-1}}{(2j-1)}$ **31.** $\displaystyle\sum_{j=1}^{\infty} \frac{(-1)^{j-1}}{j}$

The periodic decimal $.3\overline{33}$ is the sum of the following geometric series.

$$\sum_{j=1}^{\infty} \left(\frac{3}{10}\right)\left(\frac{1}{10}\right)^{j-1} = \frac{3}{10} + \frac{3}{10^2} + \frac{3}{10^3} + \cdots + \frac{3}{10^n} + \cdots$$

Since this series has $a_1 = \frac{3}{10}$ and $r = \frac{1}{10}$, $S = \lim S_n = \frac{3}{9}$ or $\frac{1}{3}$. Thus $.3\overline{33}$ is equal to the rational number $\frac{1}{3}$. Express each of the following periodic decimals as a rational number.

32. $.1\overline{11}$ **33.** $.\overline{3535}$ **34.** $.244\overline{244}$ **35.** $.\overline{157215721572}$

It can be shown that for the sequence $\{a_n\}$, if $\lim_{n \to \infty} a_n \neq 0$, then the following series diverges.

$$\sum_{j=1}^{\infty} a_j = a_1 + a_2 + \cdots + a_n + \cdots$$

Show that each of the following series diverges.

36. $\displaystyle\sum_{j=1}^{\infty} \frac{3j}{5j+1}$ **37.** $\displaystyle\sum_{j=1}^{\infty} \frac{3j-1}{3j+1}$

38. The geometric series $\displaystyle\sum_{j=1}^{\infty} a_1 \cdot r^{j-1}$ with $r = 1$ and $a_1 \neq 0$.

39. Use the limit theorems on page 316 to show that for the geometric series

$\displaystyle\sum_{j=1}^{\infty} a_1 \cdot r^{j-1}$ with nth partial sum S_n,

$$\lim_{n \to \infty} S_n = \frac{a_1}{1-r}(1 - \lim r^n) = \frac{a}{1-r} \text{ whenever } |r| < 1.$$

40. Use the ratio test to show that the following geometric series converges whenever $|r| < 1$ and diverges whenever $|r| > 1$.

$$\sum_{j=1}^{\infty} a_1 \cdot r^{j-1}$$

41. Prove that the periodic decimal $.c_1 c_2 c_3 \cdots c_p \overline{c_1 c_2 c_3 \cdots c_p}$ (where $c_1, c_2, \cdots, c_p$ are integers between 0 and 9, inclusive) can be expressed as the following rational number.

$$\frac{(c_1) \cdot (c_2) \cdot (c_3) \cdots (c_p)}{10^p - 1}$$

(HINT: See Exercises 32–35.)

8-7 Power Series

The geometric series

$$\sum_{j=1}^{\infty} ar^{j-1} = a + ar + ar^2 + \cdots + ar^{n-1} + \cdots$$

is a special case of a <u>power series</u>.

Definition: A series of the form

$$\sum_{j=0}^{\infty} a_j x^j = a_0 + a_1 x + a_2 x^2 + a_3 x^3 + \cdots + a_n x^n + \cdots$$

is called a **power series in x,** where x is a variable and $a_0, a_1, a_2, \cdots,$ $a_n \cdots$ are constants.

In working with power series, the problem is to determine for what values of the variable a power series converges. The set of values is called the **interval of convergence.**

Example 1
Find the interval of convergence of the following power series.

$$1 + 2x + 3x^2 + 4x^3 + \cdots + nx^{n-1} + \cdots$$

Solution: Use the ratio test. Since nx^{n-1} is the nth term, $(n + 1)x^n$ is the $(n + 1)$st term. Therefore,

$$\left|\frac{a_{n+1}}{a_n}\right| = \left|\frac{(n + 1)x^n}{nx^{n-1}}\right| = \left|\frac{(n + 1)x}{n}\right| \quad \longleftarrow \quad \frac{x^n}{x^{n-1}} = x$$

$$= \left|\left(\frac{n}{n} + \frac{1}{n}\right)x\right| = \left|\left(1 + \frac{1}{n}\right)x\right|$$

Thus, $\lim_{n \to \infty}\left|\frac{a_{n+1}}{a_n}\right| = \lim_{n \to \infty}\left|\left(1 + \frac{1}{n}\right)x\right| \quad \longleftarrow \quad \lim_{n \to \infty}\left(1 + \frac{1}{n}\right) = 1$

$$= |x| \quad \longleftarrow \quad L$$

By the ratio test, the power series converges when $L < 1$. Since $L = |x|$, the interval of convergence is

$$|x| < 1, \quad \text{or} \quad -1 < x < 1.$$

When L is equal to zero in the ratio test for a power series, the series converges for <u>all</u> real values of the variable, that is, for the set of real numbers.

Example 2

Find the interval of convergence for the following series.

$$1 + x + \frac{x^2}{2!} + \frac{x^3}{3!} + \cdots + \frac{x^{n-1}}{(n-1)!} + \cdots$$

Solution: Use the ratio test. Since $\frac{x^{n-1}}{(n-1)!}$ is the nth term, $\frac{x^x}{n!}$ is the $(n+1)$st term.

$$\left| \frac{a_{n+1}}{a_n} \right| = \left| \frac{\dfrac{x^n}{n!}}{\dfrac{x^{n-1}}{(n-1)!}} \right| = \left| \frac{x^n}{n!} \cdot \frac{(n-1)!}{x^{n-1}} \right| \quad \longleftarrow \quad \frac{x^n}{x^{n-1}} = x$$

$$= \left| \frac{(n-1)!\, x}{n!} \right| \quad \longleftarrow \quad \frac{(n-1)!}{n!} = \frac{1}{n}$$

$$= \left| \frac{x}{n} \right| = \frac{|x|}{n} \quad \longleftarrow \quad \text{In the series, } n \text{ is a positive integer.}$$

Thus, $\lim\limits_{n\to\infty} \left| \dfrac{a_{n+1}}{a_n} \right| = \lim\limits_{n\to\infty} \dfrac{|x|}{n} = 0 \quad \longleftarrow \quad$ **Limit Theorem 2**

Since $L = 0$, the interval of convergence is the set of real numbers.

Recall from algebra that e is the irrational number used as the base for the <u>natural logarithms</u>. The power series in Example 2 can be shown to converge to e^x where x is a real number. That is,

$$\boldsymbol{e^x = 1 + x + \frac{x^2}{2!} + \frac{x^3}{3!} + \cdots + \frac{x^{n-1}}{(n-1)!} + \cdots}$$

Example 3

Estimate the value of e to two decimal places.

Solution: Use $x = 1$ in the power series for e^x.

$$e = 1 + 1 + \frac{1^2}{2!} + \frac{1^3}{3!} + \cdots + \frac{1^{n-1}}{(n-1)!} + \cdots$$

For our purposes, it is sufficient to find the sum of the first five terms.

$$e \approx 1 + 1 + \frac{1}{2} + \frac{1}{6} + \frac{1}{24} \approx 2.71$$

Classroom Exercises

Find the (n + 1)st term.

1. $\displaystyle\sum_{j=1}^{\infty} (-1)^{j-1}(x)^{j-1}$

2. $\displaystyle\sum_{j=1}^{\infty} j(-1)^{j-1}x^{j-1}$

3. $\displaystyle\sum_{j=1}^{\infty} \frac{x^j}{j!}$

Written Exercises

a

Find the interval of convergence for each of the following series.

1. $\displaystyle\sum_{j=1}^{\infty} x^{j-1}$

2. $\displaystyle\sum_{j=1}^{\infty} \frac{x^{j-1}}{j}$

3. $\displaystyle\sum_{j=1}^{\infty} j^2 x^j$

4. $\displaystyle\sum_{j=1}^{\infty} (-1)^{j-1}x^{j-1}$

5. $\displaystyle\sum_{j=1}^{\infty} \frac{x^j}{2j-1}$

6. $\displaystyle\sum_{j=1}^{\infty} \frac{jx^j}{2^j}$

7. $\displaystyle\sum_{j=1}^{\infty} \frac{(2j-1)x^{j-1}}{2^{j-1}}$

8. $\displaystyle\sum_{j=1}^{\infty} x^{-j}$

9. $\displaystyle\sum_{j=1}^{\infty} \frac{(-1)^{j-1}x^{j-1}}{3^j}$

10. $\displaystyle\sum_{j=1}^{\infty} \frac{x^j}{j(j+1)}$

11. $\displaystyle\sum_{j=1}^{\infty} \frac{x^{j-1}}{j \cdot 5^{j-1}}$

12. $\displaystyle\sum_{j=1}^{\infty} j^2 (x+1)^j$

The power series in Exercise 4 can be shown to converge to $(1+x)^{-1}$ where $|x| < 1$. Use the first five terms of this power series to estimate the value of each of the following to two decimal places.

13. $(1.0009)^{-1}$

14. $\dfrac{1}{1.0002}$

15. $(.0003)^{-1}$

16. $\dfrac{1}{.00001}$

17. Find the interval of convergence for the following series.

$$\sum_{j=1}^{\infty} j(-1)^{j-1}x^{j-1} = 1 - 2x + 3x^2 - 4x^3 + \cdots + (-1)^{n-1}x^{n-1} + \cdots$$

The power series in Exercise 17 can be shown to converge to $(1+x)^{-2}$ where $|x| < 1$. Use the first five terms of this power series to estimate the value of each of the following to two decimal places.

18. $(1.0004)^{-2}$

19. $\left(\dfrac{1}{1.0001}\right)^2$

20. $(.0002)^{-2}$

21. $\dfrac{1}{(.0003)^2}$

Find the interval of convergence for each of the following series.

22. $\displaystyle\sum_{j=1}^{\infty} \frac{(-1)^{j+1}x^{2j-1}}{(2j-1)!}$

23. $\displaystyle\sum_{j=1}^{\infty} \frac{(-1)^{j-1}(x-1)^j}{j}$

24. $\displaystyle\sum_{j=1}^{\infty} \frac{x^{2j+1}}{2j+1}$

25. $\displaystyle\sum_{j=1}^{\infty} (-1)^j \frac{x^{2j+1}}{2j+1}$

26. $\displaystyle\sum_{j=1}^{\infty} \frac{(x-2)^j}{j^2}$

27. $\displaystyle\sum_{j=1}^{\infty} \frac{(-1)^{j-1}(x+1)^j}{j}$

8-8 Power Series and Circular Functions

There are power series representations for the circular (trigonometric) functions that can be used to approximate these functions for any radian measure x. Here are the power series for $\sin x$ and $\cos x$.

$$\sin x = x - \frac{x^3}{3!} + \frac{x^5}{5!} - \frac{x^7}{7!} + \cdots + (-1)^{n+1} \frac{x^{2n-1}}{(2n-1)!} + \cdots$$

$$\cos x = 1 - \frac{x^2}{2!} + \frac{x^4}{4!} - \frac{x^6}{6!} + \cdots + (-1)^{n+1} \frac{x^{2n-2}}{(2n-2)!} + \cdots$$

The use of a calculator simplifies the computation involved in evaluating a power series.

Example 1 Use the power series for $\sin x$ to estimate $\sin \frac{\pi}{6}$ to five decimal places. Use 3.14159 for π.

Solution: For our purposes, it is sufficient to find the sum of the first four terms of the series.

$$\sin \frac{\pi}{6} \approx \frac{\pi}{6} - \frac{\left(\frac{\pi}{6}\right)^3}{3!} + \frac{\left(\frac{\pi}{6}\right)^5}{5!} - \frac{\left(\frac{\pi}{6}\right)^7}{7!}$$

$$\approx .5235983 - .0239245 + .0003279 - .0000021$$

$$\approx .4999996 \approx .50000$$

Since $\sin \frac{\pi}{6} = .5$, this is a good approximation.

The Table of Values of the trigonometric functions in the back of the book was developed by using power series. Calculators that have trigonometric functions also use power series representations to evaluate these functions.

Recall that the power series

$$1 + x + \frac{x^2}{2!} + \frac{x^3}{3!} + \cdots + \frac{x^{n-1}}{(n-1)!} + \cdots$$

converges to e^x, where x is a real number. By replacing x in this series with ix, where $i = \sqrt{-1}$ and x is a real number, it can be shown that the resulting series converges. The expression e^{ix} is defined to be the sum of this series. This power series can be used to derive **Euler's Formula,** an important identity that relates the sine and cosine functions and the exponential function.

$$e^{ix} = 1 + ix + \frac{(ix)^2}{2!} + \frac{(ix)^3}{3!} + \frac{(ix)^4}{4!} + \cdots + \frac{(ix)^{n-1}}{(n-1)!} + \cdots$$

Use $i^2 = -1$, $i^4 = 1, \cdots$ $\longrightarrow$
$$= 1 + ix + \frac{i^2 x^2}{2!} + \frac{i^3 x^3}{3!} + \frac{i^4 x^4}{4!} + \cdots + \frac{i^{n-1} x^{n-1}}{(n-1)!} + \cdots$$

$$= 1 + ix - \frac{x^2}{2!} - \frac{ix^3}{3!} + \frac{x^4}{4!} + \frac{ix^5}{5!} + \cdots$$

Rearrange the real and imaginary terms as follows. NOTE: Rearrangements of some series that converge may produce series that do <u>not</u> converge. However, the following rearrangement is valid.

$$e^{ix} = \left(1 - \frac{x^2}{2!} + \frac{x^4}{4!} - \frac{x^6}{6!} + \cdots\right) + i\left(x - \frac{x^3}{3!} + \frac{x^5}{5!} - \frac{x^7}{7!} + \cdots\right)$$

Since $\cos x = 1 - \frac{x^2}{2!} + \frac{x^4}{4!} - \frac{x^6}{6!} + \cdots$ and $\sin x = x - \frac{x^3}{3!} + \frac{x^5}{5!} - \frac{x^7}{7!} + \cdots$

$$e^{ix} = \cos x + i \sin x \qquad \longleftarrow \text{Euler's Formula}$$

Example 2 Use Euler's Formula to evaluate $e^{i\frac{\pi}{2}}$.

Solution: $e^{ix} = \cos x + i \sin x \qquad \longleftarrow \text{Euler's Formula}$

$e^{i\frac{\pi}{2}} = \cos \frac{\pi}{2} + i \sin \frac{\pi}{2} = 0 + i \cdot 1 = i$

The following identity can be proved by using the method for deriving Euler's Formula. You are asked for the proof in the Exercises.

$$e^{-ix} = \cos x - i \sin x$$

This identity and Euler's Formula can be used to prove other identities that relate the circular functions and the exponential function.

Example 3 Prove: $\cos x = \dfrac{e^{ix} + e^{-ix}}{2}$

Solution:

$e^{ix} = \cos x + i \sin x \qquad \longleftarrow \text{Euler's Formula}$

$\dfrac{e^{-ix} = \cos x - i \sin x}{e^{ix} + e^{-ix} = 2 \cos x} \qquad \begin{array}{l} \longleftarrow \text{Identity} \\ \longleftarrow \text{Sum} \end{array}$

$\dfrac{e^{ix} + e^{-ix}}{2} = \cos x$

Classroom Exercises

Evaluate each expression.

1. $e^{i\frac{\pi}{6}}$ **2.** $e^{i\frac{\pi}{3}}$ **3.** $e^{-i\frac{\pi}{3}}$ **4.** $e^{-i\frac{\pi}{2}}$

Written Exercises

a

Use the first four terms of the power series for sin x or cos x to estimate each value. Use 3.14159 for π. Round each answer to four decimal places.

1. $\sin .2$ **2.** $\cos 1$ **3.** $\cos .1$ **4.** $\sin \frac{1}{2}$

5. $\cos 1.5$ **6.** $\sin 1$ **7.** $\cos .2$ **8.** $\sin \pi$

9. $\cos \pi$ **10.** $\cos .5$ **11.** $\sin \frac{\pi}{2}$ **12.** $\cos \frac{3\pi}{2}$

Use Euler's Formula to evaluate each expression.

13. $e^{i\pi}$ **14.** $e^{i\frac{\pi}{4}}$ **15.** $e^{2\pi i}$ **16.** $e^{\frac{3\pi}{2}i}$

17. $e^{3\pi i}$ **18.** $e^{\frac{9\pi}{4}i}$ **19.** $e^{4\pi i}$ **20.** $e^{\frac{7\pi}{2}i}$

21. Prove the following identity: $\sin x = \dfrac{e^{ix} - e^{-ix}}{2i}$ (HINT: Subtract each side of the identity $e^{-ix} = \cos x - i \sin x$ from the corresponding sides in Euler's Formula.)

b

Prove each identity.

22. $\tan x = \dfrac{e^{ix} - e^{-ix}}{i(e^{ix} + e^{-ix})}$ **23.** $\cot x = \dfrac{i(e^{ix} + e^{-ix})}{e^{ix} - e^{-ix}}$

24. $\sec x = \dfrac{2}{e^{ix} + e^{-ix}}$ **25.** $\csc x = \dfrac{2i}{e^{ix} - e^{-ix}}$

26. Prove that $e^{i(x+2k\pi)} = e^{ix}$ where k is an integer.

27. Use the ratio test to prove that the power series for $\sin x$ converges for all real values of x.

28. Repeat Exercise 27 for the power series for $\cos x$.

29. Repeat Exercise 27 for the power series for e^{ix}.

30. In the power series for e^x replace x with $-ix$. Use the ratio test to show that the resulting series converges for all real values of x.

31. Define e^{-ix} to be the sum of the series obtained in Exercise 30. Use this series and the method by which Euler's Formula was derived to derive the identity $e^{-ix} = \cos x - i \sin x$.

32. Show that the identity $e^{-ix} = \cos x - i \sin x$ can be derived by replacing x with $-x$ in Euler's Formula.

33. Recall that $a + bi = r(\cos\theta + i\sin\theta)$ where the real number θ is the argument and $r = |a + bi|$. Use Euler's Formula to show that $a + bi = re^{i\theta}$. This is called the **exponential form** of the complex number.

Write each of the following complex numbers in exponential form.

34. i 　　　　 **35.** $-i$ 　　　　 **36.** $\dfrac{1}{2} + \dfrac{\sqrt{3}}{2}i$ 　　　　 **37.** $\sqrt{2} + \sqrt{2}i$

38. 1 　　　　 **39.** -1 　　　　 **40.** $-\sqrt{2} + \sqrt{2}i$ 　　　　 **41.** $3 + 4i$

42. Use multiplication of complex numbers to prove the following.
$$(r_1 e^{i\theta_1})(r_2 e^{i\theta_2}) = r_1 r_2 e^{i(\theta_1 + \theta_2)}$$

43. Use Euler's Formula and the identity $e^{-ix} = \cos x - i \sin x$ to prove
$$e^{ix} \cdot e^{-ix} = 1.$$

44. Use division of complex numbers to prove the following.
$$\frac{r_1 e^{i\theta_1}}{r_2 e^{i\theta_2}} = \frac{r_1}{r_2} e^{i(\theta_1 - \theta_2)}$$

45. Let $z = e^{i\theta}$ be a complex number. Prove that $\bar{z} = e^{-i\theta}$.

46. Use division of complex numbers to prove that $e^{-i\theta} = \dfrac{1}{e^{i\theta}}$.

47. Use De Moivre's Theorem to prove that $(re^{i\theta})^n = r^n e^{in\theta}$ where n is an integer.

48. Use the Complex Roots Theorem to prove that if n is a positive integer and $z \neq 0$, then $z^n = re^{i\theta}$ has n roots which are given by
$$z = \sqrt[n]{r}\left(e^{\frac{i(\theta + 2k\pi)}{n}}\right) \text{ where } k = 0, 1, 2, \cdots, n-1.$$

49. Replace x with ix in the power series for $\sin x$ given on page 334. Use the ratio test to show that the resulting series converges.

50. Repeat Exercise 49 for the power series for $\cos x$.

For all real values of x define $\sin ix$ and $\cos ix$ to be the sum of the series obtained in Exercises 49 and 50, respectively. Use these definitions to prove the following identities.

51. $e^x = \cos ix - i \sin ix$ 　　　　　　 **52.** $e^{-x} = \cos ix + i \sin ix$

Use the identities obtained in Exercises 51 and 52 to prove the following identities.

53. $\cos ix = \dfrac{e^{-x} + e^x}{2}$ 　　　　　　 **54.** $\sin ix = \dfrac{e^{-x} - e^x}{2i}$

8-9 Hyperbolic Functions

The solutions of problems in the sciences and engineering are often based on functions that involve the exponential function. The following are two such functions.

$$\sinh x = \frac{e^x - e^{-x}}{2} \qquad \cosh x = \frac{e^x + e^{-x}}{2}$$

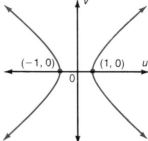

It can be shown that for each value of x, $(\cosh x, \sinh x)$ lies on the graph of the <u>unit hyperbola</u> $u^2 - v^2 = 1$ shown at the right. It is for this reason that these two functions are called the **hyperbolic sine** and the **hyperbolic cosine,** respectively.

Example 1 Find the value of each function for $x = 0$.

  a. $\sinh x$ **b.** $\cosh x$

Solutions: a. $\sinh 0 = \dfrac{e^0 - e^{-0}}{2}$ **b.** $\cosh 0 = \dfrac{e^0 + e^{-0}}{2}$

$$= \frac{1-1}{2} = 0 \qquad\qquad\qquad\qquad = \frac{1+1}{2} = 1$$

The power series for e^x can be used to find a power series expression for the sinh and cosh functions.

Example 2 Find a power series for sinh x.

Solution: Replace e^x and e^{-x} in $\sinh x = \dfrac{e^x - e^{-x}}{2}$ by their power series. To obtain a power series for e^{-x}, replace x in the power series for e^x.

$$\sinh x = \frac{e^x - e^{-x}}{2} = \frac{1}{2}e^x - \frac{1}{2}e^{-x}$$

$$\sinh x = \frac{1}{2}\left(1 + x + \frac{x^2}{2!} + \frac{x^3}{3!} + \cdots\right) - \frac{1}{2}\left(1 - x + \frac{x^2}{2!} - \frac{x^3}{3!} + \cdots\right)$$

$$\sinh x = x + \frac{x^3}{3!} + \frac{x^5}{5!} + \cdots + \frac{x^{2n-1}}{(2n-1)!} + \cdots$$

The method of Example 2 can be used to find the power series representation for $\cosh x$. You are asked to do this in the Exercises.

$$\cosh x = 1 + \frac{x^2}{2!} + \frac{x^4}{4!} + \cdots + \frac{x^{2n-2}}{(2n-2)!} + \cdots$$

The hyperbolic tangent (tanh), cotangent (coth), secant (sech), and cosecant (csch) are defined below. Many of the properties of the hyperbolic functions are analogous to those of the circular functions. You are asked to prove some of these properties in the Exercises.

$$\tanh x = \frac{e^x - e^{-x}}{e^x + e^{-x}} \qquad \coth x = \frac{e^x + e^{-x}}{e^x - e^{-x}}$$

$$\operatorname{sech} x = \frac{2}{e^x + e^{-x}} \qquad \operatorname{csch} x = \frac{2}{e^x - e^{-x}}$$

Classroom Exercises

Find the value of each function for x = 0.

1. $\tanh x$ **2.** $\coth x$ **3.** $\operatorname{sech} x$ **4.** $\operatorname{csch} x$

Written Exercises

a

1. Find a power series for $\cosh x$.

Use the definitions of the hyperbolic functions to prove each identity.

2. $\tanh x = \dfrac{\sinh x}{\cosh x}$ **3.** $\coth x = \dfrac{1}{\tanh x}$

4. $\operatorname{sech} x = \dfrac{1}{\cosh x}$ **5.** $\operatorname{csch} x = \dfrac{1}{\sinh x}$

6. $\coth x = \dfrac{\cosh x}{\sinh x}$ **7.** $\tanh = \dfrac{\operatorname{sech} x}{\operatorname{csch} x}$

8. $\cosh(-x) = \cosh x$ **9.** $\sinh(-x) = -\sinh x$

b

10. Show that the point $(\cosh x, \sinh x)$ lies on the unit hyperbola by proving the following.

$$\cosh^2 x - \sinh^2 x = 1$$

Prove each identity.

11. $\tanh^2 x + \operatorname{sech}^2 x = 1$

12. $\coth^2 x - \operatorname{csch}^2 x = 1$

13. $\sinh(x + y) = \sinh x \cosh y + \cosh x \sinh y$

14. $\cosh(x + y) = \cosh x \cosh y + \sinh x \sinh y$

15. $\cosh 2x = 2 \cosh^2 x - 1$

16. $\sinh 2x = 2 \sinh x \cosh x$

17. Use the ratio test to prove that the power series for $\sinh x$ converges for all real values of x.

18. Repeat Exercise 17 for the power series for $\cosh x$ obtained in Exercise 1.

Prove each identity where n is any integer.

19. $\sinh(x + 2n\pi i) = \sinh x$

20. $\cosh(x + 2n\pi i) = \cosh x$

21. $\tanh(x + n\pi i) = \tanh x$

22. $\coth(x + n\pi i) = \coth x$

23. $\operatorname{csch}(x + 2n\pi i) = \operatorname{csch} x$

24. $\operatorname{sech}(x + 2n\pi i) = \operatorname{sech} x$

Prove the following identities. (See Exercises 53 and 54 on page 337.)

25. $\sin ix = i \sinh x$

26. $\cos ix = \cosh x$

BASIC: POWER SERIES

The sum of a convergent series can be estimated by adding a finite number of terms. It is possible to program a computer to add a finite number of terms.

Problem: *Given a value for x, estimate sin x by adding the first 13 terms of the following power series.*

$$\sin x = x - \frac{x^3}{3!} + \frac{x^5}{5!} - \frac{x^7}{7!} + \cdots + (-1)^{n-1}\frac{x^{2n-1}}{(2n-1)!} + \cdots$$

This problem can be solved using the program shown on the following page.

```
1Ø READ X
2Ø LET S = X
3Ø LET D = 1
4Ø LET N = 2
5Ø LET E = 2*N - 1
6Ø LET D = D*(E - 1)*E
7Ø LET S = S + (-1)↑(N - 1)*(X↑E)/D
8Ø LET N = N + 1
9Ø IF N <= 13 THEN 5Ø
1ØØ PRINT "SIN(";X;") = ";S
11Ø GO TO 1Ø
12Ø DATA 1.57Ø8Ø, -.785398, 3.14159
13Ø END
```

Analysis: The strategy of the above program is to use the fact that the nth partial sum is equal to the $(n-1)$st partial sum plus the nth term. Note that the exponent, E, for the nth term is given by $2n - 1$. Thus, the denominator for the nth term, $(2n - 1)!$, can be computed by multiplying the denominator for the $(n-1)$st term times $(E - 1)$ times E.

Statement 5Ø: This computes and stores the exponent, E, for the nth term.

Statement 6Ø: This computes and stores the denominator, D, for the nth term.

Statement 7Ø: This computes the sum of the $(n-1)$st partial sum and the nth term and stores the result.

Output:
```
SIN( 1.57Ø8 )  =    1
SIN(-Ø.785398 )  =  -Ø.7Ø71Ø7
SIN( 3.14159 )  =    2.71478E-6
```
$\sin \pi = 0$
$2.71478 \times 10^{-6} \approx 0$

```
OUT OF DATA IN LINE 1Ø
```

Written Exercises

In Exercises 1–10, write a program that can be used to solve the given problem.

In Exercises 1 and 2, you are given the first term, a, and common ratio, r, of a geometric sequence $(a, ar, ar^2, ar^3, \cdots, ar^{n-1})$. You are also given a positive integer, n.

1. Compute the nth term.

2. Compute the sum of the first n terms.

3. Given the first term, a, and the common ratio, r ($|r| < 1$), of a geometric series $(a + ar + ar^2 + \cdots)$, compute the sum.

4. Given a value for x, compute $\cos x$. Use the following power series.

$$\cos x = 1 - \frac{x^2}{2!} + \frac{x^4}{4!} - \frac{x^6}{6!} + \cdots + (-1)^{n-1}\frac{x^{2n-2}}{(2n-2)!} + \cdots$$

5. Given a value for x, where $-1 \le x \le 1$, compute $\arctan x$. Use the following power series.

$$\arctan x = x - \frac{x^3}{3} + \frac{x^5}{5} - \frac{x^7}{7} + \cdots + (-1)^{n-1}\frac{x^{2n-1}}{2n-1} + \cdots$$

6. Given x, compute e^x. Use the following power series.

$$e^x = 1 + x + \frac{x^2}{2!} + \frac{x^3}{3!} + \frac{x^4}{4!} + \cdots + \frac{x^{n-1}}{(n-1)!} + \cdots$$

7. Given a value for x, compute $\sinh x$. Use the following power series.

$$\sinh x = x + \frac{x^3}{3!} + \frac{x^5}{5!} + \frac{x^7}{7!} + \cdots + \frac{x^{2n-1}}{(2n-1)!} + \cdots$$

8. Given a value for x, compute $\cosh x$. Use the following power series.

$$\cosh x = 1 + \frac{x^2}{2!} + \frac{x^4}{4!} + \frac{x^6}{6!} + \cdots + \frac{x^{2n-2}}{(2n-2)!} + \cdots$$

9. Use the following series to write a program that will approximate the value of π.

$$\frac{\pi}{4} = 1 - \frac{1}{3} + \frac{1}{5} - \frac{1}{7} + \frac{1}{9} - \cdots + \frac{(-1)^{n-1}}{(2n-1)} + \cdots$$

10. Find the limit, if it exists, of $\dfrac{\sin x}{x}$ as x approaches 0. Do this by printing x and $\dfrac{\sin x}{x}$ for $x = 1, .5, .25, .125, .0625, \cdots$, until the difference between successive values of $\dfrac{\sin x}{x}$ is less than .000005.

Chapter Objectives and Review

Objective: To know the meanings of the mathematical terms in this chapter.

1. Be sure that you know the meaning of these mathematical terms.

arithmetic sequence (p. 307)
arithmetic series (p. 322)
common difference (p. 307)
common ratio (p. 307)
convergent sequence (p. 313)
divergent sequence (p. 313)
Euler's Formula (p. 335)
general term (p. 304)
geometric sequence (p. 307)
geometric series (p. 322)
hyperbolic cosine (p. 338)

hyperbolic sine (p. 338)
index (p. 304)
interval of convergence (p. 331)
limit (p. 313)
partial sum (p. 321)
power series (p. 331)
ratio test (p. 327)
sequence (p. 304)
series (p. 321)
sum of a series (p. 327)
term (p. 304)

Objective: To use the rule or general term of a sequence to find specific terms of the sequence. (Section 8–1)

Use the rule to find the indicated term.

2. $a_n = \dfrac{1}{n+1}$; 9th term

3. $a_n = \frac{1}{2}n^2$; 8th term

Write the first five terms of each sequence.

4. $\{a_n\} = \{(-1)^n\}$

5. $\{a_n\} = \left\{\left(\dfrac{2}{n}\right)^2\right\}$

Objective: To use the formulas for the general terms of arithmetic and geometric sequences. (Section 8–2)

6. Write the sixth term of the arithmetic sequence $\{a_n\}$ given that $a_1 = 2$ and that $a_2 = 6$.

7. Write the sixth term of the geometric sequence $\{a_n\}$ given that $a_1 = 2$ and that $a_2 = 6$.

Objective: To use the definition of the limit of a sequence. (Section 8–3)

Graph the first eight terms of each sequence.

8. $\{a_n\} = \left\{2 + \dfrac{1}{n}\right\}$

9. $\{a_n\} = \{n\}$

For each sequence $\{a_n\}$, find a natural number m such that a_m and all succeeding terms differ from L by less than $\frac{1}{1,000}$.

10. $\{a_n\} = \left\{3 - \frac{1}{n}\right\}$; $L = 3$

11. $\{a_n\} = \left\{\frac{4}{n+1}\right\}$; $L = 0$

Use the definition of _limit_ to prove that the sequence $\{a_n\}$ converges to the limit L.

12. $\{a_n\} = \left\{4 + \frac{1}{n}\right\}$; $L = 4$

13. $\{a_n\} = \left\{\frac{n+1}{n+3}\right\}$; $L = 1$

Objective: To use limit theorems to find the limit of a sequence. (Section 8–4)

Find the limit of each sequence.

14. $\lim\limits_{n\to\infty} 2\left(-\frac{1}{3}\right)^n$

15. $\lim\limits_{n\to\infty} \left(3 - \frac{2}{n}\right)$

16. $\lim\limits_{n\to\infty} \left(\dfrac{3 - \frac{2}{n}}{1 - \frac{1}{n}}\right)$

17. $\lim\limits_{n\to\infty} \left[1 - \frac{1}{n}\right]\left[2 + \left(-\frac{1}{2}\right)^n\right]$

Objective: To use the definitions of series and partial sum. (Section 8–5)

Use the terms of the sequence $\{a_n\}$ and the rule $S_n = a_1 + a_2 + \cdots + a_n$ to write the first four terms of the sequence $\{S_n\}$.

18. $\{a_n\} = \left\{2 + \frac{1}{n}\right\}$

19. $\{a_n\} = \{n\}$

Identify each series as arithmetic (A) or geometric (G). Then write the series in expanded form.

20. $\sum\limits_{j=1}^{\infty} \left[-8 \cdot \left(-\frac{1}{4}\right)^{j-1}\right]$

21. $\sum\limits_{j=1}^{\infty} [50 - 5(j - 1)]$

Objective: To use the general term for the sequence of partial sums of an arithmetic or geometric series to evaluate a specific partial sum. (Section 8–5)

Evaluate the tenth partial sum of the series for each sequence.

22. $\{-3 \cdot (-1)^{n-1}\}$

23. $\{-5 + (n - 1) \cdot 3\}$

Objective: To find the sum, if it exists, of a given geometric series. (Section 8–6)

Find the sum, if it exists, of the given geometric series. Otherwise, write _diverges_.

24. $\sum\limits_{j=1}^{\infty} \frac{1}{8}(-1)^{j-1}$

25. $\sum\limits_{j=1}^{\infty} -1\left(\frac{1}{8}\right)^{j-1}$

26. $\sum\limits_{j=1}^{\infty} (.8)^{j-1}$

27. $\sum\limits_{j=1}^{\infty} (1.02)^{j-1}$

Objective: To use the ratio test to determine whether a series converges. (Section 8–6)

Use the ratio test to determine whether each of the following series converges. For each exercise write _converges_, _diverges_, or _test fails_.

28. $\displaystyle\sum_{j=1}^{\infty} \frac{1}{5j}$

29. $\displaystyle\sum_{j=1}^{\infty} \frac{1}{5^j}$

30. $\displaystyle\sum_{j=1}^{\infty} 5j$

31. $\displaystyle\sum_{j=1}^{\infty} 5^j$

Objective: To find the interval of convergence of a given power series. (Section 8–7)

Find the interval of convergence for each of the following power series.

32. $\displaystyle\sum_{j=1}^{\infty} x^j$

33. $\displaystyle\sum_{j=1}^{\infty} (-1)^j (x)^j$

34. $\displaystyle\sum_{j=1}^{\infty} \frac{x^j}{j!}$

35. $\displaystyle\sum_{j=1}^{\infty} \frac{x^j}{j}$

Objective: To use the power series for $\sin x$ and $\cos x$ to estimate values of the sine and cosine functions. (Section 8–8)

Use the first four terms of the power series for $\sin x$ or $\cos x$ to estimate each value.

36. $\sin .1$

37. $\cos (-1)$

38. $\sin .5$

39. $\cos \frac{\pi}{2}$

Objective: To use Euler's Formula to evaluate expressions. (Section 8–8)

Use Euler's Formula to evaluate each expression.

40. $e^{i\frac{3\pi}{4}}$

41. $e^{5\pi i}$

42. $e^{-i\pi}$

43. $e^{i\frac{5\pi}{6}}$

Objective: To prove identities that involve the hyperbolic functions. (Section 8–9)

Use the definitions of the hyperbolic functions to prove each identity.

44. $\cosh x - \sinh x = e^{-x}$

45. $\operatorname{csch} (-x) = -\operatorname{csch} x$

Chapter Test

1. Write the first four terms of the sequence $\{a_n\} = \left\{1 + \dfrac{2}{n}\right\}$.

2. Write the fifth term of the geometric sequence $\{a_n\}$ with $a_1 = 2$ and $a_2 = 1$.

3. For the sequence $\{a_n\} = \left\{-2 + \dfrac{1}{n}\right\}$, find a natural number m such that a_m and all succeeding terms differ from -2 by less than $\dfrac{1}{1,000}$.

4. Use the definition of limit to prove that the sequence $\left\{-2 + \dfrac{1}{n}\right\}$ converges to the limit -2.

5. Use the limit theorems to find $\lim\limits_{n\to\infty} \dfrac{\left(\frac{n-1}{n}\right)\left(-\frac{1}{2}\right)^n}{\left(-2 + \frac{1}{n}\right)}$.

6. Write the series $\sum\limits_{j=1}^{\infty} -\dfrac{1}{8}(-1)^{j-1}$ in expanded form.

7. Evaluate the tenth partial sum of the series for $\{25 - 5(n-1)\}$.

8. Find the sum, if it exists, of the geometric series $\sum\limits_{j=1}^{\infty} -\dfrac{1}{4}(-8)^{j-1}$. Otherwise, write diverges.

9. Use the ratio test to determine whether the series $\sum\limits_{j=1}^{\infty} \dfrac{4}{(j+1)!}$ converges. Write converges, diverges, or test fails.

10. Find the interval of convergence of the power series $\sum\limits_{j=1}^{\infty} \dfrac{(-1)^{j+1}x^{j+1}}{(j+1)!}$.

11. Use the first four terms of the power series for $\sin x$ to estimate $\sin(-1)$.

12. Use Euler's Formula to evaluate $e^{i\frac{3\pi}{2}}$

13. Prove: $\cosh(-x) = \cosh x$

Cumulative Review   Chapters 6–8

Write the letter of the response that best answers each question.

1. Given that $W(1.5) = (0.0707, 0.9975)$, where W is the wrapping function, find $W(\pi + 1.5)$. (Section 6–1)

    a. $(0.0707, -0.9975)$ b. $(-0.0707, 0.9975)$

    c. $(0.9975, 0.0707)$ d. $(-0.0707, -0.9975)$

2. Given that $\cos t = -.4161$, find $\cos(\pi - 2)$. (Section 6–2)

   a. $-.4161$ **b.** $.4161$ **c.** $.9093$ **d.** $-.9093$

3. Find the frequency f of $y = 2 \cos(3t - \pi)$. (Section 6–3)

   a. π **b.** $\frac{2\pi}{3}$ **c.** $\frac{3}{2\pi}$ **d.** $\frac{3}{\pi}$

4. Which is <u>not</u> a formula for simple harmonic motion? (Section 6–4)

   a. $y = \sin(t - \pi)$ **b.** $y = 3 \cos(t + \frac{\pi}{2})$

   c. $y = 3t \sin(t - \pi)$ **d.** $y = \cos t$

5. The frequency of an alternating emf is 60 cycles per second and the maximum emf is 150 volts. Write an equation for the emf of the form $E = E_m \sin \omega t$. (Section 6–5)

   a. $E = -150 \sin 60\pi t$ **b.** $E = 150 \sin 30\pi t$

   c. $E = 150 \sin 60\pi t$ **d.** $E = 150 \sin 120\pi t$

6. The frequency of a simple sound is 220 cycles per second. Which is an equation of the oscilloscope image? (Section 6–6)

   a. $y = \sin 440 \pi t$ **b.** $y = \sin 110\pi t$

   c. $y = 220 \sin \pi t$ **d.** $y = \sin 220\pi t$

7. Subtract: $(3 + 2i) - (4 - 3i)$ (Section 7–1)

   a. $-1 + 5i$ **b.** $7 + 5i$ **c.** $-1 - i$ **d.** $7 - i$

8. Multiply: $(2 - 3i)(1 - 2i)$ (Section 7–2)

   a. $8 - 7i$ **b.** $-4 + 7i$ **c.** $8 + i$ **d.** $-4 - 7i$

9. Express $2 - 2i$ in polar form. (Section 7–3)

   a. $2\sqrt{2}[\cos(-\frac{\pi}{4}) + i \sin(-\frac{\pi}{4})]$ **b.** $2\sqrt{2}(\cos\frac{\pi}{4} + i \sin\frac{\pi}{4})$

   c. $2\sqrt{2}(\cos\frac{3\pi}{4} + i \sin\frac{3\pi}{4})$ **d.** $2\sqrt{2}(\cos\frac{3\pi}{4} - i \sin\frac{3\pi}{4})$

10. Find $z_1 \cdot z_2$ and express the product in standard form. (Section 7–4)

   $z_1 = 2(\cos\frac{\pi}{3} + i \sin\frac{\pi}{3})$ $z_2 = 7(\cos\frac{2\pi}{9} + i \sin\frac{2\pi}{9})$

   a. $\frac{2}{7}(\cos\frac{\pi}{9} + i \sin\frac{\pi}{9})$ **b.** $\frac{2}{7}(\cos\frac{5\pi}{9} + i \sin\frac{5\pi}{9})$

   c. $14(\cos\frac{5\pi}{9} + i \sin\frac{5\pi}{9})$ **d.** $-2.4304 + 13.7872i$

11. Find $z_1 \div z_2$ and express the quotient in standard form. (Section 7–4)

   $z_1 = 6(\cos 75° + i \sin 75°)$ $z_2 = 2(\cos 50° + i \sin 50°)$

   a. $3 \cos 125° + 3i \sin 125°$ **b.** $.9063 + .4226i$

   c. $2.719 + 1.268i$ **d.** $3 \cos 25° + 3i \sin 25°$

12. If $z = r(\cos \theta + i \sin \theta)$, which of these represents z^5? (Section 7–5)

   a. $r^5(\cos \theta^5 + i \sin \theta^5)$ **b.** $5r(\cos 5\theta + i \sin 5\theta)$

   c. $r^5(\cos 5\theta + i \sin 5\theta)$ **d.** $5r(\cos \theta^5 + i \sin \theta^5)$

13. Evaluate $(1 - i)^4$. Express in standard form. (Section 7–5)

a. -4 b. $2 - 2i$ c. $-2i$ d. $-2 + 2i$

14. Which of the following is not a cube root of -1? (Section 7–6)

a. -1 b. $-\frac{1}{2} - i\frac{\sqrt{3}}{2}$ c. $\frac{1}{2} + i\frac{\sqrt{3}}{2}$ d. $\frac{1}{2} - i\frac{\sqrt{3}}{2}$

15. Find a_5: $\{a_n\} = \left\{3 - \frac{2n}{(n+1)^2}\right\}$. (Section 8–1)

a. 3   b. $2\frac{1}{2}$ c. $2\frac{13}{18}$ d. 0

16. Write the general term of the arithmetic sequence $\{a_n\}$ for which $a_2 = 1$ and $a_6 = -7$. (Section 8–2)

a. $5 - 2n$ b. $3 - 2n$ c. $5 - 3n$ d. $2 + 3(n - 1)$

17. For the sequence $\{a_n\} = \{\frac{4}{n^2}\}$, find the smallest natural number m such that a_m and all succeeding terms differ from 0 by less than $\frac{1}{10}$. (Section 8–3)

a. 5   b. 6   c. 7   d. 8

18. Find the limit, if it exists, of $\{\frac{n+1}{2n-1}\}$ as $n \to \infty$. (Section 8–4)

a. Does not exist.   b. 0   c. 2   d. $\frac{1}{2}$

19. Which of the following is represented by $\sum_{j=1}^{4} [-1 + (j - 1)3]$? (Section 8–5)

a. $0 + 3 + 6 + 9$ b. $-4 - 1 + 2 + 5$

c. $-1 + 2 + 5 + 8$ d. $2 + 5 + 8 + 3$

20. Find the sum, if it exists, of $\sum_{j=1}^{\infty} 2(\frac{3}{4})^{j-1}$. (Section 8–6)

a. Does not exist.   b. 8   c. 4   d. 0

21. For the series $\sum_{j=1}^{\infty} a_j$, $\lim_{n \to \infty} \frac{a_{n+1}}{a_n} = \frac{3}{2}$, what may you conclude? (Section 8–6)

a. The series converges.   b. The series diverges.

c. The series has a sum of $\frac{3}{2}$.   d. The ratio test fails.

22. Find the interval of convergence for the power series $\sum_{j=1}^{\infty} 2x^j$. (Section 8–7)

a. $0 \le x < 1$ b. $-1 < x \le 0$

c. The set of real numbers   d. $-1 < x < 1$

23. Find the value of $e^{i\frac{\pi}{2}}$. (Section 8–8)

a. -1 b. i c. 1   d. $-i$

24. Find the value of $\tanh 0$. (Section 8–9)

a. Undefined   b. 1   c. 0   d. 2

Final Cumulative Review: Chapters 1–8

Write the letter of the response that best answers each question.

1. Find the distance between $P(2, -4)$ and $Q(-2, -1)$. (Section 1–1)

    a. 25   **b.** $\sqrt{7}$ **c.** 5   **d.** $\sqrt{29}$

2. Find the measure of the angle formed by a $\frac{7}{12}$ counterclockwise rotation. (Section 1–2)

    a. $210°$ **b.** $-210°$ **c.** $105°$ **d.** $-105°$

3. Point $P(x, y)$ is on the terminal side of an angle in standard position and is r units from the origin. Which of the following is <u>false</u>? (Sections 1–3 and 1–4)

    a. $\sin \theta = \frac{y}{r}$ **b.** $\cos \theta = \frac{x}{r}$ **c.** $\cot \theta = \frac{x}{y}$ **d.** $\sec \theta = \frac{r}{y}$

4. Given that $\sin 60° = \frac{\sqrt{3}}{2}$, evaluate $\sin 300°$. (Section 1–5)

    a. $\frac{1}{2}$ **b.** $-\frac{\sqrt{3}}{2}$ **c.** $\frac{\sqrt{3}}{2}$ **d.** $-\frac{1}{2}$

5. If $\sin \theta = .9042$, where $0° < \theta < 90°$, find θ to the nearest minute. (Sections 1–6 and 1–7)

    a. $64°43'$ **b.** $65°47'$ **c.** $64°47'$ **d.** $65°43'$

6. A kite string is 60 meters long and makes an angle of $31°$ with the horizontal. The string is held one meter above the ground. Find the height of the kite to the nearest meter. (Sections 1–8 and 1–9)

    a. 52   **b.** 31   **c.** 32   **d.** 51

7. A motorboat is 12.6 kilometers due south of a lighthouse. The bearing of the boat's dock from the lighthouse is $90°$. The distance between the dock and the lighthouse is 21.3 kilometers. Find, to the nearest ten minutes, the bearing of the dock from the motorboat. (Sections 1–10 and 1–11)

    a. $53°40'$ **b.** $30°40'$ **c.** $36°20'$ **d.** $59°20'$

8. Use the table on pages 368–372 to find a four-place decimal approximation for the radian measure of an angle that measures $55°20'$. (Sections 2–1 and 2–2)

    a. .9657   **b.** .6050   **c.** .9483   **d.** .6225

9. The turntable of a record player rotates at 78 rotations per minute. Find the angular velocity of the turntable in radians per minute. (Section 2–3)

    a. 78π rad/min   **b.** 39π rad/min   **c.** 156π rad/min   **d.** 78 rad/min

10. Tell which equality is <u>false</u>. (Section 2-4)

    a. $\sin(180° - \theta) = \sin\theta$ **b.** $\cos(180° - \theta) = \cos\theta$

    c. $\sin(360° - \theta) = -\sin\theta$ **d.** $\cos(360° - \theta) = \cos\theta$

11. For the function $y = 4\sin\left(\frac{1}{2}x + \frac{\pi}{2}\right)$, find which of the following is represented by the number 4π. (Sections 2-5, 2-6, 2-7, and 2-8)

    a. The amplitude only   **b.** The period only

    c. The phase shift only   **d.** The period and the phase shift

12. Find which point is <u>not</u> on the graph of the function $y = 2\sin x - \cos x$. (Section 2-9)

    a. $(0, -1)$ **b.** $\left(\frac{\pi}{2}, 0\right)$ **c.** $(\pi, -1)$ **d.** $\left(\frac{3\pi}{2}, -2\right)$

13. Find the function that is discontinuous at 2π and has period 2π. (Section 2-10)

    a. $y = \tan x$ **b.** $y = \cot x$ **c.** $y = \sec x$ **d.** $y = \csc x$

14. Find the equality that is an identity. (Sections 3-1 and 3-2)

    a. $\tan^2\theta = \csc^2\theta - 1$ **b.** $\cot^2\theta = \sec^2\theta - 1$

    c. $\sin^2\theta \cdot \cot^2\theta = 1 - \sin^2\theta$ **d.** $\cos\theta \cdot \cot\theta = \sin\theta$

Use these four equalities for Exercise 15.

    I. $\cos 2\alpha = \cos^2\alpha - \sin^2\alpha$ **II.** $\sin 2\alpha = \sin\alpha\cos\alpha$

    III. $\tan(\alpha + \beta) = \dfrac{\tan\alpha + \tan\beta}{1 - \tan\alpha\tan\beta}$ **IV.** $\sin\dfrac{\alpha}{2} = \pm\sqrt{\dfrac{1 - \cos\alpha}{2}}$

15. Which of these equalities are identities? (Sections 3-3, 3-4, 3-5, 3-6, and 3-7)

    a. I and III only   **b.** IV only

    c. I, III, and IV only   **d.** All four

16. Express $\sin\frac{\pi}{6} + \sin\frac{\pi}{12}$ as a product. (Section 3-8)

    a. $-2\sin\frac{\pi}{24}\cos\frac{\pi}{8}$ **b.** $-2\sin\frac{\pi}{8}\cos\frac{\pi}{24}$

    c. $2\sin\frac{\pi}{8}\cos\frac{\pi}{24}$ **d.** $2\sin\frac{\pi}{24}\cos\frac{\pi}{8}$

17. When you solve a triangle that has three given sides, which of the following should you do first? (Sections 4-1, 4-2, 4-3, and 4-4)

    a. Use the Law of Sines.   **b.** Use the Law of Cosines.

    c. Consider the Ambiguous Case.   **d.** Use the Law of Tangents.

18. In $\triangle ABC$, $a = 10$, $c = 6$, and $B = 45°$. Find the area to the nearest square unit. (Section 4-5)

    a. 60   **b.** 30   **c.** 21   **d.** 42

19. In $\triangle ABC$, $a = 8$, $b = 9$, and $c = 7$. Find the area to the nearest square unit. (Section 4–6)

    a. 38   **b.** 12   **c.** 720   **d.** 27

20. Vectors OB and OC are component vectors of $\overrightarrow{OA}$ and have the origin as their initial point. Their direction angles are 150° and 235°, respectively. If $|\overrightarrow{OB}| = 8$, and $|\overrightarrow{OC}| = 5$, find $|\overrightarrow{OA}|$ to the nearest unit. (Sections 4–7 and 4–8)

    a. 8   **b.** 6   **c.** 10   **d.** 9

21. An object weighing 560 newtons rests on a plank that makes an angle of 30° with the horizontal. Find, to the nearest newton, the magnitude of the component of the weight that acts parallel to the plank. (Section 4–9)

    a. 484   **b.** 323   **c.** 485   **d.** 280

22. Evaluate arc $\sin -\frac{\sqrt{2}}{2}$. Assume that n is an integer. (Section 5–1)

    a. $\frac{5\pi}{4} + 2n$ or $\frac{7\pi}{4} + 2n\pi$ **b.** $\frac{5\pi}{4} + 2n\pi$

    c. $-\frac{\pi}{4} + 2n\pi$ **d.** $-\frac{5\pi}{4} + 2n\pi$

23. Evaluate Arc $\cos \frac{1}{2}$. (Section 5–2)

    a. $\frac{\pi}{6}$ **b.** $\frac{\pi}{4}$ **c.** $\frac{\pi}{3}$ **d.** $\frac{\pi}{2}$

24. Evaluate $\cos [$Arc $\tan (-\sqrt{3}]$. (Section 5–3)

    a. $-\frac{\sqrt{3}}{2}$ **b.** $\frac{\sqrt{3}}{2}$ **c.** $-\frac{1}{2}$ **d.** $\frac{1}{2}$

25. Solve: $2 \sin x - \cos^2 x + 2 = 0$, where $0 \le x < 2\pi$ (Section 5–4)

    a. $\{\frac{3\pi}{2}\}$ **b.** $\{0\}$ **c.** $\{\frac{\pi}{2}\}$ **d.** $\{\pi\}$

26. Which of the following refers to the circle with polar equation $r = 2 \cos \theta$? (Sections 5–5 and 5–6)

    a. Center at $(0, 1)$; radius: 2   **b.** Center at $(0, 1)$; radius: 1

    c. Center at $(1, 0)$; radius: 2   **d.** Center at $(1, 0)$; radius: 1

27. Given: $W(t) = (-0.20, 0.98)$, where W is the wrapping function Find $W(-t)$. (Section 6–1)

    a. $(-0.20, -0.98)$ **b.** $(-0.20, 0.98)$

    c. $(0.20, -0.98)$ **d.** $(0.20, 0.98)$

28. Given: $\sin 2 = .9093$ Find $\sin (\pi + 2)$. (Section 6–2)

    a. .9093   **b.** $-.9093$ **c.** .4161   **d.** $-.4161$

29. Find the frequency f of $y = -\sin (2t + \pi)$. (Section 6–3)

    a. $\frac{2}{\pi}$ **b.** $\frac{1}{\pi}$ **c.** π **d.** $\frac{\pi}{2}$

30. $y = -\sin 264\pi t$ is an equation of the oscilloscope image of a simple sound. Find the frequency of the sound. (Section 6–6)

    a. 132 cycles per second   **b.** 264 cycles per second

    c. 528 cycles per second   **d.** None of these

31. Divide: $(3 + 2i) \div (4 - 3i)$ (Sections 7–1 and 7–2)

    a. $\frac{6}{25} + \frac{17}{25}i$ **b.** $-3 - \frac{17}{2}i$ **c.** $6 + 17i$ **d.** $3 + \frac{17}{2}i$

32. Find $z_1 \div z_2$ and express the quotient in standard form. (Sections 7–3 and 7–4)

$$z_1 = 3\left(\cos \tfrac{\pi}{4} + i \sin \tfrac{\pi}{4}\right) \qquad z_2 = 2\left(\cos \tfrac{5\pi}{8} + i \sin \tfrac{5\pi}{8}\right)$$

    a. $\frac{3}{2}\left(\cos \tfrac{3\pi}{8} + i \sin \tfrac{3\pi}{8}\right)$ **b.** $\frac{3}{2}\left(\cos \tfrac{3\pi}{8} - i \sin \tfrac{3\pi}{8}\right)$

    c. $\frac{2}{3}\left(\cos \tfrac{7\pi}{8} - i \sin \tfrac{7\pi}{8}\right)$ **d.** $.5741 - 1.3859i$

33. Which of the following is a <u>false</u> statement? (Sections 7–5 and 7–6)

    a. $\left(\cos \tfrac{\pi}{4} + i \sin \tfrac{\pi}{4}\right)^4 = -1$ **b.** $(1 + i)^8 = 64$

    c. $\frac{1}{2} - i\frac{\sqrt{3}}{2}$ is a cube root of -1.   **d.** "$z^5 = 32$" has five roots.

34. Write the seventh term of the geometric sequence $\{a_n\}$, given that $a_1 = 2$ and that the common ratio r is $-\frac{1}{3}$. (Sections 8–1 and 8–2)

    a. $\frac{2}{729}$ **b.** $-\frac{2}{729}$ **c.** $-\frac{2}{2187}$ **d.** $\frac{2}{2187}$

35. Which of the following limits does not exist? (Sections 8–3 and 8–4)

    a. $\lim\limits_{n \to \infty} \frac{2000}{n}$ **b.** $\lim\limits_{n \to \infty} \left(\frac{99}{98}\right)^n$ **c.** $\lim\limits_{n \to \infty} \left(\frac{1}{5^n}\right)$ **d.** $\lim\limits_{n \to \infty} 8^{10}$

36. Which of the following is an arithmetic series? (Sections 8–5 and 8–6)

    a. $2 + 5 + 8 + \cdots$ **b.** $2 + 1 + \frac{1}{2} + \frac{1}{4} + \cdots$

    c. $2 - 2 + 2 + \cdots$ **d.** $\frac{1}{3} - \frac{1}{9} + \frac{1}{27} - \cdots$

37. Use the first four terms of the power series for $\cos x$ to estimate the value of $\cos 1$. (Sections 8–7 and 8–8)

    a. 1   **b.** $\frac{1}{2}$ **c.** $\frac{13}{24}$ **d.** $\frac{389}{720}$

38. Which of the following functions is represented by $\dfrac{e^x + e^{-x}}{2}$? (Section 8–9)

    a. $\sinh x$ **b.** $\cosh x$ **c.** $\operatorname{sech} x$ **d.** $\tanh x$

Appendix:
Logarithms of Trigonometric Functions

A-1 Logarithms

Recall that 16 is the fourth power of 2; that is,

$$2^4 = 16.$$

In this expression, 2 is the <u>base</u> and 4 is the <u>exponent</u>. It is sometimes useful to write this expression in another form. In this case, 4 is called the <u>base 2 logarithm</u> (abbreviated log) of 16; that is,

$$4 = \log_2 16.$$

Thus, a <u>logarithm is an exponent</u>.

Definition: Logarithm of a Number

If b and x are positive real numbers, then

$$y = \log_b x \quad \text{if and only if} \quad x = b^y.$$

Thus, $2^4 = 16$ can be written in <u>logarithmic form</u> as $\log_2 16 = 4$.

Example 1 Express in logarithmic form: **a.** $2^{-3} = \dfrac{1}{8}$ **b.** $27^{\frac{2}{3}} = 9$

Solutions: a. $2^{-3} = \dfrac{1}{8}$ means $\log_2 \dfrac{1}{8} = -3$. **b.** $27^{\frac{2}{3}} = 9$ means $\log_{27} 9 = \dfrac{2}{3}$.

You can also use the definition to write a logarithmic expression in <u>exponential form</u>.

Example 2 Express in exponential form: **a.** $\log_2 8 = 3$ **b.** $\log_2 .25 = -2$

Solutions: a. $\log_2 8 = 3$ means $2^3 = 8$.

b. $\log_2 .25 = -2$ means $2^{-2} = .25$.

To find the value of a variable in a logarithmic expression, first express it in exponential form.

Example 3   Find the value of the variable.

a. $\log_3 9 = y$ **b.** $\log_b 100 = 2$

Solutions: **a.** $\log_3 9 = y$ means $3^y = 9$. Thus, $y = 2$, since $3^2 = 9$.
b. $\log_b 100 = 2$ means $b^2 = 100$. Thus, $b = 10$, since $10^2 = 100$.

Recall that $b^0 = 1$ whenever $b \neq 0$. Therefore, for any non-zero base b,

$$\log_b 1 = 0.$$

Written Exercises

Express each of the following in exponential form.

1. $\log_5 25 = 2$ **2.** $\log_{10} 1000 = 3$ **3.** $\log_{100} 1 = 0$

4. $\log_5 5 = 1$ **5.** $\log_8 2 = \frac{1}{3}$ **6.** $\log_{100} 1000 = \frac{3}{2}$

7. $\log_9 27 = \frac{3}{2}$ **8.** $\log_{100} .01 = -1$ **9.** $\log_{10} .001 = -3$

Express each of the following in logarithmic form.

10. $4^2 = 16$ **11.** $2^5 = 32$ **12.** $9^1 = 9$

13. $8^0 = 1$ **14.** $25^{\frac{1}{2}} = 5$ **15.** $10^{-2} = .01$

16. $81^{\frac{3}{2}} = 729$ **17.** $10,000^{\frac{1}{2}} = 100$ **18.** $9^{-\frac{1}{2}} = \frac{1}{3}$

Find the value of the variable.

19. $\log_{10} x = 3$ **20.** $\log_4 x = \frac{1}{2}$ **21.** $\log_{16} x = \frac{3}{2}$

22. $\log_b 9 = 2$ **23.** $\log_b 16 = \frac{1}{2}$ **24.** $\log_b 9 = \frac{2}{3}$

25. $\log_2 16 = y$ **26.** $\log_{16} \frac{1}{4} = y$ **27.** $\log_{16} 1 = y$

Evaluate each expression.

28. $\log_6 6$ **29.** $\log_5 1$ **30.** $\log_{10} (10^4)$

31. $\log_{100} 10$ **32.** $\log_9 27$ **33.** $\log_{16} 8$

34. $\log_{16} 32$ **35.** $\log_{27} 9$ **36.** $\log_8 32$

37. $\log_3 \left(\frac{1}{81}\right)$ **38.** $\log_{10} (.1)$ **39.** $\log_{16} \left(\frac{1}{4}\right)$

40. $\log_{27} \left(\frac{1}{81}\right)$ **41.** $\log_b 1$ **42.** $\log_x (x^4)$

A-2 Products, Quotients, and Powers

Recall the following properties of exponents from algebra.

If a, b, x, and y are real numbers where $a > 0$ and $b > 0$, then

1. $b^x \cdot b^y = b^{x+y}$

2. $\dfrac{b^x}{b^y} = b^{x-y}$

3. $(ab)^x = a^x b^x$

4. $\left(\dfrac{a}{b}\right)^x = \dfrac{a^x}{b^x}$

5. $b^x = b^y$ if and only if $x = y$.

The fact that a logarithm is an exponent leads to several useful properties of logarithms.

Theorem A–1: Logarithm of a Product

The logarithm of a product of two positive real numbers M and N is the sum of the logarithms of the two numbers. That is,

$$\log_b (M \cdot N) = \log_b M + \log_b N.$$

Proof: Let $\quad \log_b M = x$ and $\log_b N = y$.

Then $\qquad b^x = M \quad$ and $\quad b^y = N.$ ⟵ By definition

Thus $\qquad M \cdot N = b^x \cdot b^y$

$\qquad\qquad\qquad = b^{x+y}.$ ⟵ Property **1** of exponents

Finally, $\log_b (M \cdot N) = x + y,$ ⟵ By definition

or $\qquad \log_b (M \cdot N) = \log_b M + \log_b N.$ ⟵ Substitution

Theorem A–2: Logarithm of a Quotient

The logarithm of a quotient of two positive real numbers M and N is the difference of the logarithms of the numbers. That is,

$$\log_b (M \div N) = \log_b M - \log_b N.$$

The proof of Theorem A–2 is asked for in the exercises.

Theorem A–3:   Logarithm of a Power

The logarithm of a real number N raised to an exponent a is the product of the exponent and the logarithm of that number. That is,

$$\log_b N^a = a \log_b N.$$

The proof of Theorem A–3 is asked for in the exercises.

Example 1 Express $\log_b \dfrac{x^2 \cdot y}{z^3}$ as the sum and difference of logarithms.

Solution: $\log_b \dfrac{x^2 \cdot y}{z^3} = \log_b x^2 \cdot y - \log_b z^3$ ⟵ By Theorem A–2

$= \log_b x^2 + \log_b y - \log_b z^3$ ⟵ By Theorem A–1

$= 2 \log_b x + \log_b y - 3 \log_b z$ ⟵ By Theorem A–3

Example 2 Given $\log_{10} 2 = .3010$ and $\log_{10} 3 = .4771$, find each of the following.

a. $\log_{10} 6$ **b.** $\log \dfrac{3}{2}$ **c.** $\log 16$

Solutions: a. $\log_{10} 6 = \log_{10} (2 \cdot 3)$

$= \log_{10} 2 + \log_{10} 3$ ⟵ By Theorem A–1

$= .3010 + .4771$

$= .7781$

b. $\log_{10} \dfrac{3}{2} = \log_{10} 3 - \log_{10} 2$ ⟵ By Theorem A–2

$= .4771 - .3010$

$= .1761$

c. $\log_{10} 16 = \log_{10} 2^4$

$= 4 \log_{10} 2$ ⟵ By Theorem A–3

$= 4 \cdot (.3010)$

$= 1.2040$

Written Exercises

a

Express each expression as a sum and difference of logarithms.

1. $\log_b x \cdot y \cdot z$

2. $\log_b \dfrac{2z^4}{3x^2}$

3. $\log_b \dfrac{x^{\frac{2}{3}} \cdot y^{\frac{4}{5}}}{z^{\frac{1}{2}}}$

4. $\log_b \dfrac{4}{x \cdot y^2 \cdot z^3}$

5. $\log_b \dfrac{x^4 \cdot y^{-2}}{x^{-3}}$

6. $\log_b \dfrac{a \cdot z^4}{2 \cdot x^2}$

Given $\log_{10} 2 = .3010$, $\log_{10} 3 = .4771$, $\log_{10} 5 = .6990$ and $\log_{10} 11 = 1.0414$, find each of the following.

7. $\log_{10} 22$

8. $\log_{10} 15$

9. $\log_{10} \dfrac{11}{2}$

10. $\log_{10} 12$

11. $\log_{10} 55$

12. $\log_{10} 66$

13. $\log_{10} 36$

14. $\log_{10} 30$

15. $\log_{10} 60$

16. $\log_{10} \sqrt[3]{22}$

17. $\log_{10} .3$

18. $\log_{10} \dfrac{1}{5}$

19. $\log_{10} \dfrac{24}{11}$

20. $\log_{10} \dfrac{\sqrt{11}}{5}$

21. $\log_{10} \sqrt{\dfrac{11}{5}}$

b

22. Prove Theorem A–2. (HINT: Let $\log_b M = x$ and $\log_b N = y$. Then use $\dfrac{b^x}{b^y} = b^{x-y}$.)

23. Prove Theorem A–3. (HINT: Let $\log_b N = x$.)

A-3 Common Logarithms

Logarithms with base 10 are **common logarithms.** For convenience, common logarithms are usually written as $\log x$ rather than $\log_{10} x$. Every positive <u>real number</u> has a base 10 logarithm which is the sum of an integer and a base 10 logarithm of a number between 1 and 10.

$$
\begin{aligned}
\log 576 &= \log (10^2 \cdot 5.76) \\
&= \log 10^2 + \log 5.76 \\
&= 2 + \log 5.76
\end{aligned}
$$

$$
\begin{aligned}
\log .000576 &= \log (10^{-4} \cdot 5.76) \\
&= \log 10^{-4} + \log 5.76 \\
&= -4 + \log 5.76
\end{aligned}
$$

To find log 5.76, or the logarithm of any number between 1 and 10, a four-place table of common logarithms is given on pages 365–366. Except for 1, the logarithms are approximations. For convenience, however, the symbol = will be used when writing statements that involve logarithms. Use the following portion of the table for Examples 1 and 2. Note that the decimal point is omitted from each entry. Therefore, a decimal point must be inserted before the numbers when using them in calculations.

N	0	1	2	3	4	5	6	7	8	9
5.7	7559	7566	7574	7582	7589	7597	7604	7612	7619	7627
5.8	7634	7642	7649	7657	7664	7672	7686	7686	7694	7761

Example 1 Find: a. log 5.76   b. log 5.89

Solutions: a. Find 5.7 in the N column.

   log 5.76 = .7604   ◄——— Look directly right to the column headed **6.**

b. Find 5.8 in the N column.

   log 5.89 = .7761   ◄——— Look directly right to the column headed **9.**

Since any positive number can be expressed in scientific notation (as the product of a power of 10 and as a number between 1 and 10), the table of common logarithms can be used for any positive number.

Example 2 Find: a. log 576   b. log .000576

Solutions: a. $\log 576 = \log (10^2 \cdot 5.76)$ ◄——— Write in scientific notation.

   $= \log 10^2 + \log 5.76$ ◄——— Theorem A–1

   $= 2 + \log 5.76$ ◄——— $\log 10^2 = \log 100 = 2$

   $= 2 + .7604 = 2.7604$

b. $\log .000576 = \log (10^{-4} \cdot 5.76)$ ◄——— Write in scientific notation.

   $= \log 10^{-4} + \log 5.76$

   $= -4 + .7604$

As Example 2 illustrates, every common logarithm consists of two parts. The integral part is the **characteristic.** The positive decimal portion is the **mantissa.** The mantissa is always kept positive so that it may be readily found in the table. In Example 2a, the characteristic is 2 and the mantissa is .7604.

The characteristic in Example **2b** is negative. Logarithms with negative characteristics are generally written in **standard form.** To do this, add 10 to, and subtract 10 from, the logarithm.

$$-4 + .7604 = 10 - 4 + .7604 - 10 = 6.7604 - 10$$

A table of common logarithms can be used in a second way. Given a logarithm (exponent), you can find the corresponding number N, called the **antilogarithm,** or simply antilog.

Example 3 Find the antilog, N.

a. $\log N = 2.9440$ **b.** $\log N = 4.3927 - 10$

Solutions: a. $\log N = 2.9440$

$\therefore N = 10^{2.9440}$ ←————— Definition of $\log N$

$= 10^2 \cdot 10^{.9440}$ ←————— Property of exponents

$= 100 \cdot 8.79 = 879$ ←————— $\log 8.79 = .9440$

b. First, rewrite $4.3927 - 10$ so that the mantissa stands alone.

$\log N = 4.3927 - 10$

$= 4 + .3927 - 10$

$= -6 + .3927$ ←————— Characteristic + mantissa

$\therefore N = 10^{-6 + .3927}$ ←————— Definition of $\log N$

$= 10^{-6} \cdot 10^{.3927}$

$= .000001 \cdot 2.47 = .00000247$

Written Exercises

Find the common logarithm of each number.

1. 273	**2.** 2.73	**3.** .000273	**4.** 846
5. 85300	**6.** .203	**7.** .0203	**8.** 845,000
9. 1.43	**10.** .00456	**11.** .000052	**12.** 42.30
13. 1000	**14.** .00015	**15.** 12.3	**16.** .132

In Exercises 17–32, $\log N$ is given. Find the antilog, N.

17. 3.4048	**18.** 6.9138 − 10	**19.** 2.9633	**20.** .8494
21. 9.5276 − 10	**22.** 9.5276	**23.** 3.3483	**24.** 7.7396 − 10
25. .8651	**26.** 3.7973	**27.** 3.7973 − 10	**28.** 1.0212
29. 8.9600 − 10	**30.** 5.9600	**31.** 1.0043	**32.** .9996

A-4 Computing with Logarithms

Common logarithms can be used to simplify computations by changing computations involving multiplication and division to computations involving addition and subtraction of logarithms.

Example 1

Use logarithms to find $731 \cdot 8.45$.

Solution:
$$\log (731 \cdot 8.45) = \log 731 + \log 8.45 \quad \longleftarrow \quad \log (MN) = \log M + \log N$$
$$= \log 731 + .9269 \quad \longleftarrow \quad \text{From the table}$$
$$= 2.8639 + .9269 \quad \longleftarrow \quad 731 = 7.31 \cdot 10^2. \text{ Thus, the characteristic is 2.}$$
$$= 3.7908$$
$$\therefore 731 \cdot 8.45 = 10^3 \cdot 6.18 \quad \longleftarrow \quad \text{Use the nearest table value, .7910, for the mantissa.}$$
$$= 6180$$

In Example 1, the exact answer is 6176.95. However, 6180 is as accurate as you can be, since the antilogs in the tables are given to the nearest hundredth.

Example 2

Use logarithms to find $23.7 \div 84{,}000$.

Solution:
$$\log (23.7 \div 84{,}000) = \log 23.7 - \log 84{,}000 \quad \longleftarrow \quad \text{Theorem A-2}$$
$$\log 23.7 = 1.3747; \quad \log 84{,}000 = 4.9243$$

To avoid a negative mantissa when 4.9243 is subtracted from 1.3747, add 10 to, <u>and</u> subtract 10 from, 1.3747.

$$\begin{aligned} \log 23.7 = \quad &11.3747 - 10 \quad \longleftarrow \quad 1.3747 + 10 - 10 \\ \log 84{,}000 = \quad &\underline{4.9243} \\ &6.4504 - 10 \quad \longleftarrow \quad \text{The mantissa is positive.} \end{aligned}$$

$$\therefore 23.7 \div 84{,}000 = 10^{-4} \cdot 2.82 \quad \longleftarrow \quad \text{Characteristic is } 6 - 10, \text{ or } -4.$$
$$= .000282$$

Careful organization of the calculations is especially important when there are more than two operations.

Example 3

Find N if $N = \dfrac{1.73 \cdot 2.85}{.0013}$.

Solution: $\log N = \log 1.73 + \log 2.85 - \log .0013$

$$\log 1.73 = \quad .2380$$
$$\log 2.85 = \quad \underline{.4548} \quad \longleftarrow \text{Add to the line above.}$$
$$\overline{10.6928 - 10} \quad \longleftarrow \text{To avoid a negative mantissa,}$$
$$\text{add and subtract 10.}$$
$$\log .0013 = \underline{7.1139 - 10} \quad \longleftarrow \text{Subtract from the line above.}$$
$$\log N = \overline{3.5789}$$

$$\therefore \frac{1.73 \cdot 2.85}{0.0013} = 10^3 \cdot 3.79 = 3790$$

Logarithms can also be used with the trigonometric functions. To find $\log \tan 69°$, first find $\tan 69°$ in the Table of Values of the Trigonometric Functions. Thus, $\log \tan 69° = \log 2.6051$. Then go to the Table of Common Logarithms to find $\log 2.6051$. Thus, $\log \tan 69° = \log 2.6051 = .4158$. To avoid using two tables, you can find $\log \tan 69°$ directly by using the Table of Values of the Logarithms of Trigonometric Functions on pages 372–376. Since many trigonometric functions have values smaller than one, their logarithms often have negative characteristics. To simplify matters, the negative characteristic is not included in the table. Each value in the table is assumed to have −10 added to it. Thus, $\log \tan 69° = 10.4158 - 10 = .4158$.

Example 4

Find N if $N = \dfrac{\sqrt{401} \cdot \sin 23°}{\tan 59°}$.

Solution: First write $\sqrt{401}$ as $401^{\frac{1}{2}}$. Then,

$$\log 401 = 2.6031, \tfrac{1}{2} \log 401 = 1.3016$$

$$\log \sin 23° = \quad 9.5919 - 10 \quad \longleftarrow \begin{array}{l}\text{Add} \quad 10 \text{ to the Log–Trig} \\ \text{table value, } 9.5919.\end{array}$$

$$\log (\sqrt{401} \cdot \sin 23°) = 10.8935 - 10 \quad \longleftarrow 1.3016 + (9.5919 - 10)$$
$$\log \tan 59° = \quad \underline{.2212} \quad \longleftarrow \text{Subtract.}$$
$$\log N = \overline{10.6723 - 10}, \text{ or } .6723$$

$$\therefore \frac{\sqrt{401} \cdot \sin 23°}{\tan 59°} = 10^0 \cdot 4.70 = 4.70$$

Written Exercises

a

Use logarithms to compute each answer.

1. $843 \cdot .142$

2. $93.1 \cdot .0131$

3. $9.01 \cdot .901$

4. $82,100 \cdot .0042$

5. $7.23 \cdot 8070$

6. $92,300 \cdot .000143$

7. $843 \div .42$

8. $93.1 \div 1310$

9. $9.01 \div .00901$

10. $82,100 \div .0042$

11. $7.23 \div 807$

12. $92,300 \div .0004$

13. $\dfrac{1.85 \cdot 73.8}{28,300}$

14. $\dfrac{23,800}{285 \cdot 731}$

15. $\dfrac{143}{1.23 \cdot .852}$

16. $\dfrac{800 \cdot 48,200}{.0013}$

17. $\dfrac{123 \cdot 1.23}{3.21}$

18. $\dfrac{(2.31)^3}{(5.2)^2}$

19. $\sin 35° \cdot \cos 43°$

20. $\sin 35° \div \cos 43°$

21. $\dfrac{\tan 71°}{\tan 42°}$

22. $\dfrac{1}{\sin 52°} \cdot \tan 52°$

23. $\dfrac{\sin 1° \cdot \cos 2°}{\tan 3°}$

24. $\dfrac{\tan 40°}{\sin 20° \cdot \cos 61°}$

b

In Exercises 25–29, use $\pi = 3.14$.

25. The formula for the surface area of a sphere is $A = 4\pi r^2$. The earth is approximately a sphere with radius $r = 6340$ kilometers. Find the surface area of the earth to the nearest whole number.

26. The formula for the volume of a sphere with radius r is $V = \frac{4}{3}\pi r^3$. Use the data in Exercise 25 to find the volume of the earth to the nearest whole number.

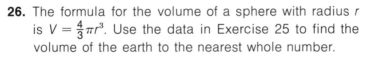

The formula for the volume of a right circular cone is $V = \frac{1}{3}\pi r^2 h$, where r is the radius of the base and h is the altitude of the cone. Find the volume to the nearest hundredth of each cone in Exercises 27–29.

27. $r = 1.23$ cm; $h = 3.54$ cm

28. $r = 8.97$ m; $h = 5.16$ m

29. $r = .13$ cm; $h = 2.31$ cm

A-5 Solving Triangles with Logarithms

Example 1

In right triangle ABC, $c = 20.3$, and $\angle A = 40°$.

a. Find a.
b. Find b.

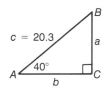

Solutions: a. $\sin 40° = \dfrac{a}{20.3}$

$a = 20.3 \cdot \sin 40°$

$\log a = \log 20.3 + \log \sin 40°$

$\quad = 1.3075 + (9.8081 - 10)$

$\quad = 11.1156 - 10$

$\quad = 1.1156$

$\therefore a = 13.0$

b. $\cos 40° = \dfrac{b}{20.3}$

$b = 20.3 \cos 40°$

$\log b = \log 20.3 + \log \cos 40°$

$\quad = 1.3075 + (9.8843 - 10)$

$\quad = 1.1918$

$\therefore b = 15.6$

Recall that in oblique triangles you can use the Law of Sines whenever you know two angles and a side.

Example 2

In $\triangle ABC$, $a = 360$, $\angle A = 43°$, and $\angle C = 110°$.
Find each of the following.

a. $\angle B$
b. b
c. c

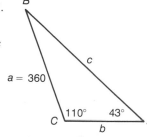

Solutions: a. $\angle B = 180° - (43° + 110°) = 27°$

b. $\dfrac{b}{\sin B} = \dfrac{a}{\sin A}$ or $b = \dfrac{360 \cdot \sin 27°}{\sin 43°}$

$\therefore \log b = \log 360 + \log \sin 27° - \log \sin 43°$

Calculations:

$\log 360 = \quad 2.5563$
$\log \sin 27° = \quad \underline{9.6570 - 10}$
$\qquad\qquad\qquad 12.2133 - 10$ ←———— $\log 360 + \log \sin 27°$
$\log \sin 43° = \quad \underline{9.8338 - 10}$ ←———— Subtract from the line above.
$\log b = \quad 2.3795$ ←———— $(\log 360 + \log \sin 27°) - \log \sin 43°$
$\therefore b = 10^2 \cdot 2.40 = 240$

c. $\dfrac{c}{\sin C} = \dfrac{a}{\sin A}$ or $c = \dfrac{360 \cdot \sin 110°}{\sin 43°}$

$\therefore \log c = \log 360 + \log \sin 110° - \log \sin 43°$

Calculations:

$$\begin{aligned}
\log 360 &= \ 2.5563 \\
\log \sin 110° &= \ \underline{9.9730 - 10} \\
&\quad\ \ 12.5293 - 10 \quad \longleftarrow \text{ } \log 360 + \log \sin 110° \\
\log \sin 43° &= \ \underline{9.8338 - 10} \quad \longleftarrow \text{ Subtract from the line above.} \\
\log c &= \ 2.6955 \quad\quad\ \longleftarrow \ (\log 360 + \log \sin 110°) - \log \sin 43° \\
\therefore c &= 10^2 \cdot 4.96 = 496
\end{aligned}$$

Written Exercises

a

Solve each right triangle ABC. Angle C is the right angle.

1. $c = 35; \ A = 40°$ **2.** $c = 7; \ B = 42°$

3. $a = 527; \ B = 29°$ **4.** $b = 80; \ B = 31°$

5. $a = 40; \ b = 62$ **6.** $a = 340; \ c = 570$

7. $b = 143; \ c = 145$ **8.** $a = 59; \ b = 59$

9. $a = 80; \ A = 24°$ **10.** $b = 430; \ B = 36°$

Solve each oblique triangle ABC.

11. $a = 359; \ A = 33°; \ B = 100°$ **12.** $b = .0456; \ A = 23°; \ C = 44°$

13. $c = 470; \ A = 70°; \ B = 75°$ **14.** $a = .0501; \ A = 39°; \ B = 40°$

15. $b = 7080; \ A = 30°; \ C = 50°$ **16.** $c = 60900; \ A = 43°; \ B = 35°$

17. $a = .27; \ A = 103°; \ B = 48°$ **18.** $b = 25.1; \ A = 121°; \ C = 37°$

19. $c = .0135; \ A = 92°; \ B = 81°$ **20.** $a = .0127; \ A = 20°; \ B = 35°$

b

21. Find, to the nearest tenth of a meter, the length of the shadow, side *AC*, cast by a tilted pole, side *BC*, where $BC = 14$ meters.

22. A diagonal of a parallelogram is 5.68 meters long and makes angles of 27° and 16° with the sides. Find the lengths of the sides to the nearest hundredth of a meter.

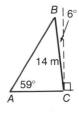

23. A vertical pole is on the side of a straight road that makes an angle of 25° with the horizontal. When the angle between the line of sight to the sun and the horizontal is 67°, the pole makes a shadow of 28 meters long extending in the uphill direction parallel to the road. Find the length of the pole to the nearest meter.

Table of Common Logarithms

N	0	1	2	3	4	5	6	7	8	9
1.0	0000	0043	0086	0128	0170	0212	0253	0294	0334	0374
1.1	0414	0453	0492	0531	0569	0607	0645	0682	0719	0755
1.2	0792	0828	0864	0899	0934	0969	1004	1038	1072	1106
1.3	1139	1173	1206	1239	1271	1303	1335	1367	1399	1430
1.4	1461	1492	1523	1553	1584	1614	1644	1673	1703	1732
1.5	1761	1790	1818	1847	1875	1903	1931	1959	1987	2014
1.6	2041	2068	2095	2122	2148	2175	2201	2227	2253	2279
1.7	2304	2330	2355	2380	2405	2430	2455	2480	2504	2529
1.8	2553	2577	2601	2625	2648	2672	2695	2718	2742	2765
1.9	2788	2810	2833	2856	2878	2900	2923	2945	2967	2989
2.0	3010	3032	3054	3075	3096	3118	3139	3160	3181	3201
2.1	3222	3243	3263	3284	3304	3324	3345	3365	3385	3404
2.2	3424	3444	3464	3483	3502	3522	3541	3560	3579	3598
2.3	3617	3636	3655	3674	3692	3711	3729	3747	3766	3784
2.4	3802	3820	3838	3856	3874	3892	3909	3927	3945	3962
2.5	3979	3997	4014	4031	4048	4065	4082	4099	4116	4133
2.6	4150	4166	4183	4200	4216	4232	4249	4265	4281	4298
2.7	4314	4330	4346	4362	4378	4393	4409	4425	4440	4456
2.8	4472	4487	4502	4518	4533	4548	4564	4579	4594	4609
2.9	4624	4639	4654	4669	4683	4698	4713	4728	4742	4757
3.0	4771	4786	4800	4814	4829	4843	4857	4871	4886	4900
3.1	4914	4928	4942	4955	4969	4983	4997	5011	5024	5038
3.2	5051	5065	5079	5092	5105	5119	5132	5145	5159	5172
3.3	5185	5198	5211	5224	5237	5250	5263	5276	5289	5302
3.4	5315	5328	5340	5353	5366	5378	5391	5403	5416	5428
3.5	5441	5453	5465	5478	5490	5502	5514	5527	5539	5551
3.6	5563	5575	5587	5599	5611	5623	5635	5647	5658	5670
3.7	5682	5694	5705	5717	5729	5740	5752	5763	5775	5786
3.8	5798	5809	5821	5832	5843	5855	5866	5877	5888	5899
3.9	5911	5922	5933	5944	5955	5966	5977	5988	5999	6010
4.0	6021	6031	6042	6053	6064	6075	6085	6096	6107	6117
4.1	6128	6138	6149	6160	6170	6180	6191	6201	6212	6222
4.2	6232	6243	6253	6263	6274	6284	6294	6304	6314	6325
4.3	6335	6345	6355	6365	6375	6385	6395	6405	6415	6425
4.4	6435	6444	6454	6464	6474	6484	6493	6503	6513	6522
4.5	6532	6542	6551	6561	6571	6580	6590	6599	6609	6618
4.6	6628	6637	6646	6656	6665	6675	6684	6693	6702	6712
4.7	6721	6730	6739	6749	6758	6767	6776	6785	6794	6803
4.8	6812	6821	6830	6839	6848	6857	6866	6875	6884	6893
4.9	6902	6911	6920	6928	6937	6946	6955	6964	6972	6981
5.0	6990	6998	7007	7016	7024	7033	7042	7050	7059	7067
5.1	7076	7084	7093	7101	7110	7118	7126	7135	7143	7152
5.2	7160	7168	7177	7185	7193	7202	7210	7218	7226	7235
5.3	7243	7251	7259	7267	7275	7284	7292	7300	7308	7316
5.4	7324	7332	7340	7348	7356	7364	7372	7380	7388	7396

Table of Common Logarithms

N	0	1	2	3	4	5	6	7	8	9
5.5	7404	7412	7419	7427	7435	7443	7451	7459	7466	7474
5.6	7482	7490	7497	7505	7513	7520	7528	7536	7543	7551
5.7	7559	7566	7574	7582	7589	7597	7604	7612	7619	7627
5.8	7634	7642	7649	7657	7664	7672	7679	7686	7694	7701
5.9	7709	7716	7723	7731	7738	7745	7752	7760	7767	7774
6.0	7782	7789	7796	7803	7810	7818	7825	7832	7839	7846
6.1	7853	7860	7868	7875	7882	7889	7896	7903	7910	7917
6.2	7924	7931	7938	7945	7952	7959	7966	7973	7980	7987
6.3	7993	8000	8007	8014	8021	8028	8035	8041	8048	8055
6.4	8062	8069	8075	8082	8089	8096	8102	8109	8116	8122
6.5	8129	8136	8142	8149	8156	8162	8169	8176	8182	8189
6.6	8195	8202	8209	8215	8222	8228	8235	8241	8248	8254
6.7	8261	8267	8274	8280	8287	8293	8299	8306	8312	8319
6.8	8325	8331	8338	8344	8351	8357	8363	8370	8376	8382
6.9	8388	8395	8401	8407	8414	8420	8426	8432	8439	8445
7.0	8451	8457	8463	8470	8476	8482	8488	8494	8500	8506
7.1	8513	8519	8525	8531	8537	8543	8549	8555	8561	8567
7.2	8573	8579	8585	8591	8597	8603	8609	8615	8621	8627
7.3	8633	8639	8645	8651	8657	8663	8669	8675	8681	8686
7.4	8692	8698	8704	8710	8716	8722	8727	8733	8739	8745
7.5	8751	8756	8762	8768	8774	8779	8785	8791	8797	8802
7.6	8808	8814	8820	8825	8831	8837	8842	8848	8854	8859
7.7	8865	8871	8876	8882	8887	8893	8899	8904	8910	8915
7.8	8921	8927	8932	8938	8943	8949	8954	8960	8965	8971
7.9	8976	8982	8987	8993	8998	9004	9009	9015	9020	9025
8.0	9031	9036	9042	9047	9053	9058	9063	9069	9074	9079
8.1	9085	9090	9096	9101	9106	9112	9117	9122	9128	9133
8.2	9138	9143	9149	9154	9159	9165	9170	9175	9180	9186
8.3	9191	9196	9201	9206	9212	9217	9222	9227	9232	9238
8.4	9243	9248	9253	9258	9263	9269	9274	9279	9284	9289
8.5	9294	9299	9304	9309	9315	9320	9325	9330	9335	9340
8.6	9345	9350	9355	9360	9365	9370	9375	9380	9385	9390
8.7	9395	9400	9405	9410	9415	9420	9425	9430	9435	9440
8.8	9445	9450	9455	9460	9465	9469	9474	9479	9484	9489
8.9	9494	9499	9504	9509	9513	9518	9523	9528	9533	9538
9.0	9542	9547	9552	9557	9562	9566	9571	9576	9581	9586
9.1	9590	9595	9600	9605	9609	9614	9619	9624	9628	9633
9.2	9638	9643	9647	9652	9657	9661	9666	9671	9675	9680
9.3	9685	9689	9694	9699	9703	9708	9713	9717	9722	9727
9.4	9731	9736	9741	9745	9750	9754	9759	9763	9768	9773
9.5	9777	9782	9786	9791	9795	9800	9805	9809	9814	9818
9.6	9823	9827	9832	9836	9841	9845	9850	9854	9859	9863
9.7	9868	9872	9877	9881	9886	9890	9894	9899	9903	9908
9.8	9912	9917	9921	9926	9930	9934	9939	9943	9948	9952
9.9	9956	9961	9965	9969	9974	9978	9983	9987	9991	9996

Table of Values of the Trigonometric Functions

θ Deg.	θ Rad.	Sin θ	Cos θ	Tan θ	Cot θ	Sec θ	Csc θ		
0° 00'	.0000	.0000	1.0000	.0000		1.000		1.5708	90° 00'
10'	.0029	.0029	1.0000	.0029	343.77	1.000	343.8	1.5679	50'
20'	.0058	.0058	1.0000	.0058	171.89	1.000	171.9	1.5650	40'
30'	.0087	.0087	1.0000	.0087	114.59	1.000	114.6	1.5621	30'
40'	.0116	.0116	.9999	.0116	85.940	1.000	85.95	1.5592	20'
50'	.0145	.0145	.9999	.0145	68.750	1.000	68.76	1.5563	10'
1° 00'	.0175	.0175	.9998	.0175	57.290	1.000	57.30	1.5533	89° 00'
10'	.0204	.0204	.9998	.0204	49.104	1.000	49.11	1.5504	50'
20'	.0233	.0233	.9997	.0233	42.964	1.000	42.98	1.5475	40'
30'	.0262	.0262	.9997	.0262	38.188	1.000	38.20	1.5446	30'
40'	.0291	.0291	.9996	.0291	34.368	1.000	34.38	1.5417	20'
50'	.0320	.0320	.9995	.0320	31.242	1.001	31.26	1.5388	10'
2° 00'	.0349	.0349	.9994	.0349	28.636	1.001	28.65	1.5359	88° 00'
10'	.0378	.0378	.9993	.0378	26.432	1.001	26.45	1.5330	50'
20'	.0407	.0407	.9992	.0407	24.542	1.001	24.56	1.5301	40'
30'	.0436	.0436	.9990	.0437	22.904	1.001	22.93	1.5272	30'
40'	.0465	.0465	.9989	.0466	21.470	1.001	21.49	1.5243	20'
50'	.0495	.0494	.9988	.0495	20.206	1.001	20.23	1.5213	10'
3° 00'	.0524	.0523	.9986	.0524	19.081	1.001	19.11	1.5184	87° 00'
10'	.0553	.0552	.9985	.0553	18.075	1.002	18.10	1.5155	50'
20'	.0582	.0581	.9983	.0582	17.169	1.002	17.20	1.5126	40'
30'	.0611	.0610	.9981	.0612	16.350	1.002	16.38	1.5097	30'
40'	.0640	.0640	.9980	.0641	15.605	1.002	15.64	1.5068	20'
50'	.0669	.0669	.9978	.0670	14.924	1.002	14.96	1.5039	10'
4° 00'	.0698	.0698	.9976	.0699	14.301	1.002	14.34	1.5010	86° 00'
10'	.0727	.0727	.9974	.0729	13.727	1.003	13.76	1.4981	50'
20'	.0756	.0756	.9971	.0758	13.197	1.003	13.23	1.4952	40'
30'	.0785	.0785	.9969	.0787	12.706	1.003	12.75	1.4923	30'
40'	.0814	.0814	.9967	.0816	12.251	1.003	12.29	1.4893	20'
50'	.0844	.0843	.9964	.0846	11.826	1.004	11.87	1.4864	10'
5° 00'	.0873	.0872	.9962	.0875	11.430	1.004	11.47	1.4835	85° 00'
10'	.0902	.0901	.9959	.0904	11.059	1.004	11.10	1.4806	50'
20'	.0931	.0929	.9957	.0934	10.712	1.004	10.76	1.4777	40'
30'	.0960	.0958	.9954	.0963	10.385	1.005	10.43	1.4748	30'
40'	.0989	.0987	.9951	.0992	10.078	1.005	10.13	1.4719	20'
50'	.1018	.1016	.9948	.1022	9.7882	1.005	9.839	1.4690	10'
6° 00'	.1047	.1045	.9945	.1051	9.5144	1.006	9.567	1.4661	84° 00'
10'	.1076	.1074	.9942	.1080	9.2553	1.006	9.309	1.4632	50'
20'	.1105	.1103	.9939	.1110	9.0098	1.006	9.065	1.4603	40'
30'	.1134	.1132	.9936	.1139	8.7769	1.006	8.834	1.4573	30'
40'	.1164	.1161	.9932	.1169	8.5555	1.007	8.614	1.4544	20'
50'	.1193	.1190	.9929	.1198	8.3450	1.007	8.405	1.4515	10'
7° 00'	.1222	.1219	.9925	.1228	8.1443	1.008	8.206	1.4486	83° 00'
10'	.1251	.1248	.9922	.1257	7.9530	1.008	8.016	1.4457	50'
20'	.1280	.1276	.9918	.1287	7.7704	1.008	7.834	1.4428	40'
30'	.1309	.1305	.9914	.1317	7.5958	1.009	7.661	1.4399	30'
40'	.1338	.1334	.9911	.1346	7.4287	1.009	7.496	1.4370	20'
50'	.1367	.1363	.9907	.1376	7.2687	1.009	7.337	1.4341	10'
8° 00'	.1396	.1392	.9903	.1405	7.1154	1.010	7.185	1.4312	82° 00'
10'	.1425	.1421	.9899	.1435	6.9682	1.010	7.040	1.4283	50'
20'	.1454	.1449	.9894	.1465	6.8269	1.011	6.900	1.4254	40'
30'	.1484	.1478	.9890	.1495	6.6912	1.011	6.765	1.4224	30'
40'	.1513	.1507	.9886	.1524	6.5606	1.012	6.636	1.4195	20'
50'	.1542	.1536	.9881	.1554	6.4348	1.012	6.512	1.4166	10'
9° 00'	.1571	.1564	.9877	.1584	6.3138	1.012	6.392	1.4137	81° 00'
		Cos θ	Sin θ	Cot θ	Tan θ	Csc θ	Sec θ	θ Rad.	θ Deg.

Table of Values of the Trigonometric Functions

θ Deg.	θ Rad.	Sin θ	Cos θ	Tan θ	Cot θ	Sec θ	Csc θ		
9° 00′	.1571	.1564	.9877	.1584	6.3138	1.012	6.392	1.4137	81° 00′
10′	.1600	.1593	.9872	.1614	6.1970	1.013	6.277	1.4108	50′
20′	.1629	.1622	.9868	.1644	6.0844	1.013	6.166	1.4079	40′
30′	.1658	.1650	.9863	.1673	5.9758	1.014	6.059	1.4050	30′
40′	.1687	.1679	.9858	.1703	5.8708	1.014	5.955	1.4021	20′
50′	.1716	.1708	.9853	.1733	5.7694	1.015	5.855	1.3992	10′
10° 00′	.1745	.1736	.9848	.1763	5.6713	1.015	5.759	1.3963	80° 00′
10′	.1774	.1765	.9843	.1793	5.5764	1.016	5.665	1.3934	50′
20′	.1804	.1794	.9838	.1823	5.4845	1.016	5.575	1.3904	40′
30′	.1833	.1822	.9833	.1853	5.3955	1.017	5.487	1.3875	30′
40′	.1862	.1851	.9827	.1883	5.3093	1.018	5.403	1.3846	20′
50′	.1891	.1880	.9822	.1914	5.2257	1.018	5.320	1.3817	10′
11° 00′	.1920	.1908	.9816	.1944	5.1446	1.019	5.241	1.3788	79° 00′
10′	.1949	.1937	.9811	.1974	5.0658	1.019	5.164	1.3759	50′
20′	.1978	.1965	.9805	.2004	4.9894	1.020	5.089	1.3730	40′
30′	.2007	.1994	.9799	.2035	4.9152	1.020	5.016	1.3701	30′
40′	.2036	.2022	.9793	.2065	4.8430	1.021	4.945	1.3672	20′
50′	.2065	.2051	.9787	.2095	4.7729	1.022	4.876	1.3643	10′
12° 00′	.2094	.2079	.9781	.2126	4.7046	1.022	4.810	1.3614	78° 00′
10′	.2123	.2108	.9775	.2156	4.6382	1.023	4.745	1.3584	50′
20′	.2153	.2136	.9769	.2186	4.5736	1.024	4.682	1.3555	40′
30′	.2182	.2164	.9763	.2217	4.5107	1.024	4.620	1.3526	30′
40′	.2211	.2193	.9757	.2247	4.4494	1.025	4.560	1.3497	20′
50′	.2240	.2221	.9750	.2278	4.3897	1.026	4.502	1.3468	10′
13° 00′	.2269	.2250	.9744	.2309	4.3315	1.026	4.445	1.3439	77° 00′
10′	.2298	.2278	.9737	.2339	4.2747	1.027	4.390	1.3410	50′
20′	.2327	.2306	.9730	.2370	4.2193	1.028	4.336	1.3381	40′
30′	.2356	.2334	.9724	.2401	4.1653	1.028	4.284	1.3352	30′
40′	.2385	.2363	.9717	.2432	4.1126	1.029	4.232	1.3323	20′
50′	.2414	.2391	.9710	.2462	4.0611	1.030	4.182	1.3294	10′
14° 00′	.2443	.2419	.9703	.2493	4.0108	1.031	4.134	1.3265	76° 00′
10′	.2473	.2447	.9696	.2524	3.9617	1.031	4.086	1.3235	50′
20′	.2502	.2476	.9689	.2555	3.9136	1.032	4.039	1.3206	40′
30′	.2531	.2504	.9681	.2586	3.8667	1.033	3.994	1.3177	30′
40′	.2560	.2532	.9674	.2617	3.8208	1.034	3.950	1.3148	20′
50′	.2589	.2560	.9667	.2648	3.7760	1.034	3.906	1.3119	10′
15° 00′	.2618	.2588	.9659	.2679	3.7321	1.035	3.864	1.3090	75° 00′
10′	.2647	.2616	.9652	.2711	3.6891	1.036	3.822	1.3061	50′
20′	.2676	.2644	.9644	.2742	3.6470	1.037	3.782	1.3032	40′
30′	.2705	.2672	.9636	.2773	3.6059	1.038	3.742	1.3003	30′
40′	.2734	.2700	.9628	.2805	3.5656	1.039	3.703	1.2974	20′
50′	.2763	.2728	.9621	.2836	3.5261	1.039	3.665	1.2945	10′
16° 00′	.2793	.2756	.9613	.2867	3.4874	1.040	3.628	1.2915	74° 00′
10′	.2822	.2784	.9605	.2899	3.4495	1.041	3.592	1.2886	50′
20′	.2851	.2812	.9596	.2931	3.4124	1.042	3.556	1.2857	40′
30′	.2880	.2840	.9588	.2962	3.3759	1.043	3.521	1.2828	30′
40′	.2909	.2868	.9580	.2994	3.3402	1.044	3.487	1.2799	20′
50′	.2938	.2896	.9572	.3026	3.3052	1.045	3.453	1.2770	10′
17° 00′	.2967	.2924	.9563	.3057	3.2709	1.046	3.420	1.2741	73° 00′
10′	.2996	.2952	.9555	.3089	3.2371	1.047	3.388	1.2712	50′
20′	.3025	.2979	.9546	.3121	3.2041	1.048	3.356	1.2683	40′
30′	.3054	.3007	.9537	.3153	3.1716	1.049	3.326	1.2654	30′
40′	.3083	.3035	.9528	.3185	3.1397	1.049	3.295	1.2625	20′
50′	.3113	.3062	.9520	.3217	3.1084	1.050	3.265	1.2595	10′
18° 00′	.3142	.3090	.9511	.3249	3.0777	1.051	3.236	1.2566	72° 00′
		Cos θ	Sin θ	Cot θ	Tan θ	Csc θ	Sec θ	θ Rad.	θ Deg.

Table of Values of the Trigonometric Functions

θ Deg.	θ Rad.	Sin θ	Cos θ	Tan θ	Cot θ	Sec θ	Csc θ		
18°00′	.3142	.3090	.9511	.3249	3.0777	1.051	3.236	1.2566	**72°00′**
10′	.3171	.3118	.9502	.3281	3.0475	1.052	3.207	1.2537	50′
20′	.3200	.3145	.9492	.3314	3.0178	1.053	3.179	1.2508	40′
30′	.3229	.3173	.9483	.3346	2.9887	1.054	3.152	1.2479	30′
40′	.3258	.3201	.9474	.3378	2.9600	1.056	3.124	1.2450	20′
50′	.3287	.3228	.9465	.3411	2.9319	1.057	3.098	1.2421	10′
19°00′	.3316	.3256	.9455	.3443	2.9042	1.058	3.072	1.2392	**71°00′**
10′	.3345	.3283	.9446	.3476	2.8770	1.059	3.046	1.2363	50′
20′	.3374	.3311	.9436	.3508	2.8502	1.060	3.021	1.2334	40′
30′	.3403	.3338	.9426	.3541	2.8239	1.061	2.996	1.2305	30′
40′	.3432	.3365	.9417	.3574	2.7980	1.062	2.971	1.2275	20′
50′	.3462	.3393	.9407	.3607	2.7725	1.063	2.947	1.2246	10′
20°00′	.3491	.3420	.9397	.3640	2.7475	1.064	2.924	1.2217	**70°00′**
10′	.3520	.3448	.9387	.3673	2.7228	1.065	2.901	1.2188	50′
20′	.3549	.3475	.9377	.3706	2.6985	1.066	2.878	1.2159	40′
30′	.3578	.3502	.9367	.3739	2.6746	1.068	2.855	1.2130	30′
40′	.3607	.3529	.9356	.3772	2.6511	1.069	2.833	1.2101	20′
50′	.3636	.3557	.9346	.3805	2.6279	1.070	2.812	1.2072	10′
21°00′	.3665	.3584	.9336	.3839	2.6051	1.071	2.790	1.2043	**69°00′**
10′	.3694	.3611	.9325	.3872	2.5826	1.072	2.769	1.2014	50′
20′	.3723	.3638	.9315	.3906	2.5605	1.074	2.749	1.1985	40′
30′	.3752	.3665	.9304	.3939	2.5386	1.075	2.729	1.1956	30′
40′	.3782	.3692	.9293	.3973	2.5172	1.076	2.709	1.1926	20′
50′	.3811	.3719	.9283	.4006	2.4960	1.077	2.689	1.1897	10′
22°00′	.3840	.3746	.9272	.4040	2.4751	1.079	2.669	1.1868	**68°00′**
10′	.3869	.3773	.9261	.4074	2.4545	1.080	2.650	1.1839	50′
20′	.3898	.3800	.9250	.4108	2.4342	1.081	2.632	1.1810	40′
30′	.3927	.3827	.9239	.4142	2.4142	1.082	2.613	1.1781	30′
40′	.3956	.3854	.9228	.4176	2.3945	1.084	2.595	1.1752	20′
50′	.3985	.3881	.9216	.4210	2.3750	1.085	2.577	1.1723	10′
23°00′	.4014	.3907	.9205	.4245	2.3559	1.086	2.559	1.1694	**67°00′**
10′	.4043	.3934	.9194	.4279	2.3369	1.088	2.542	1.1665	50′
20′	.4072	.3961	.9182	.4314	2.3183	1.089	2.525	1.1636	40′
30′	.4102	.3987	.9171	.4348	2.2998	1.090	2.508	1.1606	30′
40′	.4131	.4014	.9159	.4383	2.2817	1.092	2.491	1.1577	20′
50′	.4160	.4041	.9147	.4417	2.2637	1.093	2.475	1.1548	10′
24°00′	.4189	.4067	.9135	.4452	2.2460	1.095	2.459	1.1519	**66°00′**
10′	.4218	.4094	.9124	.4487	2.2286	1.096	2.443	1.1490	50′
20′	.4247	.4120	.9112	.4522	2.2113	1.097	2.427	1.1461	40′
30′	.4276	.4147	.9100	.4557	2.1943	1.099	2.411	1.1432	30′
40′	.4305	.4173	.9088	.4592	2.1775	1.100	2.396	1.1403	20′
50′	.4334	.4200	.9075	.4628	2.1609	1.102	2.381	1.1374	10′
25°00′	.4363	.4226	.9063	.4663	2.1445	1.103	2.366	1.1345	**65°00′**
10′	.4392	.4253	.9051	.4699	2.1283	1.105	2.352	1.1316	50′
20′	.4422	.4279	.9038	.4734	2.1123	1.106	2.337	1.1286	40′
30′	.4451	.4305	.9026	.4770	2.0965	1.108	2.323	1.1257	30′
40′	.4480	.4331	.9013	.4806	2.0809	1.109	2.309	1.1228	20′
50′	.4509	.4358	.9001	.4841	2.0655	1.111	2.295	1.1199	10′
26°00′	.4538	.4384	.8988	.4877	2.0503	1.113	2.281	1.1170	**64°00′**
10′	.4567	.4410	.8975	.4913	2.0353	1.114	2.268	1.1141	50′
20′	.4596	.4436	.8962	.4950	2.0204	1.116	2.254	1.1112	40′
30′	.4625	.4462	.8949	.4986	2.0057	1.117	2.241	1.1083	30′
40′	.4654	.4488	.8936	.5022	1.9912	1.119	2.228	1.1054	20′
50′	.4683	.4514	.8923	.5059	1.9768	1.121	2.215	1.1025	10′
27°00′	.4712	.4540	.8910	.5095	1.9626	1.122	2.203	1.0996	**63°00′**
		Cos θ	Sin θ	Cot θ	Tan θ	Csc θ	Sec θ	θ Rad.	θ Deg.

Tables **369**

Table of Values of the Trigonometric Functions

θ Deg.	θ Rad.	Sin θ	Cos θ	Tan θ	Cot θ	Sec θ	Csc θ		
27° 00′	.4712	.4540	.8910	.5095	1.9626	1.122	2.203	1.0996	63° 00′
10′	.4741	.4566	.8897	.5132	1.9486	1.124	2.190	1.0966	50′
20′	.4771	.4592	.8884	.5169	1.9347	1.126	2.178	1.0937	40′
30′	.4800	.4617	.8870	.5206	1.9210	1.127	2.166	1.0908	30′
40′	.4829	.4643	.8857	.5243	1.9074	1.129	2.154	1.0879	20′
50′	.4858	.4669	.8843	.5280	1.8940	1.131	2.142	1.0850	10′
28° 00′	.4887	.4695	.8829	.5317	1.8807	1.133	2.130	1.0821	62° 00′
10′	.4916	.4720	.8816	.5354	1.8676	1.134	2.118	1.0792	50′
20′	.4945	.4746	.8802	.5392	1.8546	1.136	2.107	1.0763	40′
30′	.4974	.4772	.8788	.5430	1.8418	1.138	2.096	1.0734	30′
40′	.5003	.4797	.8774	.5467	1.8291	1.140	2.085	1.0705	20′
50′	.5032	.4823	.8760	.5505	1.8165	1.142	2.074	1.0676	10′
29° 00′	.5061	.4848	.8746	.5543	1.8040	1.143	2.063	1.0647	61° 00′
10′	.5091	.4874	.8732	.5581	1.7917	1.145	2.052	1.0617	50′
20′	.5120	.4899	.8718	.5619	1.7796	1.147	2.041	1.0588	40′
30′	.5149	.4924	.8704	.5658	1.7675	1.149	2.031	1.0559	30′
40′	.5178	.4950	.8689	.5696	1.7556	1.151	2.020	1.0530	20′
50′	.5207	.4975	.8675	.5735	1.7437	1.153	2.010	1.0501	10′
30° 00′	.5236	.5000	.8660	.5774	1.7321	1.155	2.000	1.0472	60° 00′
10′	.5265	.5025	.8646	.5812	1.7205	1.157	1.990	1.0443	50′
20′	.5294	.5050	.8631	.5851	1.7090	1.159	1.980	1.0414	40′
30′	.5323	.5075	.8616	.5890	1.6977	1.161	1.970	1.0385	30′
40′	.5352	.5100	.8601	.5930	1.6864	1.163	1.961	1.0356	20′
50′	.5381	.5125	.8587	.5969	1.6753	1.165	1.951	1.0327	10′
31° 00′	.5411	.5150	.8572	.6009	1.6643	1.167	1.942	1.0297	59° 00′
10′	.5440	.5175	.8557	.6048	1.6534	1.169	1.932	1.0268	50′
20′	.5469	.5200	.8542	.6088	1.6426	1.171	1.923	1.0239	40′
30′	.5498	.5225	.8526	.6128	1.6319	1.173	1.914	1.0210	30′
40′	.5527	.5250	.8511	.6168	1.6212	1.175	1.905	1.0181	20′
50′	.5556	.5275	.8496	.6208	1.6107	1.177	1.896	1.0152	10′
32° 00′	.5585	.5299	.8480	.6249	1.6003	1.179	1.887	1.0123	58° 00′
10′	.5614	.5324	.8465	.6289	1.5900	1.181	1.878	1.0094	50′
20′	.5643	.5348	.8450	.6330	1.5798	1.184	1.870	1.0065	40′
30′	.5672	.5373	.8434	.6371	1.5697	1.186	1.861	1.0036	30′
40′	.5701	.5398	.8418	.6412	1.5597	1.188	1.853	1.0007	20′
50′	.5730	.5422	.8403	.6453	1.5497	1.190	1.844	.9977	10′
33° 00′	.5760	.5446	.8387	.6494	1.5399	1.192	1.836	.9948	57° 00′
10′	.5789	.5471	.8371	.6536	1.5301	1.195	1.828	.9919	50′
20′	.5818	.5495	.8355	.6577	1.5204	1.197	1.820	.9890	40′
30′	.5847	.5519	.8339	.6619	1.5108	1.199	1.812	.9861	30′
40′	.5876	.5544	.8323	.6661	1.5013	1.202	1.804	.9832	20′
50′	.5905	.5568	.8307	.6703	1.4919	1.204	1.796	.9803	10′
34° 00′	.5934	.5592	.8290	.6745	1.4826	1.206	1.788	.9774	56° 00′
10′	.5963	.5616	.8274	.6787	1.4733	1.209	1.781	.9745	50′
20′	.5992	.5640	.8258	.6830	1.4641	1.211	1.773	.9716	40′
30′	.6021	.5664	.8241	.6873	1.4550	1.213	1.766	.9687	30′
40′	.6050	.5688	.8225	.6916	1.4460	1.216	1.758	.9657	20′
50′	.6080	.5712	.8208	.6959	1.4370	1.218	1.751	.9628	10′
35° 00′	.6109	.5736	.8192	.7002	1.4281	1.221	1.743	.9599	55° 00′
10′	.6138	.5760	.8175	.7046	1.4193	1.223	1.736	.9570	50′
20′	.6167	.5783	.8158	.7089	1.4106	1.226	1.729	.9541	40′
30′	.6196	.5807	.8141	.7133	1.4019	1.228	1.722	.9512	30′
40′	.6225	.5831	.8124	.7177	1.3934	1.231	1.715	.9483	20′
50′	.6254	.5854	.8107	.7221	1.3848	1.233	1.708	.9454	10′
36° 00′	.6283	.5878	.8090	.7265	1.3764	1.236	1.701	.9425	54° 00′
		Cos θ	Sin θ	Cot θ	Tan θ	Csc θ	Sec θ	θ Rad.	θ Deg.

Table of Values of the Trigonometric Functions

θ Deg.	θ Rad.	Sin θ	Cos θ	Tan θ	Cot θ	Sec θ	Csc θ		
36° 00'	.6283	.5878	.8090	.7265	1.3764	1.236	1.701	.9425	54° 00'
10'	.6312	.5901	.8073	.7310	1.3680	1.239	1.695	.9396	50'
20'	.6341	.5925	.8056	.7355	1.3597	1.241	1.688	.9367	40'
30'	.6370	.5948	.8039	.7400	1.3514	1.244	1.681	.9338	30'
40'	.6400	.5972	.8021	.7445	1.3432	1.247	1.675	.9308	20'
50'	.6429	.5995	.8004	.7490	1.3351	1.249	1.668	.9279	10'
37° 00'	.6458	.6018	.7986	.7536	1.3270	1.252	1.662	.9250	53° 00'
10'	.6487	.6041	.7969	.7581	1.3190	1.255	1.655	.9221	50'
20'	.6516	.6065	.7951	.7627	1.3111	1.258	1.649	.9192	40'
30'	.6545	.6088	.7934	.7673	1.3032	1.260	1.643	.9163	30'
40'	.6574	.6111	.7916	.7720	1.2954	1.263	1.636	.9134	20'
50'	.6603	.6134	.7898	.7766	1.2876	1.266	1.630	.9105	10'
38° 00'	.6632	.6157	.7880	.7813	1.2799	1.269	1.624	.9076	52° 00'
10'	.6661	.6180	.7862	.7860	1.2723	1.272	1.618	.9047	50'
20'	.6690	.6202	.7844	.7907	1.2647	1.275	1.612	.9018	40'
30'	.6720	.6225	.7826	.7954	1.2572	1.278	1.606	.8988	30'
40'	.6749	.6248	.7808	.8002	1.2497	1.281	1.601	.8959	20'
50'	.6778	.6271	.7790	.8050	1.2423	1.284	1.595	.8930	10'
39° 00'	.6807	.6293	.7771	.8098	1.2349	1.287	1.589	.8901	51° 00'
10'	.6836	.6316	.7753	.8146	1.2276	1.290	1.583	.8872	50'
20'	.6865	.6338	.7735	.8195	1.2203	1.293	1.578	.8843	40'
30'	.6894	.6361	.7716	.8243	1.2131	1.296	1.572	.8814	30'
40'	.6923	.6383	.7698	.8292	1.2059	1.299	1.567	.8785	20'
50'	.6952	.6406	.7679	.8342	1.1988	1.302	1.561	.8756	10'
40° 00'	.6981	.6428	.7660	.8391	1.1918	1.305	1.556	.8727	50° 00'
10'	.7010	.6450	.7642	.8441	1.1847	1.309	1.550	.8698	50'
20'	.7039	.6472	.7623	.8491	1.1778	1.312	1.545	.8668	40'
30'	.7069	.6494	.7604	.8541	1.1708	1.315	1.540	.8639	30'
40'	.7098	.6517	.7585	.8591	1.1640	1.318	1.535	.8610	20'
50'	.7127	.6539	.7566	.8642	1.1571	1.322	1.529	.8581	10'
41° 00'	.7156	.6561	.7547	.8693	1.1504	1.325	1.524	.8552	49° 00'
10'	.7185	.6583	.7528	.8744	1.1436	1.328	1.519	.8523	50'
20'	.7214	.6604	.7509	.8796	1.1369	1.332	1.514	.8494	40'
30'	.7243	.6626	.7490	.8847	1.1303	1.335	1.509	.8465	30'
40'	.7272	.6648	.7470	.8899	1.1237	1.339	1.504	.8436	20'
50'	.7301	.6670	.7451	.8952	1.1171	1.342	1.499	.8407	10'
42° 00'	.7330	.6691	.7431	.9004	1.1106	1.346	1.494	.8378	48° 00'
10'	.7359	.6713	.7412	.9057	1.1041	1.349	1.490	.8348	50'
20'	.7389	.6734	.7392	.9110	1.0977	1.353	1.485	.8319	40'
30'	.7418	.6756	.7373	.9163	1.0913	1.356	1.480	.8290	30'
40'	.7447	.6777	.7353	.9217	1.0850	1.360	1.476	.8261	20'
50'	.7476	.6799	.7333	.9271	1.0786	1.364	1.471	.8232	10'
43° 00'	.7505	.6820	.7314	.9325	1.0724	1.367	1.466	.8203	47° 00'
10'	.7534	.6841	.7294	.9380	1.0661	1.371	1.462	.8174	50'
20'	.7563	.6862	.7274	.9435	1.0599	1.375	1.457	.8145	40'
30'	.7592	.6884	.7254	.9490	1.0538	1.379	1.453	.8116	30'
40'	.7621	.6905	.7234	.9545	1.0477	1.382	1.448	.8087	20'
50'	.7650	.6926	.7214	.9601	1.0416	1.386	1.444	.8058	10'
44° 00'	.7679	.6947	.7193	.9657	1.0355	1.390	1.440	.8029	46° 00'
10'	.7709	.6967	.7173	.9713	1.0295	1.394	1.435	.7999	50'
20'	.7738	.6988	.7153	.9770	1.0235	1.398	1.431	.7970	40'
30'	.7767	.7009	.7133	.9827	1.0176	1.402	1.427	.7941	30'
40'	.7796	.7030	.7112	.9884	1.0117	1.406	1.423	.7912	20'
50'	.7825	.7050	.7092	.9942	1.0058	1.410	1.418	.7883	10'
45° 00'	.7854	.7071	.7071	1.0000	1.0000	1.414	1.414	.7854	45° 00'
		Cos θ	Sin θ	Cot θ	Tan θ	Csc θ	Sec θ	θ Rad.	θ Deg.

Table of Values of the Logarithms of Trigonometric Functions

θ Deg.	θ Rad.	Log Sin θ	Log Cos θ	Log Tan θ	Log Cot θ		
0° 00′	.0000	—	10.0000	—	12.5363	1.5708	90° 00′
10′	.0029	7.4637	10.0000	7.4637	12.2352	1.5679	50′
20′	.0058	7.7648	10.0000	7.7648	12.0591	1.5650	40′
30′	.0087	7.9408	10.0000	7.9409	12.0591	1.5621	30′
40′	.0116	8.0658	10.0000	8.0658	11.9342	1.5592	20′
50′	.0145	8.1627	10.0000	8.1627	11.8373	1.5563	10′
1° 00′	.0175	8.2419	9.9999	8.2419	11.7581	1.5533	89° 00′
10′	.0204	8.3088	9.9999	8.3089	11.6911	1.5504	50′
20′	.0233	8.3668	9.9999	8.3669	11.6331	1.5475	40′
30′	.0262	8.4179	9.9999	8.4181	11.5819	1.5446	30′
40′	.0291	8.4637	9.9998	8.4638	11.5362	1.5417	20′
50′	.0320	8.5050	9.9998	8.5053	11.4947	1.5388	10′
2° 00′	.0349	8.5428	9.9997	8.5431	11.4569	1.5359	88° 00′
10′	.0378	8.5776	9.9997	8.5779	11.4221	1.5330	50′
20′	.0407	8.6097	9.9996	8.6101	11.3899	1.5301	40′
30′	.0436	8.6397	9.9996	8.6401	11.3599	1.5272	30′
40′	.0465	8.6677	9.9995	8.6682	11.3318	1.5243	20′
50′	.0495	8.6940	9.9995	8.6945	11.3055	1.5213	10′
3° 00′	.0524	8.7188	9.9994	8.7194	11.2806	1.5184	87° 00′
10′	.0553	8.7423	9.9993	8.7429	11.2571	1.5155	50′
20′	.0582	8.7645	9.9993	8.7652	11.2348	1.5126	40′
30′	.0611	8.7857	9.9992	8.7865	11.2135	1.5097	30′
40′	.0640	8.8059	9.9991	8.8067	11.1933	1.5068	20′
50′	.0669	8.8251	9.9990	8.8261	11.1739	1.5039	10′
4° 00′	.0698	8.8436	9.9989	8.8446	11.1554	1.5010	86° 00′
10′	.0727	8.8613	9.9989	8.8624	11.1376	1.4981	50′
20′	.0756	8.8783	9.9988	8.8795	11.1205	1.4952	40′
30′	.0785	8.8946	9.9987	8.8960	11.1040	1.4923	30′
40′	.0814	8.9104	9.9986	8.9118	11.0882	1.4893	20′
50′	.0844	8.9256	9.9985	8.9272	11.0728	1.4864	10′
5° 00′	.0873	8.9403	9.9983	8.9420	11.0580	1.4835	85° 00′
10′	.0902	8.9545	9.9982	8.9563	11.0437	1.4806	50′
20′	.0931	8.9682	9.9981	8.9701	11.0299	1.4777	40′
30′	.0960	8.9816	9.9980	8.9836	11.0164	1.4748	30′
40′	.0989	8.9945	9.9979	8.9966	11.0034	1.4719	20′
50′	.1018	9.0070	9.9977	9.0093	10.9907	1.4690	10′
6° 00′	.1047	9.0192	9.9976	9.0216	10.9784	1.4661	84° 00′
10′	.1076	9.0311	9.9975	9.0336	10.9664	1.4632	50′
20′	.1105	9.0426	9.9973	9.0453	10.9547	1.4603	40′
30′	.1134	9.0539	9.9972	9.0567	10.9433	1.4573	30′
40′	.1164	9.0648	9.9971	9.0678	10.9322	1.4544	20′
50′	.1193	9.0755	9.9969	9.0786	10.9214	1.4515	10′
7° 00′	.1222	9.0859	9.9968	9.0891	10.9109	1.4486	83° 00′
10′	.1251	9.0961	9.9966	9.0995	10.9005	1.4457	50′
20′	.1280	9.1060	9.9964	9.1096	10.8904	1.4428	40′
30′	.1309	9.1157	9.9963	9.1194	10.8806	1.4399	30′
40′	.1338	9.1252	9.9961	9.1291	10.8709	1.4370	20′
50′	.1367	9.1345	9.9959	9.1385	10.8615	1.4341	10′
8° 00′	.1396	9.1436	9.9958	9.1478	10.8522	1.4312	82° 00′
10′	.1425	9.1525	9.9956	9.1569	10.8431	1.4283	50′
20′	.1454	9.1612	9.9954	9.1658	10.8342	1.4254	40′
30′	.1484	9.1697	9.9952	9.1745	10.8255	1.4224	30′
40′	.1513	9.1781	9.9950	9.1831	10.8169	1.4195	20′
50′	.1542	9.1863	9.9948	9.1915	10.8085	1.4166	10′
9° 00′	.1571	9.1943	9.9946	9.1997	10.8003	1.4137	81° 00′
		Log Cos θ	Log Sin θ	Log Cot θ	Log Tan θ	θ Rad.	θ Deg.

The tables give the logarithms increased by 10. In each case, 10 should be subtracted.

Table of Values of the Logarithms of Trigonometric Functions

θ Deg.	θ Rad.	Log Sin θ	Log Cos θ	Log Tan θ	Log Cot θ		
9° 00'	.1571	9.1943	9.9946	9.1997	10.8003	1.4137	81° 00'
10'	.1600	9.2022	9.9944	9.2078	10.7922	1.4108	50'
20'	.1629	9.2100	9.9942	9.2158	10.7842	1.4079	40'
30'	.1658	9.2176	9.9940	9.2236	10.7764	1.4050	30'
40'	.1687	9.2251	9.9938	9.2313	10.7687	1.4021	20'
50'	.1716	9.2324	9.9936	9.2389	10.7611	1.3992	10'
10° 00'	.1745	9.2397	9.9934	9.2463	10.7537	1.3963	80° 00'
10'	.1774	9.2468	9.9931	9.2536	10.7464	1.3934	50'
20'	.1804	9.2538	9.9929	9.2609	10.7391	1.3904	40'
30'	.1833	9.2606	9.9927	9.2680	10.7320	1.3875	30'
40'	.1862	9.2674	9.9924	9.2750	10.7250	1.3846	20'
50'	.1891	9.2740	9.9922	9.2819	10.7181	1.3817	10'
11° 00'	.1920	9.2806	9.9919	9.2887	10.7113	1.3788	79° 00'
10'	.1949	9.2870	9.9917	9.2953	10.7047	1.3759	50'
20'	.1978	9.2934	9.9914	9.3020	10.6980	1.3730	40'
30'	.2007	9.2997	9.9912	9.3085	10.6915	1.3701	30'
40'	.2036	9.3058	9.9909	9.3149	10.6851	1.3672	20'
50'	.2065	9.3119	9.9907	9.3212	10.6788	1.3643	10'
12° 00'	.2094	9.3179	9.9904	9.3275	10.6725	1.3614	78° 00'
10'	.2123	9.3238	9.9901	9.3336	10.6664	1.3584	50'
20'	.2153	9.3296	9.9899	9.3397	10.6603	1.3555	40'
30'	.2182	9.3353	9.9896	9.3458	10.6542	1.3526	30'
40'	.2211	9.3410	9.9893	9.3517	10.6483	1.3497	20'
50'	.2240	9.3466	9.9890	9.3576	10.6424	1.3468	10'
13° 00'	.2269	9.3521	9.9887	9.3634	10.6366	1.3439	77° 00'
10'	.2298	9.3575	9.9884	9.3691	10.6309	1.3410	50'
20'	.2327	9.3629	9.9881	9.3748	10.6252	1.3381	40'
30'	.2356	9.3682	9.9878	9.3804	10.6196	1.3352	30'
40'	.2385	9.3734	9.9875	9.3859	10.6141	1.3323	20'
50'	.2414	9.3786	9.9872	9.3914	10.6086	1.3294	10'
14° 00'	.2443	9.3837	9.9869	9.3968	10.6032	1.3265	76° 00'
10'	.2473	9.3887	9.9866	9.4021	10.5979	1.3235	50'
20'	.2502	9.3937	9.9863	9.4074	10.5926	1.3206	40'
30'	.2531	9.3986	9.9859	9.4127	10.5873	1.3177	30'
40'	.2560	9.4035	9.9856	9.4178	10.5822	1.3148	20'
50'	.2589	9.4083	9.9853	9.4230	10.5770	1.3119	10'
15° 00'	.2618	9.4160	9.9849	9.4281	10.5719	1.3090	75° 00'
10'	.2647	9.4177	9.9846	9.4331	10.5669	1.3061	50'
20'	.2676	9.4223	9.9843	9.4381	10.5619	1.3032	40'
30'	.2705	9.4269	9.9839	9.4430	10.5570	1.3003	30'
40'	.2734	9.4314	9.9836	9.4479	10.5521	1.2974	20'
50'	.2763	9.4359	9.9832	9.4527	10.5473	1.2945	10'
16° 00'	.2793	9.4403	9.9828	9.4575	10.5425	1.2915	74° 00'
10'	.2822	9.4447	9.9825	9.4622	10.5378	1.2886	50'
20'	.2851	9.4491	9.9821	9.4669	10.5331	1.2857	40'
30'	.2880	9.4533	9.9817	9.4716	10.5284	1.2828	30'
40'	.2909	9.4576	9.9814	9.4762	10.5238	1.2799	20'
50'	.2938	9.4618	9.9810	9.4808	10.5192	1.2770	10'
17° 00'	.2967	9.4659	9.9806	9.4853	10.5147	1.2741	73° 00'
10'	.2996	9.4700	9.9802	9.4898	10.5102	1.2712	50'
20'	.3025	9.4741	9.9798	9.4943	10.5057	1.2683	40'
30'	.3054	9.4781	9.9794	9.4987	10.5013	1.2654	30'
40'	.3083	9.4821	9.9790	9.5031	10.4969	1.2625	20'
50'	.3113	9.4861	9.9786	9.5075	10.4925	1.2595	10'
18° 00'	.3142	9.4900	9.9782	9.5118	10.4882	1.2566	72° 00'
		Log Cos θ	Log Sin θ	Log Cot θ	Log Tan θ	θ Rad.	θ Deg.

Table of Values of the Logarithms of Trigonometric Functions

θ Deg.	θ Rad.	Log Sin θ	Log Cos θ	Log Tan θ	Log Cot θ		
18° 00'	.3142	9.4900	9.9782	9.5118	10.4882	1.2566	72° 00'
10'	.3171	9.4939	9.9778	9.5161	10.4839	1.2537	50'
20'	.3200	9.4977	9.9774	9.5203	10.4797	1.2508	40'
30'	.3229	9.5015	9.9770	9.5245	10.4755	1.2479	30'
40'	.3258	9.5052	9.9765	9.5287	10.4713	1.2450	20'
50'	.3287	9.5090	9.9761	9.5329	10.4671	1.2421	10'
19° 00'	.3316	9.5126	9.9757	9.5370	10.4630	1.2392	71° 00'
10'	.3345	9.5163	9.9752	9.5411	10.4589	1.2363	50'
20'	.3374	9.5199	9.9748	9.5451	10.4549	1.2334	40'
30'	.3403	9.5235	9.9743	9.5491	10.4509	1.2305	30'
40'	.3432	9.5270	9.9739	9.5531	10.4469	1.2275	20'
50'	.3462	9.5306	9.9734	9.5571	10.4429	1.2246	10'
20° 00'	.3491	9.5341	9.9730	9.5611	10.4389	1.2217	70° 00'
10'	.3520	9.5375	9.9725	9.5650	10.4350	1.2188	50'
20'	.3549	9.5409	9.9721	9.5689	10.4311	1.2159	40'
30'	.3578	9.5443	9.9716	9.5727	10.4273	1.2130	30'
40'	.3607	9.5477	9.9711	9.5766	10.4234	1.2101	20'
50'	.3636	9.5510	9.9706	9.5804	10.4196	1.2072	10'
21° 00'	.3665	9.5543	9.9702	9.5842	10.4158	1.2043	69° 00'
10'	.3694	9.5576	9.9697	9.5879	10.4121	1.2014	50'
20'	.3723	9.5609	9.9692	9.5917	10.4083	1.1985	40'
30'	.3752	9.5641	9.9687	9.5954	10.4046	1.1956	30'
40'	.3782	9.5673	9.9682	9.5991	10.4009	1.1926	20'
50'	.3811	9.5704	9.9677	9.6028	10.3972	1.1897	10'
22° 00'	.3840	9.5736	9.9672	9.6064	10.3936	1.1868	68° 00'
10'	.3869	9.5767	9.9667	9.6100	10.3900	1.1839	50'
20'	.3898	9.5798	9.9661	9.6136	10.3864	1.1810	40'
30'	.3927	9.5828	9.9656	9.6172	10.3828	1.1781	30'
40'	.3956	9.5859	9.9651	9.6208	10.3792	1.1752	20'
50'	.3985	9.5889	9.9646	9.6243	10.3757	1.1723	10'
23° 00'	.4014	9.5919	9.9640	9.6279	10.3721	1.1694	67° 00'
10'	.4043	9.5948	9.9635	9.6314	10.3686	1.1665	50'
20'	.4072	9.5978	9.9629	9.6348	10.3652	1.1636	40'
30'	.4102	9.6007	9.9624	9.6383	10.3617	1.1606	30'
40'	.4131	9.6036	9.9618	9.6417	10.3583	1.1577	20'
50'	.4160	9.6065	9.9613	9.6452	10.3548	1.1548	10'
24° 00'	.4189	9.6093	9.9607	9.6486	10.3514	1.1519	66° 00'
10'	.4218	9.6121	9.9602	9.6520	10.3480	1.1490	50'
20'	.4247	9.6149	9.9596	9.6553	10.3447	1.1461	40'
30'	.4276	9.6177	9.9590	9.6587	10.3413	1.1432	30'
40'	.4305	9.6205	9.9584	9.6620	10.3380	1.1403	20'
50'	.4334	9.6232	9.9579	9.6654	10.3346	1.1374	10'
25° 00'	.4363	9.6259	9.9573	9.6687	10.3313	1.1345	65° 00'
10'	.4392	9.6286	9.9567	9.6720	10.3280	1.1316	50'
20'	.4422	9.6313	9.9561	9.6752	10.3248	1.1286	40'
30'	.4451	9.6340	9.9555	9.6785	10.3215	1.1257	30'
40'	.4480	9.6366	9.9549	9.6817	10.3183	1.1228	20'
50'	.4509	9.6392	9.9543	9.6850	10.3150	1.1199	10'
26° 00'	.4538	9.6418	9.9537	9.6882	10.3118	1.1170	64° 00'
10'	.4567	9.6444	9.9530	9.6914	10.3086	1.1141	50'
20'	.4596	9.6470	9.9524	9.6946	10.3054	1.1112	40'
30'	.4625	9.6495	9.9518	9.6977	10.3023	1.1083	30'
40'	.4654	9.6521	9.9512	9.7009	10.2991	1.1054	20'
50'	.4683	9.6546	9.9505	9.7040	10.2960	1.1025	10'
27° 00'	.4712	9.6570	9.9499	9.7072	10.2928	1.0996	63° 00'
		Log Cos θ	Log Sin θ	Log Cot θ	Log Tan θ	θ Rad.	θ Deg.

Table of Values of the Logarithms of Trigonometric Functions

θ Deg.	θ Rad.	Log Sin θ	Log Cos θ	Log Tan θ	Log Cot θ		
27°00'	.4712	9.6570	9.9499	9.7072	10.2928	1.0996	63°00'
10'	.4741	9.6595	9.9492	9.7103	10.2897	1.0966	50'
20'	.4771	9.6620	9.9486	9.7134	10.2866	1.0937	40'
30'	.4800	9.6644	9.9479	9.7165	10.2835	1.0908	30'
40'	.4829	9.6668	9.9473	9.7196	10.2804	1.0879	20'
50'	.4858	9.6692	9.9466	9.7226	10.2774	1.0850	10'
28°00'	.4887	9.6716	9.9459	9.7257	10.2743	1.0821	62°00'
10'	.4916	9.6740	9.9453	9.7287	10.2713	1.0792	50'
20'	.4945	9.6763	9.9446	9.7317	10.2683	1.0763	40'
30'	.4974	9.6787	9.9439	9.7348	10.2652	1.0734	30'
40'	.5003	9.6810	9.9432	9.7378	10.2622	1.0705	20'
50'	.5032	9.6833	9.9425	9.7408	10.2592	1.0676	10'
29°00'	.5061	9.6856	9.9418	9.7438	10.2562	1.0647	61°00'
10'	.5091	9.6878	9.9411	9.7467	10.2533	1.0617	50'
20'	.5120	9.6901	9.9404	9.7497	10.2503	1.0588	40'
30'	.5149	9.6923	9.9397	9.7526	10.2474	1.0559	30'
40'	.5178	9.6946	9.9390	9.7556	10.2444	1.0530	20'
50'	.5207	9.6968	9.9383	9.7585	10.2415	1.0501	10'
30°00'	.5236	9.6990	9.9375	9.7614	10.2386	1.0472	60°00'
10'	.5265	9.7012	9.9368	9.7644	10.2356	1.0443	50'
20'	.5294	9.7033	9.9361	9.7673	10.2327	1.0414	40'
30'	.5323	9.7055	9.9353	9.7701	10.2299	1.0385	30'
40'	.5352	9.7076	9.9346	9.7730	10.2270	1.0356	20'
50'	.5381	9.7097	9.9338	9.7759	10.2241	1.0327	10'
31°00'	.5411	9.7118	9.9331	9.7788	10.2212	1.0297	59°00'
10'	.5440	9.7139	9.9323	9.7816	10.2184	1.0268	50'
20'	.5469	9.7160	9.9315	9.7845	10.2155	1.0239	40'
30'	.5498	9.7181	9.9308	9.7873	10.2127	1.0210	30'
40'	.5527	9.7201	9.9300	9.7902	10.2098	1.0181	20'
50'	.5556	9.7222	9.9292	9.7930	10.2070	1.0152	10'
32°00'	.5585	9.7242	9.9284	9.7958	10.2042	1.0123	58°00'
10'	.5614	9.7262	9.9276	9.7986	10.2014	1.0094	50'
20'	.5643	9.7282	9.9268	9.8014	10.1986	1.0065	40'
30'	.5672	9.7302	9.9260	9.8042	10.1958	1.0036	30'
40'	.5701	9.7322	9.9252	9.8070	10.1930	1.0007	20'
50'	.5730	9.7342	9.9244	9.8097	10.1903	.9977	10'
33°00'	.5760	9.7361	9.9236	9.8125	10.1875	.9948	57°00'
10'	.5789	9.7380	9.9228	9.8153	10.1847	.9919	50'
20'	.5818	9.7400	9.9219	9.8180	10.1820	.9890	40'
30'	.5847	9.7419	9.9211	9.8208	10.1792	.9861	30'
40'	.5876	9.7438	9.9203	9.8235	10.1765	.9832	20'
50'	.5905	9.7457	9.9194	9.8263	10.1737	.9803	10'
34°00'	.5934	9.7476	9.9186	9.8290	10.1710	.9774	56°00'
10'	.5963	9.7494	9.9177	9.8317	10.1683	.9745	50'
20'	.5992	9.7513	9.9169	9.8344	10.1656	.9716	40'
30'	.6021	9.7531	9.9160	9.8371	10.1629	.9687	30'
40'	.6050	9.7550	9.9151	9.8398	10.1602	.9657	20'
50'	.6080	9.7568	9.9142	9.8425	10.1575	.9628	10'
35°00'	.6109	9.7586	9.9134	9.8452	10.1548	.9599	55°00'
10'	.6138	9.7604	9.9125	9.8479	10.1521	.9570	50'
20'	.6167	9.7622	9.9116	9.8506	10.1494	.9541	40'
30'	.6196	9.7640	9.9107	9.8533	10.1467	.9512	30'
40'	.6225	9.7657	9.9098	9.8559	10.1441	.9483	20'
50'	.6254	9.7675	9.9089	9.8586	10.1414	.9454	10'
36°00'	.6283	9.7692	9.9080	9.8613	10.1387	.9425	54°00'
		Log Cos θ	Log Sin θ	Log Cot θ	Log Tan θ	θ Rad.	θ Deg.

Table of Values of the Logarithms of Trigonometric Functions

θ Deg.	θ Rad.	Log Sin θ	Log Cos θ	Log Tan θ	Log Cot θ		
36° 00′	.6283	9.7692	9.9080	9.8613	10.1387	.9425	54° 00′
10′	.6312	9.7710	9.9070	9.8639	10.1361	.9396	50′
20′	.6341	9.7727	9.9061	9.8666	10.1334	.9367	40′
30′	.6370	9.7744	9.9052	9.8692	10.1308	.9338	30′
40′	.6400	9.7761	9.9042	9.8718	10.1282	.9308	20′
50′	.6429	9.7778	9.9033	9.8745	10.1255	.9279	10′
37° 00′	.6458	9.7795	9.9023	9.8771	10.1229	.9250	53° 00′
10′	.6487	9.7811	9.9014	9.8797	10.1203	.9221	50′
20′	.6516	9.7828	9.9004	9.8824	10.1176	.9192	40′
30′	.6545	9.7844	9.8995	9.8850	10.1150	.9163	30′
40′	.6574	9.7861	9.8985	9.8876	10.1124	.9134	20′
50′	.6603	9.7877	9.8975	9.8902	10.1098	.9105	10′
38° 00′	.6632	9.7893	9.8965	9.8928	10.1072	.9076	52° 00′
10′	.6661	9.7910	9.8955	9.8954	10.1046	.9047	50′
20′	.6690	9.7926	9.8945	9.8980	10.1020	.9018	40′
30′	.6720	9.7941	9.8935	9.9006	10.0994	.8988	30′
40′	.6749	9.7957	9.8925	9.9032	10.0968	.8959	20′
50′	.6778	9.7973	9.8915	9.9058	10.0942	.8930	10′
39° 00′	.6807	9.7989	9.8905	9.9084	10.0916	.8901	51° 00′
10′	.6836	9.8004	9.8895	9.9110	10.0890	.8872	50′
20′	.6865	9.8020	9.8884	9.9135	10.0865	.8843	40′
30′	.6894	9.8035	9.8874	9.9161	10.0839	.8814	30′
40′	.6923	9.8050	9.8864	9.9187	10.0813	.8785	20′
50′	.6952	9.8066	9.8853	9.9212	10.0788	.8756	10′
40′ 00′	.6981	9.8081	9.8843	9.9238	10.0762	.8727	50° 00′
10′	.7010	9.8096	9.8832	9.9264	10.0736	.8698	50′
20′	.7039	9.8111	9.8821	9.9289	10.0711	.8668	40′
30′	.7069	9.8125	9.8810	9.9315	10.0685	.8639	30′
40′	.7098	9.8140	9.8800	9.9341	10.0659	.8610	20′
50′	.7127	9.8155	9.8789	9.9366	10.0634	.8581	10′
41° 00′	.7156	9.8169	9.8778	9.9392	10.0608	.8552	49° 00′
10′	.7185	9.8184	9.8767	9.9417	10.0583	.8523	50′
20′	.7214	9.8198	9.8756	9.9443	10.0557	.8494	40′
30′	.7243	9.8213	9.8745	9.9468	10.0532	.8465	30′
40′	.7272	9.8227	9.8733	9.9494	10.0506	.8436	20′
50′	.7301	9.8241	9.8722	9.9519	10.0481	.8407	10′
42° 00′	.7330	9.8255	9.8711	9.9544	10.0456	.8378	48° 00′
10′	.7359	9.8269	9.8699	9.9570	10.0430	.8348	50′
20′	.7389	9.8283	9.8688	9.9595	10.0405	.8319	40′
30′	.7418	9.8297	9.8676	9.9621	10.0379	.8290	30′
40′	.7447	9.8311	9.8665	9.9646	10.0354	.8261	20′
50′	.7476	9.8324	9.8653	9.9671	10.0329	.8232	10′
43° 00′	.7505	9.8338	9.8641	9.9697	10.0303	.8203	47° 00′
10′	.7534	9.8351	9.8629	9.9722	10.0278	.8174	50′
20′	.7563	9.8365	9.8618	9.9747	10.0253	.8145	40′
30′	.7592	9.8378	9.8606	9.9772	10.0228	.8116	30′
40′	.7621	9.8391	9.8594	9.9798	10.0202	.8087	20′
50′	.7650	9.8405	9.8582	9.9823	10.0177	.8058	10′
44° 00′	.7679	9.8418	9.8569	9.9848	10.0152	.8029	46° 00′
10′	.7709	9.8431	9.8557	9.9874	10.0126	.7999	50′
20′	.7738	9.8444	9.8545	9.9899	10.0101	.7970	40′
30′	.7767	9.8457	9.8532	9.9924	10.0076	.7941	30′
40′	.7796	9.8469	9.8520	9.9949	10.0051	.7912	20′
50′	.7825	9.8482	9.8507	9.9975	10.0025	.7883	10′
45° 00′	.7854	9.8495	9.8495	10.0000	10.0000	.7854	45° 00′
		Log Cos θ	Log Sin θ	Log Cot θ	Log Tan θ	θ Rad.	θ Deg.

Table of Squares, Cubes, Square and Cube Roots

No.	Squares	Cubes	Square Roots	Cube Roots	No.	Squares	Cubes	Square Roots	Cube Roots
1	1	1	1.000	1.000	51	2,601	132,651	7.141	3.708
2	4	8	1.414	1.260	52	2,704	140,608	7.211	3.733
3	9	27	1.732	1.442	53	2,809	148,877	7.280	3.756
4	16	64	2.000	1.587	54	2,916	157,464	7.348	3.780
5	25	125	2.236	1.710	55	3,025	166,375	7.416	3.803
6	36	216	2.449	1.817	56	3,136	175,616	7.483	3.826
7	49	343	2.646	1.913	57	3,249	185,193	7.550	3.849
8	64	512	2.828	2.000	58	3,364	195,112	7.616	3.871
9	81	729	3.000	2.080	59	3,481	205,379	7.681	3.893
10	100	1,000	3.162	2.154	60	3,600	216,000	7.746	3.915
11	121	1,331	3.317	2.224	61	3,721	226,981	7.810	3.936
12	144	1,728	3.464	2.289	62	3,844	238,328	7.874	3.958
13	169	2,197	3.606	2.351	63	3,969	250,047	7.937	3.979
14	196	2,744	3.742	2.410	64	4,096	262,144	8.000	4.000
15	225	3,375	3.873	2.466	65	4,225	274,625	8.062	4.021
16	256	4,096	4.000	2.520	66	4,356	287,496	8.124	4.041
17	289	4,913	4.123	2.571	67	4,489	300,763	8.185	4.062
18	324	5,832	4.243	2.621	68	4,624	314,432	8.246	4.082
19	361	6,859	4.359	2.668	69	4,761	328,509	8.307	4.102
20	400	8,000	4.472	2.714	70	4,900	343,000	8.367	4.121
21	441	9,261	4.583	2.759	71	5,041	357,911	8.426	4.141
22	484	10,648	4.690	2.802	72	5,184	373,248	8.485	4.160
23	529	12,167	4.796	2.844	73	5,329	389,017	8.544	4.179
24	576	13,824	4.899	2.884	74	5,476	405,224	8.602	4.198
25	625	15,625	5.000	2.924	75	5,625	421,875	8.660	4.217
26	676	17,576	5.099	2.962	76	5,776	438,976	8.718	4.236
27	729	19,683	5.196	3.000	77	5,929	456,533	8.775	4.254
28	784	21,952	5.292	3.037	78	6,084	474,552	8.832	4.273
29	841	24,389	5.385	3.072	79	6,241	493,039	8.888	4.291
30	900	27,000	5.477	3.107	80	6,400	512,000	8.944	4.309
31	961	29,791	5.568	3.141	81	6,561	531,441	9.000	4.327
32	1,024	32,768	5.657	3.175	82	6,724	551,368	9.055	4.344
33	1,089	35,937	5.745	3.208	83	6,889	571,787	9.110	4.362
34	1,156	39,304	5.831	3.240	84	7,056	592,704	9.165	4.380
35	1,225	42,875	5.916	3.271	85	7,225	614,125	9.220	4.397
36	1,296	46,656	6.000	3.302	86	7,396	636,056	9.274	4.414
37	1,369	50,653	6.083	3.332	87	7,569	658,503	9.327	4.431
38	1,444	54,872	6.164	3.362	88	7,744	681,472	9.381	4.448
39	1,521	59,319	6.245	3.391	89	7,921	704,969	9.434	4.465
40	1,600	64,000	6.325	3.420	90	8,100	729,000	9.487	4.481
41	1,681	68,921	6.403	3.448	91	8,281	753,571	9.539	4.498
42	1,764	74,088	6.481	3.476	92	8,464	778,688	9.592	4.514
43	1,849	79,507	6.557	3.503	93	8,649	804,357	9.644	4.531
44	1,936	85,184	6.633	3.530	94	8,836	830,584	9.695	4.547
45	2,025	91,125	6.708	3.557	95	9,025	857,375	9.747	4.563
46	2,116	97,336	6.782	3.583	96	9,216	884,736	9.798	4.579
47	2,209	103,823	6.856	3.609	97	9,409	912,673	9.849	4.595
48	2,304	110,592	6.928	3.634	98	9,604	941,192	9.899	4.610
49	2,401	117,649	7.000	3.659	99	9,801	970,299	9.950	4.626
50	2,500	125,000	7.071	3.684	100	10,000	1,000,000	10.000	4.642

Glossary

The following definitions and statements reflect the usage of terms in this textbook.

Amplitude When M and m are the maximum and minimum values, respectively, of a periodic function, then the amplitude of the function is $\frac{1}{2}(M - m)$. (Page 88)

Angle An angle is a rotation of a ray about its endpoint from an initial position to a terminal position. (Page 8)

Angular Velocity The angular displacement per unit of time of an object in rotary motion is the angular velocity. (Page 70)

Arithmetic Means The terms between any two given terms of an arithmetic sequence are called arithmetic means. (Page 310)

Arithmetic Sequence If each term of a sequence is obtained by adding a constant d (called the common difference) to the preceding term, the sequence is called an arithmetic sequence. (Page 307)

Asymptotes Lines that a curve nears but does not intersect are called asymptotes. (Page 109)

Circular Functions Given the wrapping function, $W(t) = (x, y)$, the cosine function, $\cos t$, maps t onto x, and the sine function, $\sin t$, maps t onto y. In symbols, $\cos t = x$ and $\sin t = y$. These functions are called circular functions. (Page 252)

Common Logarithms Logarithms with base 10 are common logarithms. (Page 357)

Complex Number A complex number is a number of the form $a + bi$, where a and b are real numbers and $i = \sqrt{-1}$. (Page 276)

Complex Roots Theorem If n is a positive integer and $z \neq 0$, then $z^n = r(\cos \theta + i \sin \theta)$ has n roots which are given by
$$z = \sqrt[n]{r}\left[\cos\left(\frac{\theta}{n} + \frac{k \cdot 360°}{n}\right) + i \sin\left(\frac{\theta}{n} + \frac{k \cdot 360°}{n}\right)\right],$$
where $k = 0, 1, 2, \ldots, n - 1$. (Page 296)

Conditional Equation A conditional equation is an equation that is true for some, but not all, permissible replacements of the variable. (Page 226)

Cosecant Ratio Let $P(x, y)$ be any point (not the origin) on the terminal side of an angle θ in standard position, and let r be the radius vector. Then, cosecant θ ($\csc \theta$) $= \frac{r}{y}$, $y \neq 0$. (Page 17)

Cosine Ratio Let $P(x, y)$ be any point (not the origin) on the terminal side of an angle in standard position with measure θ and let r be the radius vector. Then, cosine θ ($\cos \theta$) $= \frac{x}{r}$. (Page 13)

Cotangent Ratio Let $P(x, y)$ be any point (not the origin) on the terminal side of an angle θ in standard position, and let r be the radius vector. Then, cotangent θ ($\cot \theta$) $= \frac{x}{y}$, $y \neq 0$. (Page 17)

Coterminal Angles Angles in standard position whose terminal sides coincide are called coterminal angles. (Page 10)

Cycle One complete execution of an event that is repeated over and over again is called a cycle. (Page 257)

De Moivre's Theorem If $z = r(\cos \theta + i \sin \theta)$, then $z^n = r^n (\cos n\theta + i \sin n\theta)$, where n is a positive integer. (Page 292)

Distance Formula The distance between two points $P_1(x_1, y_1)$ and $P_2(x_2, y_2)$ is
$$P_1P_2 = \sqrt{(x_2 - x_1)^2 + (y_2 - y_1)^2}. \text{ (Page 3)}$$

Divergent Sequence A sequence that does not converge to a limit is called a divergent sequence. (Page 313)

Domain In a relation or a function, the set of first elements is the domain. (Page 3)

Equivalent Vectors Vectors with the same magnitude and direction are called <u>equivalent vectors</u>. (Page 188)

Even Function Any function having the property that for each x in the domain, $-x$ is in the domain and $f(-x) = f(x)$, is called an <u>even function</u>. (Page 75)

Frequency The reciprocal of the period of a periodic function is the <u>frequency</u> of the function. (Page 97)

Function A <u>function</u> is a relation such that no two of its ordered pairs have the same first element. (Page 3)

Geometric Means The terms between any two given terms of a geometric sequence are called <u>geometric means</u>. (Page 310)

Geometric Sequence If each term of a sequence is obtained by multiplying the preceding term by some fixed number r (called the common ratio), the sequence is called a <u>geometric sequence</u>. (Page 307)

Heron's Formula In any triangle ABC with sides of lengths a, b, and c, the area K is given by the formula $K = \sqrt{s(s-a)(s-b)(s-c)}$, where $s = \dfrac{a+b+c}{2}$. (Page 185)

Inverse Relation Every relation has an <u>inverse relation</u> which is formed by interchanging the elements of each ordered pair in the relation. (Page 210)

Law of Cosines In any triangle, the square of a side is equal to the sum of the squares of the other sides minus twice the product of these sides and the cosine of the included angle. (Page 168)

Law of Sines In any triangle, the ratio of the length of a side to the sine of the angle opposite that side is the same for each side—angle pair. (Page 165)

Law of Tangents In any triangle ABC,

$$\frac{a-b}{a+b} = \frac{\tan\left(\dfrac{A-B}{2}\right)}{\tan\left(\dfrac{A+B}{2}\right)}; \frac{c-a}{c+a} = \frac{\tan\left(\dfrac{C-A}{2}\right)}{\tan\left(\dfrac{C+A}{2}\right)};$$

and $\dfrac{b-c}{b+c} = \dfrac{\tan\left(\dfrac{B-C}{2}\right)}{\tan\left(\dfrac{B+C}{2}\right)}$. (Page 179)

Limit For a sequence $\{a_n\}$ and a real number L, if, corresponding to any positive number d (no matter how small), a natural number m can be found such that a_m and all succeeding terms differ from L by less than d, then the sequence converges to the <u>limit L</u>. (Page 313)

Logarithm of a Number If b and x are positive real numbers, then $y = \log_b x$ if and only if $x = b^y$. (Page 353)

N^{th} Partial Sum The expression $S_n = \displaystyle\sum_{j=1}^{n} a_j$ is the <u>nth partial sum</u> of the series. (Page 321)

Odd Function Any function having the property that for each x in the domain, $-x$ is also in the domain and $f(-x) = -f(x)$, is called an <u>odd function</u>. (Page 75)

Periodic Function If there is a smallest positive number p such that $f(p+x) = f(x)$ for every x in the domain of f, then p is the <u>period</u> of the function f and f is a <u>periodic function</u>. (Page 83)

Phase Shift The <u>phase shift</u> for functions of the form $y = A \cos B(x - C)$ and $y = A \sin B(x - C)$ is C units to the right if $C > 0$ or $|C|$ units to the left if $C < 0$. (Page 101)

Polar Form of a Complex Number $r(\cos\theta + i \sin\theta)$ is the <u>polar</u> or <u>trigonometric form</u> of the complex number $a + bi$. (Page 285)

Power Series A series of the form
$$\sum_{j=0}^{\infty} a_j x^j = a_0 + a_1 x + a_2 x^2 + \ldots + a_n x^n + \ldots$$
is called a <u>power series</u> in x, where x is a variable and $a_0, a_1, a_2, \ldots, a_n \ldots$ are constants. (Page 331)

Pythagorean Theorem In any right triangle, if the length of the hypotenuse is c and the lengths of the legs are a and b, then $a^2 + b^2 = c^2$. (Page 2)

Quadrantal Angle An angle whose terminal

side coincides with the x axis or the y axis is a quadrantal angle. (Page 9)

Radius Vector In trigonometry, the distance r from the origin to a point P on the terminal side of an angle in standard position is the radius vector. (Page 13)

Range In a relation or function, the set of second elements is the range. (Page 3)

Reference Angle The reference angle of a given angle θ is the positive acute angle determined by the x axis and the terminal side of the given angle. (Page 23)

Reference Triangle When you construct a perpendicular from any point P on the terminal side of an angle to the x axis, a right triangle is formed. This triangle is called a reference triangle. (Page 13)

Relation A relation is a set of ordered pairs. (Page 3)

Rotary Motion The motion of a body turning about an axis is called rotary motion. (Page 70)

Secant Ratio Let $P(x, y)$ be any point (not the origin) on the terminal side of an angle θ in standard position, and let r be the radius vector. Then secant θ (sec θ) $= \dfrac{r}{x}$, $x \neq 0$. (Page 17)

Sequence A sequence is a function whose domain is the set of positive integers. The numbers in the range of the function are the terms of the sequence. When the rule for the terms of a sequence is written in symbols, it is called the general term of the sequence. (Page 304)

Sequence of Partial Sums The expression $\{S_n\} = \left\{ \displaystyle\sum_{j=1}^{n} a_j \right\}$ is called the sequence of partial sums of the series. (Page 321)

Series The expression $\displaystyle\sum_{j=1}^{\infty} a_j$ is called a series for the sequence $\{a_n\}$. (Page 321)

Sine Ratio Let $P(x, y)$ be any point (not the origin) on the terminal side of an angle in standard position with measure θ and let r be the radius vector. Then sine θ (sin θ) $= \dfrac{y}{r}$. (Page 13)

Standard Form of a Complex Number The form $a + bi$ is called the standard form of a complex number. (Page 276)

Standard Position An angle is in standard position in the coordinate plane when its vertex is at the origin and its initial side coincides with the positive x axis. (Page 9)

Sum of the Series When the sequence of partial sums of the series $\displaystyle\sum_{j=1}^{\infty} a_j$ converges to a limit S, the series converges, and S is called the sum of the series. In symbols, $\displaystyle\sum_{j=1}^{\infty} a_j = S$ or $S = a_1 + a_2 + \ldots + a_n + \ldots$. (Page 327)

Tangent Ratio Let $P(x, y)$ be any point (not the origin) on the terminal side of an angle in standard position with measure θ and let r be the radius vector. Then tangent θ (tan θ) $= \dfrac{y}{x}$, $x \neq 0$. (Page 13)

Unit Circle A circle whose center is the origin and whose radius is one is a unit circle. (Page 248)

Vector A line segment with both direction and magnitude is called a vector. (Page 188)

Wrapping Function The wrapping function maps the set of real numbers onto the points of the unit circle. (Page 249)

Index

Boldfaced numerals indicate the pages that contain formal or informal definitions.

The answers to the odd-numbered problems in the Written Exercises, Chapter Objectives and Reviews, and Cumulative Reviews, are given below and on the pages that follow.

There are four exercise categories that are designed to be self-checking: Review Capsule exercises, Classroom Exercises, Review exercises, and Chapter Tests. For that reason, answers are provided for all of the problems in these four categories.

CHAPTER 1 TRIGONOMETRIC FUNCTIONS

PAGE 5 CLASSROOM EXERCISES

1. Domain: $\{-2, 0, 5\}$; Range: $\{0, 1, 3, 6\}$; Not a function   2. Domain: $\{-1, 2\}$; Range: $\{1, 4, 7, 9\}$; Not a function   3. Domain: $\{1, 2, 5, 8\}$; Range: $\{3, 4, 6, 13\}$; Function   4. Domain: $\{-2, 2, 4, 6\}$; Range: $\{-3, -1, 3\}$; Function   5. Domain: $\{-1, 1, 2, 3\}$; Range: $\{-1, 1, 3, 4\}$; Not a function
6. Domain: $\{x : -1 \leq x \leq 1, x \in R\}$; Range: $\{x : -1 \leq x \leq 1, x \in R\}$; Not a function   7. Domain: $\{3\}$; Range: R; Not a function   8. Domain: $\{x : 0 \leq x \leq 2, x \in R\}$; Range: $\{x : 0 \leq x \leq 4, x \in R\}$; Function

PAGES 5-7 WRITTEN EXERCISES

1. $\sqrt{2}$ 3. 4   5. $2\sqrt{3}$ 7. $\sqrt{10}$ 9. $4\sqrt{5}$ 11. $\sqrt{58}$ 13. Domain: $\{-1, 0, 1, 2, 3\}$; Range: $\{-1, 0, 1, 2, 3\}$; Function   15. Domain: $\{-5, -2, 0, 2, 5\}$; Range: $\{-5, -2, 0, 2, 5\}$; Function   Answers for Exercises 17-22 will vary.

17.
x	−1	0	1	2	3
y	−5	−3	−1	1	3

19.
x	−2	−1	0	1	2
y	−8	−1	0	1	8

21.
x	−5	−3	$-2\frac{1}{2}$	−1	0	1
y	$-\frac{1}{3}$	−1	−2	1	$\frac{1}{2}$	$\frac{1}{3}$

23. 0   25. 3   27. 3 and −3   29. 0 and 2   31. 1 and −1   33. Domain: R; Range: $\{-2\}$
35. Domain: R; Range: $\{x : x \geq 0, x \in R\}$ 37. $r = 10; \frac{y}{r} = \frac{8}{10} = \frac{4}{5}; \frac{x}{r} = \frac{6}{10} = \frac{3}{5}; \frac{y}{x} = \frac{8}{6} = \frac{4}{3}$
39. $r = 20; \frac{y}{r} = \frac{16}{20} = \frac{4}{5}; \frac{x}{r} = \frac{12}{20} = \frac{3}{5}; \frac{y}{x} = \frac{16}{12} = \frac{4}{3}$ 41. PQ = 35 m; RP = $35\sqrt{2}$ m   43. 84 m; $42\sqrt{3}$ m

PAGES 10-11 CLASSROOM EXERCISES

1. 180°   2. −270°   3. −240°   4. 120°   5. 750°   6. −420°   7. 215°   8. −865°

1.

$\frac{1}{4}(360°)$
$= 90°$

3.

5.

7.

9.

11.

13.

15.

17. −660° 19. 780° 21. 270° 23. −1110° 25. −510° 27. I 29. II 31. II 33. IV

35. III 37. IV 39. I 41. II 43. 30° 45. 0° 47. 280° 49. 310° 51. 5 53. $x = -\frac{\sqrt{2}}{2}$ and
$y = \frac{\sqrt{2}}{2}$ 55. $x = -\frac{3}{2}\sqrt{2}$ and $y = -\frac{3}{2}\sqrt{2}$ 57. For all angles in standard position with measure $\theta < 0°$
which lie in Quadrant I, θ is between $-360° - n \cdot 360°$ and $-270° - n \cdot 360°$ or $-(n + 1) \cdot 360° < \theta <$
$-270° - n \cdot 360°$. Quadrant II: $-270° - n \cdot 360° < \theta < -180° - n \cdot 360°$. Quadrant III: $-180°$
$- n \cdot 360° < \theta < -90° - n \cdot 360°$. Quadrant IV: $-90° - n \cdot 360° < \theta < -n \cdot 360°$, where n is a whole
number.

PAGE 12 REVIEW CAPSULE FOR SECTION 1-3

1. $2\sqrt{2}$ 2. $5\sqrt{5}$ 3. $2\sqrt{13}$ 4. $\frac{2\sqrt{2}}{5}$ 5. $\frac{\sqrt{6}}{4}$ 6. $\frac{3\sqrt{7}}{14}$ 7. $-\frac{2\sqrt{5}}{15}$

PAGE 15 CLASSROOM EXERCISES

1. $\sin \theta = \frac{3}{5}$; $\cos \theta = \frac{4}{5}$; $\tan \theta = \frac{3}{4}$ 2. $\sin \theta = \frac{3}{5}$; $\cos \theta = -\frac{4}{5}$; $\tan \theta = -\frac{3}{4}$ 3. $\sin \theta = -\frac{3}{5}$; $\cos \theta = -\frac{4}{5}$;
$\tan \theta = \frac{3}{4}$ 4. $\sin \theta = -\frac{3}{5}$; $\cos \theta = \frac{4}{5}$; $\tan \theta = -\frac{3}{4}$

PAGES 15-16 WRITTEN EXERCISES

A sketch is shown for Exercise 1 only. Exercises 2-12 are done in a similar manner.

1. $\frac{12}{13}$, $\frac{5}{13}$, $\frac{12}{5}$

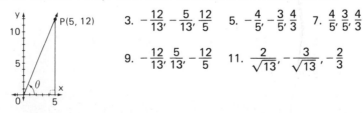

3. $-\frac{12}{13}$, $-\frac{5}{13}$, $\frac{12}{5}$ 5. $-\frac{4}{5}$, $-\frac{3}{5}$, $\frac{4}{3}$ 7. $\frac{4}{5}$, $\frac{3}{5}$, $\frac{4}{3}$

9. $-\frac{12}{13}$, $\frac{5}{13}$, $-\frac{12}{5}$ 11. $\frac{2}{\sqrt{13}}$, $-\frac{3}{\sqrt{13}}$, $-\frac{2}{3}$

CHAPTER 1

13. $\cos \theta = -\dfrac{\sqrt{3}}{2}$; $\tan \theta = -\dfrac{1}{\sqrt{3}}$ 15. $\sin \theta = \dfrac{3}{\sqrt{34}}$; $\cos \theta = -\dfrac{5}{\sqrt{34}}$ 17. $\sin \theta = \dfrac{4}{\sqrt{41}}$; $\cos \theta = \dfrac{5}{\sqrt{41}}$

19. $\sin \theta = \dfrac{\sqrt{2}}{2}$; $\tan \theta = -1$ A sketch is shown for Ex. 21. Exercises 22-28 are done in a similar manner.

21. $\sin \theta = \dfrac{5}{13}$; $\cos \theta = -\dfrac{12}{13}$; $\tan \theta = -\dfrac{5}{12}$

23. $\dfrac{3}{\sqrt{13}}$; $-\dfrac{2}{\sqrt{13}}$; $-\dfrac{3}{2}$

25. $\dfrac{7}{17\sqrt{2}}$; $-\dfrac{23}{17\sqrt{2}}$; $-\dfrac{7}{23}$

27. $\dfrac{3}{5}, \dfrac{4}{5}, \dfrac{3}{4}$

P(−12, 5)

29. $\sin \theta = -\dfrac{4}{5}$; $\tan \theta = \dfrac{4}{3}$ 31. $\sin \theta = -\dfrac{4}{5}$; $\cos \theta = \dfrac{3}{5}$ 33. $\sin \theta = \dfrac{1}{\sqrt{2}}$ or $\dfrac{\sqrt{2}}{2}$; $\cos \theta = -\dfrac{1}{\sqrt{2}}$ or $-\dfrac{\sqrt{2}}{2}$

35. $\cos \theta = \dfrac{4\sqrt{6}}{11}$; $\tan \theta = -\dfrac{5}{4\sqrt{6}}$ or $-\dfrac{5\sqrt{6}}{24}$ 37. +, +, −, −   39. +, −, +, −   41. III or IV

43. II or IV   45. II   47. IV   49. III   51. IV

PAGE 19   CLASSROOM EXERCISES

1. $\csc \theta = \dfrac{5}{3}$, $\sec \theta = \dfrac{5}{4}$, $\cot \theta = \dfrac{4}{3}$ 2. $\csc \theta = \dfrac{5}{3}$, $\sec \theta = -\dfrac{5}{4}$, $\cot \theta = -\dfrac{4}{3}$ 3. $\csc \theta = -\dfrac{5}{3}$, $\sec \theta = -\dfrac{5}{4}$,
$\cot \theta = \dfrac{4}{3}$ 4. $\csc \theta = -\dfrac{5}{3}$, $\sec \theta = \dfrac{5}{4}$, $\cot \theta = -\dfrac{4}{3}$

PAGES 19-20   WRITTEN EXERCISES

1. $\csc \theta = -\sqrt{2}$; $\sec \theta = -\sqrt{2}$; $\cot \theta = 1$ 3. $\sqrt{2}$; $-\sqrt{2}$; -1 5. $\dfrac{\sqrt{34}}{3}$; $\dfrac{\sqrt{34}}{5}$; $\dfrac{5}{3}$ 7. $-\dfrac{\sqrt{34}}{3}$; $-\dfrac{\sqrt{34}}{5}$; $\dfrac{5}{3}$

9. $-\dfrac{\sqrt{29}}{5}$; $\dfrac{\sqrt{29}}{2}$; $-\dfrac{2}{5}$ 11. $\sqrt{5}$; $-\dfrac{\sqrt{5}}{2}$; -2 13. $\sin \theta = -\dfrac{\sqrt{15}}{4}$; $\cos \theta = \dfrac{1}{4}$; $\tan \theta = -\sqrt{15}$; $\csc \theta = -\dfrac{4}{\sqrt{15}}$;

$\cot \theta = -\dfrac{1}{\sqrt{15}}$ 15. $\sin \theta = \dfrac{2}{\sqrt{29}}$; $\cos \theta = -\dfrac{5}{\sqrt{29}}$; $\csc \theta = \dfrac{\sqrt{29}}{2}$; $\sec \theta = -\dfrac{\sqrt{29}}{5}$; $\cot \theta = -\dfrac{5}{2}$ 17. $\sin \theta = -\dfrac{2}{3}$;

$\cos \theta = -\dfrac{\sqrt{5}}{3}$; $\tan \theta = \dfrac{2}{\sqrt{5}}$; $\sec \theta = -\dfrac{3}{\sqrt{5}}$; $\cot \theta = \dfrac{\sqrt{5}}{2}$ 19. $\sin \theta = -\dfrac{1}{2}$; $\cos \theta = -\dfrac{3}{2\sqrt{3}}$; $\tan \theta = \dfrac{\sqrt{3}}{3}$;

$\csc \theta = -2$; $\cot \theta = \dfrac{3}{\sqrt{3}}$ 21. +, +, −, −   23. +, −, +, −   25. $\dfrac{5}{\sqrt{6}}$ 27. $\dfrac{1}{2}$ 29. $-\dfrac{7}{6}$ 31. 1   33. 1

35. II or IV   37. I   39. II   41. II; $\cos \theta = -\dfrac{4}{5}$; $\csc \theta = \dfrac{5}{3}$; $\cot \theta = -\dfrac{4}{3}$; $\sec \theta = \dfrac{5}{3}$ 43. I or IV;

$\sin \theta = \dfrac{20}{29}$; $\tan \theta = \dfrac{20}{21}$; $\csc \theta = \dfrac{29}{20}$; $\cot \theta = \dfrac{21}{20}$ 45. II or IV; $\sin \theta = \dfrac{11}{61}$; $\cos \theta = -\dfrac{60}{61}$; $\csc \theta = \dfrac{61}{11}$;

$\sec \theta = -\dfrac{61}{60}$

PAGE 20   REVIEW CAPSULE FOR SECTION 1-5

1. $b = 2\sqrt{3}$; $a = 2$ 2. $c = 2$; $b = \sqrt{3}$ 3. $a = \dfrac{1}{2}$; $c = 1$ 4. $b = 1$, $a = 1$ 5. $b = \dfrac{1}{\sqrt{2}}$; $c = 1$
6. $a = 2$; $c = 2\sqrt{2}$

PAGE 24   CLASSROOM EXERCISES

1. 60°; $\sin 120° = \dfrac{\sqrt{3}}{2}$; $\cos 120° = -\dfrac{1}{2}$; $\tan 120° = -\sqrt{3}$ 2. 30°; $\sin 150° = \dfrac{1}{2}$; $\cos 150° = -\dfrac{\sqrt{3}}{2}$;

TRIGONOMETRIC FUNCTIONS

$\tan 150° = -\dfrac{1}{\sqrt{3}}$ 3. $45°$; $\sin 135° = \dfrac{\sqrt{2}}{2}$; $\cos 135° = -\dfrac{\sqrt{2}}{2}$; $\tan 135° = -1$ 4. $45°$; $\sin 315° = -\dfrac{\sqrt{2}}{2}$;

$\cos 315° = \dfrac{\sqrt{2}}{2}$; $\tan 315° = -1$

PAGES 24-26   WRITTEN EXERCISES

1. $\sin 210° = -\dfrac{1}{2}$; $\cos 210° = -\dfrac{\sqrt{3}}{2}$; $\tan 210° = \dfrac{1}{\sqrt{3}}$ 3. $-\dfrac{\sqrt{3}}{2}$; $-\dfrac{1}{2}$; $\sqrt{3}$ 5. $\sec 0° = 1$; $\csc 0°$ is

undefined; $\cot 0°$ is undefined   7. -1; undefined; undefined   9. $180° - 135° = 45°$ 11. $60°$ 13. $12°$

15. $59°$ 17. $53°$ 19. $71°$ 21. $30°$; $\sec 30° = \dfrac{2}{\sqrt{3}}$; $\csc 30° = 2$; $\cot 30° = \sqrt{3}$ 23. $30°$; $-\dfrac{2}{\sqrt{3}}$;

-2; $\sqrt{3}$ 25. $45°$; $\sec 45° = \sqrt{2}$; $\csc 45° = \sqrt{2}$; $\cot 45° = 1$ 27. $45°$; $-\sqrt{2}$; $-\sqrt{2}$; 1 29. $60°$;

$\sec 60° = 2$; $\csc 60° = \dfrac{2}{\sqrt{3}}$; $\cot 60° = \dfrac{1}{\sqrt{3}}$ 31. $60°$; -2, $-\dfrac{2}{\sqrt{3}}$, $\dfrac{1}{\sqrt{3}}$

	$\sin\theta$	$\cos\theta$	$\tan\theta$	$\cot\theta$	$\sec\theta$	$\csc\theta$
33.	0	1	0	undefined	1	undefined
35.	$\dfrac{1}{\sqrt{2}}$	$\dfrac{1}{\sqrt{2}}$	1	1	$\sqrt{2}$	$\sqrt{2}$
37.	1	0	undefined	0	undefined	1
39.	$\dfrac{1}{\sqrt{2}}$	$-\dfrac{1}{\sqrt{2}}$	-1	-1	$-\sqrt{2}$	$\sqrt{2}$
41.	0	-1	0	undefined	-1	undefined
43.	$-\dfrac{1}{\sqrt{2}}$	$-\dfrac{1}{\sqrt{2}}$	1	1	$-\sqrt{2}$	$-\sqrt{2}$
45.	-1	0	undefined	0	undefined	-1
47.	$-\dfrac{1}{\sqrt{2}}$	$\dfrac{1}{\sqrt{2}}$	-1	-1	$\sqrt{2}$	$-\sqrt{2}$
49.	0	1	0	undefined	1	undefined

A sketch is given for Exercise 51.
Exercises 52-64 are done in a
similar manner.

51.

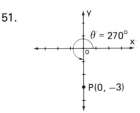

$\theta = 270°$

$P(0, -3)$

	$\sin\theta$	$\cos\theta$	$\tan\theta$	$\csc\theta$	$\sec\theta$	$\cot\theta$	θ
51.	-1	0	undefined	-1	undefined	0	$270°$
53.	1	0	undefined	1	undefined	0	$90°$
55.	$-\dfrac{1}{2}$	$-\dfrac{\sqrt{3}}{2}$	$\dfrac{1}{\sqrt{3}}$	-2	$-\dfrac{2}{\sqrt{3}}$	$\sqrt{3}$	$210°$
57.	$-\dfrac{1}{2}$	$\dfrac{\sqrt{3}}{2}$	$-\dfrac{1}{\sqrt{3}}$	-2	$\dfrac{2}{\sqrt{3}}$	$-\sqrt{3}$	$330°$

	$\sin\theta$	$\cos\theta$	$\tan\theta$	$\csc\theta$	$\sec\theta$	$\cot\theta$	θ
59.	$\dfrac{\sqrt{3}}{2}$	$-\dfrac{1}{2}$	$-\sqrt{3}$	$\dfrac{2}{\sqrt{3}}$	-2	$-\dfrac{1}{\sqrt{3}}$	$120°$
61.	$-\dfrac{\sqrt{3}}{2}$	$\dfrac{1}{2}$	$-\sqrt{3}$	$-\dfrac{2}{\sqrt{3}}$	2	$-\dfrac{1}{\sqrt{3}}$	$300°$
63.	$-\dfrac{1}{\sqrt{2}}$	$-\dfrac{1}{\sqrt{2}}$	1	$-\sqrt{2}$	$-\sqrt{2}$	1	$225°$

65. $x = \dfrac{\sqrt{3}}{2}$; $y = \dfrac{1}{2}$ 67. $x = 0$; $y = 5$ 69. $x = -\dfrac{1}{\sqrt{2}}$; $y = -\dfrac{1}{\sqrt{2}}$ 71. $x = 7$; $y = 0$ 73. $x = 2$; $y = -2\sqrt{3}$
75. $x = -\dfrac{\sqrt{3}}{2}$; $y = -\dfrac{1}{2}$ 77. $120°$ 79. $330°$ 81. $360°$

PAGE 26   REVIEW

1. $\sqrt{85}$ 2. $\sqrt{109}$ 3. Function   4. Domain: $\{-2, -1, 0, 1, 2\}$; Range: $\{-2, -1, 0\}$ 5. $-630°$
6. $240°$ 7. $\dfrac{3}{5}$ 8. $-\dfrac{4}{5}$ 9. $-\dfrac{3}{4}$ 10. $-\dfrac{4}{3}$ 11. $-\dfrac{5}{4}$ 12. $\dfrac{5}{3}$ 13. $-\dfrac{12}{13}$ 14. $\dfrac{5}{12}$ 15. $\dfrac{12}{5}$ 16. $-\dfrac{13}{12}$
17. $-\dfrac{13}{5}$ 18. 0 19. -1 20. $\dfrac{1}{2}$ 21. 1 22. -2

PAGE 30   CLASSROOM EXERCISES

1. .9063   2. .8450   3. .7907   4. 1.1237   5. 1.026   6. 1.011   7. $-\tan 22°$ 8. $\cos 35°$
9. $-\sin 88°$ 10. $\tan 35°$ 11. $\cot 60°$ 12. $-\sec 25°$ 13. $\sin 41° 50'$ 14. $-\tan 56° 20'$
15. $-\cos 82°$ 16. $-\csc 20° 15'$ 17. $-\cot 60° 40'$ 18. $-\sec 40°$ 19. $-\tan 79° 40'$
20. $\cos 20° 50'$ 21. $-\sin 80°$

PAGE 31   WRITTEN EXERCISES

1. .6157   3. .5169   5. 1.499   7. 2.254   9. .0058   11. 343.8   13. 4.8430   15. .7092
17. 34.368   19. .9995   21. 1.019   23. 1.4106   25. .7050   27. 3.326   29. $-.5299$ 31. .9159
33. -1.086 35. .0787   37. $-.9628$ 39. .9528   41. 1.9074   43. -1.012 45. -7.337
47. 49.104   49. $-.8541$ 51. -1.003 53. .3057   55. 1.1106   57. 1.272   59. $-.2035$ 61. 1.390
63. -1.169 65. $328° 20'$ 67. $214° 30'$ or $325° 30'$ 69. $234° 20'$ 71. $154° 50'$ or $205° 10'$

PAGE 34   CLASSROOM EXERCISES

1. $\sin 14° 12'$ 2. $\cos 34° 45'$ 3. $-\tan 49° 37'$ 4. $-\csc 33° 12'$

PAGE 34   WRITTEN EXERCISES

1. .4266   3. .4081   5. 1.352   7. .6211   9. .2979   11. 4.995   13. -1.3960 15. $-.1010$

17. −2.805   19. .8946   21. −3.1911   23. −5.778   25. .3948   27. −1.4696   29. 228° 26′ or 311° 34′   31. 138° 14′   33. 190° 19′   35. 36° 07′   37. 222° 55′

PAGE 34   REVIEW CAPSULE FOR SECTION 1-8

1. 66°   2. 57° 40′   3. 21° 30′   4. 45° 10′

PAGE 36   CLASSROOM EXERCISES

1. $\frac{a}{c}$ 2. $\frac{c}{a}$ 3. $\frac{b}{c}$ 4. $\frac{c}{b}$ 5. sin B   6. cos A   7. tan B

PAGE 37   WRITTEN EXERCISES

1. a = 18; A = 36° 52′; B = 53° 08′   3. c = 20; A = 45°; B = 45°   5. a = 1.7 or 2; A = 60°; B = 30°
7. a = 60; B = 36° 52′; A = 53° 08′   9. b = 120; A = 44° 46′; B = 45° 14′   11. b = 17; A = 41° 16′;
B = 48° 44′   13. B = 45°; a = 17; c = 24   15. B = 30°; c = 2; b = 1   17. A = 27° 30′; b = 27; a = 14
19. A = 17° 25′; b = 10,903; c = 11,427   21. A = 27° 07′; a = 34, b = 66

PAGE 37   REVIEW

1. .2868   2. −.7934   3. −3.4495   4. 23°   5. 68° 30′   6. 23° 20′   7. .7830   8. 23° 28′
9. b = 10; A = 67° 23′; B = 22° 37′   10. 8.7

PAGES 42-46   WRITTEN EXERCISES

1. 150 m   3. 1006 m   5. 2756 m   7. 207 m   9. 150 m   11. 22° 40′   13. 38° 40′   15. 4 m
17. 17 m   19. 258 m   21. 28 m   23. 12 m

PAGES 47-49   WRITTEN EXERCISES

1. 64 m   3. 168 m   5. 36 m   7. 182 m   9. 18 m

PAGES 51-52   WRITTEN EXERCISES

1. 270°   3. 297°   5. 48° 50′   7. 4.7 kilometers   9. 296.7 kilometers   11. 244° 40′   13. 13
kilometers per hour   15. 194.4 kilometers

PAGES 53-54   WRITTEN EXERCISES

1. $\frac{1}{3}$ 3. $\frac{1}{4}$ 5. 122 cm   7. 74 in.   9. $\frac{1}{3}$ 11. 250 cm   13. AB = 189 in.; CD = 168 in.

1. $(X - 3) \uparrow 2$ 3. $(Q + R)/4$ 5. $2*(L + W)$ 7. $-7 + B/(C - D)$ 9. $SQR((X\uparrow2 - A)/(5*Z))$

11.
```
1Ø READ D
2Ø LET S = 180 − D
3Ø PRINT "SUPPLEMENT ="; S
4Ø GO TO 1Ø
5Ø DATA ...
6Ø END
```

13.
```
1Ø READ R
2Ø LET A = 3.14159 * R↑2
3Ø PRINT "AREA ="; A
4Ø GO TO 1Ø
5Ø DATA ...
6Ø END
```

15.
```
1Ø READ A, B
2Ø LET X = .5 * A * B
3Ø PRINT "AREA ="; X
4Ø GO TO 1Ø
5Ø DATA ...
6Ø END
```

PAGES 57-59   CHAPTER OBJECTIVES AND REVIEW

3. $3\sqrt{5}$ 5. Domain: R; Range: $\{1\}$ 7. 1 and −1   9. 135°   11. $-\frac{3}{5}$ 13. $-\frac{3}{4}$ 15. $\frac{5}{4}$ 17. $\frac{5}{12}$

19. $\frac{13}{12}$ 21. 1   23. $\frac{1}{\sqrt{3}}$ 25. $-\frac{1}{\sqrt{2}}$ 27. .4279   29. −2.1283   31. 55° 40′   33. .6728   35. 15

37. 10.7   39. 207 meters   41. 303°

PAGE 60   CHAPTER TEST

1. True   2. False   3. True   4. False   5. False   6. 5 and −5   7. −660°   8. 5   9. $\frac{5}{13}$ 10. $\frac{13}{5}$

11. $-\frac{5}{12}$ 12. $-\frac{13}{12}$ 13. $-\frac{15}{17}$ 14. $\frac{15}{8}$ 15. $-\frac{17}{15}$ 16. $-\frac{17}{8}$ 17. −.6655   18. 58° 35′   19. a = 8;

B = 62°; A = 28°   20. 48°; 42°

CHAPTER 2   GRAPHS OF TRIGONOMETRIC FUNCTIONS

PAGE 63   CLASSROOM EXERCISES

1. 30°, $\frac{\pi}{6}$; 45°, $\frac{\pi}{4}$; 60°, $\frac{\pi}{3}$; 90°, $\frac{\pi}{2}$; 120°, $\frac{2\pi}{3}$; 135°, $\frac{3\pi}{4}$; 150°, $\frac{5\pi}{6}$; −30°, $-\frac{\pi}{6}$; −60°, $-\frac{\pi}{3}$

PAGES 64-65   WRITTEN EXERCISES

1. 1.5 radians   3. .5 radian   5. .4 radian   7. 2π 9. $-\frac{\pi}{2}$ 11. $-\frac{\pi}{3}$ 13. $\frac{2\pi}{3}$ 15. $\frac{4\pi}{3}$ 17. $\frac{11\pi}{72}$

19. $\frac{11\pi}{144}$ 21. $\frac{\pi}{360}$ 23. 60°   25. −270°   27. 30°   29. 120°   31. 210°   33. 720°   35. 450°

37. 45°   39. −300°   41. 1170°

43.    45. 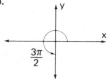   47. 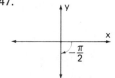   49.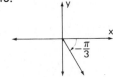

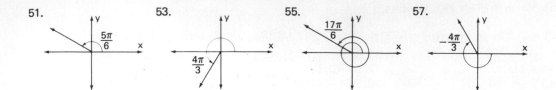

51. $\frac{5\pi}{6}$ **53.** $\frac{4\pi}{3}$ **55.** $\frac{17\pi}{6}$ **57.** $-\frac{4\pi}{3}$

59. $\frac{1}{10}$ **61.** $\frac{\pi}{5}$ **63.** $\frac{2\pi}{3}$ **65.** $\frac{3\pi}{2}$ **67.** π **69.** $-\frac{13\pi}{6}$ **71.** $\frac{7\pi}{4}$ **73.** 0   **75.** $-\frac{13\pi}{4}$ **77.** $\frac{\pi}{2}$ **79.** III

81. IV   **83.** Quadrant I: $-2\pi - n \cdot 2\pi < x < -\frac{3\pi}{2} - n \cdot 2\pi$, or $-(n+1)2\pi < x < -\frac{3\pi}{2} - n \cdot 2\pi$;

Quadrant II: $-\frac{3\pi}{2} - n \cdot 2\pi < x < -\pi - n \cdot 2\pi$; Quadrant III: $-\pi - n \cdot 2\pi < x < -\frac{\pi}{2} - n \cdot 2\pi$;

Quadrant IV: $-\frac{\pi}{2} - n \cdot 2\pi < x < -n \cdot 2\pi$

PAGE 68   CLASSROOM EXERCISES

1. .2123   **2.** .9512   **3.** .1047   **4.** .0640   **5.** 37° 20′   **6.** 60° 20′   **7.** 86°   **8.** 9° 40′   **9.** 18 cm

PAGES 68-69   WRITTEN EXERCISES

1. .2036   **3.** .3142   **5.** 1.4224   **7.** .1571   **9.** 7° 50′   **11.** 12° 20′   **13.** 71° 30′   **15.** 33°   **17.** $\frac{20\pi}{3}$
19. 9π **21.** $\frac{88\pi}{5}$ **23.** 24π **25.** $\frac{65\pi}{9}$ **27.** 14π **29.** π **31.** $\frac{784\pi}{9}$ **33.** $\frac{65\pi}{3}$ **35.** 1768 mm

PAGE 72   CLASSROOM EXERCISES

1. $\frac{2\pi}{3}$ **2.** $\frac{5\pi}{3}$ **3.** $\frac{7\pi}{6}$ **4.** $\frac{\pi}{6}$ **5.** π **6.** $\frac{3\pi}{2}$ **7.** $\frac{11\pi}{6}$ **8.** $\frac{4\pi}{3}$

PAGES 72-74   WRITTEN EXERCISES

1. $\frac{2\pi}{3}$ **3.** $\frac{4\pi}{5}$ **5.** $\frac{2\pi}{9}$ rad/sec   **7.** $\frac{4\pi}{15}$ rad/sec   **9.** 120π rad/sec   **11.** 26.6 cm   **13.** 2π radians per

minute   **15.** $\frac{\pi}{360}$ radians per minute   **17.** $.76\pi$ meters per second   **19.** π meters per second   **21.** $\frac{7\pi}{3}$

meters per second   **23.** 600π cm/sec   **25.** 120π rad/sec   **27.** 528π km/hr   **29.** 420π cm/sec

31. 420π cm/sec

PAGE 75   REVIEW CAPSULE FOR SECTION 2-4

1. HA   **2.** SAS   **3.** ASA

PAGE 78   CLASSROOM EXERCISES

1. −.8660   **2.** .8660   **3.** −.8660   **4.** .5000   **5.** −.5000   **6.** −.5000

1. $-\frac{1}{2}$ 3. $\frac{\sqrt{3}}{2}$ 5. $-.7880$ 7. $.8000$ 9. $-.8000$ 11. $.6000$ 13. $-.6000$ 15. $-.4226$

17. $-.4226$ 19. $.9063$ 21. $-.9063$ 23. $-.3827$ 25. $-.3827$ 27. $.8090$ 29. $-.8090$

31. $-.3090$ 33. $-.3090$ 35. $.9511$ 37. $-.9511$ 39. $-.2588$ 41. $-.2588$ 43. $.9659$

45. $-.9659$ 47. $\cos(360° - \theta) = \cos[180° + (180° - \theta)]$; Let $\alpha = 180° - \theta$. Then $\cos(360° - \theta)$

$= \cos(180° + \alpha) = -\cos\alpha = -\cos(180° - \theta) = -(-\cos\theta) = \cos\theta$.   49. $\cos(360° + \theta)$

$= \cos[180° + (180° + \theta)]$; Let $\alpha = 180° + \theta$. Then $\cos(360° + \theta) = \cos(180° + \alpha) = -\cos\alpha$

$= -\cos(180° + \theta) = -(-\cos\theta) = \cos\theta$.   51. Using the diagram for Exercise 48, $\cos(180° - \theta)$

$= \frac{x'}{r} = \frac{-x}{r} = -\frac{x}{r} = -\cos\theta$.   53. $-\frac{12}{13}$, if θ is in Quadrant I; $\frac{12}{13}$, if θ is in Quadrant IV.   55. $-\frac{12}{13}$, if θ is in

Quadrant I; $\frac{12}{13}$, if θ is in Quadrant IV.   57. even   59. neither   61. odd   63. even   65. $x = 0$

67.

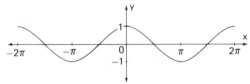

PAGE 84   CLASSROOM EXERCISES

1. c   2. a   3. b   4. d

PAGES 85-86   WRITTEN EXERCISES

1.

x	0	$\frac{\pi}{4}$	$\frac{\pi}{2}$	$\frac{3\pi}{4}$	π	$\frac{5\pi}{4}$	$\frac{3\pi}{2}$	$\frac{7\pi}{4}$	2π
$x - 4\pi$	-4π	$-\frac{15\pi}{4}$	$-\frac{7\pi}{2}$	$-\frac{13\pi}{4}$	-3π	$-\frac{11\pi}{4}$	$-\frac{5\pi}{2}$	$-\frac{9\pi}{4}$	-2π
sin x or sin (x − 4π)	0	0.71	1	0.71	0	−0.71	−1	−0.71	0

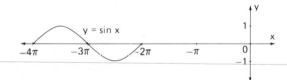

3.

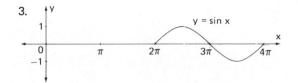

5.

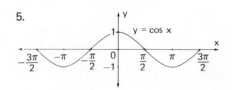

7.

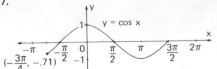

9.

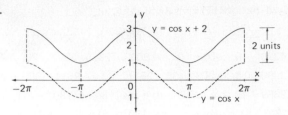

11.

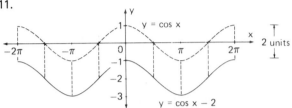

13. 2   15. 2   17. $\frac{\pi}{2}$

19. $\frac{1}{2}$ 21. $-\frac{1}{2}$ 23. $\frac{\pi}{4}$ or $\frac{5\pi}{4}$

25. $0 < x < \frac{\pi}{4}$ and $\frac{5\pi}{4} < x < 2\pi$

27. $\frac{\pi}{2}$

29.

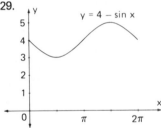

31.

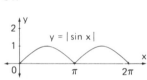

33.

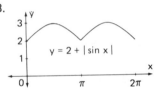

35.

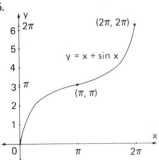

37.

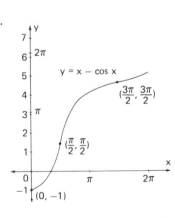

39. -3

41. 1

PAGE 87   REVIEW

1. $1\frac{2}{3}$, or 1.7 to the nearest tenth   2. 1.8   3. $-\frac{\pi}{4}$ 4. $\frac{\pi}{3}$ 5. π 6. $-\frac{3\pi}{4}$ 7. 60°   8. $-15°$

9. 135°   10. 30°   11. .5992   12. .2763   13. 1.3701   14. 1.0501   15. 41° 30′   16. 29° 30′
17. 70° 00′   18. 88° 50′   19. 20π   20. 219.9 m   21. .6000   22. .8660, −.8660, −.8660, .8660
23.

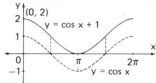

PAGES 90-91   CLASSROOM EXERCISES

1. $\sin \pi = 0$, $-\sin \pi = 0$; $\sin \frac{3\pi}{2} = -1$, $-\sin \frac{3\pi}{2} = 1$ 2. $\frac{1}{2} \sin \frac{\pi}{2} + 4 = 4\frac{1}{2}$; $\sin \pi = 0$, $\frac{1}{2} \sin \pi + 4 = 4$;

$\frac{1}{2} \sin \frac{3\pi}{2} + 4 = 3\frac{1}{2}$ 3. $\cos 0 = 1$, $-\cos 0 = -1$; $\cos \frac{\pi}{3} = \frac{1}{2}$, $-\cos \frac{\pi}{3} = -\frac{1}{2}$; $\cos \frac{\pi}{2} = 0$, $-\cos \frac{\pi}{2} = 0$; $\cos \pi = -1$,

$-\cos \pi = 1$ 4. $\cos 0 = 1$, $\frac{1}{3} \cos 0 - 2 = -\frac{5}{3}$; $\cos \frac{\pi}{3} = \frac{1}{2}$, $\frac{1}{3} \cos \frac{\pi}{3} - 2 = -\frac{11}{6}$; $\cos \frac{\pi}{2} = 0$, $\frac{1}{3} \cos \frac{\pi}{2} - 2 = -2$;

$\cos \pi = -1$, $\frac{1}{3} \cos \pi - 2 = -\frac{7}{3}$ 5. 3   6. 2   7. 2   8. $\frac{1}{2}$

PAGES 91-92   WRITTEN EXERCISES

1. Amplitude: 5

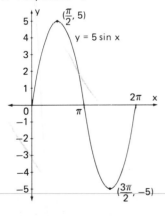

3. Amplitude: 2

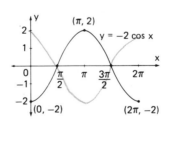

5. Amplitude: 1

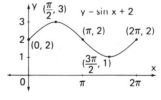

7. Amplitude: 1

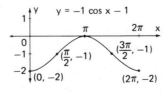

9. $\frac{1}{2}$; b

11. 2; d

GRAPHS OF TRIGONOMETRIC FUNCTIONS

395

13.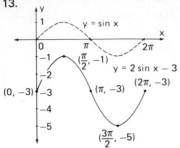

15.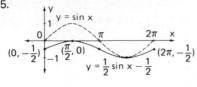

17. y = 3 cos x and
y = −3 cos x + 6

19. y = 10 cos x and
y = −10 cos x

21.

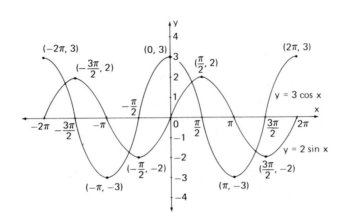

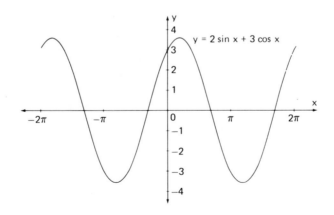

23. 3.6

1. 1   2. −.71   3. 0   4. 1   5. 0   6. .87   7. −.87   8. 0   9. 1, $\frac{2\pi}{3}$ 10. 1; 6π 11. 4; $\frac{\pi}{2}$

PAGES 96-97   WRITTEN EXERCISES

1. Period: $\frac{2\pi}{3}$

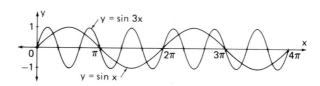

3. Period: 6π

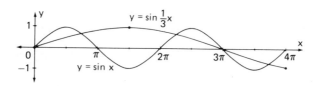

5. Period: $\frac{4\pi}{3}$

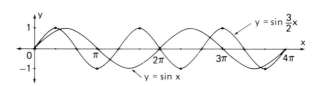

7. Period: π

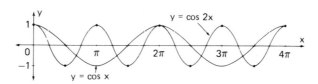

9. Period: 6π

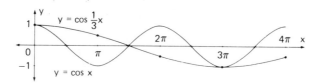

11. Period: 3π

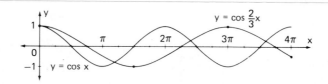

GRAPHS OF TRIGONOMETRIC FUNCTIONS

13. 2π; 2; d   15. π; 1; e   17. π; 1; b   19. 2; 6π 21. $\frac{1}{18}$; 36π 23. $\frac{4}{5}$; π 25. y = 3 cos 2x;

y = -3 cos 2x; y = 3 cos (-2x); and y = -3 cos (-2x)   27. y = 12 cos 6x, y = -12 cos 6x;

y = 12 cos (-6x); and y = -12 cos (-6x)   29. y = 4 sin 6x; y = -4 sin 2x. Answers vary.

31. y = 7 sin 3x; y = -7 sin 3x. Answers vary.   33. y = 2 sin $(-\frac{2}{5}x)$ and y = -2 sin $(\frac{2}{5}x)$ 35. $\frac{1}{4\pi}$

37. $\frac{40}{\pi}$ 39. $\frac{10}{\pi}$

PAGE 97   REVIEW CAPSULE FOR SECTION 2-8

1. $2[x - (-1)]$ 2. $4(x - 1)$ 3. $2(x - 1)$ 4. $4(x - \frac{1}{2})$ 5. $3(x - \frac{2}{3})$ 6. $3[x - (-\frac{2}{3})]$ 7. $2(x - \frac{\pi}{2})$
8. $5[x - (-\frac{\pi}{10})]$

PAGE 101   CLASSROOM EXERCISES

1. $\frac{\pi}{2}$ units to the left   2. π units to the right   3. 2π units to the left   4. $\frac{\pi}{6}$ units to the right

PAGES 102-103   WRITTEN EXERCISES

1. Amplitude: 2; Period: 2π;   3. Amplitude: $\frac{1}{2}$; Period: 2π;   5. Amplitude: 1; Period: $\frac{2\pi}{3}$;

Phase shift: $\frac{\pi}{4}$ units to the left   Phase shift: $\frac{\pi}{4}$ units to the left   Phase shift: $\frac{\pi}{4}$ units to the right

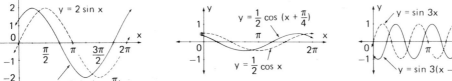
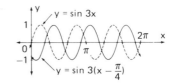

7. Amplitude: 2; Period: π;

Phase shift: $\frac{\pi}{2}$ units to the right

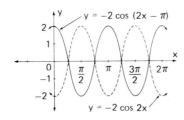

9. $\frac{1}{2}$; $\frac{2\pi}{3}$; $\frac{2\pi}{3}$ units to the right   11. 3; 2π; $\frac{3\pi}{2}$ units

to the right   13. $\frac{3}{4}$; $\frac{2\pi}{3}$; $\frac{\pi}{3}$ units to the left   15. 4;

π; π units to the left   17. 2; π; $\frac{\pi}{4}$ units to the left,

b   19. 2; $\frac{2\pi}{3}$; $\frac{\pi}{6}$ units to the left, a   21. y = 5 sin

$(x - \frac{\pi}{3})$; y = -5 sin $(x - \frac{\pi}{3})$; y = 5 sin $(-x + \frac{\pi}{3})$, or

y = -5 sin $(-x + \frac{\pi}{3})$ 23. y = $\frac{2}{3}$ sin 8$(x + \frac{\pi}{8})$;

y = $-\frac{2}{3}$ sin 8$(x + \frac{\pi}{8})$; y = $\frac{2}{3}$ sin $[-8(x + \frac{\pi}{8})]$, or

y = $-\frac{2}{3}$ sin $[-8(x + \frac{\pi}{8})]$ 25. y = $\frac{7}{3}$ cos $\frac{12}{5}(x + \pi)$;

y = $-\frac{7}{3}$ cos $\frac{12}{5}(x + \pi)$; y = $\frac{7}{3}$ cos $[-\frac{12}{5}(x + \pi)]$, or

y = $-\frac{7}{3}$ cos $[-\frac{12}{5}(x + \pi)]$

27. Amplitude: 2; Period: π; Phase shift: $\frac{\pi}{6}$ units to the left

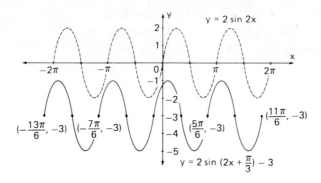

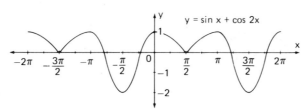

$(-\frac{13\pi}{6}, -3)$ $(-\frac{7\pi}{6}, -3)$ $(\frac{5\pi}{6}, -3)$ $(\frac{11\pi}{6}, -3)$

$y = 2\sin(2x + \frac{\pi}{3}) - 3$

29.

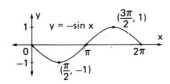

$y = \sin x + \cos 2x$

PAGE 103   REVIEW

1. Amplitude = 1

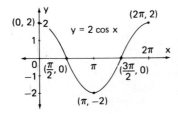

$(\frac{3\pi}{2}, 1)$ $y = -\sin x$ $(\frac{\pi}{2}, -1)$

2. Amplitude = 2

$(0, 2)$ $y = 2\cos x$ $(2\pi, 2)$ $(\frac{\pi}{2}, 0)$ $(\frac{3\pi}{2}, 0)$ $(\pi, -2)$

3. Amplitude = 4

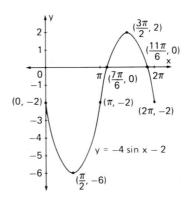

$(\frac{3\pi}{2}, 2)$ $(\frac{11\pi}{6}, 0)$ $(\frac{7\pi}{6}, 0)$ $(\pi, -2)$ $(2\pi, -2)$ $(0, -2)$ $y = -4\sin x - 2$ $(\frac{\pi}{2}, -6)$

4. Amplitude = $\frac{1}{2}$

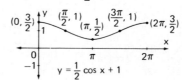

$(0, \frac{3}{2})$ $(\frac{\pi}{2}, 1)$ $(\pi, \frac{1}{2})$ $(\frac{3\pi}{2}, 1)$ $(2\pi, \frac{3}{2})$

$y = \frac{1}{2}\cos x + 1$

GRAPHS OF TRIGONOMETRIC FUNCTIONS

5.

x	0	$\frac{\pi}{6}$	$\frac{\pi}{3}$	$\frac{\pi}{2}$	$\frac{2\pi}{3}$	$\frac{5\pi}{6}$	π	$\frac{7\pi}{6}$	$\frac{4\pi}{3}$	$\frac{3\pi}{2}$	$\frac{5\pi}{3}$	$\frac{11\pi}{6}$	2π
3x	0	$\frac{\pi}{2}$	π	$\frac{3\pi}{2}$	2π	$\frac{5\pi}{2}$	3π	$\frac{7\pi}{2}$	4π	$\frac{9\pi}{2}$	5π	$\frac{11\pi}{2}$	6π
cos 3x	1	0	−1	0	1	0	−1	0	1	0	−1	0	1

Period = $\frac{2\pi}{3}$

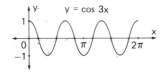

6.

x	0	$\frac{\pi}{2}$	π	$\frac{3\pi}{2}$	2π
$\frac{1}{3}x$	0	$\frac{\pi}{6}$	$\frac{\pi}{3}$	$\frac{\pi}{2}$	$\frac{2\pi}{3}$
$\sin\frac{1}{3}x$	0	$\frac{1}{2}$	$\frac{\sqrt{3}}{2}\approx .9$	1	$\frac{\sqrt{3}}{2}\approx .9$

Period = 6π

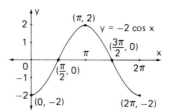

7.

x	0	$\frac{\pi}{8}$	$\frac{\pi}{4}$	$\frac{3\pi}{8}$	$\frac{\pi}{2}$	$\frac{5\pi}{8}$	$\frac{3\pi}{4}$	$\frac{7\pi}{8}$	π	$\frac{9\pi}{8}$	$\frac{5\pi}{4}$	$\frac{11\pi}{8}$	$\frac{3\pi}{2}$	$\frac{13\pi}{8}$	$\frac{7\pi}{4}$	$\frac{15\pi}{8}$	2π
4x	0	$\frac{\pi}{2}$	π	$\frac{3\pi}{2}$	2π	$\frac{5\pi}{2}$	3π	$\frac{7\pi}{2}$	4π	$\frac{9\pi}{2}$	5π	$\frac{11\pi}{2}$	6π	$\frac{13\pi}{2}$	7π	$\frac{15\pi}{2}$	8π
$\frac{1}{4}\sin 4x$	0	$\frac{1}{4}$	0	$-\frac{1}{4}$	0	$\frac{1}{4}$	0	$-\frac{1}{4}$	0	$\frac{1}{4}$	0	$-\frac{1}{4}$	0	$\frac{1}{4}$	0	$-\frac{1}{4}$	0

Period = $\frac{\pi}{2}$

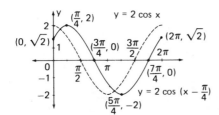

8. Period = 2π

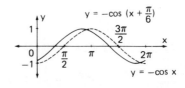

9. Amplitude: 2; Period: 2π;

Phase shift: $\frac{\pi}{4}$ units to the right

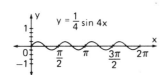

10. Amplitude: 1; Period: 2π

Phase shift: $\frac{\pi}{6}$ units to the left

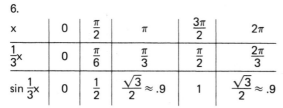

1. 2; 1   2. $\frac{\sqrt{3}}{2}$; $\sqrt{3}$; $\frac{\pi}{3}$; $\frac{1}{2}$; $\sqrt{3}+\frac{1}{2}$ 3. $\frac{\sqrt{2}}{2}$; $\sqrt{2}$; $\frac{\pi}{2}$; 0; $\sqrt{2}$ 4. $\frac{1}{2}$; 1; $\frac{2\pi}{3}$; $-\frac{1}{2}$; $\frac{1}{2}$

1.
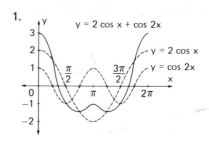

3.
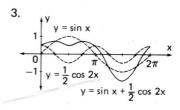

5.
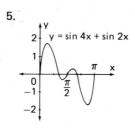

7.
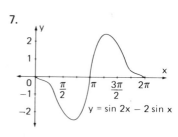

9.
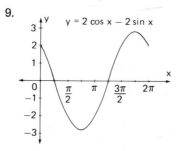

11.
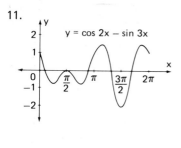

13.
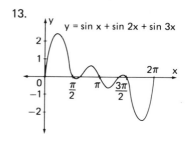

15.
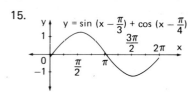

17.

$y = \frac{1}{x}\sin x$

1. −.58   2. −1.7   3. −.58   4. .58   5. 2   6. 2

GRAPHS OF TRIGONOMETRIC FUNCTIONS

1. $\frac{\pi}{3}; \frac{2\pi}{3}$ units to the right, b   2. 2π; 2π units to the right, c   3. π; $\frac{3\pi}{2}$ units to the left, d   4. $\frac{\pi}{2}; \frac{\pi}{2}$ units to the right, a

PAGES 112-113   WRITTEN EXERCISES

1. 2π; $\frac{\pi}{2}$ units to the right   3. π; $\frac{\pi}{2}$ units to the left   5. $\frac{\pi}{3}; \frac{\pi}{3}$ units to the right   7. π; $\frac{\pi}{8}$ units to the right

9. $\frac{2\pi}{3}$; π units to the right   11. $-\frac{3\pi}{2}; -\frac{\pi}{2}; \frac{\pi}{2}; \frac{3\pi}{2}$ 13. $-\frac{3\pi}{2}; -\frac{\pi}{2}; \frac{\pi}{2}; \frac{3\pi}{2}$

15.

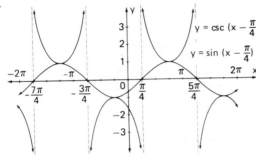

17.

Period: $\frac{\pi}{2}$;

Phase shift: 0

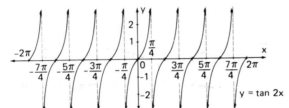

19.

Period: $\frac{2\pi}{3}$;

Phase shift: 0

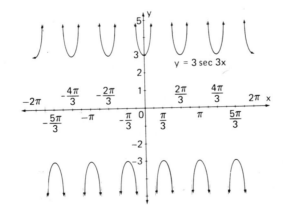

21.

Period: π;

Phase shift: $\frac{\pi}{2}$ units to the right

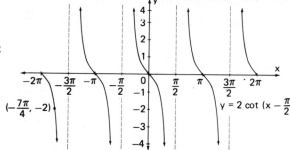

$\left(-\frac{7\pi}{4}, -2\right)$

$y = 2 \cot \left(x - \frac{\pi}{2}\right)$

23.

Period: $\frac{2\pi}{3}$;

Phase shift: $\frac{\pi}{3}$ units to the left

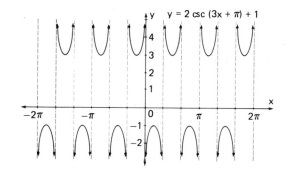

$y = 2 \csc (3x + \pi) + 1$

25. $\tan (\pi + \theta) = \dfrac{\sin (\pi + \theta)}{\cos (\pi + \theta)} = \dfrac{-\sin \theta}{-\cos \theta} = \tan \theta$ **27.** $|\cot x| = \left| \dfrac{\cos x}{\sin x} \right| \geq \left| \dfrac{\cos x}{1} \right| = |\cos x|$

29. $|\cot x| = \left| \dfrac{\cos x}{\sin x} \right| = \left| \cos x \cdot \dfrac{1}{\sin x} \right| = |\cos x \cdot \csc x| \leq |1 \cdot \csc x| = |\csc x|$

PAGE 115 WRITTEN EXERCISES

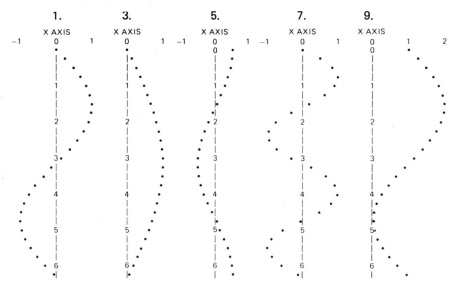

GRAPHS OF TRIGONOMETRIC FUNCTIONS

3. 1.7 radians 5. $-\dfrac{\pi}{6}$ 7. $\dfrac{7\pi}{6}$ 9. $-18°$ 11. $-150°$ 13. 1.1810 15. .2240 17. $62°\ 40'$

19. $75°\ 30'$ 21. 838.8 cm 23. .6000 25. .8000 27. .5000 29. $-.5000$ 31. $-.8660$

33. .8660 35. See Example 3 on page 83.

37.

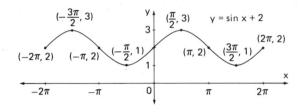

39. Amplitude: $\dfrac{1}{2}$

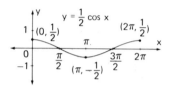

41. Amplitude: $\dfrac{1}{2}$

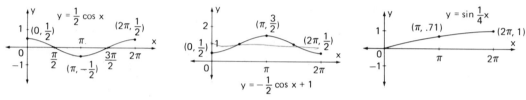

43. Period: 8π

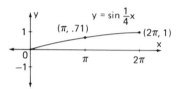

45. Period: 4π

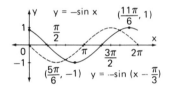

47. Amplitude: 1;

　period: 2π;

　phase shift: $\dfrac{\pi}{3}$ units to

　　　　the right

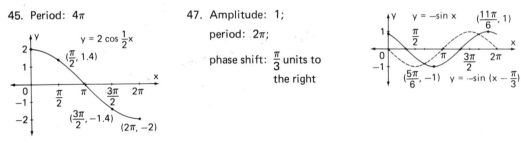

49. Amplitude: $\dfrac{1}{2}$;

　period: $\dfrac{2\pi}{4} = \dfrac{\pi}{2}$;

　phase shift: 2π units

　　　to the left

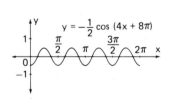

51.

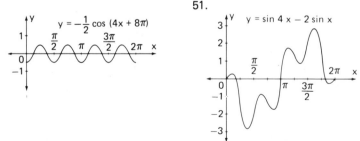

53.

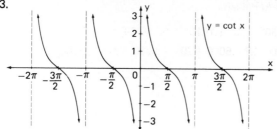

55.

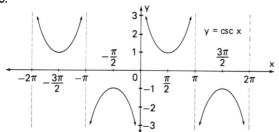

PAGE 119 CHAPTER TEST

1. 3   2. $-\dfrac{7\pi}{6}$ 3. 240°   4. .9832

5. 33° 00'   6. 120π rad/sec

7. 18.8 cm   8. −.4226   9. −.8000

10.

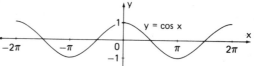

11.

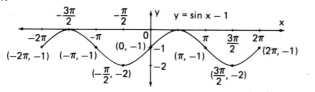

12. Amplitude: $\dfrac{1}{2}$

14. Period: $\dfrac{2\pi}{\frac{1}{4}} = 8\pi$

13.

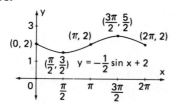

15.

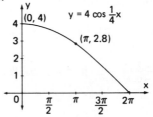

GRAPHS OF TRIGONOMETRIC FUNCTIONS

16. Amplitude: $\frac{1}{4}$; period: $\frac{\pi}{2}$; phase shift: 2π units to the left

17.

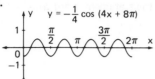

18.

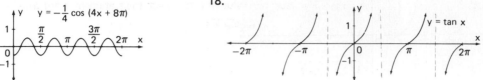

19.

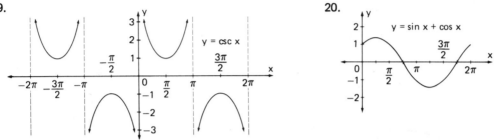

20.

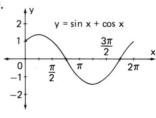

PAGES 120-122 CUMULATIVE REVIEW: CHAPTERS 1 AND 2

1. d 3. b 5. a 7. d 9. a 11. c 13. a 15. c 17. d 19. a 21. c 23. b 25. a 27. c

CHAPTER 3 TRIGONOMETRIC IDENTITIES

PAGE 126 CLASSROOM EXERCISES

1. $\dfrac{\sin^2 \theta}{\cos^2 \theta} - 2\dfrac{\sin \theta}{\cos \theta}$, or $\dfrac{\sin \theta}{\cos \theta}\Big(\dfrac{\sin \theta}{\cos \theta} - 2\Big)$ 2. $\tan^2 \theta - 2 \tan \theta$, or $\tan \theta\,(\tan \theta - 2)$ 3. $\dfrac{1}{\cot^2 \theta} - \dfrac{2}{\cot \theta}$,
or $\dfrac{1 - 2 \cot \theta}{\cot^2 \theta}$

PAGES 126-127 WRITTEN EXERCISES

1. $\csc \theta = \dfrac{r}{y}$, $\therefore \csc \theta = \dfrac{r}{r \sin \theta}$, $\csc \theta = \dfrac{1}{\sin \theta}$ 3. $\cot \theta = \dfrac{x}{y} = \dfrac{1}{\frac{y}{x}} = \dfrac{1}{\tan \theta}$ 5. $1 + \cot^2 \theta = 1 + \dfrac{x^2}{y^2} = \dfrac{y^2 + x^2}{y^2}$

$= \dfrac{r^2}{y^2} = \csc^2 \theta$ 7. $\dfrac{1 - 2 \cos^2 \theta}{(1 - \cos^2 \theta)^2}$ 9. $1 - 2 \cos^2 \theta$ 11. $\dfrac{1 - \sin^2 \theta}{\sin \theta}$ 13. $\sin^2 \theta + \dfrac{1}{\sin \theta}$ 15. $\dfrac{\tan^2 \theta}{\tan^2 \theta + 1}$

17. $\tan \theta + \dfrac{1}{\tan \theta}$ 19. $\pm\sqrt{1 - \cos^2 \theta}$ 21. $\pm\dfrac{\sqrt{1 - \cos^2 \theta}}{\cos \theta}$ 23. $\dfrac{\csc \theta \sec \theta}{\sec \theta + \csc \theta}$ 25. Let $\theta = \dfrac{\pi}{3}$,

$\sin \theta + \cos \theta = \dfrac{\sqrt{3}}{2} + \dfrac{1}{2} \neq 1$. Since there is at least one value of θ for which the statement is not true,

$\sin \theta + \cos \theta = 1$ is not an identity. 27. $\csc \theta = \pm\dfrac{\sqrt{\tan^2 \theta + 1}}{\tan \theta}$; $\sin \theta = \pm\dfrac{\tan \theta}{\sqrt{\tan^2 \theta + 1}}$;

$\sec \theta = \pm\sqrt{1 + \tan^2 \theta}$; $\cos \theta = \pm\dfrac{1}{\sqrt{1 + \tan^2 \theta}}$; $\cot \theta = \dfrac{1}{\tan \theta}$ 29. $\sin \theta = \pm\dfrac{\sqrt{\sec^2 \theta - 1}}{\sec \theta}$;

$\csc \theta = \pm\dfrac{\sec \theta}{\sqrt{\sec^2 \theta - 1}}$; $\cos \theta = \dfrac{1}{\sec \theta}$; $\tan \theta = \pm\sqrt{\sec^2 \theta - 1}$; $\cot \theta = \pm\dfrac{1}{\sqrt{\sec^2 \theta - 1}}$

PAGE 128   REVIEW CAPSULE FOR SECTION 3-2

1. $\dfrac{a - b}{b}$ 2. $\dfrac{\tan \theta - \sec \theta}{\sec \theta}$ 3. $\dfrac{2b - x^3}{x^2}$ 4. $\dfrac{2 \cos \theta - \sin^3 \theta}{\sin^2 \theta}$ 5. $\dfrac{1}{\cos \theta}$, or $\sec \theta$ 6. $\dfrac{ad + 2b}{bd}$

7. $\sin \theta + 2 \tan \theta$ 8. $\dfrac{3x - 1}{x^2 - 4}$ 9. $\dfrac{3 \csc \theta - 1}{\csc^2 \theta - 4}$

PAGE 130   CLASSROOM EXERCISES

1. $1 - \cos^2 \theta$ 2. $\dfrac{1}{1 - \sin^2 \theta}$ 3. $1 + \tan^2 \theta$ $\bigg|$ $\dfrac{1}{1 - \sin^2 \theta}$ 4. $\dfrac{1}{\cot^2 \theta + 1}$ $\bigg|$ $1 - \cos^2 \theta$

$\quad\;\; \sin^2 \theta$ $\qquad\quad \dfrac{1}{\cos^2 \theta}$ $\qquad\quad\;\; \sec^2 \theta$ $\bigg|$ $\dfrac{1}{\cos^2 \theta}$ $\qquad\quad\;\; \dfrac{1}{\csc^2 \theta}$ $\bigg|$ $\sin^2 \theta$

$\quad\;\; \dfrac{1}{\csc^2 \theta}$ $\qquad\quad \sec^2 \theta$ $\qquad\quad\;\; \sec^2 \theta = \sec^2 \theta$ $\qquad\quad\;\; \sin^2 \theta = \sin^2 \theta$

PAGES 130-131   WRITTEN EXERCISES

1. $\sin \theta \cdot \sec \theta$ 3. $\dfrac{1 - \cos^2 \theta}{\cos^2 \theta}$ 5. $\sec^2 \theta - 1$ 7. $\dfrac{\cos^2 \theta}{\sin \theta} + \sin \theta$ 9. $\dfrac{\cot \theta}{\cos \theta}$

$\;\sin \theta \cdot \dfrac{1}{\cos \theta}$ $\quad\;\; \dfrac{\sin^2 \theta}{\cos^2 \theta}$ $\qquad\quad\; \tan^2 \theta$ $\qquad\quad \dfrac{\cos^2 \theta + \sin^2 \theta}{\sin \theta}$ $\qquad\; \dfrac{\cos \theta}{\sin \theta} \cdot \dfrac{1}{\cos \theta}$

$\;\tan \theta$ $\qquad\quad\;\; \tan^2 \theta$ $\qquad\qquad\qquad\qquad\qquad \dfrac{1}{\sin \theta}$ $\qquad\qquad\; \dfrac{1}{\sin \theta}$

$\qquad\qquad\qquad\qquad\qquad\qquad\qquad\qquad\qquad\qquad\quad \csc \theta$ $\qquad\qquad\quad \csc \theta$

11. $\csc^2 \theta \cdot \tan^2 \theta - 1$ 13. $\csc^4 \theta - \cot^4 \theta$ 15. $(1 - \tan \theta)^2$

$\;\; \dfrac{1}{\sin^2 \theta} \cdot \dfrac{\sin^2 \theta}{\cos^2 \theta} - 1$ $\qquad (\csc^2 \theta - \cot^2 \theta)(\csc^2 \theta + \cot^2 \theta)$ $\qquad 1 - 2 \tan \theta + \tan^2 \theta$

$\;\; \dfrac{1}{\cos^2 \theta} - 1$ $\qquad\qquad 1 \cdot (\csc^2 \theta + \cot^2 \theta)$ $\qquad\qquad\qquad (1 + \tan^2 \theta) - 2 \tan \theta$

$\;\; \sec^2 \theta - 1$ $\qquad\qquad\quad \csc^2 \theta + \cot^2 \theta$ $\qquad\qquad\qquad\qquad\; \sec^2 \theta - 2 \tan \theta$

$\;\; \tan^2 \theta$

17. $\qquad\qquad\quad \dfrac{\cot \theta}{\cos \theta} + \dfrac{\sec \theta}{\cot \theta}$ $\bigg|$ $\sec^2 \theta \cdot \csc \theta$ $\qquad\qquad \dfrac{\cos^2 \theta + \sin^2 \theta}{\sin \theta \cdot \cos^2 \theta}$ $\bigg|$

$\dfrac{\cos \theta}{\sin \theta} \cdot \dfrac{1}{\cos \theta} + \dfrac{1}{\cos \theta} \cdot \dfrac{1}{\dfrac{\cos \theta}{\sin \theta}}$ $\qquad\quad \dfrac{1}{\cos^2 \theta} \cdot \dfrac{1}{\sin \theta}$ $\qquad\qquad \dfrac{1}{\sin \theta \cdot \cos^2 \theta}$

$\qquad\qquad\qquad \dfrac{1}{\sin \theta} + \dfrac{\sin \theta}{\cos^2 \theta}$ $\qquad\qquad\qquad\qquad\qquad \dfrac{1}{\cos^2 \theta} \cdot \dfrac{1}{\sin \theta} = \dfrac{1}{\cos^2 \theta} \cdot \dfrac{1}{\sin \theta}$

TRIGONOMETRIC IDENTITIES

19. $\sec\theta - \tan\theta \cdot \sin\theta = \dfrac{1}{\cos\theta} - \dfrac{\sin\theta}{\cos\theta} \cdot \sin\theta = \dfrac{1 - \sin^2\theta}{\cos\theta} = \dfrac{\cos^2\theta}{\cos\theta} = \cos\theta$ 21. $\dfrac{1 - \sin^2\theta}{1 + \tan^2\theta} = \dfrac{\cos^2\theta}{\sec^2\theta} \cdot$

$= \cos^4\theta$ 23. $\dfrac{\sin\theta + \tan\theta}{1 + \sec\theta} = \dfrac{\sin\theta + \dfrac{\sin\theta}{\cos\theta}}{1 + \dfrac{1}{\cos\theta}} = \dfrac{\dfrac{\sin\theta\cos\theta + \sin\theta}{\cos\theta}}{\dfrac{\cos\theta + 1}{\cos\theta}} = \dfrac{\sin\theta(\cos\theta + 1)}{\cos\theta} = \cos\theta + 1$

25. $\sec\theta - \tan\theta\sin\theta = \dfrac{1}{\cos\theta} - \dfrac{\sin^2\theta}{\cos\theta} = \dfrac{1 - \sin^2\theta}{\cos\theta} = \dfrac{\cos^2\theta}{\cos\theta} = \cos\theta = \dfrac{1}{\sec\theta}$ 27. $\dfrac{\sec\theta \cdot \sin\theta \cdot}{\tan\theta + \cot\theta}$

$= \dfrac{\dfrac{1}{\cos\theta} \cdot \sin\theta}{\dfrac{\sin\theta}{\cos\theta} + \dfrac{\cos\theta}{\sin\theta}} = \dfrac{\dfrac{\sin\theta}{\cos\theta}}{\dfrac{\sin^2\theta + \cos^2\theta}{\cos\theta\sin\theta}} = \sin^2\theta$ 29. $\cos^2\theta - \sin^2\theta = 1 - \sin^2\theta - \sin^2\theta = 1 - 2\sin^2\theta$

31. $\dfrac{\sec^2\theta}{\sec^2\theta - 1} = \dfrac{\dfrac{1}{\cos^2\theta}}{\dfrac{1}{\cos^2\theta} - 1} = \dfrac{1}{\sin^2\theta} = \csc^2\theta$ 33. $(\sin\theta + \cos\theta)^2 + (\sin\theta - \cos\theta)^2 = \sin^2\theta$

$+ 2\sin\theta\cos\theta + \cos^2\theta + \sin^2\theta - 2\sin\theta\cos\theta + \cos^2\theta = \sin^2\theta + \cos^2\theta + \sin^2\theta + \cos^2\theta = 1 + 1 = 2$

35. $\dfrac{\tan\theta - 1}{\tan\theta + 1} = \dfrac{\dfrac{1}{\cot\theta} - 1}{\dfrac{1}{\cot\theta} + 1} = \dfrac{\dfrac{1 - \cot\theta}{\cot\theta}}{\dfrac{1 + \cot\theta}{\cot\theta}} = \dfrac{1 - \cot\theta}{1 + \cot\theta}$ 37. $\dfrac{\cos\theta + 1}{\sin^3\theta} = \dfrac{\cos\theta + 1}{\sin\theta(1 - \cos^2\theta)} =$

$\dfrac{1 + \cos\theta}{\sin\theta(1 - \cos\theta)(1 + \cos\theta)} = \dfrac{1}{\sin\theta(1 - \cos\theta)} = \dfrac{1}{\sin\theta} \cdot \dfrac{1}{1 - \cos\theta} = \dfrac{\csc\theta}{1 - \cos\theta}$ 39. $\dfrac{\tan\theta}{\sec\theta} + \dfrac{\cot\theta}{\csc\theta}$

$= \dfrac{\tan\theta\csc\theta + \cot\theta\sec\theta}{\sec\theta\csc\theta} = \dfrac{\dfrac{\sin\theta}{\cos\theta} \cdot \dfrac{1}{\sin\theta} + \dfrac{\cos\theta}{\sin\theta} \cdot \dfrac{1}{\cos\theta}}{1} = \sin\theta + \cos\theta$ 41. $\dfrac{2\tan\theta}{1 + \tan^2\theta} = \dfrac{2\tan\theta}{\sec^2\theta}$

$= \dfrac{\dfrac{2\sin\theta}{\cos\theta}}{\dfrac{1}{\cos^2\theta}} = 2\sin\theta\cos\theta$ 43. $\dfrac{1 - \cos^6\theta}{\sin^2\theta} = \dfrac{(1 - \cos^6\theta)(1 + \cos^2\theta + \cos^4\theta)}{1 - \cos^2\theta} = 1 + \cos^2\theta + \cos^4\theta$

45. $\dfrac{\sin^3\theta + \cos^3\theta}{\sin\theta + \cos\theta} = \dfrac{(\sin\theta + \cos\theta)(\sin^2\theta - \sin\theta\cos\theta + \cos^2\theta)}{\sin\theta + \cos\theta} = 1 - \sin\theta\cos\theta$

47. $\dfrac{1 - 2\sin\theta - 3\sin^2\theta}{\cos^2\theta} = \dfrac{(1 + \sin\theta)(1 - 3\sin\theta)}{1 - \sin^2\theta} = \dfrac{(1 + \sin\theta)(1 - 3\sin\theta)}{(1 + \sin\theta)(1 - \sin\theta)} = \dfrac{1 - 3\sin\theta}{1 - \sin\theta}$

49. $\sec x + \tan x = \dfrac{1}{\cos x} + \dfrac{\sin x}{\cos x} = \dfrac{1 + \sin x}{\cos x} = \dfrac{1 + \sin x}{\cos x} \cdot \dfrac{\cos x}{\cos x} = \dfrac{\cos x(1 + \sin x)}{\cos^2 x} =$

$\dfrac{\cos x(1 + \sin x)}{1 - \sin^2 x} = \dfrac{\cos x}{1 - \sin x}$

PAGE 131   REVIEW CAPSULE FOR SECTION 3-3

1. $2\sqrt{3}$ 2. $5\sqrt{2}$ 3. $12\sqrt{3}$ 4. $7\sqrt{2}$ 5. $\sqrt{2}$ 6. $\sqrt{6}$ 7. $9\sqrt{2}$ 8. 36

PAGE 135   CLASSROOM EXERCISES

1. $\dfrac{\sqrt{3}}{2}$ 2. $-\dfrac{\sqrt{2}}{2}$ 3. $\frac{1}{4}(\sqrt{6} - \sqrt{2})$ 4. $-\frac{1}{4}(\sqrt{6} + \sqrt{2})$

1. $\frac{1}{4}(\sqrt{6} - \sqrt{2})$ 3. $-\frac{1}{4}(\sqrt{2} + \sqrt{6})$ 5. $\frac{1}{4}(\sqrt{6} + \sqrt{2})$ 7. $\frac{1}{4}(\sqrt{6} + \sqrt{2})$ 9. $-\frac{16}{65}; \frac{56}{65}$ 11. $-\frac{16}{65}; \frac{56}{65}$

13. $\frac{297}{425}; \frac{87}{425}$ 15. $\cos \theta = \cos (-\theta) = \cos [(90° - \theta) - 90°] = \cos (90° - \theta) \cos 90° + \sin (90° - \theta)$

$\sin 90° = \cos (90° - \theta) \cdot 0 + \sin (90° - \theta) \cdot 1 = \sin (90° - \theta)$ 17. $\cos (180° + \theta) = \cos 180° \cos \theta$

$- \sin 180° \sin \theta = -1 \cdot \cos \theta - 0 \cdot \sin \theta = -\cos \theta$ 19. $\cos (\frac{3\pi}{2} - \theta) = \cos \frac{3\pi}{2} \cos \theta + \sin \frac{3\pi}{2} \sin \theta$

$= 0 \cdot \cos \theta + (-1) \sin \theta = -\sin \theta$ 21. $\cos (2\pi - \theta) = \cos 2\pi \cos \theta + \sin 2\pi \sin \theta = 1 \cdot \cos \theta + 0 \cdot \sin \theta$

$= \cos \theta$ 23. $\frac{\sqrt{2}}{2} \cos \theta - \frac{\sqrt{2}}{2} \sin \theta$ 25. $\frac{1}{2} \cos \theta + \frac{\sqrt{3}}{2} \sin \theta$ 27. $\frac{\sqrt{3}}{2} \cos \theta + \frac{1}{2} \sin \theta$ 29. $\cos (\alpha + \beta)$

$- \cos (\alpha - \beta) = (\cos \alpha \cos \beta - \sin \alpha \sin \beta) - (\cos \alpha \cos \beta + \sin \alpha \sin \beta) = -2 \sin \alpha \sin \beta$ 31. $\cos (-\alpha)$

$= \cos \alpha = \cos (360° + \alpha) = \cos (360° - (-\alpha)) = \cos 360° \cos (-\alpha) + \sin 360° \sin (-\alpha) = 1 \cdot \cos (-\alpha)$

$+ 0 \cdot \sin (-\alpha) = \cos (-\alpha)$

PAGE 136   REVIEW CAPSULE FOR SECTION 3-4
1. 40°   2. 70°   3. 27°   4. 78°   5. 38° 40'   6. 72° 20'

PAGE 139   CLASSROOM EXERCISES

1. $\frac{\pi}{2} - \frac{\pi}{3}$ 2. $90° + 45°$ 3. $\frac{\pi}{4} + \frac{\pi}{6}$ 4. $225° - 30°$ 5. $\frac{1}{2}$ 6. $\frac{\sqrt{2}}{2}$ 7. $\frac{1}{4}(\sqrt{6} + \sqrt{2})$ 8. $\frac{1}{4}(\sqrt{2} - \sqrt{6})$

PAGES 139-140   WRITTEN EXERCISES

1. $\frac{1}{4}(\sqrt{6} - \sqrt{2})$ 3. $\frac{1}{4}(\sqrt{2} - \sqrt{6})$ 5. $\frac{1}{4}(\sqrt{2} - \sqrt{6})$ 7. $-\frac{\sqrt{2}}{2}$ 9. $\frac{63}{65}; -\frac{33}{65}$ 11. $\frac{63}{65}; \frac{33}{65}$ 13. $-\frac{304}{425};$

$\frac{416}{425}$ 15. $\sin (\frac{\pi}{2} + \theta) = \sin \frac{\pi}{2} \cos \theta + \cos \frac{\pi}{2} \sin \theta = 1 \cdot \cos \theta + 0 \cdot \sin \theta = \cos \theta$ 17. $\sin (\pi - \theta)$

$= \sin \pi \cos \theta - \cos \pi \sin \theta = 0 \cdot \cos \theta - (-1)(\sin \theta) = \sin \theta$ 19. $\sin (270° - \theta) = \sin 270° \cos \theta$

$- \cos 270° \sin \theta = (-1)(\cos \theta) - 0 \cdot \sin \theta = -\cos \theta$ 21. $\sin (2\pi - \theta) = \sin 2\pi \cos \theta - \cos 2\pi \sin \theta$

$= 0 \cdot \cos \theta - 1 \cdot \sin \theta = -\sin \theta$ 23. $\frac{\sqrt{2}}{2} \sin \theta + \frac{\sqrt{2}}{2} \cos \theta$ 25. $\frac{1}{2} \sin \theta - \frac{\sqrt{3}}{2} \cos \theta$ 27. $\frac{\sqrt{3}}{2} \sin \theta$

$- \frac{1}{2} \cos \theta$ 29. $2 \sin (\frac{\pi}{4} + \theta) \sin (\theta - \frac{\pi}{4}) = 2(\frac{\sqrt{2}}{2} \cos \theta + \frac{\sqrt{2}}{2} \sin \theta)(\frac{\sqrt{2}}{2} \sin \theta - \frac{\sqrt{2}}{2} \cos \theta)$

$= 2(\frac{\sqrt{2}}{2})^2 (\cos \theta + \sin \theta)(\sin \theta - \cos \theta) = 1 \cdot (\cos \theta \sin \theta - \cos^2 \theta + \sin^2 \theta - \sin \theta \cos \theta) = \sin^2 \theta - \cos^2 \theta$

31. $\sin (\alpha + \beta) \cdot \sin (\alpha - \beta) = (\sin \alpha \cos \beta + \cos \alpha \sin \beta)(\sin \alpha \cos \beta - \cos \alpha \sin \beta) = (\sin \alpha \cos \beta)^2$

$- \sin \alpha \cos \alpha \sin \beta \cos \beta + \sin \alpha \cos \alpha \sin \beta \cos \beta - (\cos \alpha \sin \beta)^2 = \sin^2 \alpha \cos^2 \beta - \cos^2 \alpha \sin^2 \beta$

$= \sin^2 \alpha (1 - \sin^2 \beta) - (1 - \sin^2 \alpha) \sin^2 \beta = \sin^2 \alpha - \sin^2 \alpha \sin^2 \beta - \sin^2 \beta + \sin^2 \beta \sin^2 \alpha = \sin^2 \alpha - \sin^2 \beta$

PAGE 140   REVIEW

1. $\sin \theta = \frac{y}{r} \therefore r \sin \theta = y, \cos \theta = \frac{x}{r} \therefore r \cos \theta = x; \cot \theta = \frac{x}{y}$ then $\cot \theta = \frac{r \cos \theta}{r \sin \theta}$ and $\cot \theta = \frac{\cos \theta}{\sin \theta}$

2. $\sec^2 \theta - \tan^2 \theta = (\frac{r}{x})^2 - (\frac{y}{x})^2 = \frac{r^2 - y^2}{x^2} = \frac{x^2}{x^2} = 1 \therefore \sec^2 \theta - \tan^2 \theta = 1$

TRIGONOMETRIC IDENTITIES

3.

$$\frac{\cos\theta - \sin\theta}{\cos\theta}$$

$$\frac{\cos\theta}{\cos\theta} - \frac{\sin\theta}{\cos\theta}$$

$$1 - \tan\theta$$

4.

$$\tan\theta + \cot\theta$$

$$\frac{\sin\theta}{\cos\theta} + \frac{\cos\theta}{\sin\theta}$$

$$\frac{\sin^2\theta + \cos^2\theta}{\cos\theta\sin\theta}$$

$$\frac{1}{\cos\theta\sin\theta}$$

$$\frac{1}{\cos\theta} \cdot \frac{1}{\sin\theta}$$

$$\sec\theta\csc\theta$$

5.

$$\frac{\cot\theta + 1}{\cot\theta}$$

$$\frac{\cot\theta}{\cot\theta} + \frac{1}{\cot\theta}$$

$$1 + \tan\theta$$

6.

$$\tan\theta\,(\tan\theta + \cot\theta)$$

$$\tan^2\theta + \tan\theta\cot\theta$$

$$\tan^2\theta + 1$$

$$\sec^2\theta$$

7. $-\frac{1}{4}(\sqrt{2} + \sqrt{6})$ 8. $\cos(3\pi - \theta) = \cos 3\pi\cos\theta + \sin 3\pi\sin\theta = -1 \cdot \cos\theta + 0 \cdot \sin\theta = -\cos\theta$

9. $-\frac{16}{65}$ 10. $-\frac{1}{4}(\sqrt{6} + \sqrt{2})$ 11. $\sin(\pi + \theta) = \sin\pi\cos\theta + \cos\pi\sin\theta = 0 \cdot \cos\theta + (-1)\sin\theta$

$= -\sin\theta$ 12. $-\frac{33}{65}$

PAGE 143   CLASSROOM EXERCISES

1. $\sqrt{3} - 2$ 2. -1 3. $2 + \sqrt{3}$ 4. $2 - \sqrt{3}$

PAGE 144   WRITTEN EXERCISES

1. $-2 + \sqrt{3}$ 3. $2 + \sqrt{3}$ 5. $\sqrt{3} - 2$ 7. $\sqrt{3} - 2$ 9. $-\tan\theta$ 11. $-\tan\theta$ 13. $-\frac{63}{16}, -\frac{33}{56}$

15. $-\frac{63}{16}, -\frac{33}{56}$ 17. $-\frac{304}{297}, \frac{416}{87}$ 19. $\frac{1 + \tan\theta}{1 - \tan\theta}$ 21. $\frac{\sqrt{3} - \tan\theta}{1 + \sqrt{3}\tan\theta}$ 23. $\frac{3\tan\theta - \sqrt{3}}{3 + \sqrt{3}\tan\theta}$

25. $\tan(3\pi - \theta) = \frac{\tan 3\pi - \tan\theta}{1 + \tan 3\pi\tan\theta} = \frac{0 - \tan\theta}{1 + 0} = -\tan\theta$ 27. $-\frac{\cot\alpha\cot\beta + 1}{\cot\alpha - \cot\beta}$

PAGE 144   REVIEW CAPSULE FOR SECTION 3-6

1. $\sin\alpha\cos\beta + \cos\alpha\sin\beta$ 2. $\sin\alpha\cos\beta - \cos\alpha\sin\beta$ 3. $\cos\alpha\cos\beta - \sin\alpha\sin\beta$ 4. $\cos\alpha\cos\beta$
$+ \sin\alpha\sin\beta$ 5. $\frac{\tan\alpha + \tan\beta}{1 - \tan\alpha\tan\beta}$ 6. $\frac{\tan\alpha - \tan\beta}{1 + \tan\alpha\tan\beta}$

PAGE 147   CLASSROOM EXERCISES

1. $\frac{24}{25}$ 2. $\frac{7}{25}$ 3. $\frac{24}{7}$

PAGES 147-148   WRITTEN EXERCISES

1. $\frac{24}{25}$ 3. $-\frac{24}{7}$ 5. $\frac{4}{5}$ 7. $-\frac{120}{169}$ 9. $-\frac{120}{119}$ 11. $0.6300; -0.7766; -0.8112$

13.

$$\frac{\sin 2\alpha}{1 + \cos 2\alpha}$$

$$\frac{2\sin\alpha\cos\alpha}{1 + (2\cos^2\alpha - 1)}$$

$$\frac{2\sin\alpha\cos\alpha}{2\cos^2\alpha}$$

$$\frac{\sin\alpha}{\cos\alpha}$$

$$\tan\alpha$$

15.

$$\frac{2\tan\alpha}{1 + \tan^2\alpha}$$

$$\frac{2\tan\alpha}{\sec^2\alpha}$$

$$\frac{2\frac{\sin\alpha}{\cos\alpha}}{\frac{1}{\cos^2\alpha}}$$

$$2\sin\alpha\cos\alpha$$

$$\sin 2\alpha$$

17.

$$4\csc^2 2\alpha \quad\bigg|\quad \sec^2\alpha\,(1 + \cot^2\alpha)$$

$$\frac{4}{\sin^2 2\alpha} \quad\bigg|\quad \sec^2\alpha\csc^2\alpha$$

$$\frac{4}{(2\sin\alpha\cos\alpha)^2}$$

$$\frac{1}{\sin^2\alpha\cos^2\alpha} \quad\bigg|$$

$$\csc^2\alpha\sec^2\alpha \;=\; \csc^2\alpha\sec^2\alpha$$

19.

$$\frac{1 + \cos 2\alpha}{\sin 2\alpha}$$

$$\frac{1 + (2\cos^2\alpha - 1)}{2\sin\alpha\cos\alpha}$$

$$\frac{2\cos^2\alpha}{2\sin\alpha\cos\alpha}$$

$$\frac{\cos\alpha}{\sin\alpha}$$

$$\cot\alpha$$

21.

$$\frac{\cot^2 \alpha - 1}{\csc^2 \alpha}$$

$$\frac{\cot^2 \alpha}{\csc^2 \alpha} - \frac{1}{\csc^2 \alpha}$$

$$\frac{\dfrac{\cos^2 \alpha}{\sin^2 \alpha}}{\dfrac{1}{\sin^2 \alpha}} - \sin^2 \alpha$$

$$\cos^2 \alpha - \sin^2 \alpha$$

$$\cos 2\alpha$$

23.

$$(\sin \alpha - \cos \alpha)^2 \quad \bigg| \quad \sec^2 \alpha - \tan^2 \alpha - \sin 2\alpha$$

$$\sin^2 \alpha - 2 \sin \alpha \cos \alpha + \cos^2 \alpha \quad \bigg| \quad 1 - \sin 2\alpha$$

$$(\sin^2 \alpha + \cos^2 \alpha) - 2 \sin \alpha \cos \alpha \quad \bigg|$$

$$1 - \sin 2\alpha \ = \ 1 - \sin 2\alpha$$

25.

$$\frac{\sec^2 \alpha}{2 - \sec^2 \alpha}$$

$$\frac{\dfrac{1}{\cos^2 \alpha}}{2 - \dfrac{1}{\cos^2 \alpha}}$$

$$\frac{1}{2 \cos^2 \alpha - 1}$$

$$\frac{1}{\cos 2\alpha}$$

$$\sec 2\alpha$$

27.

$$\frac{\sec^2 \alpha \csc^2 \alpha}{\csc^2 \alpha - \sec^2 \alpha}$$

$$\frac{\dfrac{1}{\cos^2 \alpha} \cdot \dfrac{1}{\sin^2 \alpha}}{\dfrac{1}{\sin^2 \alpha} - \dfrac{1}{\cos^2 \alpha}}$$

$$\frac{\dfrac{1}{\cos^2 \alpha - \sin^2 \alpha}}{}$$

$$\frac{1}{\cos 2\alpha}$$

$$\sec 2\alpha$$

29.

$$\sin 3\alpha$$

$$\sin (2\alpha + \alpha)$$

$$\sin 2\alpha \cos \alpha + \cos 2\alpha \sin \alpha$$

$$(2 \sin \alpha \cos \alpha) \cos \alpha + (2 \cos^2 \alpha - 1) \sin \alpha$$

$$4 \sin \alpha \cos^2 \alpha - \sin \alpha$$

$$4 \sin \alpha (1 - \sin^2 \alpha) - \sin \alpha$$

$$3 \sin \alpha - 4 \sin^3 \alpha$$

31.

$$\tan 3\alpha$$

$$\tan (2\alpha + \alpha)$$

$$\frac{\tan 2\alpha + \tan \alpha}{1 - \tan 2\alpha \tan \alpha}$$

$$\frac{\dfrac{2 \tan \alpha}{1 - \tan^2 \alpha} + \tan \alpha}{1 - (\dfrac{2 \tan \alpha}{1 - \tan^2 \alpha}) \tan \alpha}$$

$$\frac{2 \tan \alpha + \tan \alpha (1 - \tan^2 \alpha)}{(1 - \tan^2 \alpha) - 2 \tan^2 \alpha}$$

$$\frac{3 \tan \alpha - \tan^3 \alpha}{1 - 3 \tan^2 \alpha}$$

33. 6 m

35.

$$\sin 4\alpha$$

$$2 \sin 2\alpha \cos 2\alpha$$

$$2(2 \sin \alpha \cos \alpha)(2 \cos^2 \alpha - 1)$$

$$4 \sin \alpha \cos \alpha (2 \cos^2 \alpha - 1)$$

PAGE 148   REVIEW CAPSULE FOR SECTION 3-7

1. $\dfrac{\pi}{3}$ 2. 2π 3. $-\dfrac{3\pi}{4}$ 4. $-\dfrac{9\pi}{2}$ 5. 270°   6. −330°   7. 630°   8. −675°

PAGE 151   CLASSROOM EXERCISES

1. I   2. I   3. II   4. II   5. $-\dfrac{1}{4}\sqrt{14}$

TRIGONOMETRIC IDENTITIES

1. $\frac{1}{2}\sqrt{2-\sqrt{2}}$ 3. $\sqrt{2}-1$ 5. $-\frac{1}{2}\sqrt{2-\sqrt{3}}$ 7. $\frac{1}{2}\sqrt{2+\sqrt{2}}$ 9. $\sqrt{2}+1$ 11. $-\frac{1}{2}\sqrt{2+\sqrt{3}}$

13. $\frac{1}{2}\sqrt{2-\sqrt{2-\sqrt{3}}}$ 15. $2\sqrt{2-\sqrt{2}}-\sqrt{2+\sqrt{2}}\sqrt{2-\sqrt{2}}-1$ 17. $-\frac{5}{13}$ 19. $\frac{3\sqrt{13}}{13}$

21. $-\frac{3}{2}$ 23. $-\frac{4}{5}$ 25. $-\frac{2\sqrt{5}}{5}$ 27. $-\frac{1}{2}$ 29. $-\frac{5}{13}$ 31. $-\frac{3\sqrt{13}}{13}$ 33. $-\frac{3}{2}$ 35. $-\frac{4}{5}$ 37. $\frac{2\sqrt{5}}{5}$

39. $-\frac{1}{2}$ 41. $\tan\frac{\alpha}{2} = \pm\sqrt{\frac{1-\cos\alpha}{1+\cos\alpha}} \cdot \sqrt{\frac{1-\cos\alpha}{1-\cos\alpha}} = \pm\frac{1-\cos\alpha}{\sqrt{1-\cos^2\alpha}} = \pm\frac{1-\cos\alpha}{\sqrt{\sin^2\alpha}} = \frac{1-\cos\alpha}{\sin\alpha}$ The $\pm$

sign is not needed because $\tan\frac{\alpha}{2}$ and $\sin\alpha$ have the same sign.   43. $2\sin\frac{\alpha}{2}\cos\frac{\alpha}{2}$

$= 2\left(\pm\sqrt{\frac{1-\cos\alpha}{2}}\right)\left(\pm\sqrt{\frac{1+\cos\alpha}{2}}\right) = 2\left(\pm\sqrt{\frac{1-\cos^2\alpha}{4}}\right) = \pm\sqrt{\sin^2\alpha} = \sin\alpha$ 45. $\dfrac{2\tan\frac{\alpha}{2}}{1+\tan^2\frac{\alpha}{2}}$

$\dfrac{\frac{2\sin\alpha}{1+\cos\alpha}}{1+(\frac{\sin\alpha}{1+\cos\alpha})^2} = \frac{2\sin\alpha(1+\cos\alpha)}{(1+\cos\alpha)^2+\sin^2\alpha} = \frac{2\sin\alpha(1+\cos\alpha)}{1+2\cos\alpha+(\cos^2\alpha+\sin^2\alpha)} = \frac{2\sin\alpha(1+\cos\alpha)}{2+2\cos\alpha}$

$= \frac{\sin\alpha(1+\cos\alpha)}{1+\cos\alpha} = \sin\alpha$ 47. $\pm\frac{\sqrt{2+2\cos\alpha}}{\sin\alpha} = \pm\frac{\sqrt{2(1+\cos\alpha)}}{\sin\alpha} \cdot \frac{\sqrt{1-\cos\alpha}}{\sqrt{1-\cos\alpha}} = \pm\frac{\sqrt{2(1-\cos^2\alpha)}}{\sin\alpha\sqrt{1-\cos\alpha}}$

$= \pm\frac{\sqrt{2\sin^2\alpha}}{\sin\alpha\sqrt{1-\cos\alpha}} = \frac{\sqrt{2}}{\pm\sqrt{1-\cos\alpha}} = \frac{1}{\pm\sqrt{\frac{1-\cos\alpha}{2}}} = \frac{1}{\sin\frac{\alpha}{2}} = \csc\frac{\alpha}{2}$

49. $\dfrac{1-\tan\frac{\alpha}{2}}{1+\tan\frac{\alpha}{2}} = \dfrac{1-\dfrac{\sin\frac{\alpha}{2}}{\cos\frac{\alpha}{2}}}{1+\dfrac{\sin\frac{\alpha}{2}}{\cos\frac{\alpha}{2}}} = \dfrac{\cos\frac{\alpha}{2}-\sin\frac{\alpha}{2}}{\cos\frac{\alpha}{2}+\sin\frac{\alpha}{2}} \cdot \dfrac{\cos\frac{\alpha}{2}-\sin\frac{\alpha}{2}}{\cos\frac{\alpha}{2}-\sin\frac{\alpha}{2}} = \dfrac{\cos^2\frac{\alpha}{2}-2\sin\frac{\alpha}{2}\cos\frac{\alpha}{2}+\sin^2\frac{\alpha}{2}}{\cos^2\frac{\alpha}{2}-\sin^2\frac{\alpha}{2}}$

$= \dfrac{(\sin^2\frac{\alpha}{2}+\cos^2\frac{\alpha}{2})-2\sin\frac{\alpha}{2}\cos\frac{\alpha}{2}}{\cos^2\frac{\alpha}{2}-\sin^2\frac{\alpha}{2}} = \dfrac{1-\sin\alpha}{\cos\alpha}$ 51. $\alpha \approx 223° \pm n \cdot 360°$

PAGE 153   REVIEW CAPSULE FOR SECTION 3-8

1. $x = 4; y = 1$ 2. $x = 2; y = -1$ 3. $x = 11; y = 8$

PAGE 155   CLASSROOM EXERCISES

1. $\frac{1}{2}\sin\frac{7\pi}{12}+\frac{1}{2}\sin\frac{\pi}{12}$ 2. $-2\sin\frac{7\pi}{24}\sin\frac{\pi}{24}$

1. $\cos 72° + \cos 8°$ 3. $\sin \frac{3\pi}{10} - \sin \frac{\pi}{10}$ 5. $\sin 4x + \sin 2x$ 7. $\cos 7x + \cos 3x$ 9. $-\frac{1}{2}\cos 14x$
$+\frac{1}{2}\cos 6x$ 11. $\frac{1}{2}\sin 12x - \frac{1}{2}\sin 4x$ 13. $\frac{1}{2}\cos 6x + \frac{1}{2}\cos 2x$ 15. $-2\sin 37° \sin 14°$ 17. $2\cos 87°$
$\sin 44°$ 19. $2\sin \frac{7\pi}{24}\cos \frac{\pi}{24}$ 21. $2\cos \frac{1}{2}\cos \frac{1}{4}$ 23. $2\cos(\frac{2\pi}{3})\sin(-\frac{\pi}{4})$ 25. $\cos(\alpha + \beta) + \cos(\alpha - \beta)$
$= (\cos \alpha \cos \beta - \sin \alpha \sin \beta) + (\cos \alpha \cos \beta + \sin \alpha \sin \beta) = 2\cos \alpha \cos \beta$

27. $\dfrac{\cos 7t + \cos 5t}{\sin 7t - \sin 5t} = \dfrac{2\cos 6t \cos t}{2\cos 6t \sin t} = \dfrac{\cos t}{\sin t} = \dfrac{\csc t}{\sec t}$ 29. $\dfrac{\sin 4x + \sin 2x}{\cos 4x + \cos 2x} = \dfrac{2\sin 3x \cos x}{2\cos 3x \cos x} = \dfrac{\sin 3x}{\cos 3x}$

$= \tan 3x = \dfrac{1}{\cot 3x}$ 31. $\dfrac{\sin 3x - \sin x}{\cos 3x + \cos x} = \dfrac{2\cos 2x \sin x}{2\cos 2x \cos x} = \dfrac{\sin x}{\cos x} = \tan x$ 33. $\dfrac{\sin \alpha - \sin \beta}{\sin \alpha + \sin \beta}$

$= \dfrac{2\cos \frac{\alpha + \beta}{2}\sin \frac{\alpha - \beta}{2}}{2\sin \frac{\alpha + \beta}{2}\cos \frac{\alpha - \beta}{2}} = \dfrac{\dfrac{\sin \frac{\alpha - \beta}{2}}{\cos \frac{\alpha - \beta}{2}}}{\dfrac{\sin \frac{\alpha + \beta}{2}}{\cos \frac{\alpha + \beta}{2}}} = \dfrac{\tan \frac{\alpha - \beta}{2}}{\tan \frac{\alpha + \beta}{2}}$ 35. $\dfrac{\sin 5t - \sin 3t}{\sin 5t + \sin 3t} = \dfrac{2\cos 4t \sin t}{2\sin 4t \cos t} = \dfrac{\dfrac{\sin t}{\cos t}}{\dfrac{\sin 4t}{\cos 4t}} = \dfrac{\tan t}{\tan 4t}$

1. $\sin^2 x(\csc^2 x - 1) = \sin^2 x(\cot^2 x) = \sin^2 x\left(\dfrac{1}{\tan^2 x}\right) = \sin^2 x\left(\dfrac{\cos^2 x}{\sin^2 x}\right) = \cos^2 x$ 3. $\sec^2 x + \csc^2$

$= \sec^2 x\left(1 + \dfrac{\csc^2 x}{\sec^2 x}\right) = \sec^2 x\left(1 + \dfrac{\cos^2 x}{\sin^2 x}\right) = \sec^2 x(1 + \cot^2 x)(1 + \cot^2 x) = \sec^2 x \csc^2 x$

5. $\dfrac{\sec x}{\cot x + \tan x} = \dfrac{1}{\cos x\left(\dfrac{\cos x}{\sin x} + \dfrac{\sin x}{\cos x}\right)} = \dfrac{1}{\dfrac{\cos^2 x}{\sin x} + \sin x} = \dfrac{1}{\sin x(\tan^2 x + 1)} = \dfrac{1}{\sin x \sec^2 x} = \dfrac{\sin^2 x}{\sin x}$

$= \sin x$ 7. $\dfrac{\cos x - \cos 2x}{\sin 2x + \sin x} = \dfrac{2\sin \frac{3x}{2}\sin \frac{x}{2}}{2\sin \frac{3x}{2}\cos \frac{x}{2}} = \tan \frac{x}{2}$ 9. $\sin 2x \tan x = 2\sin x \cos x \tan x = 2\sin^2 x$

11. $\tan^2 x - \sin^2 x = \tan^2 x\left(1 - \dfrac{\sin^2 x}{\tan^2 x}\right) = \tan^2 x\left(1 = \sin^2 x\left(\dfrac{\cos^2 x}{\sin^2 x}\right)\right) = \tan^2 x(1 - \cos^2 x)$

$= \tan^2 x \sin^2 x$ 13. $(\tan x + \cot x)^2 = \tan^2 x + 2\tan x \cot x + \cot^2 x = \tan^2 x + 2 + \cot^2 x = \tan^2 x$
$+ 1 + 1 + \cot^2 x = \sec^2 x + \csc^2 x$ 15. $\cos x \cot x \cos x + 1 = \left(\dfrac{1}{\sin x}\right)\left(\dfrac{\cos x}{\sin x}\right)\cos x + 1 = \dfrac{\cos^2 x}{\sin^2 x} + 1$

$= \cot^2 x + 1 = \csc^2 x$ 17. $\dfrac{\cos x + \sin x}{\sec x + \csc x} = \cos x\left(\dfrac{1 + \tan x}{\sec x + \csc x}\right) = \dfrac{\cos x \sin x(\csc x + \sec x)}{\sec x + \csc x}$

$= \cos x \sin x$ 19. $\dfrac{\sin x + \tan x}{\csc x + \cot x} = \dfrac{\sin x + \dfrac{\sin x}{\cos x}}{\csc x + \cot x} = \dfrac{\sin x(1 + \sec x)}{\csc x + \cot x} = \dfrac{\sin x \tan x(\cot x + \csc x)}{\csc x + \cot x}$

$= \sin x \tan x$ 21. $\sin^2 x \sec^2 x + \sin^2 x \csc^2 x = \sin^2 x\left(\dfrac{1}{\cos^2 x}\right) + \sin^2 x\left(\dfrac{1}{\sin^2 x}\right) = \tan^2 x + 1 = \sec^2 x$

23. $\tan^2 x - \cot^2 x = \tan^2 x + 1 - 1 - \cot^2 x = (\tan^2 x + 1) - (1 + \cot^2 x) = \sec^2 x - \csc^2 x$

25. $\cot x - \sec x \csc x(1 - 2\sin^2 x) = \cot x - \sec x \csc x(\cos 2x) = \cot x - \sec x \csc x(\cos^2 x - \sin^2 x)$

$= \cot x - \cos x \csc x + \sin x \sec x = \cot x - \cot x + \tan x = \tan x$

27. $\dfrac{1 + \cot x}{\csc x} = \dfrac{1 + \dfrac{\cos x}{\sin x}}{\dfrac{1}{\sin x}}\left(\dfrac{\dfrac{\sin x}{\cos x}}{\dfrac{\sin x}{\cos x}}\right) = \dfrac{\dfrac{\sin x}{\cos x} + 1}{\dfrac{1}{\cos x}} = \dfrac{\tan x + 1}{\sec x}$

29. $\dfrac{2\tan x}{1 + \tan^2 x} = \dfrac{2\tan x}{\sec^2 x} = \dfrac{2\sin x}{\cos x} \cdot \cos^2 x = 2\sin x \cos x$ 31. $\dfrac{1 - 2\cos^2 x}{\sin x \cos x} = \dfrac{-\cos^2 x}{\sin x \cos x} = \dfrac{\sin^2 x - \cos^2 x}{\sin x \cos x}$

$= \dfrac{\sin x \cos x \left(\dfrac{\sin^2 x}{\sin x \cos x} - \dfrac{\cos^2 x}{\sin x \cos x}\right)}{\sin x \cos x} = \dfrac{\sin x}{\cos x} - \dfrac{\cos x}{\sin x} = \tan x - \cot x$

33. $\dfrac{3\cos x + \cos 3x}{3\sin x - \sin 3x} = \dfrac{3\cos x + \cos(2x + x)}{3\sin x - \sin(2x + x)} = \dfrac{3\cos x + \cos 2x \cos x - \sin 2x \sin x}{3\sin x - \sin 2x \cos x - \cos 2x \sin x}$

$= \dfrac{\cos x(3 + \cos 2x - 2\sin^2 x)}{\sin x(3 - 2\cos^2 x - \cos 2x)} = \dfrac{\cos x}{\sin x}\left(\dfrac{3 + \cos^2 X - 3\sin^2 x}{3 - 3\cos^2 x + \sin^2 x}\right) = \cot x \left(\dfrac{3 + \cos^2 x - 3(1 - \cos^2 x)}{3(1 - \cos^2 x) + \sin^2 x}\right)$

$= \dfrac{\cot x(4\cos^2 x)}{4\sin^2 x} = \cot^3 x$ 35. $\dfrac{\sin x}{1 + \cos x} = \dfrac{\sin(2)\left(\dfrac{x}{2}\right)}{1 + \cos(2)\left(\dfrac{x}{2}\right)} = \dfrac{2\sin\dfrac{x}{2}\cos\dfrac{x}{2}}{1 + \cos^2\left(\dfrac{x}{2}\right) - \sin^2\left(\dfrac{x}{2}\right)} = \dfrac{2\sin\dfrac{x}{2}\cos\dfrac{x}{2}}{2\cos^2\left(\dfrac{x}{2}\right)}$

$= \dfrac{\sin\dfrac{x}{2}}{\cos\dfrac{x}{2}} = \tan\dfrac{x}{2}$ 37. $2\cos^2\left(\dfrac{x}{2}\right) - \cos x = 2\left(\dfrac{1 + \cos x}{2}\right) - \cos x = 1 + \cos x - \cos x = 1$ 39. $\cos 4x$

$= 2\cos^2 2x - 1 = \dfrac{2}{\sec^2 2x} - 1 = \dfrac{2 - \sec^2 2x}{\sec^2 2x}$ 41. $\sin 3x = \sin(2x + x) = \sin 2x \cos x + \cos 2x \sin x$

$= 2\sin x \cos^2 x + (1 - 2\sin^2 x)\sin x = 2\sin x \cos^2 x + \sin x - 2\sin^3 x = 2\sin x(1 - \sin^2 x) + \sin x$

$- 2\sin^3 x = 2\sin x - 2\sin^2 x + \sin x - 2\sin^3 x = 3\sin x - 4\sin^3 x$ 43. $\csc^2\left(\dfrac{x}{2}\right) - 1 = \cot^2\left(\dfrac{x}{2}\right)$

$= \dfrac{1}{\tan^2\left(\dfrac{x}{2}\right)} = \dfrac{1}{\dfrac{1 - \cos x}{1 + \cos x}} = \dfrac{1 + \cos x}{1 - \cos x}$ 45. $\sec x \sec y \sin(x - y) = \dfrac{\sin(x - y)}{\cos x \cos y} = \dfrac{\sin x \cos y - \cos x \sin y}{\cos x \cos y}$

$= \dfrac{\sin x \cos y}{\cos x \cos y} - \dfrac{\cos x \sin y}{\cos x \cos y} = \dfrac{\sin x}{\cos x} - \dfrac{\sin y}{\cos y} = \tan x - \tan y$ 47. $\dfrac{\sin 5x - \sin 3x}{\cos 5x - \cos 3x} = \dfrac{-2\cos 4x \sin x}{2\sin 4x \sin x}$

$= -\cot 4x$ 49. $\dfrac{\cos 2x - \cos 6x}{\cos 2x + \cos 6x} = \dfrac{-(\cos 6x - \cos 2x)}{\cos 6x + \cos 2x} = \dfrac{2\sin 4x \sin 2x}{2\cos 4x \cos 2x} = \dfrac{\tan 4x}{\cot 2x}$ 51. $\sin(x + y)$

$\cdot \sin(x - y) = (\sin x \cos y + \cos x \sin y)(\sin x \cos y - \cos x \sin y) = \sin^2 x \cos^2 y - \cos^2 x \sin^2 y$

$= \sin^2 x(1 - \sin^2 y) - (1 - \sin^2 x)\sin^2 y = \sin^2 x - \sin^2 x \sin^2 y - \sin^2 y + \sin^2 x \sin^2 y = \sin^2 x$

$- \sin^2 y$ 53. $\dfrac{2\sin x \cos x}{1 + \cos^2 x - \sin^2 x} = \dfrac{2\sin x \cos x}{\cos^2 x + \cos^2 x} = \dfrac{2\sin x \cos x}{2\cos^2 x} = \dfrac{\sin x}{\cos x} = \tan x$ 55. $\dfrac{\sin x}{1 + \cos x}$

$= \dfrac{2\sin\dfrac{x}{2}\cos\dfrac{x}{2}}{1 + 2\cos^2\left(\dfrac{x}{2}\right) - 1} = \dfrac{\sin\dfrac{x}{2}}{\cos\dfrac{x}{2}} = \tan\dfrac{x}{2}$ 57. $\cos 6x = \cos(3x + 3x) = \cos 3x \cos 3x - \sin 3x \sin 3x$

$= \cos^2 3x - \sin^2 3x$ 59. $\dfrac{\sin 3x + \sin x}{\sin 6x - \sin 2x} = \dfrac{2\sin 2x \cos x}{2\cos 4x \sin 2x} = \cos x \sec 4x$ 61. $\sin(x + y + z)$

$= \sin[(x + y) + z] = \sin(x + y)\cos z + \cos(x + y)\sin z = (\sin x \cos y + \cos x \sin y)\cos z + (\cos x \cos y$

$- \sin x \sin y)\sin z$

3. $\sin \theta = \frac{y}{r}$ and $\csc \theta = \frac{r}{y}$; Then $\sin \theta = \frac{y}{r} \cdot \frac{\frac{1}{y}}{\frac{1}{y}} = \frac{1}{\frac{r}{y}} = \frac{1}{\csc \theta}$; Thus, $\sin \theta = \frac{1}{\csc \theta}$ 5. $\tan \theta = \frac{y}{x}$, $\cot \theta = \frac{x}{y}$; $\therefore$

$\tan \theta = \frac{y}{x} \cdot \frac{\frac{1}{y}}{\frac{1}{y}} = \frac{1}{\frac{x}{y}}$; and $\tan \theta = \frac{1}{\cot \theta}$ 7. $\sec^2 \theta - \sin^2 \theta - \cos^2 \theta = \sec^2 \theta - (\sin^2 \theta + \cos^2 \theta) = \sec^2 \theta - 1$

$= \tan^2 \theta$ 9. $\sin^2 \theta + \frac{1}{\sec^2 \theta}$ $\tan \theta \cot \theta$ 11. $2 \sin^2 x - 1 = 2(1 - \cos^2 x) - 1 = 2 - 2\cos^2 x$

$\sin^2 \theta + \cos^2 \theta$ $\frac{\sin \theta}{\cos \theta} \cdot \frac{\cos \theta}{\sin \theta}$ $- 1 = 1 - 2\cos^2 x$

$1 = 1$

13. $-\frac{1}{4}(\sqrt{2} + \sqrt{6})$ 15. $\cos(\theta - \pi) = \cos \theta \cos \pi + \sin \theta \sin \pi = (\cos \theta)(-1) + \sin \theta \cdot 0 = -\cos \theta$

17. $\frac{1}{4}(\sqrt{6} + \sqrt{2})$ 19. $\sin(180° - \theta) = \sin 180° \cos \theta - \cos 180° \sin \theta = 0 \cdot \cos \theta - (-1) \sin \theta = \sin \theta$

21. $\frac{33}{65}$ 23. $\tan(2\pi - \theta) = \frac{\tan 2\pi - \tan \theta}{1 + \tan 2\pi \tan \theta} = \frac{0 - \tan \theta}{1 + 0 \cdot \tan \theta} = -\tan \theta$ 25. $\frac{24}{25}$ 27. $\frac{24}{7}$

29. $\frac{1}{2}\sqrt{2 + \sqrt{3}}$ 31. $2 \sin \frac{\alpha}{2} \cos \frac{\alpha}{2} = 2\left(\pm \sqrt{\frac{1 - \cos \alpha}{2}} \sqrt{\frac{1 + \cos \alpha}{2}}\right) = \pm \sqrt{1 - \cos \alpha} \sqrt{1 + \cos \alpha}$

$= \pm \sqrt{1 - \cos^2 \alpha} = \pm \sqrt{\sin^2 \alpha} = \sin \alpha$ 33. $2 \sin \frac{3\pi}{20} \cos \frac{\pi}{20}$ 35. $\frac{\sin 4\alpha + \sin 2\alpha}{\cos 4\alpha + \cos 2\alpha} = \frac{2 \sin 3\alpha \cos \alpha}{2 \cos 3\alpha \cos \alpha}$

$= \frac{\sin 3\alpha}{\cos 3\alpha} = \tan 3\alpha = \frac{1}{\cot 3\alpha}$ 37. $\left(\frac{1 + \tan \theta}{1 - \tan \theta}\right)^2 = \frac{1 + 2 \tan \theta + \tan^2 \theta}{1 - 2 \tan \theta + \tan^2 \theta} = \frac{\sec^2 \theta + 2 \tan \theta}{\sec^2 \theta - 2 \tan \theta}$

$= \frac{\sec^2 \theta (1 + 2 \sin \theta \cos \theta)}{\sec^2 \theta (1 - 2 \sin \theta \cos \theta)} = \frac{1 + 2 \sin \theta \cos \theta}{1 - 2 \sin \theta \cos \theta} = \frac{1 + \sin 2\theta}{1 - \sin 2\theta}$

PAGE 162   CHAPTER TEST

1. $\sin \theta = \frac{y}{r} \therefore r \sin \theta = y$; $\cos \theta = \frac{x}{r} \therefore r \cos \theta = x$; $\cot \theta = \frac{x}{y}$, then $\cot \theta = \frac{r \cos \theta}{r \sin \theta}$ and $\cot \theta = \frac{\cos \theta}{\sin \theta}$ 2. $\tan \theta$

$= \frac{y}{x}$ and $\sec \theta = \frac{r}{x}$; $\therefore \frac{\tan \theta}{\sec \theta} = \frac{y}{x} \div \frac{r}{x} = \frac{y}{x} \cdot \frac{x}{r} = \frac{y}{r}$ and $\frac{y}{r} = \sin \theta$. Thus, $\frac{\tan \theta}{\sec \theta} = \sin \theta$.   3. $\cos \theta \cdot \csc \theta = \cos \theta \cdot \frac{1}{\sin \theta}$

$= \frac{\cos \theta}{\sin \theta} = \cot \theta$ 4. $\csc^2 \theta \tan^2 \theta - 1 = \frac{1}{\sin^2 \theta} \cdot \frac{\sin^2 \theta}{\cos^2 \theta} - 1 = \frac{1}{\cos^2 \theta} - 1 = \tan^2 \theta$

5.

$\frac{\tan \theta}{1 - \cos^2 \theta}$	$\sec \theta \cdot \csc \theta$
$\frac{\frac{\sin \theta}{\cos \theta}}{\sin^2 \theta}$	$\frac{1}{\cos \theta} \cdot \frac{1}{\sin \theta}$
$\frac{1}{\cos \theta \sin \theta}$	$= \frac{1}{\cos \theta \sin \theta}$

6.

$\csc \theta + \cot \theta$
$\frac{1}{\sin \theta} + \frac{\cos \theta}{\sin \theta}$
$\frac{1 + \cos \theta}{\sin \theta}$

7.

	$\frac{2 \tan \theta}{1 + \tan^2 \theta}$	$2 \sin \theta \cos \theta$
	$\frac{1 - \tan^2 \theta}{1 - \tan^2 \theta} \cdot \frac{2 \tan \theta}{1 + \tan^2 \theta}$	$\sin 2\theta$
	$\frac{1 - \tan^2 \theta}{1 + \tan^2 \theta} \cdot \frac{2 \tan \theta}{1 - \tan^2 \theta}$	
	$\frac{\cos^2 \theta}{\cos^2 \theta} \cdot \frac{1 - \tan^2 \theta}{1 + \tan^2 \theta} \cdot \tan 2\theta$	
	$\frac{\cos^2 \theta - \sin^2 \theta}{\cos^2 \theta + \sin^2 \theta} \cdot \tan 2\theta$	
	$\frac{\cos 2\theta}{1} \cdot \frac{\sin 2\theta}{\cos 2\theta}$	
	$\sin 2\theta \quad = \quad \sin 2\theta$	

8.

$1 - \cos 2\beta$	$\sin 2\beta \cdot \tan \beta$
$1 - (2 \cos^2 \beta - 1)$	$2 \sin \beta \cos \beta \cdot \frac{\sin \beta}{\cos \beta}$
$2 - 2 \cos^2 \beta$	
$2(1 - \cos^2 \beta)$	
$2 \sin^2 \beta \quad = \quad 2 \sin^2 \beta$	

9. $\frac{1}{4}(\sqrt{6}-\sqrt{2})$ 10. $\frac{1}{4}(\sqrt{6}+\sqrt{2})$ 11. $\frac{1}{4}(\sqrt{6}-\sqrt{2})$ 12. $2-\sqrt{3}$ 13. $\cos(\alpha+\beta)=-\frac{12}{65}$;

$\sin(\alpha-\beta)=-\frac{33}{65}$ 14. $\cos(270°+\theta)=\cos 270°\cos\theta-\sin 270°\sin\theta=0\cdot\cos\theta-(-1)\sin\theta=\sin\theta$

15. $\sin 2\alpha=.6692$; $\cos 2\alpha=.7431$; $\tan 2\alpha=.9006$ 16. $\frac{161}{289}$ 17. $\frac{1}{4}$ 18. $2\cos\frac{3\pi}{20}\cos\frac{\pi}{20}$ 19. $\cos 95°$

$+\cos 55°$ 20. $\sin(\alpha+\beta)-\sin(\alpha-\beta)=(\sin\alpha\cos\beta+\cos\alpha\sin\beta)-(\sin\alpha\cos\beta-\cos\alpha\sin\beta)$

$=2\cos\alpha\sin\beta$

CHAPTER 4   TRIGONOMETRY AND TRIANGLES

PAGE 166   CLASSROOM EXERCISES

1. $95°$ 2. 6.5 3. 7.1

PAGES 166-167   WRITTEN EXERCISES

1. $C=67°$; $a=15.4$; $b=10.9$ 3. $A=9°$; $b=42.0$; $c=49.0$ 5. $C=60°$; $a=24$; $c=20.8$ 7. $C=64°$;

$b=19.1$; $c=17.8$ 9. $A=100°\ 30'$; $a=46.7$; $c=28.7$ 11. No   13. $\frac{a-b}{b}=\frac{\sin A-\sin B}{\sin B}$ can be written as

$\frac{a}{b}-\frac{b}{b}=\frac{\sin A}{\sin B}-\frac{\sin B}{\sin B}$ or $\frac{a}{b}-1=\frac{\sin A}{\sin B}-1$ or $\frac{a}{b}=\frac{\sin A}{\sin B}$ or $\frac{a}{\sin A}=\frac{b}{\sin B}$, which is the Law of Sines.

15. 52 km   17. 6 cm   19. $\dfrac{\sin A-\sin B}{\sin A+\sin B}=\dfrac{\frac{a}{b}\sin B-\sin B}{\frac{a}{b}\sin B+\sin B}=\dfrac{\sin B(\frac{a}{b}-1)}{\sin B(\frac{a}{b}+1)}=\dfrac{a-b}{a+b}$

PAGE 168   REVIEW CAPSULE FOR SECTION 4-2

1. $c=\frac{b^2-3}{2b}$ 2. $b=a+3cd$ 3. $d=\frac{x-2-b^2}{3}$ 4. $b=\pm\sqrt{x+y-a^2}$ 5. $x=\frac{b^2+c^2-a^2}{2bc}$

6. $\cos C=\frac{a^2+b^2-c^2}{2ab}$

PAGE 171   CLASSROOM EXERCISES

1. If $c=13$, then $\cos C=\frac{(-13)^2+8^2+9^2}{2\cdot8\cdot9}<0$, which implies the largest angle is obtuse. (Cosine is

negative in Quadrant II.)   2. $A=83°$; $B=41°$; $C=56°$

PAGES 171-172   WRITTEN EXERCISES

1. $c=5.3$; $A=37°\ 20'$; $B=102°\ 40'$ 3. $b=24.9$; $A=3°\ 40'$; $C=5°\ 20'$ 5. $c=7.2$; $B=46°\ 10'$;

$A=73°\ 50'$ 7. $A=21°\ 50'$; $C=38°\ 20'$; $B=119°\ 50'$ 9. $A=90°$; $B=53°\ 10'$; $C=36°\ 50'$

11. $A=29°$; $B=46°\ 40'$; $C=104°\ 20'$ 13. $c=13.5$; $A=13°\ 30'$; $B=127°\ 30'$ 15. $A=48°$;

$B=33°\ 50'$; $C=98°\ 10'$ 17. If $C=90°$, $\cos C=0$. Thus, the equation becomes $c^2=a^2+b^2$, which is

the Pythagorean Theorem.   19. 173 m   21. 34 cm

23.

$1+\cos C$	$\dfrac{(a+b+c)(a+b-c)}{2ab}$	$1+\dfrac{a^2+b^2-c^2}{2ab}$ = $1+\dfrac{a^2+b^2-c^2}{2ab}$
$1+\dfrac{a^2+b^2-c^2}{2ab}$	$\dfrac{(a+b)^2-c^2}{2ab}$	
	$\dfrac{a^2+2ab+b^2-c^2}{2ab}$	

25. Let ℓ_1 and ℓ_2 represent the lengths of the sides of a parallelogram and let d_1 and d_2 represent the lengths of the diagonals of a

parallelogram. $\cos \theta = \dfrac{\ell_1^2 + \ell_2^2 - d_1^2}{2(\ell_1)(\ell_2)}$, $\cos (180° - \theta) = \dfrac{\ell_1^2 + \ell_2^2 - d_2^2}{2(\ell_1)(\ell_2)}$

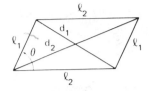

Since $\cos (180° - \theta) = -\cos \theta$, $\dfrac{\ell_1^2 + \ell_2^2 - d_1^2}{2(\ell_1)(\ell_2)} = -\dfrac{\ell_1^2 + \ell_2^2 - d_2^2}{2(\ell_1)(\ell_2)}$ and

$\ell_1^2 + \ell_2^2 - d_1^2 = -\ell_1^2 - \ell_2^2 + d_2^2$ and $\ell_1^2 + \ell_2^2 + \ell_1^2 + \ell_2^2 = d_1^2 + d_2^2$

PAGE 176   CLASSROOM EXERCISES

1. None   2. Two   3. One   4. None

PAGE 177   WRITTEN EXERCISES

1. None   3. One   5. One   7. None   9. None   11. None   13. One   15. None   17. One
19. None   21. None   23. One   25. None   27. One   29. c = 8.4; B = 47°; A = 71°   31. B = 35°;
C = 104°; c = 22.2   33. B = 90°; C = 60°; c = 24.2   35. A = 24°; B = 64°; C = 92°   37. A = 49°;
C = 99°; c = 13.0   39. EV_2 is about 2.4×10^8 km. EV_1 is about 4.5×10^7 km.

PAGE 180   CLASSROOM EXERCISES

1. 155° 30'   2. 3° 30'   3. 70.5 meters

PAGES 180-181   WRITTEN EXERCISES

1. A = 63° 40'; B = 44° 20'; c = 67.9   3. B = 34° 40'; C = 22° 20'; a = 26.5   5. A = 39° 40'; B = 23° 20';
c = 304.3   7. B = 149°; C = 19°; a = 15.3   9. B = 63° 10'; C = 33° 50'; a = 8.9   11. 1403 meters
13. c = 76 m; 49° and 33°

PAGE 181   REVIEW

1. C = 63°; a = 4.1; b = 5.3   2. A = 33°; b = 5.2; c = 10.8   3. c = 11.4; A = 22°; B = 38°   4. b = 15.1;
A = 78°; C = 64°   5. Two   6. None   7. Two   8. One   9. C = 36°; A = 44°; a = 7.1   10. B = 62°;
C = 73°; c = 10.8 or B = 118°; C = 17°; c = 3.3   11. A = 41° 30' or 42°; B = 83°; c = 12.2   12. B = 18°;
C = 32°; a = 57.0

PAGE 183   CLASSROOM EXERCISES

1. 26,900 square units   2. 4564 m^2

PAGES 183-184   WRITTEN EXERCISES

1. 700 cm^2 3. 376 square units   5. 2641 km^2 7. 1009 cm^2 9. 67,417 cm^2 11. $K = \frac{1}{2}ac \sin B$
$= \frac{1}{2}(\sin A \frac{c}{\sin C}) c \sin B = \dfrac{c^2 \sin A \sin B}{2 \sin C}$ 13. $K = \frac{1}{2}bc \sin A = \frac{1}{2}b(\sin C \frac{b}{\sin B}) \sin A = \dfrac{b^2 \sin A \sin C}{2 \sin B}$
15. 3258 m^2 17. $a \cos B + b \cos A = a(\dfrac{a^2 + c^2 - b^2}{2ac}) + b(\dfrac{b^2 + c^2 - a^2}{2bc}) = \frac{1}{2}(\dfrac{a^2}{c} + c - \dfrac{b^2}{c} + \dfrac{b^2}{c} + c - \dfrac{a^2}{c}) = c$

19. From Exercise 12, $K = \dfrac{a^2 \sin B \sin C}{2 \sin A} = \dfrac{1}{2}a^2 \dfrac{\sin B \sin C}{\sin [180° - (B + C)]} = \dfrac{1}{2}a^2 \dfrac{\sin B \sin C}{\sin (B + C)}$ 21. $\dfrac{\pi}{2}$

23. $\dfrac{4}{3}(4\pi - 3\sqrt{3})$

PAGE 187   CLASSROOM EXERCISES

1. 3.9   2. 27.7

PAGE 187   WRITTEN EXERCISES

1. 15   3. 1   5. 4   7. 1625   9. 133,675   11. 3.3   13. 6.2   15. 96 cm²   17. $K = \sqrt{x^2 - 1}$
19. 4.8

PAGE 190   CLASSROOM EXERCISES

1. $\overrightarrow{OB}$ 2. $\overrightarrow{OC}$ 3. $\overrightarrow{OA}$ 4. $\overrightarrow{BA}$ 5. $\overrightarrow{OA}$

PAGES 190-191   WRITTEN EXERCISES

1. 1.4   3. 5   5. 9.4   7. 12.0   9. $|\overrightarrow{OB}| = 3$; $|\overrightarrow{OC}| = 4$ 11. $|\overrightarrow{OB}| = 3$; $|\overrightarrow{OC}| = 8$ 13. 306° 50′
15. 216° 50′   17. 53° 10′   19. 112° 40′   21. Draw $\overrightarrow{BA}$ equivalent to $\overrightarrow{OC}$. Thus, $m\angle ABO = 90°$,

$\cos \alpha = \dfrac{|\overrightarrow{OB}|}{|\overrightarrow{OA}|}$, and $|\overrightarrow{OB}| = |\overrightarrow{OA}| \cos \alpha$.

PAGE 191   REVIEW

1. 170   2. 57   3. 9   4. 32   5. 79   6. 463   7. 7.2   8. 146° 20′

PAGE 194   WRITTEN EXERCISES

1. 5.3   3. 17.6   5. 248 km   7. 259°   9. 77° 10′

PAGES 196-197   WRITTEN EXERCISES

1. 126 N   3. 67°   5. 106 N   7. 268 N   9. 84°   11. 2.5 meters

PAGES 199-200   WRITTEN EXERCISES

1. 482 km/hr   3. 40 km/hr   5. 15°   7. $|\overrightarrow{AH}| = 325$ km/hr; $|\overrightarrow{HC}| = 87$ km/hr   9. 13 km/hr
11. 4 km/hr

PAGE 202   WRITTEN EXERCISES

1.
```
10  READ A, B, C1
20  LET D1=C1*3.14159 / 180
30  LET K = .5 * A * B * SIN(D1)
40  PRINT "AREA ="; K
50  GO TO 10
60  DATA . . .
70  END
```

3.
```
10  READ A, B, C
20  LET S = (A + B + C)/2
30  LET K = SQR(S*(S—A)*(S—B)*(S—C))
40  PRINT "AREA ="; K
50  GO TO 10
60  DATA . . .
70  END
```

5.
```
10   READ C, A1
20   LET B1 = 90 − A1
30   LET D1 = A1*3.14159/180
40   LET A = C * SIN(D1)
50   LET B = C * COS(D1)
60   PRINT "THE OTHER ACUTE ANGLE =";B1
70   PRINT "THE LEGS =";A; " AND"; B
80   PRINT
90   GO TO 10
100  DATA . . .
110  END
```

7.
```
10   READ A, B
20   LET A1 = ATN(A/B)
30   LET D1 = A1 * 180/3.14159
40   LET B1 = 90 − D1
50   LET C = A/SIN(A1)
60   PRINT "ACUTE ANGLES =";D1;" AND";B1
70   PRINT "HYPOTENUSE ="; C
80   PRINT
90   GO TO 10
100  DATA . . .
110  END
```

9. SAA case
```
10   READ A,A1, B1
20   LET C1 = 180−A1−B1
30   PRINT "ANGLES ARE:"; A1;B1;C1
40   LET A1 = A1*3.14159/180
50   LET B1 = B1*3.14159/180
60   LET C1 = C1*3.14159/180
70   LET B = A*SIN(B1)/SIN(A1)
80   LET C = A*SIN(C1)/SIN(A1)
90   PRINT "SIDES ARE:";A;B;C
100  GO TO 10
110  DATA 16.5, 38, 54, 224, 84,
            21.1, 75.36, 18, 32
120  END
```

11. SSS case
```
10   READ A,B,C
20   LET T = (B↑2+C↑2−A↑2)/(2*B*C)
30   LET A1 = ATN(SQR(1−T↑2)/T)
40   LET S = B*SIN(A1)/A
50   LET B1 = ATN(S/SQR(1−S↑2))
60   LET A1 = A1*180/3.14159
70   LET B1 = B1*180/3.14159
80   PRINT "ANGLES ARE:";A1;B1;
            180−A1−B1
90   PRINT
100  GO TO 10
110  DATA 4, 5, 7, 25, 40, 58, 8, 9, 13
120  END
```

PAGES 203-205 CHAPTER OBJECTIVES AND REVIEW

3. $A = 51°$; $b = 4.5$; $c = 14.3$ 5. $A = 12°$; $b = 109.3$; $c = 110.5$ 7. $A = 54°$; $B = 80°$; $C = 46°$
9. $A = 47°$; $B = 56°$; $C = 77°$ 11. None   13. None   15. One   17. One   19. $C = 25°$; $A = 106°$;
$a = 20.4$ 21. $A = 64°$; $B = 30°$; $C = 86°$ 23. $B = 29°$; $C = 45°$; $a = 49.6$ 25. 96   27. 155   29. 238
31. 0.2   33. 238°   35. 57 km   37. 475 km/hr

PAGE 206 CHAPTER TEST

1. 15.4   2. 5.3   3. 28.5 or 6.2   4. 76°   5. 16   6. 27   7. 3   8. 60°   9. 5   10. 291 N

PAGES 207-208 CUMULATIVE REVIEW: CHAPTERS 3 and 4

1. c   3. d   5. b   7. b   9. c   11. b   13. a   15. c   17. b   19. d

CHAPTER 5 INVERSE FUNCTIONS/EQUATIONS

PAGE 213 CLASSROOM EXERCISES

1. Yes   2. No   3. $y = \frac{1}{2}x$ 4. $x = 4$ 5. $y = -\frac{1}{3}x$ 6. $y = x + 1$ 7. $\frac{\pi}{6}$; $\frac{5\pi}{6}$ 8. π; $-\pi$ 9. $-\pi$; 0; π
10. $\frac{\pi}{6}$; $-\frac{\pi}{6}$

WRITTEN EXERCISES (The "I" in Exercises 21-31 represents the set of integers.)

1. $\{0, 1, 2\}$

3. $\{(1, 0), (4, 1), (4, 2)\}$

5. No

7.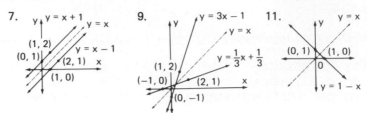

9.

11.

13.

15. The functions in Exercises 10, 12 and 14 have inverses that are not functions.   17. $x = -1$ 19. $y = \arcsin x$

21. $\{y : y = \frac{\pi}{2} + 2n\pi, n \in I\}$ 23. $\{y : y = 2n\pi, n \in I\}$

25. $\{y : y = \frac{3\pi}{2} + 2n\pi, n \in I\}$ 27. $\{y : y = -\frac{\pi}{6} + 2n\pi$ or

$y = -\frac{5\pi}{6} + 2n\pi, n \in I\}$ 29. $\{y : y = \frac{\pi}{6} + 2n\pi$ or $y = -\frac{\pi}{6} + 2n\pi, n \in I\}$

31. $\{y : y = \frac{2\pi}{3} + 2n\pi$ or $y = -\frac{2\pi}{3} + 2n\pi, n \in I\}$

PAGE 217   CLASSROOM EXERCISES

1. $45°$ 2. $45°$ 3. $\frac{\sqrt{3}}{2}$ 4. $\frac{1}{2}$

PAGES 217-219   WRITTEN EXERCISES

1. Domain: $-1 \leq x \leq 1$; Range: $-\frac{\pi}{2} \leq y \leq \frac{\pi}{2}$ 3. $0 \leq x \leq 1$ 5. $0 < x < 1$

7.

9.

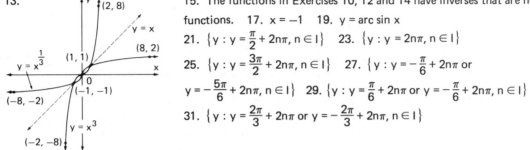

11. $30°$ 13. $45°$ 15. $135°$ 17. 0 19. 0 21. $-\frac{\pi}{2}$

23. $\frac{\pi}{3}$ 25. $-\frac{\pi}{3}$ 27. $-\frac{\pi}{6}$ 29. $\frac{1}{2}$ 31. $\frac{\sqrt{3}}{2}$ 33. $\frac{1}{2}$

35. 1 37. $\frac{\sqrt{2}}{2}$ 39. 1 41. 0 43. $\frac{\sqrt{3}}{2}$ 45. $-\frac{\sqrt{3}}{2}$

47. $\frac{4}{5}$ 49. $-\frac{12}{13}$ 51. $\frac{4}{3}$ 53. 1 55. $-\frac{5}{4}$ 57. $\frac{\sqrt{15}}{4}$

59. $\frac{1}{2}$ 61. $\frac{1}{a}$ 63. $-\pi \leq y \leq 0$ 65. Let

$y =$ Arc cos $(-\sqrt{3})$, cos $y = -\sqrt{3}$. This is impossible since $-1 \leq \cos y \leq 1$.   67. $0 \leq x \leq \pi$

69. The inverse is a function.

INVERSE FUNCTIONS/EQUATIONS

1. $-1 < y < 1$ 2. All real numbers   3. $y \geq 1$ 4. $y \leq -1, y \geq 1$, or $|y| \geq 1$

PAGE 222   CLASSROOM EXERCISES (The "I" in Exercises 1 and 2 represents the set of integers.)

1. $180° + n \cdot 360°, n \in I$ 2. $135° + n \cdot 180°, n \in I$ 3. $60°$ 4. $\frac{\pi}{6} + n\pi, n \in I$ 5. $\frac{\pi}{3}$ 6. $\frac{5\pi}{6}$ 7. $\frac{7}{24}$

PAGES 222-223   WRITTEN EXERCISES (The "I" in Exercises 1-9 represents the set of integers.)

1. $\{y : y = 45° + n \cdot 180°, n \in I\}$ 3. $\{y : y = 30° + n \cdot 180°, n \in I\}$ 5. $\{y : y = 30° + n \cdot 360°$ or $y = 330° + n \cdot 360°, n \in I\}$ 7. $\{y : y = \frac{\pi}{2} + n\pi, n \in I\}$ 9. $\{y : y = \frac{7\pi}{6} + 2n\pi$ or $y = \frac{11\pi}{6} + 2n\pi, n \in I\}$

11. $\frac{\pi}{3}$ 13. $-\frac{\pi}{6}$ 15. $\frac{\pi}{2}$ 17. $-\frac{\sqrt{2}}{2}$ 19. $\frac{5}{13}$ 21. -3 23. $\frac{\sqrt{15}}{4}$ 25. $\frac{\sqrt{13}}{2}$ 27. $\frac{336}{527}$ 29. $-\frac{3}{2}$

31. $\tan [\text{Arc tan } (\frac{1}{2}) + \text{Arc tan } (\frac{1}{3})] = \dfrac{\tan [\text{Arc tan } (\frac{1}{2})] + \tan [\text{Arc tan } (\frac{1}{3})]}{1 - \tan [\text{Arc tan } (\frac{1}{2})] \tan [\text{Arc tan } (\frac{1}{3})]} = \dfrac{\frac{1}{2} + \frac{1}{3}}{1 - (\frac{1}{2})(\frac{1}{3})} = 1.$ Since

$\tan [\text{Arc tan } (\frac{1}{2}) + \text{Arc tan } (\frac{1}{3})] = 1,$ Arc tan $(\frac{1}{2}) + $ Arc tan $(\frac{1}{3}) = \frac{\pi}{4}.$ 33. $0 < x < \pi$ 35. $-\frac{\pi}{2} \leq x \leq \frac{\pi}{2}, x \neq 0$

37. $x \geq 1$ or $x \leq -1$ 39. $\tan [\text{Arc tan } b - \text{Arc tan } a] = \dfrac{\tan (\text{Arc tan } b) - \tan (\text{Arc tan } a)}{1 + \tan (\text{Arc tan } b) \tan (\text{Arc tan } a)} = \dfrac{b - a}{1 + ba}$

$= \frac{b - a}{1 + ab}$ 41. Let $\alpha = $ Arc tan $\frac{1}{3}$, $\beta = $ Arc tan $\frac{1}{7}$; 2 Arc tan $\frac{1}{3} + $ Arc tan $\frac{1}{7} = 2\alpha + \beta$; Thus, tan $(2\alpha + \beta)$

$= \dfrac{\tan 2\alpha + \tan \beta}{1 - \tan 2\alpha \tan \beta} = \dfrac{\frac{2 \tan \alpha}{1 - \tan^2 \alpha} + \tan \beta}{1 - \frac{2 \tan \alpha}{1 - \tan^2 \alpha} \tan \beta} = \dfrac{2 \tan \alpha + \tan \beta - \tan^2 \alpha \tan \beta}{1 - \tan^2 \alpha - 2 \tan \alpha \tan \beta} = \dfrac{2(\frac{1}{3}) + \frac{1}{7} - (\frac{1}{9})(\frac{1}{7})}{1 - \frac{1}{9} - 2(\frac{1}{3})(\frac{1}{7})} = 1;$ Since

Arc tan $(2\alpha + \beta) = 1,$ $2\alpha + \beta = 2$ Arc tan $\frac{1}{3} + $ Arc tan $\frac{1}{7} = \frac{\pi}{4}.$ 43. sin [Arc cos x + Arc sin x]

$= \sin (\text{Arc cos } x) \cos (\text{Arc sin } x) + \cos (\text{Arc cos } x) \sin (\text{Arc sin } x) = \sqrt{1 - x^2} \sqrt{1 - x^2} + x^2 = 1;$ Since

$\sin [\text{Arc cos } x + \text{Arc sin } x] = 1,$ Arc cos $x + $ Arc sin $x = \frac{\pi}{2}.$

PAGE 229   CLASSROOM EXERCISES (The "I" in Exercise 1 represents the set of integers.)

1. $\{x : x = \frac{\pi}{3} + 2n\pi$ or $x = \frac{2\pi}{3} + 2n\pi, n \in I\},$ or $\{x : x = 60° + n \cdot 360°$ or $x = 120° + n \cdot 360°, n \in I\}$

2. $\{\frac{\pi}{4}, \frac{5\pi}{4}\},$ or $\{45°, 225°\}$ 3. $\{0, \pi\},$ or $\{0°, 180°\}$ 4. $\{\frac{\pi}{4}\},$ or $\{45°\}$ 5. $\{\frac{\pi}{4}, \frac{3\pi}{4}\},$ or $\{45°, 135°\}$

6. $\{\pm\frac{\pi}{3}, \pm\frac{\pi}{2}\},$ or $\{\pm60°, \pm90°\}$

PAGES 229-230   WRITTEN EXERCISES (The "I" in Exercises 1-5 represents the set of integers.)

1. $\{x : x = \frac{4\pi}{3} + 2n\pi$ or $x = \frac{5\pi}{3} + 2n\pi, n \in I\},$ or $\{x : x = 240° + n \cdot 360°$ or $x = 300° + n \cdot 360°, n \in I\}$

3. $\{x : x = \frac{\pi}{4} + 2n\pi$ or $x = \frac{7\pi}{4} + 2n\pi, n \in I\},$ or $\{x : x = 45° + n \cdot 360°$ or $x = 315° + n \cdot 360°, n \in I\}$

5. $\{x : x = \frac{5\pi}{6} + 2n\pi$ or $x = \frac{7\pi}{6} + 2n\pi, n \in I\},$ or $\{x : x = 150° + n \cdot 360°$ or $x = 210° + n \cdot 360°, n \in I\}$

7. $\{\frac{\pi}{6}, \frac{5\pi}{6}, \frac{7\pi}{6}, \frac{11\pi}{6}\},$ or $\{30°, 150°, 210°, 330°\}$ 9. $\{0\},$ or $\{0°\}$ 11. $\{0, \frac{\pi}{6}, \frac{5\pi}{6}, \pi\},$ or

$\{0°, 30°, 150°, 180°\}$ 13. $\{0, \pi\},$ or $\{0°, 180°\}$ 15. $\{\frac{\pi}{4}, \frac{5\pi}{4}\},$ or $\{45°, 225°\}$ 17. $\{\frac{\pi}{6}, \frac{5\pi}{6}, \frac{7\pi}{6}, \frac{11\pi}{6}\},$

or $\{30°, 150°, 210°, 330°\}$ 19. No solution. 21. $\{0\}$, or $\{0°\}$ 23. $\left\{\frac{\pi}{6}, \frac{5\pi}{6}\right\}$, or $\{30°, 150°\}$

25. $\left\{\frac{4\pi}{3}, \frac{5\pi}{3}\right\}$, or $\{240°, 300°\}$ 27. $\left\{\frac{\pi}{3}, \frac{5\pi}{3}\right\}$, or $\{60°, 300°\}$ 29. $\left\{\frac{\pi}{2}, \frac{2\pi}{3}, \frac{4\pi}{3}, \frac{3\pi}{2}\right\}$, or

$\{90°, 120°, 240°, 270°\}$ 31. $\left\{0, \frac{\pi}{2}, \pi, \frac{3\pi}{2}\right\}$, or $\{0°, 90°, 180°, 270°\}$

33. $\{109° \ 30', 120°, 240°, 250° \ 30'\}$, or $\left\{\frac{1.825\pi}{3}, \frac{2\pi}{3}, \frac{4\pi}{3}, \frac{4.175\pi}{3}\right\}$ 35. $\{26° \ 30', 116° \ 30', 206° \ 30',$

$296° \ 30'\}$, or $\left\{\frac{1.325\pi}{9}, \frac{5.825\pi}{9}, \frac{10.325\pi}{9}, \frac{14.825\pi}{9}\right\}$ 37. $\{54°, 90°, 126°, 198°, 270°, 342°\}$, or

$\left\{\frac{3\pi}{10}, \frac{\pi}{2}, \frac{7\pi}{10}, \frac{11\pi}{10}, \frac{3\pi}{2}, \frac{19\pi}{10}\right\}$ 39. $\left\{0, \frac{\pi}{2}, \frac{2\pi}{3}, \pi, \frac{4\pi}{3}, \frac{3\pi}{2}\right\}$, or $\{0°, 90°, 120°, 180°, 240°, 270°\}$

41. $\left\{x : x = 0, \pi, \frac{\pi}{9} + \frac{2n\pi}{3}, \frac{5\pi}{9} + \frac{2n\pi}{3}, n = 0, 1, 2\right\}$, or $\{x : x = 0°, 180°, 20° + n \cdot 120°,$

$100° + n \cdot 120°, n = 0, 1, 2\}$

PAGE 230 REVIEW (The "I" in Exercises 4-6, 10, 13-16 represents the set of integers.)

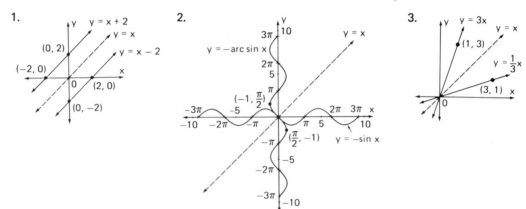

4. $\{y : y = 225° + n \cdot 360°$ or $y = 315° + n \cdot 360°, n \in I\}$ 5. $\{y : y = 90° + n \cdot 360°, n \in I\}$

6. $\{y : y = 30° + n \cdot 360°$ or $y = 330° + n \cdot 360°, n \in I\}$ 7. 0 8. $\frac{\sqrt{2}}{2}$ 9. $\frac{4}{5}$ 10. $\{y : y = \frac{\pi}{3} + n\pi,$

$n \in I\}$, or $\{y : y = 60° + n \cdot 180°, n \in I\}$ 11. 0, or 0° 12. 4 13. $\{x : x = \frac{\pi}{3} + 2n\pi$ or

$x = \frac{2\pi}{3} + 2n\pi, n \in I\}$, or $\{x : x = 60° + n \cdot 360°$ or $x = 120° + n \cdot 360°, n \in I\}$ 14. $\{x : x = \frac{\pi}{3} + 2n\pi$ or

$x = \frac{2\pi}{3} + 2n\pi, n \in I\}$, or $\{x : x = 60° + n \cdot 360°$ or $x = 120° + n \cdot 360°, n \in I\}$ 15. $\{x : x = 0 + n\pi$ or

$x = \frac{\pi}{4} + n\pi, n \in I\}$, or $\{x : x = 0° + n \cdot 180°$ or $x = 45° + n \cdot 180°, n \in I\}$ 16. $\{x : x = \frac{\pi}{2} + 2n\pi$ or

$x = \frac{7\pi}{6} + 2n\pi$ or $x = \frac{11\pi}{6} + 2n\pi, n \in I\}$ or $\{x : x = 90° + n \cdot 360°$ or $x = 210° + n \cdot 360°$ or x

$= 330° + n \cdot 360°, n \in I\}$

PAGE 230 REVIEW CAPSULE FOR SECTION 5-5

1. π 2. $\frac{3\pi}{2}$ 3. -4π 4. $\frac{4\pi}{3}$ 5. $\frac{5\pi}{3}$ 6. $\frac{2\pi}{3}$ 7. $-\frac{\pi}{3}$ 8. $\frac{5\pi}{6}$ 9. $-60°$ 10. $72°$ 11. $150°$

12. $-720°$ 13. $105°$ 14. $630°$ 15. $-135°$ 16. $405°$

1. $(1, 0°)$ or $(1, 0)$ 2. $(1, 90°)$ or $(1, \frac{\pi}{2})$ 3. $(1, 180°)$ or $(1, \pi)$ 4. $(1, 270°)$ or $(1, \frac{3\pi}{2})$

Answers to Exercises 1 and 3 will vary.

1.   3. 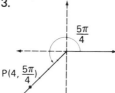 5. 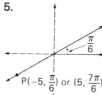 7.

9. $(\sqrt{2}, \frac{\pi}{4})$ or $(\sqrt{2}, 45°)$ 11. $(9, \pi)$ or $(9, 180°)$ 13. $(2, -\frac{\pi}{3})$ or $(2, -60°)$ 15. $(4, \frac{2\pi}{3})$ or $(4, 120°)$

17. $(-2\sqrt{2}, 2\sqrt{2})$ 19. $(0, -5)$ 21. $(\frac{\sqrt{3}}{2}, \frac{1}{2})$ 23. $(-\frac{9}{2}, -\frac{9\sqrt{3}}{2})$ 25. $r = -4 \csc \theta$ 27. $\theta = \frac{\pi}{4}$, or $45°$

29. $r^2 \cos \theta \sin \theta = 16$ 31. $r = 4$ 33. $x^2 + y^2 - 2x = 0$ 35. $x^2 + y^2 = 9$ 37. $x^2 + y^2 = 2y + 2x$

39. $x = 4$ 41. $y = 0$ 43. Using the given information and Equations 1 and 2 on page 233,
$x_1 = r \cos \theta$, $y_1 = r \sin \theta$ and $x_2 = r \cos (-\theta) = r \cos \theta$, $y_2 = r \sin (-\theta) = -r \sin \theta$; thus, $x_1 = x_2$ and $y_1 = -y_2$.

1. Circle with radius 2 units, center at the pole.   2. Straight line that includes the pole. Slope is 1.
3. Circle with radius 3 units, center on the line $\theta = 90°$ (y axis), 3 units from the pole.   4. A vertical line that includes the point (2, 0).

1. A circle of radius 1 with center at (0, 0).
5. A circle of radius $\frac{1}{2}$ with center at $(\frac{1}{2}, 0)$.
7. A circle of radius $\frac{3}{2}$ with center at $(-\frac{3}{2}, 0)$.
11. Same graph as Exercise 9.
13.

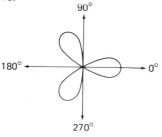

3.

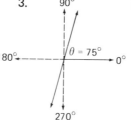

9.

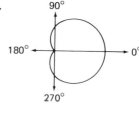

15. Five-leaved rose with petals symmetric about the 18° line, the 90° line, the 162° line, the 234° line and the 306° line.

17.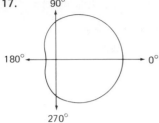

19.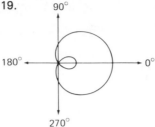

21.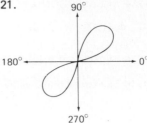

23.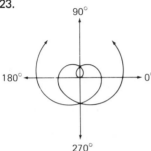

25.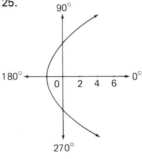

27. This is the equation of a hyperbola. Selected polar coordinates: $(-5\frac{1}{3}, 0°)$, $(-16, 60°)$, $(16, 90°)$, $(5\frac{1}{3}, 120°)$, $(3\frac{1}{5}, 180°)$, $(16, 270°)$ $(-16, 300°)$, $(-5\frac{1}{3}, 360°)$

29. Let $P(r, \theta)$ be a point on the circle with center at $C(r_1, \theta_1)$ and radius a. Draw $\overline{OC}$, $\overline{OP}$, and $\overline{PC}$. Let $\theta - \theta_1$ represent the measure of $\angle POC$. $OP = r$, $OC = r_1$, and $\overline{CP}$ is a radius of the circle, with length a. Thus, using the Law of Cosines in $\triangle POC$, $a^2 = r^2 + r_1^2 - 2rr_1 \cos(\theta - \theta_1)$. **31.** Both graphs are the same. See Example 4 on page 238.

33.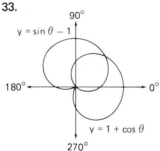

None of the intersection points provides a simultaneous solution of the two equations.

PAGES 241-242 CHAPTER OBJECTIVES AND REVIEW (The "I" in Ex. 7, 9, 19 represents the set of integers.)

3. $y = \arcsin x$ See·Example 1, Section 5-2   **5.** $y = \arccos x$ See Example 2, Section 5-1

7. $\{y : y = n \cdot 180°, n \in I\}$ **9.** $\{y : y = 60° + n \cdot 360° \text{ or } y = 300° + n \cdot 360°, n \in I\}$ **11.** $\frac{\pi}{4}$ **13.** 1

15. 0   **17.** 150°   **19.** $\{x : x = n\pi \text{ or } \frac{\pi}{6} + 2n\pi \text{ or } \frac{5\pi}{6} + 2n\pi, n \in I\}$, or $\{x : x = n \cdot 180° \text{ or } 30° + n \cdot 360°$ or $150° + n \cdot 360°, n \in I\}$

21.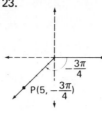

23.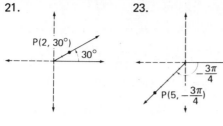

25. $(-\frac{3\sqrt{3}}{2}, \frac{3}{2})$ **27.** $(10, -\frac{\pi}{3})$, or $(10, -60°)$

29. $r \cos \theta = 5$ **31.** $r = 4 \tan \theta \sec \theta$

33. 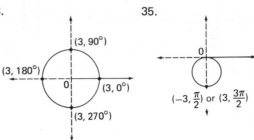   **35.**

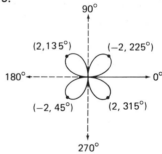

$(-3, \frac{\pi}{2})$ or $(3, \frac{3\pi}{2})$

PAGE 243   CHAPTER TEST   (The "I" in Exercises 3, 5, 7, 8 represents the set of integers.)

1. $y = \frac{1}{4}x + 3$ **2.** See Example 2, on page 211.   **3.** $\{y : y = \frac{7\pi}{6} + 2n\pi \text{ or } y = \frac{11\pi}{6} + 2n\pi, n \in I\}$, or
$\{y : y = 210° + n \cdot 360° \text{ or } y = 330° + n \cdot 360°, n \in I\}$ **4.** 0   **5.** $\{y : y = \pm \frac{2\pi}{3} + 2n\pi, n \in I\}$, or
$\{y : y = \pm 120° + n \cdot 360°, n \in I\}$ **6.** $\frac{5}{2}$ **7.** $\{x : x = \frac{2\pi}{3} + n\pi, n \in I\}$, or $\{x : x = 120° + n \cdot 180°, n \in I\}$
8. $\{x : x = \frac{\pi}{4} + n\frac{\pi}{2}, n \in I\}$ or $\{x : x = 45° + n \cdot 90°, n \in I\}$ **9.** $(\sqrt{2}, \frac{3\pi}{4})$ or $(\sqrt{2}, 135°)$ **10.** $(4, \frac{2\pi}{3})$ or
$(4, 120°)$ **11.** $(0, -2)$ **12.** $(-\frac{3\sqrt{2}}{4}, \frac{3\sqrt{2}}{4})$

13. 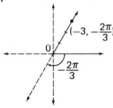   **14.** $r = 9 \cot \theta \csc \theta$ **15.**

PAGES 244-246   CUMULATIVE REVIEW: CHAPTERS 1-5

1. c   **3.** c   **5.** b   **7.** b   **9.** d   **11.** a   **13.** c   **15.** d   **17.** c   **19.** d   **21.** a   **23.** c   **25.** d   **27.** b
29. c   **31.** b   **33.** b

CHAPTER 6   CIRCULAR FUNCTIONS AND APPLICATIONS

PAGE 250   CLASSROOM EXERCISES

1. No   **2.** Yes   **3.** No   **4.** Yes   **5.** 2π **6.** $(\frac{\sqrt{2}}{2}, -\frac{\sqrt{2}}{2})$ **7.** $(-\frac{\sqrt{2}}{2}, \frac{\sqrt{2}}{2})$ **8.** $(-\frac{\sqrt{2}}{2}, -\frac{\sqrt{2}}{2})$
9. $(\frac{\sqrt{2}}{2}, \frac{\sqrt{2}}{2})$

1. I   3. IV   5. IV   7. Quadrantal between quadrants III and IV   9. $\frac{1}{2} + 2\pi, \frac{1}{2} + 4\pi$ 11. $2.1 - 2\pi$,

$2.1 + 2\pi$ 13. $(-\frac{4}{5}, -\frac{3}{5}), (\frac{4}{5}, \frac{3}{5}), (\frac{4}{5}, -\frac{3}{5}), (-\frac{4}{5}, \frac{3}{5})$ 15. $(\frac{2}{3}, -\frac{\sqrt{5}}{3}), (-\frac{2}{3}, \frac{\sqrt{5}}{3}), (-\frac{2}{3}, -\frac{\sqrt{5}}{3}), (\frac{2}{3}, \frac{\sqrt{5}}{3})$

17. $(\frac{\sqrt{13}}{\sqrt{14}}, \frac{1}{\sqrt{14}}), (-\frac{\sqrt{13}}{\sqrt{14}}, -\frac{1}{\sqrt{14}}), (-\frac{\sqrt{13}}{\sqrt{14}}, \frac{1}{\sqrt{14}}), (\frac{\sqrt{13}}{\sqrt{14}}, -\frac{1}{\sqrt{14}})$ 19. $(-1, 0)$ 21. $(-1, 0)$

23. Decreases   25. Increases   27. Decreases   29. Increases   31. ±0.9908   33. $(1, 0)$ 35. $(0, 1)$

37. $(-1, 0)$ 39. $(0, -1)$ 41. $(1, 0)$ 43. $(0, 1)$ 45. Yes, 8.

PAGE 254   CLASSROOM EXERCISES

1. $\frac{4}{5}$ 2. $-\frac{4}{5}$ 3. $-\frac{1}{3}$ 4. $\frac{1}{3}$ 5. $-\sec t$ 6. $\csc t$

PAGES 255-256   WRITTEN EXERCISES

1. The origin; 1   3. Their values are coordinates of points on the unit circle.   5. The set of real numbers

between −1 and 1 inclusive.   7. $(x, -y)$ 9. $t + 2\pi; t + 2\pi$ 11. 2π 13. $(-0.5403, -0.8415)$

15. .8415   17. −.8415   19. $\cos t = \frac{4}{5}$ 21. $\cos t = -\frac{3}{5}$ 23. $\tan(\pi + t) = \tan t$ 25. $-\cot t$

27. $\sec t$ 29. $-\csc t$ 31. $Q(\cos\frac{\pi}{6}, -\sin\frac{\pi}{6})$ 33. $\sin\frac{\pi}{6} = \frac{1}{2}; \cos\frac{\pi}{6} = \frac{\sqrt{3}}{2}$ 35. $-\frac{\sqrt{3}}{2}$ 37. $-\frac{\sqrt{3}}{2}$

39. $\frac{\sqrt{3}}{2}$ 41. $\frac{2}{\sqrt{3}}$, or $\frac{2\sqrt{3}}{3}$

PAGE 256   REVIEW CAPSULE FOR SECTION 6-3

1. b   2. d   3. c   4. a

5.
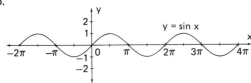

6.
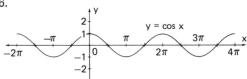

7.
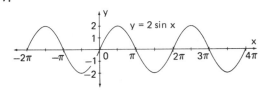

8.
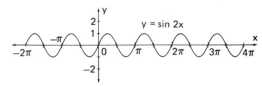

9. 2π 10. 2π 11. π 12. 4π 13. 2π 14. π 15. π 16. 4π

1. $\frac{1}{4\pi}$ cycles per unit   2. $\frac{2}{\pi}$ cycles per unit   3. $\frac{1}{\pi}$ cycles per unit   4. $\frac{6}{\pi}$ cycles per unit   5. $\frac{2}{\pi}$ cycles per unit   6. $\frac{1}{2\pi}$ cycles per unit

PAGES 259-260 WRITTEN EXERCISES

1. $\frac{1}{2\pi}$ 3. $\frac{1}{\pi}$ 5. $\frac{3}{2\pi}$ cycles per unit   7. $\frac{2}{\pi}$ cycles per unit   9. $y = \sin \pi t$, or $y = \sin(-\pi t)$ 11. $y = \sin 2t$, or $y = \sin(-2t)$ 13. $y = \cos 4t$, or $y = \cos(-4t)$ 15. $y = \cos \frac{\pi}{2}t$, or $y = \cos(-\frac{\pi}{2}t)$ 17. $f = \frac{4}{15}$ or .267 cycles per second   19. $y = \pm\frac{2}{3}\sin 8(t + \frac{\pi}{8})$

PAGES 263-264 WRITTEN EXERCISES

1. 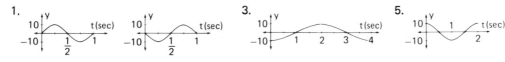   3. 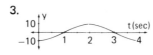   5.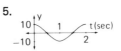

Answers for Exercises 7, 9, 11, 13, 15, 17, and 19 may vary.   7. Using the first graph in Exercise 1, the equation is $y = 10 \sin 2\pi t$. Another solution is $y = -10 \sin 2\pi(t - \frac{1}{2})$. Using the second graph, two possible solutions are $y = -10 \sin 2\pi t$, or $y = 10 \sin 2\pi(t - \frac{1}{2})$.   9. $y = 10 \sin \frac{\pi}{2}(t - 1)$, or $y = -10 \sin \frac{\pi}{2}(t + 1)$

11. $y = 10 \sin \pi(t + \frac{1}{2})$, or $y = -10 \sin \pi(t - \frac{1}{2})$

13.

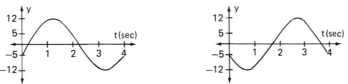

$y = 12 \sin \frac{\pi}{2}(t - \frac{5}{18})$ or $y = -12 \sin \frac{\pi}{2}(t + \frac{31}{18})$ $y = 12 \sin \frac{\pi}{2}(t - \frac{31}{18})$ or $y = -12 \sin \frac{\pi}{2}(t + \frac{5}{18})$

15. First graph: $y = 10 \cos 2\pi(t - \frac{1}{4})$ or $y = -10 \cos 2\pi(t + \frac{1}{4})$ Second graph: $y = -10 \cos 2\pi(t - \frac{1}{4})$ or $y = 10 \cos 2\pi(t + \frac{1}{4})$ 17. $y = -10 \cos \frac{\pi}{2}t$ or $y = 10 \cos \frac{\pi}{2}(t \pm 2)$ 19. $y = 10 \cos \pi t$ or $y = -10 \cos \pi(t \pm 1)$

21. 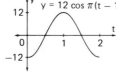   23. All three graphs are the same.   25. 0 cycles per second   27. 0 cycles per second   29. $\pm 20\pi$ cycles per second   31. $\frac{1}{2}$ sec.   33. $\frac{1}{2}$ sec.   35. $y = 2 \sin 4\pi(t - \frac{1}{8})$

PAGES 267-268 WRITTEN EXERCISES

1. f = 45 cycles per second   3. E = 0 volts   5. E = 220 sin 120πt   7. E = 100 sin 124πt   9. E = 0 volts

11. 0 volts   13. E = 100 sin 250πt   15. E = 120 sin 90πt   17. I = 15 sin 120πt

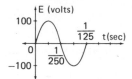

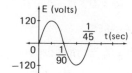

PAGES 270-271 WRITTEN EXERCISES

1. y = ± sin 1056πt

3. y = ± sin 264πt

5.

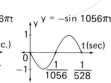

7.
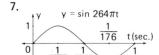
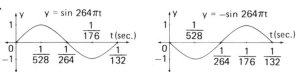

9.

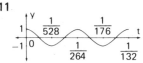

11
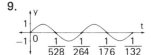

13. y = − cos 528πt

15.

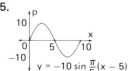

17. n = 25

PAGES 273-274 CHAPTER OBJECTIVES AND REVIEW

3. $(-\frac{\sqrt{2}}{2}, \frac{\sqrt{2}}{2})$ 5. $(\frac{\sqrt{2}}{2}, \frac{\sqrt{2}}{2})$ 7. .4617   9. −tan t   11. $f = \frac{1}{\pi}$ cycles per unit   13. $\frac{1}{2}$ cycle per unit

15. y = cos 10πt or y = cos (−10πt)   17. Answers may vary. $y = 6 \sin \frac{\pi}{4}(t - 2)$ 19. $E = 100\sqrt{3}$ or about 173.2 volts

PAGE 274 CHAPTER TEST

1. $(-\frac{3}{5}, -\frac{4}{5}); (\frac{3}{5}, \frac{4}{5}); (\frac{3}{5}, -\frac{4}{5}); (-\frac{3}{5}, \frac{4}{5})$ 2. $(\frac{\sqrt{15}}{4}, -\frac{1}{4}); (-\frac{\sqrt{15}}{4}, \frac{1}{4}); (-\frac{\sqrt{15}}{4}, -\frac{1}{4}); (\frac{\sqrt{15}}{4}, \frac{1}{4})$ 3. −.4618

4. −.8870   5. cot t   6. $\frac{3}{2}$ cycles per unit   7. $\frac{6}{\pi}$ cycles per unit   8. y = ± sin 3t

9.

10. One equation is $y = -20 \sin \frac{2\pi}{3}(t - \frac{3}{4})$. Another equation is $y = 20 \cos \frac{2\pi}{3}t$.

Answers may vary.

428   CIRCULAR FUNCTIONS AND APPLICATIONS

CHAPTER 7 COMPLEX NUMBERS

1. $3 - 3i$ 2. $-i$ 3. $3 - 2i$ 4. $-8 + 4i$ 5. $-1 + 8i$ 6. $16 - 7i$

1. $a = 4, b = -7$ 3. $a = 0, b = 3$ 5. $a = \sqrt{3}, b = 1$ 7. $a = 1, b = \sqrt{3}$ 9. $a = 0, b = 1 + \sqrt{3}$
11. $12 + 9i$ 13. $-3 - i$ 15. $-3 + 7i$ 17. 0 19. $3 - 6i$ 21. $-5 + 4i$ 23. $-7i$ 25. 0
27. $-10 + 9i$ 29. $7 + 5i$ 31. $1 - i$ 33. $a = 4, b = 0$ 35. $a = -5, b = 1 + 3\sqrt{2}$ 37. $c + di = 0 + 0i$
39. $(a + bi) + (c + di) = (a + c) + (b + d)i = (c + a) + (d + b)i = (c + di) + (a + bi)$, using the Commutative
Property for Addition of real numbers.

1. $23 + 27i$ 2. $16 - 11i$ 3. $3 - 4i$ 4. $-3 - 2i$ 5. $1 + 2i$ 6. $-2 + i$ 7. $-\frac{1}{25} - \frac{7}{25}i$ 8. $-\frac{7}{5} + \frac{11}{5}i$

1. $5 + 5i$ 3. $6 - 10i$ 5. $-17 - 17i$ 7. 45 9. $12 + 6i$ 11. 0 13. $3 - 4i$ 15. $2 + \frac{11}{4}i$ 17. 53
19. 5 21. 144 23. $4 + 2\sqrt{2}$ 25. $1 - \frac{1}{2}i$ 27. $2 - i$ 29. $-\frac{31}{61} + \frac{25}{61}i$ 31. $3 - i$ 33. $1 + i$
35. $\frac{2}{13} + \frac{3}{13}i$ 37. $-4 + 3i$ 39. $\frac{3}{5} - i$ 41. $13 + 4i$ 43. $i^5 = i; i^6 = -1; i^7 = -i; i^8 = 1; i^9 = i; i^{10} = -1;$
$i^{11} = -i; i^{12} = 1$ 45. $f(x + 4) = i^{x+4} = i^x \cdot i^4 = i^x \cdot 1 = i^x = f(x)$; Thus $f(x) = i^x$ is periodic with period 4.
47. $(a + bi) \cdot (\frac{a}{a^2 + b^2} - \frac{b}{a^2 + b^2}i) = \frac{a^2}{a^2 + b^2} - \frac{ab}{a^2 + b^2}i + \frac{ab}{a^2 + b^2}i - \frac{b^2i^2}{a^2 + b^2} = \frac{a^2 - b^2i^2}{a^2 + b^2} = \frac{a^2 + b^2}{a^2 + b^2} = 1$
49. Let $a + bi$ and $c + di$ represent two complex numbers. Then, $(a + bi)(c + di) = (ac - bd) + (ad + bc)i$.
Since the real numbers are closed with respect to multiplication and addition, $(ac - bd)$ and $(ad + bc)$ are
real numbers. Thus, $(ac - bd) + (ad + bc)i$ is a complex number and the set of complex numbers is closed
with respect to multiplication.   51. $[(a + bi) \cdot (c + di)] \cdot (e + fi) = [(ac - bd) + (ad + bc)i] (e + fi)$
$= (ac - bd)e + (ac - bd)fi + (ad + bc)ei + (ad + bc)fi^2 = (ac - bd)e - (ad + bc)f + [(ac - bd)f + (ad + bc)e] i$
$= (ace - bde - adf - bcf) + (acf - bdf + ade + bce)i$; $(a + bi) \cdot [(c + di) \cdot (e + fi)] = (a + bi) \cdot [(ce - df)$
$+ (cf + de)i] = a(ce - df) + a(cf + de)i + b(ce - df)i + b(cf + de)i^2 = a(ce - df) - b(cf + de) + [a(cf + de)$
$+ b(ce - df)] i = ace - adf - bcf - bde + (acf + ade + bce - bdf)i$ or $(ace - bde - adf - bcf)$
$+ (acf - bdf + ade + bce)i$. Thus, $[(a + bi) \cdot (c + di)] \cdot (e + fi) = (a + bi) \cdot [(c + di) \cdot (e + fi)]$ and
multiplication of complex numbers is associative.

1. $(2, 6)$ 2. $(0, -3)$ 3. $(-1, 0)$ 4. $(3, 2)$ 5. $4 + 3i$ 6. $3 + 0i$ 7. $0 + 4i$ 8. $-\sqrt{2} - i$

1. $|z| = \sqrt{2}; \theta = \frac{3\pi}{4}$, or 135°   3. $|z| = 1; \theta = \frac{\pi}{6}$, or 30°   5. $|z| = 2; \theta = \frac{\pi}{4}$, or 45°

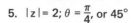

7. $|z| = 3; \theta = \frac{3\pi}{2}$, or 270°   9. $|z| = 6; \theta = \frac{2\pi}{3}$, or 120°   11. $|z| = 3\sqrt{2}; \theta = \frac{\pi}{4}$, or 45°

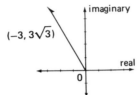

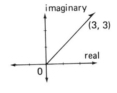

13. $|z| = \sqrt{2}; \theta = \frac{\pi}{4}$, or 45°   15. $|z| = 2; \theta = \frac{7\pi}{6}$, or 210°   17. $2 + 2i; r = 2\sqrt{2}; \theta = \frac{\pi}{4}$, or 45°

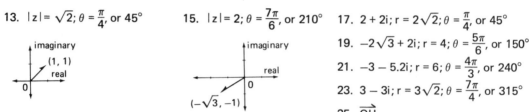

19. $-2\sqrt{3} + 2i; r = 4; \theta = \frac{5\pi}{6}$, or 150°

21. $-3 - 5.2i; r = 6; \theta = \frac{4\pi}{3}$, or 240°

23. $3 - 3i; r = 3\sqrt{2}; \theta = \frac{7\pi}{4}$, or 315°

25. $\overrightarrow{\text{OH}}$

27. $3\sqrt{2}(\cos \frac{3\pi}{4} + i \sin \frac{3\pi}{4})$ or $3\sqrt{2}(\cos 135° + i \sin 135°)$ 29. $4(\cos \frac{\pi}{6} + i \sin \frac{\pi}{6})$ or $4(\cos 30° + i \sin 30°)$

31. $2(\cos \frac{5\pi}{3} + i \sin \frac{5\pi}{3})$ or $2(\cos 300° + i \sin 300°)$ 33. $\sqrt{6}(\cos \frac{5\pi}{4} + i \sin \frac{5\pi}{4})$ or $\sqrt{6}(\cos 225°$

$+ i \sin 225°)$ 35. $3(\cos \pi + i \sin \pi)$ or $3(\cos 180° + i \sin 180°)$ 37. $2\sqrt{3}(\cos \frac{5\pi}{6} + i \sin \frac{5\pi}{6})$ or

$2\sqrt{3}(\cos 150° + i \sin 150°)$ 39. $4(\cos \frac{\pi}{3} + i \sin \frac{\pi}{3})$ or $4(\cos 60° + i \sin 60°)$ 41. $2(\cos \frac{2\pi}{3} + i \sin \frac{2\pi}{3})$ or

$2(\cos 120° + i \sin 120°)$ 43. $\frac{3}{2} + \frac{3\sqrt{3}}{2}i$ 45. $\frac{\sqrt{3}}{2} - \frac{1}{2}i$ 47. $0 + 4i$ 49. $\sqrt{3} + i$ 51. $|a|$

53. 12 or −12

55. −4 and its conjugate have the same graph.   57.

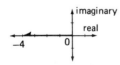

59. θ is given as the argument of a complex number, which shall be called $a + bi$. Let θ_1 represent the argument of its conjugate $a - bi$ or $a + (-b)i$. For $a > 0$, $\theta_1 = $ Arc tan $(\frac{-b}{a}) = -$Arc tan $\frac{b}{a} = -\theta$. For

COMPLEX NUMBERS

$a < 0$, $\theta_1 = \pi + \text{Arc tan} \left(-\frac{b}{a}\right) = \pi - \text{Arc tan} \frac{b}{a} = -\theta$. For $a = 0$ and $b > 0$, $\theta_1 = \frac{3\pi}{2} = -\frac{\pi}{2} = -\theta$. For $a = 0$ and $b < 0$, $\theta_1 = \frac{\pi}{2} = -\frac{3\pi}{2} = -\theta$. Thus, $-\theta$ is the argument of the conjugate of a complex number with argument θ.

PAGE 287 REVIEW

1. $15 - 2i$ 2. $-2 - i$ 3. $25 - 5i$ 4. $3 - 16i$ 5. 89 6. $\frac{5}{2} - \frac{19}{2}i$ 7. $-\frac{109}{61} - \frac{21}{61}i$ 8. $66 + 37i$

9. $|z| = \sqrt{2^2 + 2^2} = 2\sqrt{2}$; $\theta = \text{Arc tan} \frac{2}{2} = \frac{\pi}{4}$, or $45°$; $z = 2\sqrt{2}(\cos \frac{\pi}{4} + i \sin \frac{\pi}{4})$ or $2\sqrt{2}(\cos 45° + i \sin 45°)$

10. $|z| = 2\sqrt{2}$, $\theta = \frac{5\pi}{3}$, or $300°$, $z = 2\sqrt{2}(\cos \frac{5\pi}{3} + i \sin \frac{5\pi}{3})$ or $2\sqrt{2}(\cos 300° + i \sin 300°)$

11. $|z| = 3$; $\theta = \frac{\pi}{2}$, or $90°$; $z = 3(\cos \frac{\pi}{2} + i \sin \frac{\pi}{2})$ or $3(\cos 90° + i \sin 90°)$ 12. $|z| = \sqrt{2}$; $\theta = \frac{3\pi}{4}$, or $135°$; $z = \sqrt{2}(\cos \frac{3\pi}{4} + i \sin \frac{3\pi}{4})$ or $\sqrt{2}(\cos 135° + i \sin 135°)$

PAGE 291 CLASSROOM EXERCISES

1. $r = 10$, $\theta = 48°$ 2. $r = 4$, $\theta = 70°$ 3. $r = 2$, $\theta = 50°$ 4. $r = 9$, $\theta = 30°$

PAGES 291-292 WRITTEN EXERCISES

1. $0 + 24i$ 3. $-5 - 5\sqrt{3}i$ 5. $40.9584 + 38.192i$ 7. $0 + 3i$ 9. $\frac{\sqrt{2}}{3} - \frac{\sqrt{2}}{3}i$ 11. $-\frac{3}{4} + \frac{3\sqrt{3}}{4}i$;
$4.596 + 3.8568i$ 13. $0.3451 - 1.2879i$; $2.8977 - 0.7764i$ 15. $-1 - i$; $1 + i$ 17. $\frac{1}{z} = (\cos 0 + i \sin 0)$
$\div [r(\cos \theta + i \sin \theta)] = \frac{1}{r}[\cos (-\theta) + i \sin (-\theta)] = \frac{1}{r}(\cos \theta - i \sin \theta) = \frac{1}{r}[\cos (-\theta) + i \sin (-\theta)] \cdot \frac{r}{r} = \frac{1}{r^2}\bar{z}$
19. $z \div \bar{z} = r(\cos \theta + i \sin \theta) \div [r[\cos (-\theta) + i \sin (-\theta)]] = \frac{r}{r}[\cos (\theta - (-\theta)) + i \sin (\theta - (-\theta))] = \cos 2\theta + i \sin 2\theta$ 21. Let $z_1 = r_1(\cos \theta + i \sin \theta)$ and $z_2 = r_2(\cos \theta + i \sin \theta)$. Then $z_1 \div z_2$
$= \frac{r_1}{r_2}[\cos (\theta - \theta) + i \sin (\theta - \theta)] = \frac{r_1}{r_2}(\cos 0 + i \sin 0) = \frac{r_1}{r_2}(1 + 0i) = \frac{r_1}{r_2}$, which is a real number.

PAGE 294 CLASSROOM EXERCISES

1. $128 + 128i$ 2. i 3. $2 + 2\sqrt{3}i$

PAGES 294-295 WRITTEN EXERCISES

1. $-8 + 0i$ 3. $0 + 27i$ 5. $0 - \frac{1}{64}i$ 7. $32\sqrt{2} - 32\sqrt{2}i$ 9. $-64 + 0i$ 11. $0 + 8i$ 13. $-\frac{1}{4} + 0i$
15. $16 + 16\sqrt{3}i$ 17. $0 + i$ 19. $-8 - 8\sqrt{3}i$ 21. $\cos 2\theta = 2 \cos^2 \theta - 1$; $\sin 2\theta = 2 \cos \theta \sin \theta$
23. $\cos 5\theta = \cos^5 \theta - 10 \cos^3 \theta \sin^2 \theta + 5 \cos \theta \sin^4 \theta$; $\sin 5\theta = 5 \cos^4 \theta \sin \theta - 10 \cos^2 \theta \sin^3 \theta + \sin^5 \theta$

25.

27. Let $P_n = z^n = r^n(\cos n\theta + i \sin n\theta)$, $n \in N$. Thus $P_1 = z^1 = r^1 (\cos 1 \cdot \theta + i \sin 1 \cdot \theta) = r(\cos \theta + i \sin \theta)$. This is true by definition. Next, assume P_n is true for $n = k$, and show that P_n is true for $n = k + 1$. Thus, $z^{k+1} = z^k \cdot z^1 = r^k (\cos k\theta + i \sin k\theta) \cdot r(\cos \theta + i \sin \theta)$ $= r^k \cdot r(\cos (k\theta + \theta) + i \sin (k\theta + \theta))$. So, $z^{k+1} = r^{k+1}(\cos [(k + 1)\theta] + i \sin [(k + 1)\theta])$. Therefore, P_{k+1} is true whenever P_k is true, and hence P_n is true for all natural numbers.

CHAPTER 7

431

1. ± 13 2. $\pm 2\sqrt[4]{2}$ 3. -3 4. $3\sqrt[3]{3}$ 5. $z = \pm 2\sqrt{13}$ 6. $z = -5$ 7. $z = \pm 3\sqrt[4]{2}$ 8. $z = -2\sqrt[5]{2}$

1. $1 + 0i, -\frac{1}{2} + \frac{\sqrt{3}}{2}i, -\frac{1}{2} - \frac{\sqrt{3}}{2}i$ 2. $3, -\frac{3}{2} + \frac{3\sqrt{3}}{2}i, -\frac{3}{2} - \frac{3\sqrt{3}}{2}i$

1. $z_0 = 2(\cos 45° + i \sin 45°) = \sqrt{2} + \sqrt{2}i$;
$z_1 = 2(\cos 225° + i \sin 225°) = -\sqrt{2} - \sqrt{2}i$

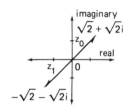

3. $z_0 = 2(\cos 30° + i \sin 30°) = \sqrt{3} + i$;
$z_1 = 2(\cos 150° + i \sin 150°) = -\sqrt{3} + i$;
$z_2 = 2(\cos 270° + i \sin 270°) = 0 - 2i$

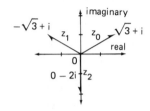

5. $z_0 = 3(\cos 60° + i \sin 60°) = \frac{3}{2} + \frac{3\sqrt{3}}{2}i$;
$z_1 = 3(\cos 180° + i \sin 180°) = -3 + 0i$;
$z_2 = 3(\cos 300° + i \sin 300°) = \frac{3}{2} - \frac{3\sqrt{3}}{2}i$

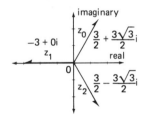

7. $z_0 = 3(\cos 45° + i \sin 45°) = \frac{3\sqrt{2}}{2} + \frac{3\sqrt{2}}{2}i$;
$z_1 = 3(\cos 135° + i \sin 135°) = -\frac{3\sqrt{2}}{2} + \frac{3\sqrt{2}}{2}i$;
$z_2 = 3(\cos 225° + i \sin 225°) = -\frac{3\sqrt{2}}{2} - \frac{3\sqrt{2}}{2}i$;
$z_3 = 3(\cos 315° + i \sin 315°) = \frac{3\sqrt{2}}{2} - \frac{3\sqrt{2}}{2}i$

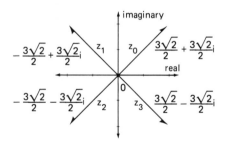

9. $z_0 = 2(\cos 30° + i \sin 30°) = \sqrt{3} + i$;
$z_1 = 2(\cos 120° + i \sin 120°) = -1 + \sqrt{3}i$;
$z_2 = 2(\cos 210° + i \sin 210°) = -\sqrt{3} - i$;
$z_3 = 2(\cos 300° + i \sin 300°) = 1 - \sqrt{3}i$

11. $z_0 = \cos 0° + i \sin 0° = 1 + 0i$;
$z_1 = \cos 90° + i \sin 90° = 0 + i$;
$z_2 = \cos 180° + i \sin 180° = -1 + 0i$;
$z_3 = \cos 270° + i \sin 270° = 0 - i$

13. $z_0 = \cos 30° + i \sin 30° = \frac{\sqrt{3}}{2} + \frac{1}{2}i$; $z_1 = \cos 90° + i \sin 90° = 0 + i$; $z_2 = \cos 150° + i \sin 150° = -\frac{\sqrt{3}}{2}$

$+ \frac{1}{2}i$; $z_3 = \cos 210° + i \sin 210° = -\frac{\sqrt{3}}{2} - \frac{1}{2}i$; $z_4 = \cos 270° + i \sin 270° = 0 - i$; $z_5 = \cos 330° + i \sin 330°$

$= \frac{\sqrt{3}}{2} - \frac{1}{2}i$ 15. $z_0 = \cos 0° + i \sin 0° = 1 + 0i$; $z_1 = \cos 30° + i \sin 30° = \frac{\sqrt{3}}{2} + \frac{1}{2}i$; $z_2 = \cos 60° + i \sin 60°$

$= \frac{1}{2} + \frac{\sqrt{3}}{2}i$; $z_3 = \cos 90° + i \sin 90° = 0 + i$; $z_4 = \cos 120° + i \sin 120° = -\frac{1}{2} + \frac{\sqrt{3}}{2}i$; $z_5 = \cos 150°$

$+ i \sin 150° = -\frac{\sqrt{3}}{2} + \frac{1}{2}i$; $z_6 = \cos 180° + i \sin 180° = -1 + 0i$; $z_7 = \cos 210° + i \sin 210° = -\frac{\sqrt{3}}{2} - \frac{1}{2}i$;

$z_8 = \cos 240° + i \sin 240° = -\frac{1}{2} - \frac{\sqrt{3}}{2}i$; $z_9 = \cos 270° + i \sin 270° = 0 - i$; $z_{10} = \cos 300° + i \sin 300°$

$= \frac{1}{2} - \frac{\sqrt{3}}{2}i$; $z_{11} = \cos 330° + i \sin 330° = \frac{\sqrt{3}}{2} - \frac{1}{2}i$ 17. $z_0 = \sqrt[5]{2}(\cos 60° + i \sin 60°)$ $= 0.5744$

$+ 0.9948i$; $z_1 = \sqrt[5]{2}(\cos 132° + i \sin 132°)$ $= -0.7686 + 0.8536i$; $z_2 = \sqrt[5]{2}(\cos 204° + i \sin 204°)$

$= -1.0493 - 0.4672i$; $z_3 = \sqrt[5]{2}(\cos 276° + i \sin 276°)$ $= 0.1200 - 1.1424i$; $z_4 = \sqrt[5]{2}(\cos 348°$

$+ i \sin 348°) = 1.1235 - 0.2388i$ 19. $z_0 = 2(\cos 60° + i \sin 60°) = 1 + \sqrt{3}i$; $z_1 = 2(\cos 180° + i \sin 180°)$

$= -2 + 0i$; $z_2 = 2(\cos 300° + i \sin 300°) = 1 - \sqrt{3}i$ 21. $z_0 = \cos 60° + i \sin 60° = \frac{1}{2} + \frac{\sqrt{3}}{2}i$; $z_1 = \cos 180°$

$+ i \sin 180° = -1 + 0i$; $z_2 = \cos 300° + i \sin 300° = \frac{1}{2} - \frac{\sqrt{3}}{2}i$ 23. $z_0 = \sqrt[3]{4}(\cos 40° + i \sin 40°)$;

$z_1 = \sqrt[3]{4}(\cos 160° + i \sin 160°)$; $z_2 = \sqrt[3]{4}(\cos 280° + i \sin 280°)$

PAGE 300   WRITTEN EXERCISES

1.	3.	5.
10 READ A, B, C, D	10 READ A, B, C, D	10 READ R, T
20 LET R = A + C	20 LET R = A − C	20 LET A = T * 3.14159 / 180
30 LET S = B + D	30 LET S = B − D	30 LET X = R * COS(A)
40 PRINT "SUM ="; R;"+"; S;"I"	40 PRINT "DIFFERENCE ="; R;"+"; S;"I"	40 LET Y = R * SIN(A)
50 GO TO 10	50 GO TO 10	50 PRINT "("; X;", "; Y;")"
60 DATA . . .	60 DATA . . .	60 GO TO 10
70 END	70 END	70 DATA . . .
		80 END

PAGES 301-302   CHAPTER OBJECTIVES AND REVIEW

3. $7 + 8i$ 5. $-20 - 3i$ 7. $\frac{1}{5} + \frac{19}{10}i$ 9. $148 + 0i$ 11. $z = 2(\cos \frac{5\pi}{6} + i \sin \frac{5\pi}{6})$, or $z = 2(\cos 150°$

$+ i \sin 150°)$ 13. $z = \cos 300° + i \sin 300°$, or $z = \cos \frac{5\pi}{3} + i \sin \frac{5\pi}{3}$ 15. $45 + 0i$ 17. $-\frac{1}{9} + 0i$

19.
$\frac{1}{18} - \frac{\sqrt{3}}{18}i$

21. $-4 - 4i$

23. $z_0 = 2(\cos 45° + i \sin 45°) = \sqrt{2} + \sqrt{2}i$;

$z_1 = 2(\cos 135° + i \sin 135°) = -\sqrt{2} + \sqrt{2}i$;

$z_2 = 2(\cos 225° + i \sin 225°) = -\sqrt{2} - \sqrt{2}i$;

$z_3 = 2(\cos 315° + i \sin 315°) = \sqrt{2} - \sqrt{2}i$

25. $z_0 = \sqrt[3]{2}(\cos 100° + i \sin 100°) = -0.2188 + 1.2409i$

$z_1 = \sqrt[3]{2}(\cos 220° + i \sin 220°) = -0.9652 - 0.8099i$

$z_2 = \sqrt[3]{2}(\cos 340° + i \sin 340°) = 1.1840 - 0.4309i$

1. $1 + 2i$ 2. $1 + 7i$ 3. $-34 - 12i$ 4. 6   5. $2(\cos 300° + i \sin 300°)$, or $2(\cos \frac{5\pi}{3} + i \sin \frac{5\pi}{3})$

6. $0 + 6i$ 7. $-\frac{\sqrt{2}}{4} + \frac{\sqrt{2}}{4}i$ 8. $16 + 0i$

9.

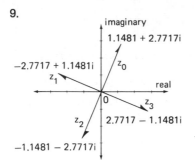

10. $z_0 = \sqrt[4]{4}\,(\cos 30° + i \sin 30°) = 1.2246 + 0.707i$;

$z_1 = \sqrt[4]{4}\,(\cos 120° + i \sin 120°) = -0.707 + 1.2246i$;

$z_2 = \sqrt[4]{4}\,(\cos 210° + i \sin 210°) = -1.2246 - 0.707i$;

$z_3 = \sqrt[4]{4}\,(\cos 300° + i \sin 300°) = 0.707 - 1.2246i$

CHAPTER 8   SEQUENCES AND SERIES

1. $a_n = 3n + 1$ 2. $a_n = 2n - 1$ 3. $a_n = -n + 1$ 4. 2, 4, 6   5. $\frac{1}{3}, \frac{2}{3}, 1$ 6. $\frac{1}{9}, \frac{1}{3}, 1$

1. $a_n = 3n - 1$ 3. $a_n = \frac{1}{n}$ 5. $a_n = 5n - 8$ 7. 1, 2, 3, 4, 5   9. 1, 8, 27, 64, 125   11. $-1, -\frac{1}{2}, -\frac{1}{3}, -\frac{1}{4}$,

$-\frac{1}{5}$ 13. $a_1 = (-1)^0 = 1, a_2 = (-1)^1 = -1, a_3 = (-1)^2 = 1, a_4 = (-1)^3 = -1, a_5 = (-1)^4 = 1$ 15. $1, \frac{1}{4}$,

$\frac{1}{9}, \frac{1}{16}, \frac{1}{25}$ 17. $\frac{3}{5}, \frac{4}{7}, \frac{5}{9}, \frac{6}{11}, \frac{7}{13}$ 19. $\frac{1}{7}$ 21. 10   23. 81   25. 252   27. -1 29. $\frac{1}{243}$ 31. 1, 3, 6, 10

33. 0, 4, 18, 48   35. $3, \frac{5}{3}, \frac{7}{5}, \frac{9}{7}$ 37. $1, 1, \frac{1}{2}, \frac{1}{6}$ 39. $\frac{1}{2}, \frac{1}{4}, \frac{1}{8}, \frac{1}{16}$ 41. $-6, 11, \frac{16}{3}, \frac{21}{5}$

43. $-\frac{\sqrt{2}}{2}, 1, -\frac{\sqrt{2}}{2}, 0$ 45. $1, -\frac{1}{2}, \frac{1}{6}, -\frac{1}{24}$ 47. $\frac{\cos x}{x^2 + 1}, \frac{\cos 2x}{x^2 + 4}, \frac{\cos 3x}{x^2 + 9}, \frac{\cos 4x}{x^2 + 16}$ 49. $1, -\frac{x^2}{2}, \frac{x^4}{24}, -\frac{x^6}{720}$

51. $1, ix, -\frac{x^2}{2}, -\frac{ix^3}{6}$ 53. $4, -4, -4, 4, 4, -4$ 55. $b_i = a_i + (i - 1) \cdots (i - i) \cdots (i - k) = a_i + (i - 1)$

$\cdots 0 \cdots (i - k) = a_i + 0 = a_i$, for $1 \le i \le k$. Thus, the first k terms of the sequence $\{a_n\}$ are identical to

the first k terms of the sequence $\{b_n\}$.

1. 6, 11, 16, 21   2. $-32, -34, -36, -38$ 3. 20b, 15b, 10b, 5b   4. 66   5. $21\frac{1}{2}$ 6. $-4, 8, -16, 32$

7. $\frac{1}{2}, 1, 2, 4$ 8. $2, 2\sqrt{2}, 4, 4\sqrt{2}$ 9. -128 10. $\frac{3}{32}$

1. 5, 8, 11, 14   3. $-5, -7, -9, -11$ 5. $50, 0, -50, -100$ 7. 23   9. -8 11. $36\frac{1}{2}$ 13. 6, 18, 54,

162   15. $-\frac{1}{2}, \frac{1}{4}, -\frac{1}{8}, \frac{1}{16}$ 17. $-\frac{10}{3}, \frac{20}{9}, -\frac{40}{27}, \frac{80}{81}$ 19. 4374   21. $-\frac{8192}{59049}$ 23. $\frac{16}{9}$ 25. $a_n = 4n - 10$

27. $a_n = nx - n + 2$ 29. $a_n = 2^{\frac{n}{2}}$ 31. $a_n = 6(-4)^{3-n}$ 33. $12, -18, 27, -\dfrac{81}{2}$ 35. y^6, y^9, y^{12}, y^{15}

37. $-1, 2$ 39. $30, 26, 22, 18, 14$ 41. $-\dfrac{1}{2}$ 43. 10 45. 31 47. 112 49. $a_n = n$ 51. $a_n = 2n - 1$
53. $\dfrac{x+y}{2}$ 55. $\sqrt{xy}$ 57. About 444 cycles/sec. 59. $a_{n+1} = a_n + d$ 61. Let r = the common ratio

of the geometric sequence with general term a_n. Then, $b_j = (a_j)^2$ and $b_{j+1} = (a_{j+1})^2$. Thus, $\dfrac{b_{j+1}}{b_j}$

$= \dfrac{(a_{j+1})^2}{(a_j)^2} = \dfrac{(a_{j+1})(a_{j+1})}{(a_j)(a_j)} = r \cdot r = r^2$ and b_n is the general term of a geometric sequence with common ratio r^2.

63. Let d represent the common difference between the terms of the arithmetic sequence $\{a_n\}$. Then
$b_{j+1} - b_j = ca_{j+1} - ca_j = c(a_{j+1} - a_j) = cd$. Thus, $\{b_n\}$ is an arithmetic sequence with common difference
= cd.

PAGE 311   REVIEW CAPSULE FOR SECTION 8-3

1. $3500 < n$ 2. $9997 < n$ 3. $n < -\dfrac{197}{99}$ 4. $n > 20$ or $n < -20$

PAGE 314   CLASSROOM EXERCISES

1. $\dfrac{1}{n+4}$ 2. $\dfrac{1}{n}$ 3. $\dfrac{1}{100}$ 4. 1

PAGES 314-315   WRITTEN EXERCISES

1.

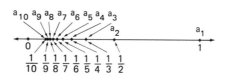

The graphs for Exercises 3, 5, 7, 9, 11, 13, 15, 17, 19, 21, and 23 are done in a similar manner.

3. $-5, -2\dfrac{1}{2}, -1\dfrac{2}{3}, -1\dfrac{1}{4}, -1, -\dfrac{5}{6}, -\dfrac{5}{7}, -\dfrac{5}{8}, -\dfrac{5}{9}, -\dfrac{1}{2}$ 5. $0, \dfrac{1}{2}, \dfrac{2}{3}, \dfrac{3}{4}, \dfrac{4}{5}, \dfrac{5}{6}, \dfrac{6}{7}, \dfrac{7}{8}, \dfrac{8}{9}, \dfrac{9}{10}$ 7. $8, 6\dfrac{1}{2}, 6, 5\dfrac{3}{4}, 5\dfrac{3}{5}, 5\dfrac{1}{2}$,

$5\dfrac{3}{7}, 5\dfrac{3}{8}, 5\dfrac{1}{3}, 5\dfrac{3}{10}$ 9. $4, 1, \dfrac{4}{9}, \dfrac{1}{4}, \dfrac{4}{25}, \dfrac{1}{9}, \dfrac{4}{49}, \dfrac{1}{16}, \dfrac{4}{81}, \dfrac{1}{25}$ 11. $\dfrac{7}{2}, \dfrac{7}{4}, \dfrac{7}{6}, \dfrac{7}{8}, \dfrac{7}{10}, \dfrac{7}{12}, \dfrac{1}{2}, \dfrac{7}{16}, \dfrac{7}{18}, \dfrac{7}{20}$ 13. $1, 2, 3, 4,$

$5, 6, 7, 8, 9, 10$ 15. $1, 4, 9, 16, 25, 36, 49, 64, 81, 100$ 17. $-1, 1, -1, 1, -1, 1, -1, 1, -1, 1$

19. $-1, \dfrac{1}{2}, -\dfrac{1}{3}, \dfrac{1}{4}, -\dfrac{1}{5}, \dfrac{1}{6}, -\dfrac{1}{7}, \dfrac{1}{8}, -\dfrac{1}{9}, \dfrac{1}{10}$ 21. $0, 2, 0, 2, 0, 2, 0, 2, 0, 2$ 23. $\dfrac{1}{4}, \dfrac{2}{5}, \dfrac{1}{2}, \dfrac{4}{7}, \dfrac{5}{8}, \dfrac{2}{3}, \dfrac{7}{10}, \dfrac{8}{11}, \dfrac{3}{4}, \dfrac{10}{13}$

25. $m \geq 10{,}001$ 33. L = 0   The graphs for Exercises 35,
27. $m \geq 30{,}001$ 37, and 39 are done in a
29. $m \geq 10{,}001$ similar manner.
31. $m \geq 20{,}000$

35. $\dfrac{2}{3}, \dfrac{4}{9}, \dfrac{8}{27}, \dfrac{16}{81}, \dfrac{32}{243}, \dfrac{64}{729}$; L = 0   37. $2, 1, \dfrac{1}{2}, \dfrac{1}{4}, \dfrac{1}{8}, \dfrac{1}{16}$; L = 0   39. $2\dfrac{2}{3}, 3\dfrac{1}{9}, 2\dfrac{26}{27}, 3\dfrac{1}{81}, 2\dfrac{242}{243}, 3\dfrac{1}{729}$; L = 3

41. $\left|1 - \frac{n+1}{n+3}\right| = \frac{2}{n+3}.$ Therefore, $\left|1 - \frac{n+1}{n+3}\right| < d$ is equivalent to $\frac{2}{n+3} < d$ which is true whenever

$\frac{2}{d} < n + 3$ or $\frac{2}{d} - 3 < n$. Thus, for any positive number d, you can choose m to be any natural number

greater than $\frac{2}{d} - 3$. Then $\left|1 - \frac{n+1}{n+3}\right|$ will be less than d whenever $n \geq m$. Therefore, $\lim\limits_{n \to \infty} \frac{n+1}{n+3} = 1.$
Exercises 43, 45, 47, 49, 51, 53, and 55 are done in a similar manner. The value of m is given for each
exercise. 43. $m > \frac{1}{d} - 1$ 45. $m > \frac{1}{d}$ 47. $m > \frac{1}{d}$ 49. $m > \frac{1}{d}$ 51. $m > \frac{2}{\sqrt{d}}$ 53. $m > \frac{1}{d} - 1$

55. Choose m to be any natural number greater than $- \log d$. 57. $\left\{2 + \frac{1}{n}\right\} = \left\{3, 2\frac{1}{2}, 2\frac{1}{3}, 2\frac{1}{4}, \ldots,\right.$

$\left.2 + \frac{1}{n}, \ldots\right\}$ The terms appear to converge on 2. Proof: $\left|2 - (2 + \frac{1}{n})\right| = \left|-\frac{1}{n}\right| = \frac{1}{n}$ Therefore

$\left|2 - (2 + \frac{1}{n})\right| < d$ is equivalent to $\frac{1}{n} < d$, which is true whenever $\frac{1}{d} < n$. Thus for any positive number d,

you can choose m to be any natural number greater than $\frac{1}{d}$. Then $\left|2 - (2 + \frac{1}{n})\right|$ will be less than d

whenever $n \geq m$. Therefore, $\lim\limits_{n \to \infty} 2 + \frac{1}{n} = 2.$ 59. $L = 0, m > \frac{1}{d} - 1$ Proof is similar to Ex. 57.

PAGE 318   CLASSROOM EXERCISES

1. Limit Theorem 3 2. Limit Theorem 3 3. Limit Theorem 2 4. Limit Theorem 2 5. 0 6. 0
7. 0 8. 0

PAGES 318-319   WRITTEN EXERCISES

1. 0 3. 0 5. No limit exists. 7. 0 9. 0 11. 12 13. 0 15. $\frac{3}{2}$ 17. 3 19. 0 21. 0 23. 6
25. $\frac{3}{5}$ 27. No limit exists. 29. 0 31. $\lim\limits_{n \to \infty} a_1 \cdot r^{n-1} = \lim\limits_{n \to \infty} a_1 \cdot r^{-1} \cdot r^n = \lim\limits_{n \to \infty} a_1 \cdot \lim\limits_{n \to \infty} r^{-1} \cdot \lim\limits_{n \to \infty} r^n$
$= a_1 \cdot r^{-1} \cdot 0 = 0$

PAGE 319   REVIEW

1. $-1, -2, -3, -4, -5$ 2. $\frac{1}{2}, \frac{1}{4}, \frac{1}{8}, \frac{1}{16}, \frac{1}{32}$ 3. $\frac{78125}{729}$ 4. $m \geq 101$ 5. $m \geq 299$ 6. $\left|1 - \frac{n}{n+2}\right| < d$

is equivalent to $\frac{2}{n+2} < d, \frac{2}{d} < n + 2, \frac{2}{d} - 2 < n.$ Thus, for any positive number d, you can choose m to be

any natural number greater than $\frac{2}{d} - 2$. Then $\left|1 - \frac{n}{n+2}\right|$ will be less than d whenever $n \geq m$. Therefore,

$\lim\limits_{n \to \infty} \frac{n}{n+2} = 1.$ 7. $m > \frac{6}{d}$ (The proof is similar to Exercise 6.) 8. 0 9. 1

PAGE 323   CLASSROOM EXERCISES

1. G 2. A 3. $1 + \frac{1}{2} + \frac{1}{3} + \frac{1}{4}$ 4. $2 + 1\frac{1}{2} + 1\frac{1}{3} + 1\frac{1}{4} + 1\frac{1}{5}$

PAGES 324-325   WRITTEN EXERCISES

1. $-1, 0, -1, 0, -1$ 3. $1, 5, 14, 30, 55$ 5. $1, 4, 9, 16, 25$ 7. $2, 6, 12, 20, 30$ 9. $1, 2\frac{1}{2}, 4\frac{1}{6}, 5\frac{11}{12}, 7\frac{43}{60}$
11. A; $2 + 6 + 10 + 14 + \cdots + (2 + (n-1) \cdot 4 + \cdots$ 13. G; $2 + 6 + 18 + 54 + \cdots + 2(3)^{n-1} + \cdots$

15. A; $-3 - 5 - 7 - 9 - \cdots - (3 + 2(n-1)) - \cdots$ 17. G; $4 - 4 + 4 - 4 + \cdots + (4 \cdot (-\frac{1}{1})^{n-1}) + \cdots$

19. 265   21. -84 23. 325   25. $\frac{255}{64}$ 27. $\frac{1533}{256}$ 29. -340 31. 0   33. $-n^2 - 2n$

35. $3(1 - (\frac{1}{3})^n)$ 37. $50n^2 - 100n$ 39. The sequence of positive integers is $\{1 + 1(n-1)\}$. Thus,

$S_n = \frac{n(2(1) + (n-1)1)}{2} = \frac{n(n+1)}{2}$.   41. The sequence of n positive odd integers is $\{1 + 2(n-1)\}$. Thus,

$S_n = \frac{n(2 \cdot 1 + (n-1)2)}{2} = n^2$.   43. 1316   45. $S_n - S_{n-1} = \sum\limits_{j=1}^{n} a_j - \sum\limits_{j=1}^{n-1} a_j = (a_1 + a_2 + \cdots + a_{n-1} + a_n)$

$- (a_1 + a_2 + \cdots + a_{n-1}) = a_n$. Thus, $S_n - S_{n-1} = a_n$.   47. $\sum\limits_{j=1}^{n} c \cdot a_j = c \cdot a_1 + c \cdot a_2 + \cdots + c \cdot a_n$

$= c(a_1 + a_2 + \cdots + a_n) = c \cdot \sum\limits_{j=1}^{n} a_j$

PAGE 329   CLASSROOM EXERCISES

1. Converges   2. Converges   3. Diverges   4. Diverges   5. Diverges   6. Converges

PAGES 329-330   WRITTEN EXERCISES

1. 3   3. 12   5. $\frac{500}{6}$ 7. $\frac{35}{9}$ 9. Diverges   11. $312\frac{1}{2}$ 13. 18   15. Diverges   17. Test fails

19. Converges   21. Converges   23. Diverges   25. Test fails   27. Converges   29. $-.19$ 31. $.78$

33. $\frac{35}{99}$ 35. $\frac{1572}{9999}$ 37. $\lim\limits_{n\to\infty} \frac{3n-1}{3n+1} = \lim\limits_{n\to\infty} \frac{3 - \frac{1}{n}}{3 + \frac{1}{n}} = \frac{\lim\limits_{n\to\infty} 3 - \lim\limits_{n\to\infty} \frac{1}{n}}{\lim\limits_{n\to\infty} 3 + \lim\limits_{n\to\infty} \frac{1}{n}} = \frac{3-0}{3+0} = 1$. Since $\lim\limits_{n\to\infty} a_n \neq 0$,

this series diverges.   39. $\lim\limits_{n\to\infty} S_n = \lim\limits_{n\to\infty} \frac{a_1}{1-r}(1 - r^n) = \frac{a_1}{1-r} \lim\limits_{n\to\infty} (1 - r^n) = \frac{a_1}{1-r}(\lim\limits_{n\to\infty} 1 - \lim\limits_{n\to\infty} r^n)$

$= \frac{a_1}{1-r}(1-0) = \frac{a_1}{1-r}$ whenever $|r| < 1$.   41. $c_1 c_2 c_3 \cdots c_p c_1 c_2 c_3 \cdots c_p = \sum\limits_{j=1}^{\infty} \frac{c_1 c_2 c_3 \cdots c_p}{10^p}(\frac{1}{10^p})^{j-1}$

Since this series has $a_1 = \frac{c_1 c_2 c_3 \cdots c_p}{10^p}$ and $r = \frac{1}{10^p}$, $S = \lim\limits_{n\to\infty} S_n = \frac{\frac{c_1 c_2 c_3 \cdots c_p}{10^p}}{1 - \frac{1}{10^p}} = \frac{\frac{c_1 c_2 c_3 \cdots c_p}{10^p}}{\frac{10^p - 1}{10^p}}$

$= \frac{c_1 c_2 c_3 \cdots c_p}{10^p - 1}$

PAGE 333   CLASSROOM EXERCISES

1. $(-1)^n x^n$ 2. $(n+1)(-1)^n x^n$ 3. $\frac{x^{n+1}}{(n+1)!}$

PAGE 333   WRITTEN EXERCISES

1. $|x| < 1$ 3. $|x| < 1$ 5. $|x| < 1$ 7. $|x| < 2$ 9. $|x| < 3$ 11. $|x| < 5$ 13. 1.00   15. 5.00
17. $|x| < 1$ 19. 1.00   21. 14.99   23. $|x - 1| < 1$ or $0 < x < 2$ 25. $|x| < 1$ 27. $|x + 1| < 1$
or $-2 < x < 0$

PAGE 336   CLASSROOM EXERCISES

1. $\frac{\sqrt{3}}{2} + \frac{1}{2}i$ 2. $\frac{1}{2} + \frac{\sqrt{3}}{2}i$ 3. $\frac{1}{2} - \frac{\sqrt{3}}{2}i$ 4. $-i$

1. .1987 3. .9950 5. .0701 7. .9801 9. −1.2114 11. .9998 13. −1 15. 1 17. −1 19. 1

21. $e^{ix} - e^{-ix} = (\cos x + i \sin x) - (\cos x - i \sin x)$, $e^{ix} - e^{-ix} = 2i \sin x$, and $\sin x = \dfrac{e^{ix} - e^{-ix}}{2i}$

23. $\cot x = \dfrac{1}{\tan x} = \dfrac{1}{\dfrac{e^{ix} - e^{-ix}}{i(e^{ix} + e^{-ix})}} = \dfrac{i(e^{ix} + e^{-ix})}{e^{ix} - e^{-ix}}$ 25. $\csc x = \dfrac{1}{\sin x} = \dfrac{1}{\dfrac{e^{ix} - e^{-ix}}{2i}} = \dfrac{2i}{e^{ix} - e^{-ix}}$

27. $\lim\limits_{n \to \infty} \left| \dfrac{a_{n+1}}{a_n} \right| = \lim\limits_{n \to \infty} \left| \dfrac{\dfrac{(-1)^{n+2} x^{2n+1}}{(2n+1)!}}{\dfrac{(-1)^{n+1} x^{2n-1}}{(2n-1)!}} \right| = \lim\limits_{n \to \infty} \left| \dfrac{(-1) x^2}{(2n+1)2n} \right| = x^2 \lim\limits_{n \to \infty} \dfrac{1}{(2n+1)2n} = x^2 \cdot 0 = 0.$ Thus,

the series converges for all values of x. 29. $e^{ix} = 1 + ix + \dfrac{(ix)^2}{2!} + \dfrac{(ix)^3}{3!} + \cdots + \dfrac{(ix)^{n-1}}{(n-1)!} + \cdots$ therefore

$\lim\limits_{n \to \infty} \left| \dfrac{a_{n+1}}{a_n} \right| = \lim\limits_{n \to \infty} \left| \dfrac{\dfrac{(ix)^n}{n!}}{\dfrac{(ix)^{n-1}}{(n-1)!}} \right| = \lim\limits_{n \to \infty} \left| \dfrac{ix}{n} \right| = i \, | \, x \, | \lim\limits_{n \to \infty} \dfrac{1}{n} = i \, | \, x \, | \cdot 0 = 0.$ Thus, the series converges

for all values of x. 31. $e^{-ix} = 1 - ix + \dfrac{i^2 x^2}{2!} - \dfrac{i^3 x^3}{3!} + \dfrac{i^4 x^4}{4!} - \dfrac{i^5 x^5}{5!} + \dfrac{i^6 x^6}{6!} - \dfrac{i^7 x^7}{7!} + \cdots + \dfrac{(-1)^{n-1} i^{n-1} x^{n-1}}{(n-1)!}$

$+ \cdots = 1 - ix - \dfrac{x^2}{2!} + \dfrac{ix^3}{3!} + \dfrac{x^4}{4!} - \dfrac{ix^5}{5!} - \dfrac{x^6}{6!} + \dfrac{ix^7}{7!} + \cdots + \dfrac{(-1)^{n-1} i^{n-1} x^{n-1}}{(n-1)!} + \cdots = (1 - \dfrac{x^2}{2!} + \dfrac{x^4}{4!} - \dfrac{x^6}{6!} + \cdots)$

$- i(x - \dfrac{x^3}{3!} + \dfrac{x^5}{5!} - \dfrac{x^7}{7!} + \cdots) = \cos x - i \sin x$ 33. $a + bi = r(\cos \theta + i \sin \theta) = re^{i\theta}$, using Euler's Formula

35. $e^{\frac{3\pi}{2}i}$ 37. $2e^{\frac{\pi}{4}i}$ 39. $e^{\pi i}$ 41. $5e^{i \, \text{Arc tan} \frac{4}{3}}$ 43. $e^{ix} \cdot e^{-ix} = (\cos x + i \sin x)(\cos x - i \sin x)$

$= (\cos x + i \sin x)(\cos (-x) + i \sin (-x)) = \cos (x - x) + i \sin (x - x) = \cos 0 + i \sin 0 = 1 + i \cdot 0 = 1$

45. $z = e^{i\theta} = \cos \theta + i \sin \theta$; $\bar{z} = \cos (-\theta) + i \sin (-\theta) = e^{i(-\theta)} = e^{-i\theta}$ 47. $(re^{i\theta})^n = [r(\cos \theta + i \sin \theta)]^n$

$= r^n [\cos (n\theta) + i \sin (n\theta)] = r^n e^{in\theta}$ 49. $\sin x = x - \dfrac{x^3}{3!} + \dfrac{x^5}{5!} - \dfrac{x^7}{7!} + \cdots + \dfrac{(-1)^{n+1} x^{2n-1}}{(2n-1)!} + \cdots$ and

$\sin (ix) = ix - \dfrac{(ix)^3}{3!} + \dfrac{(ix)^5}{5!} - \dfrac{(ix)^7}{7!} + \cdots + \dfrac{(-1)^{n+1} (ix)^{2n-1}}{(2n-1)!} + \cdots; \lim\limits_{n \to \infty} \left| \dfrac{a_{n+1}}{a_n} \right|$

$= \lim\limits_{n \to \infty} \left| \dfrac{\dfrac{(-1)^{n+2} (ix)^{2n+1}}{(2n+1)!}}{\dfrac{(-1)^{n+1} (ix)^{2n-1}}{(2n-1)!}} \right| = \lim\limits_{n \to \infty} \left| \dfrac{(-1)(ix)^2}{(2n+1)(2n)} \right| = \lim\limits_{n \to \infty} \left| \dfrac{(-1)(-1)x^2}{(2n+1)2n} \right| = x^2 \lim\limits_{n \to \infty} \dfrac{1}{(2n+1)2n}$

$= x^2 \cdot 0 = 0.$ Thus, the series converges for all values of x. 51. $\cos ix - i \sin ix = [1 - \dfrac{(ix)^2}{2!} + \dfrac{(ix)^4}{4!}$

$- \dfrac{(ix)^6}{6!} + \cdots + \dfrac{(-1)^{n+1} (ix)^{2n-2}}{(2n-2)!} + \cdots] - i[ix - \dfrac{(ix)^3}{3!} + \dfrac{(ix)^5}{5!} - \dfrac{(ix)^7}{7!} + \cdots + \dfrac{(-1)^{n+1} (ix)^{2n-1}}{(2n-1)!} + \cdots]$

$= 1 - i^2 x - \dfrac{i^2 x^2}{2!} + \dfrac{i^4 x^3}{3!} + \dfrac{i^4 x^4}{4!} - \dfrac{i^6 x^5}{5!} - \dfrac{i^6 x^6}{6!} + \dfrac{i^8 x^7}{7!} + \cdots = 1 + x + \dfrac{x^2}{2!} + \dfrac{x^3}{3!} + \dfrac{x^4}{4!} + \dfrac{x^5}{5!} + \dfrac{x^6}{6!} + \dfrac{x^7}{7} + \cdots = e^x$

53. $e^{-x} + e^x = (\cos ix + i \sin ix) + (\cos ix - i \sin ix)$, $e^{-x} + e^x = 2 \cos ix$ and $\cos ix = \dfrac{e^{-x} + e^x}{2}$

1. 0 2. Undefined 3. 1 4. Undefined

PAGES 339-340 WRITTEN EXERCISES

1. $1 + \dfrac{x^2}{2!} + \dfrac{x^4}{4!} + \cdots + \dfrac{x^{2n-2}}{(2n-2)!} + \cdots$ 3. $\coth x = \dfrac{1}{\tanh x} = \dfrac{1}{\dfrac{e^x - e^{-x}}{e^x + e^{-x}}} = \dfrac{e^x + e^{-x}}{e^x - e^{-x}} = \coth x$

5. $\operatorname{csch} x = \dfrac{1}{\sinh x} = \dfrac{1}{\dfrac{e^x - e^{-x}}{2}} = \dfrac{2}{e^x - e^{-x}} = \operatorname{csch} x$ 7. $\tanh x = \dfrac{\operatorname{sech} x}{\operatorname{csch} x} = \dfrac{\dfrac{2}{e^x + e^{-x}}}{\dfrac{2}{e^x - e^{-x}}} = \dfrac{e^x - e^{-x}}{e^x + e^{-x}} = \tanh x$

9. $\sinh(-x) = \dfrac{e^{-x} - e^{-(-x)}}{2} = \dfrac{e^{-x} - e^x}{2} = -\left(\dfrac{e^x - e^{-x}}{2}\right) = -\sinh x$ 11. $\tanh^2 x + \operatorname{sech}^2 x = \left(\dfrac{e^x - e^{-x}}{e^x + e^{-x}}\right)^2$

$+ \left(\dfrac{2}{e^x + e^{-x}}\right)^2 = \dfrac{e^{2x} + e^{-2x} - 2}{e^{2x} + e^{-2x} + 2} + \dfrac{4}{e^{2x} + e^{-2x} + 2} = \dfrac{e^{2x} + e^{-2x} - 2 + 4}{e^{2x} + e^{-2x} + 2} = \dfrac{e^{2x} + e^{-2x} + 2}{e^{2x} + e^{-2x} + 2} = 1$

13. $\sinh(x + y) = \sinh x \cosh y + \cosh x \sinh y = \left(\dfrac{e^x - e^{-x}}{2}\right)\left(\dfrac{e^y + e^{-y}}{2}\right) + \left(\dfrac{e^x + e^{-x}}{2}\right)\left(\dfrac{e^y - e^{-y}}{2}\right)$

$= \dfrac{e^{x+y} + e^{x-y} - e^{-x+y} - e^{-x-y}}{4} + \dfrac{e^{x+y} - e^{x-y} + e^{-x+y} - e^{-x-y}}{4} = \dfrac{e^{x+y} - e^{-x-y}}{2} = \sinh(x + y)$

15. $\cosh 2x = 2\cosh^2 x - 1 = 2\left(\dfrac{e^x + e^{-x}}{2}\right)^2 - 1 = \dfrac{1}{2}(e^{2x} + e^{-2x} + 2) - 1 = \dfrac{e^{2x} + e^{-2x}}{2} = \cosh 2x$

17. $\sinh x = x + \dfrac{x^3}{3!} + \dfrac{x^5}{5!} + \cdots + \dfrac{x^{2n-1}}{(2n-1)!} + \cdots;\ \lim\limits_{n\to\infty}\left|\dfrac{a_{n+1}}{a_n}\right| = \lim\limits_{n\to\infty}\left|\dfrac{\dfrac{x^{2n+1}}{(2n+1)!}}{\dfrac{x^{2n-1}}{(2n-1)!}}\right| = \lim\limits_{n\to\infty}\left|\dfrac{x^2}{(2n+1)2n}\right|$

$= x^2 \lim\limits_{n\to\infty} \dfrac{1}{(2n+1)2n} = x^2 \cdot 0 = 0$. Thus, the series converges for all values of x.

19. $\sinh(x + 2n\pi i) = \dfrac{e^{x+2n\pi i} - e^{-x-2n\pi i}}{2} = \dfrac{1}{2}(e^x e^{2n\pi i} - e^{-x} \cdot e^{-2n\pi i}) = \dfrac{e^x}{2}(\cos 2n\pi + i \sin 2n\pi)$

$- \dfrac{e^{-x}}{2}(\cos 2n\pi - i \sin 2n\pi) = \dfrac{e^x}{2}(1 + i \cdot 0) - \dfrac{e^{-x}}{2}(1 - i \cdot 0) = \dfrac{e^x - e^{-x}}{2} = \sinh x$

21. $\tanh(x + n\pi i) = \dfrac{e^{x+n\pi i} - e^{-x-n\pi i}}{e^{x+n\pi i} + e^{-x-n\pi i}} = \dfrac{e^x \cdot e^{n\pi i} - e^{-x} e^{-n\pi i}}{e^x e^{n\pi i} + e^{-x} e^{-n\pi i}}$

$= \dfrac{e^x(\cos n\pi + i \sin n\pi) - e^{-x}(\cos n\pi - i \sin n\pi)}{e^x(\cos n\pi + i \sin n\pi) + e^{-x}(\cos n\pi - i \sin n\pi)} = \dfrac{e^x((-1)^n + i \cdot 0) - e^{-x}((-1)^n - i \cdot 0)}{e^x((-1)^n + i \cdot 0) + e^{-x}((-1)^n - i \cdot 0)}$

$= \dfrac{e^x - e^{-x}}{e^x + e^{-x}} = \tanh x$ 23. $\operatorname{csch}(x + 2n\pi i) = \dfrac{2}{e^{x+2n\pi i} - e^{-x-2n\pi i}} = \dfrac{2}{e^x e^{2n\pi i} - e^{-x} e^{-2n\pi i}}$

$= \dfrac{2}{e^x(\cos 2n\pi + i \sin 2n\pi) - e^{-x}(\cos 2n\pi - i \sin 2n\pi)} = \dfrac{2}{e^x(1 + i \cdot 0) - e^{-x}(1 - i \cdot 0)} = \dfrac{2}{e^x - e^{-x}} = \operatorname{csch} x$

25. $\sin ix = \dfrac{e^{-x} - e^x}{2i} = \dfrac{i}{i}\left(\dfrac{e^{-x} - e^x}{2i}\right) = -i\left(\dfrac{e^{-x} - e^x}{2}\right) = i\left(\dfrac{e^x - e^{-x}}{2}\right) = i \sinh x$

1.
```
10  READ A,R,N
20  LET T=A*R↑(N−1)
30  PRINT "TERM";N;"=";T
40  GO TO 10
50  DATA 2,2,10,6,1,100,10,−.5,20
60  END
```

3.
```
10  READ A,R
20  IF R>=1 THEN 60
30  IF R<=−1 THEN 60
40  PRINT "INFINITE SUM =";A/(1−R)
50  GO TO 10
60  PRINT "NO INFINITE SUM EXISTS."
70  GO TO 10
80  DATA 2,2,10,.5,100,−.25,50,−3
90  END
```

5.
```
10  READ X
20  IF X>1 THEN 110
30  IF X<−1 THEN 110
40  LET S = X
50  LET I = 3
60  LET S = S−X↑I/I+X↑(I+2)/(I+2)
70  LET I = I+4
80  IF I <= 23 THEN 60
90  PRINT "ARCTAN(";X;") = ";S
100  GO TO 10
110  PRINT "SERIES DOES NOT WORK FOR X =";X
120  GO TO 10
130  DATA .5,−1.2,−.6,.785398,1.5
140  END
```

7.
```
10  READ X
20  LET S = X
30  LET D = 1
40  LET N = 2
50  LET E = 2*N − 1
60  LET D = D*(E − 1)*E
70  LET S = S + X↑E/D
80  LET N = N+1
90  IF N <= 13 THEN 50
100  PRINT "SINH(";X;") = ";S
110  GO TO 10
120  DATA 0,1.57080,−.785398,3.14159
130  END
```

9.
```
10  LET S = 1
20  LET I = 3
30  LET S = S − 1/I + 1/(I+2)
40  LET I = I + 4
50  IF I <= 1E6 THEN 30
60  PRINT "PI =";S*4
70  END
```

PAGES 343-345 CHAPTER OBJECTIVES AND REVIEW

3. 32 **5.** 4, 1, $\frac{4}{9}$, $\frac{1}{4}$, $\frac{4}{25}$ **7.** 486 **9.** **11.** 4000

$$a_1 \quad a_2 \quad a_3 \quad a_4 \quad a_5 \quad a_6 \quad a_7 \quad a_8$$

$$0 \quad 1 \quad 2 \quad 3 \quad 4 \quad 5 \quad 6 \quad 7 \quad 8 \quad 9 \quad 10$$

13. $\left| 1 - \frac{n+1}{n+3} \right| = \left| \frac{n+3-(n+1)}{n+3} \right| = \left| \frac{2}{n+3} \right| = \frac{2}{n+3}$. Therefore, $\left| 1 - \frac{n+1}{n+3} \right| < d$ is equivalent to

$\frac{2}{n+3} < d$, which is true whenever $\frac{2}{d} < n + 3$ or $\frac{2}{d} - 3 < n$. Thus, for any positive number d, you can choose

SEQUENCES AND SERIES

m to be any natural number greater than $\frac{2}{d} - 3$. Thus, $\left| 1 - \frac{n+1}{n+3} \right|$ will be less than d whenever $n \geq m$.

Therefore, $\lim\limits_{n\to\infty} \frac{n+1}{n+3} = 1$.   15. 3   17. 2   19. 1, 3, 6, 10   21. A; $50 + 45 + 40 + \cdots + (50 - 5(n-1))$

$+ \cdots$ 23. 85   25. $-\frac{8}{7}$ 27. Diverges   29. Converges   31. Diverges   33. $|x| < 1$ 35. $|x| < 1$

37. .5403   39. $-.0009$ 41. -1 43. $-\frac{\sqrt{3}}{2} + \frac{1}{2}i$ 45. csch $(-x) = \frac{2}{e^{-x} - e^{-(-x)}} = \frac{2}{e^{-x} - e^{x}}$

$= -(\frac{2}{e^{x} - e^{-x}}) = -$csch x

PAGE 346   CHAPTER TEST

1. 3, 2, $1\frac{2}{3}$, $1\frac{1}{2}$ 2. $\frac{1}{8}$ 3. $m \geq 1001$ 4. $\left| -2 - (-2 + \frac{1}{n}) \right| = \left| -\frac{1}{n} \right| = \frac{1}{n}$. Therefore, $\left| -2 - (-2 + \frac{1}{n}) \right|$

$< d$ is equivalent to $\frac{1}{n} < d$ which is true whenever $\frac{1}{d} < n$. Thus, for any positive number d, you can choose

m to be any natural number greater than $\frac{1}{d}$. Then, $\left| -2 - (-2 + \frac{1}{n}) \right|$ will be less than d whenever $n \geq m$.

Therefore, $\lim\limits_{n\to\infty} (-2 + \frac{1}{n}) = -2$.   5. 0   6. $-\frac{1}{8} + \frac{1}{8} - \frac{1}{8} + \cdots + [-\frac{1}{8}(-1)^{n-1}] + \cdots$ 7. 25   8. Diverges

9. Converges   10. The set of real numbers.   11. $-.8415$ 12. $-i$ 13. cosh $(-x) = \frac{e^{-x} + e^{-(-x)}}{2}$

$= \frac{e^{-x} + e^{x}}{2} = \frac{e^{x} + e^{-x}}{2} = $ cosh x

PAGE 346-348   CUMULATIVE REVIEW: CHAPTERS 6-8

1. d   3. c   5. d   7. a   9. a   11. c   13. a   15. c   17. c   19. c   21. b   23. b

PAGES 349-352   FINAL CUMULATIVE REVIEW: CHAPTERS 1-8

1. c   3. d   5. a   7. d   9. c   11. b   13. d   15. c   17. b   19. d   21. d   23. c   25. a   27. a
29. b   31. a   33. b   35. b   37. d

APPENDIX

PAGE 355   WRITTEN EXERCISES

1. $5^2 = 25$ 3. $100^0 = 1$ 5. $8^{\frac{1}{3}} = 2$ 7. $9^{\frac{3}{2}} = 27$ 9. $10^{-3} = .001$ 11. $\log_2 32 = 5$ 13. $\log_8 1 = 0$

15. $\log_{10} .01 = -2$ 17. $\log_{10,000} 100 = \frac{1}{2}$ 19. 1000   21. 64   23. 256   25. $2^y = 16$, $y = 4$ 27. 0

29. 0   31. $\frac{1}{2}$ 33. $\frac{3}{4}$ 35. $\frac{2}{3}$ 37. -4 39. $-\frac{1}{2}$ 41. 0

PAGE 358   WRITTEN EXERCISES

1. $\log_b x + \log_b y + \log_b z$ 3. $\frac{2}{3}\log_b x + \frac{4}{5}\log_b y - \frac{1}{2}\log_b z$ 5. $7\log_b x - 2\log_b y$ 7. 1.3424

9. .7404   11. 1.7404   13. 1.5562   15. 1.7781   17. $-.5229$ 19. .3387   21. .1712   23. Let

$\log_b N = x$. Then $N = b^x$. Thus, $N^a = (b^x)^a = b^{ax}$. Finally $\log_b N^a = ax$ or $\log_b N^a = a \log_b N$.

PAGE 360   WRITTEN EXERCISES

1. 2.4362   3. 6.4362 − 10   5. 4.9309   7. 8.3075 − 10   9. .1553   11. 5.7160 − 10   13. 3.0000
15. 1.0899   17. 2540   19. 919   21. .337   23. 2230   25. 7.33   27. .000000627   29. .0912
31. 10.1

PAGE 363   WRITTEN EXERCISES

1. 120   3. 8.12   5. 58,300   7. 2010   9. 1000   11. .00896   13. .00483   15. 136   17. 47.1
19. .420   21. 3.23   23. .333   25. 5.05×10^8 km^2 27. 5.61 cm^3 29. .04 cm^3

PAGE 365   WRITTEN EXERCISES

1. B = 50°; a = 22.5; b = 26.8   3. A = 61°; c = 603; b = 292   5. A = 32° 50′; B = 57° 10′; c = 73.8
7. B = 80° 30′; A = 9° 30′; a = 24.0   9. B = 66°; c = 197; b = 180   11. C = 47°; b = 649; c = 482
13. C = 35°; a = 770; b = 791   15. B = 100°; a = 3590; c = 5510   17. C = 29°; b = .206; c = .134
19. C = 7°; a = .111; b = .109   21. 9.8 meters   23. 72 m

A 6
B 7
C 8
D 9
E 0
F 1
G 2
H 3
I 4
J 5